Advances in Acoustics

The book series Advances in Acoustics, endorsed by the Acoustic Society of China, publishes both monographs and edited volumes. It covers all research fields of acoustics, including physical acoustics, underwater acoustics, solid-state acoustics, environmental acoustics, detection acoustics, power ultrasound, ultrasonic electronics, audio engineering, biomedical ultrasonic engineering, speech and hearing acoustics, and musical acoustics. By bridging theories, experiments, and practical applications, this series serves as an invaluable resource for researchers, engineers, and students engaged in the diverse aspects of acoustics.

Xiaozhou Liu

Nonlinear Sound Waves in Solids

Science Press
Beijing

Springer

Xiaozhou Liu
Nanjing University
Nanjing, Jiangsu, China

ISSN 3091-3373 ISSN 3091-3381 (electronic)
Advances in Acoustics
ISBN 978-981-96-5252-5 ISBN 978-981-96-5253-2 (eBook)
https://doi.org/10.1007/978-981-96-5253-2

Jointly published with China Science Publishing & Media Ltd. (Science Press)
The print edition is not for sale in China (Mainland). Customers from China (Mainland) please order the print book from: China Science Publishing & Media Ltd. (Science Press).
ISBN of the Co-Publisher's edition: 978-7-03-067573-6

The original submitted manuscript has been translated into English. The translation was done using artificial intelligence. A subsequent revision was performed by the author(s) to further refine the work and to ensure that the translation is appropriate concerning content and scientific correctness. It may, however, read stylistically different from a conventional translation.

This Springer imprint is published by the registered company Springer Nature Singapore Pte Ltd.
The registered company address is: 152 Beach Road, #21-01/04 Gateway East, Singapore 189721, Singapore

Preface

Nonlinear acoustics refers to the branch of acoustics that studies the nonlinear phenomena in the propagation process of large amplitude sound waves. When the amplitude of the sound wave is large enough, the nonlinear terms of the motion equation, continuity equation, and medium state equation describing the sound wave process cannot be ignored, resulting in a nonlinear sound wave equation. At this time, the propagation speed of sound disturbance is related to the strength of the disturbance, resulting in waveform distortion, sound saturation, nonlinear interaction between sound waves, and a series of nonlinear phenomena during the propagation of sound waves. Since the late 1950s, high-frequency high-power ultrasonic waves have increasingly appeared and been utilized, such as ultrasonic treatment and detection of industrial materials, detection of underwater targets by ultrasonic waves, propagation of strong noise, and applications of ultrasonic waves in medicine, all involving the propagation and interaction of finite amplitude sound waves, which do not follow the laws of linear acoustics.

The research of nonlinear acoustics involves nonlinear parameter arrays, strong noise nonlinearity, period-doubling bifurcation and chaos, solitons, nonlinearity in biological media, etc. Since the 1980s, the impact of nonlinear effects has been increasingly valued, the development of nonlinear acoustics has been accelerating, and its applications have been expanding. Today, nonlinear acoustics has permeated various occasions and covered all fields. For example, to study lattice structures and dislocation phenomena in crystals, nonlinear acoustic techniques can measure the relationship between temperature and higher-order elastic constants of different solids; there is a nonlinear interaction between sound waves and light waves, and in the field of acousto-optics, this interaction can be used to more effectively detect the internal structure of objects; in low-temperature research, nonlinear acoustics affects ammonia, copper, and some high-temperature superconductors, and when the temperature changes, the change in nonlinear parameters is more sensitive than the change in linear parameters; in biophysics, nonlinearity affects the tissue structure of organisms from biological tissues to molecular-level amino acids and proteins, and nonlinear acoustic effects have been widely used in lithotripsy technology and soft tissue treatment; in chemical physics, the nonlinear parameters of different mixed liquids and different purities have been measured; in the field of geophysics, nonlinear acoustics can be used to study the nonlinear effects of stratum materials and cracks, and it also has applications in geology and seismology.

In fluid media, there is only one elastic constant (such as the compression coefficient), and there is only one nonlinear parameter, namely B/A, while in the nonlinear acoustics of solids, there are two linear elastic constants for isotropic solids, and there are three nonlinear elastic constants (third-order elastic constants), and the number of anisotropic nonlinear elastic parameters varies with the structure. The third-order or higher-order elastic constants are macroscopic parameters characterizing the properties of solids, which are not only closely related to the structure of solids, but will also become new parameters for nondestructive evaluation of defects and fatigue. The nonlinearity of solids is divided into classical nonlinearity and non-classical nonlinearity. The nonlinearity in solids is due to the non-symmetry of the interatomic potential curve in the crystal (that is, the interaction force is not linearly related to the atomic deviation from the equilibrium position), which causes the non-harmonic nature of lattice vibration. Non-classical nonlinearity is now generally believed to be caused by the microscopic characteristics of solid materials, such as cracks, fissures, particles, etc., which affect the mechanical properties, and on a macroscopic level, stress and strain appear as hysteresis phenomena.

The use of sound waves for non-destructive testing of materials, based on their classical and non-classical nonlinear characteristics in solids, has developed into a mature system after many years of development. By studying the changes in nonlinear acoustic parameters during the propagation of sound waves in solid materials, we can obtain information about the mechanical properties and microstructure of solid materials. Experimental research shows that fatigued metal materials have nonlinear acoustic characteristics, and nonlinear acoustic parameters are more sensitive to material fatigue than linear acoustic parameters; sound waves show significant nonlinearity in porous materials, and have strong nonlinearity even when the porosity of the material is very small; rocks, concrete and other nonlinear mesoscopic elastic solid materials show non-classical nonlinear acoustic phenomena that are different from classical nonlinear acoustics; in recent years, phononic crystals and metamaterials have become hot research topics, mainly focusing on their linear aspects, but research shows that the nonlinearity of these materials cannot be ignored and can be utilized, and asymmetric control of sound waves can be achieved using prepressurized nonlinear particle chains. By performing nonlinear analysis on metamaterials periodically paralleled on ordinary waveguides with Helmholtz resonators, it is found that the structural parameters of the Helmholtz resonators can be reasonably adjusted to effectively suppress the occurrence of shock waves.

This book cannot cover all aspects of nonlinear sound waves in solids. The book mainly introduces the following aspects of nonlinear sound waves in solids:

Chapter 1 introduces the basics of nonlinear sound waves in solids, including nonlinear elasticity theory, nonlinear propagation and interaction of elastic waves in solids, and briefly introduces the sound nonlinearity at the solid interface contact, nonlinear Lamb waves in plates, and nonlinear sound propagation in non-uniform media;

Chapter 2 discusses the nonlinear effects and acoustic memory phenomena of sound waves in crystals and ceramic materials, introduces the interaction of collinear elastic waves propagating along the pure mode direction of crystals, proposes a calculation method for the third-order effective elastic parameters of crystals, introduces the nonlinear propagation of sound waves in lithium tantalate, PIN-PMN-PT crystals and ceramic materials, and introduces the acoustic memory phenomenon in lithium niobate crystals;

Chapter 3 discusses the nonlinear propagation of sound waves in porous materials, elaborates on the nonlinear propagation characteristics of sound waves in micro-porous viscoelastic materials, and the effect of dissipation on the high-order harmonics of porous media;

Chapter 4 discusses the non-classical nonlinear propagation of sound waves in solids with cracks, introduces the Preisach-Mayergoyz (PM) model and the nonlinear resonance acoustic spectrum method, establishes the nonlinear vibration equation through the hysteresis effect generated by the nonlinear stress-strain relationship of micro-cracks in a one-dimensional rod, introduces the inversion and positioning of single and multiple cracks, and the two-dimensional non-classical nonlinear signal time reversal imaging method;

Chapter 5 discusses the nonlinear theory of sound waves in concrete, introduces experimental research on concrete under different degrees of damage, concrete with different water contents, concrete under sulfate attack, and concrete with different mix ratios, and their non-classical nonlinear characteristics;

Chapter 6 discusses the nonlinear properties of structures with the same mass and different spring stiffness, predicts the nonlinear characteristics of this structure based on the dispersion curve of this periodic structure, introduces a non-reciprocal structure designed based on this linear and nonlinear periodic structure, realizes the asymmetric transmission of elastic waves, and simulates the linear and nonlinear acoustics of metamaterials;

Chapter 7 discusses the metamaterial structure of periodically cascaded small tubes with holes and vibrating thin plates in the sound wave guide. In the case of considering nonlinearity, this structure can obtain the second harmonic excited in reverse; it also introduces the method of obtaining high-efficiency reverse second harmonic excitation by using quasi-phase matching in the nonlinear complementary medium of acoustic metamaterials.

The author has compiled some insights from more than thirty years of work in solid nonlinear acoustics into this book, dedicated to readers, to stimulate readers' interest in the study of nonlinear acoustics, and thus promote the development of nonlinear acoustics in our country. Special thanks to the author's mentor, Prof. Xiufen Gong, for her multifaceted guidance and meticulous help. Professors Jianchun Cheng, Wenhua Jiang, Xiaojun Liu and Dong Zhang of the Acoustics Research Institute of Nanjing University have provided many valuable suggestions and advice, for which we are also grateful. Thanks to the support and help of Academician Shining Zhu of the

School of Physics at Nanjing University, Prof. Wenwu Cao of Pennsylvania State University, USA, Prof. Mingxi Ding of Chongqing University. Prof. Fenghua Li, Jun Yang, Xiuming Wang of the Institute of Acoustics, Chinese of Academy of Sciences. Thanks to the encouragement and support of the leaders and teachers of the School of Physics at Nanjing University. Associate Professor Jiehui Liu of the research group, master's students Dao Zhou, Tingting Zhang, Yulin Feng, Xiaohui Wang, Li Quan, Benbiao Luo, Jinlin Zhu, postdoctoral fellow Lue Zhang, etc., participated in some of the work, for which we express our thanks. During the writing process, the author strives to be serious and rigorous, but there are still shortcomings, and we warmly welcome criticism and correction from readers.

This book has received support from National Key Research and Development Program of China (2020YFA0211400, 2017YFA0303702, 2016YFF0203000, 2012CB921504, 2011CB707902), the Key Program of the National Natural Science Foundation of China (11834008), the National Natural Science Foundation of China (12174192, 11774167, 11474160, 11274166, 11074122, 10674066), and the State Key Laboratory of Acoustics, Chinese Academy of Sciences. We express our gratitude here.

Nanjing, China
December 2024

Xiaozhou Liu

Introduction

This book systematically introduces the basic principles and analysis methods of nonlinear acoustics in solids, with a focus on the theory and application of nonlinear acoustics in ultrasonic nondestructive testing, metamaterials, etc. The main content includes: the basics of solid nonlinear sound waves, the nonlinear effects and acoustic memory phenomena of sound waves in crystals and ceramic materials, the nonlinear propagation of sound waves in porous materials, the non-classical nonlinear propagation of sound waves in solids with cracks, the nonlinear theory and experimental research of sound waves in concrete, the nonlinear propagation of sound waves in mass-spring periodic structures, and the reverse excitation and enhancement of second harmonics in metamaterials. This book can promote the development of nonlinear acoustic theory and the development of nonlinear acoustics in nondestructive testing, metamaterials, and other fields.

This book can be used as a textbook for senior undergraduates and graduate students majoring in acoustics, nondestructive testing, etc., in higher education institutions, and can also be used as a reference book for graduate students and engineering technicians engaged in acoustics, nondestructive testing, and other fields.

Contents

1 Fundamentals of Nonlinear Propagation of Sound Waves in Solids 1
1.1 Nonlinear Elastic Theory Equation 2
1.1.1 Deformation 2
1.1.2 Elastic Properties of Isotropic Solids 4
1.1.3 Nonlinear Motion Equations 7
1.2 Nonlinear Propagation of Elastic Waves in Solids 10
1.3 Interaction of Elastic Waves in Solids 12
1.4 Nonlinear Acoustics at Solid Interfaces 21
1.4.1 Mechanism of High Order Harmonic Generation 21
1.4.2 Theoretical Model 22
1.5 Nonlinear Lamb Waves in Plates 25
1.5.1 Nonlinear Lamb Waves 25
1.5.2 Using the Mode Expansion Analysis Method [32] 26
1.6 Nonlinear Sound Propagation in Non-uniform Media 30
1.6.1 Theory 30
1.6.2 Calculation Results and Discussion 31
References 43

2 Nonlinear Effects and Acoustic Memory Phenomena of Sound Waves in Crystals and Ceramic Materials 45
2.1 Nonlinear Interactions of Sound Waves in Crystals 45
2.1.1 Theory 45
2.2 Interaction of Collinear Acoustic Waves Propagating Along Pure Mode Directions in Crystals 62
2.2.1 Basic Equation 63
2.2.2 Generation of Second Harmonic and Mixing Waves 67
2.2.3 Effective Second-Order and Third-Order Elastic Constants of Crystals 72

2.3 Study of Pure Mode Axes of PMN-PT Crystals with Different PT Contents ... 80
2.3.1 Pure Mode of PMN-PT Crystals ... 81
2.3.2 Energy Velocity of Elastic Waves Represented by the Slowness Surface ... 82
2.3.3 Specific Analysis of the Propagation Speed and Pure Mode of PMN-PT Crystal ... 83
2.4 Study on the Acoustic Characteristics of Lithium Tantalate and PIN-PMN-PT Crystals ... 87
2.4.1 Experimental Method ... 91
2.5 Ceramic Phase Transition and Nonlinear Effects ... 106
2.5.1 High-Order Nonlinear Changes Caused by Polarization in PZT ... 110
2.6 Acoustic Memory Phenomenon in Crystals ... 115
2.6.1 Experimental Method ... 117
2.6.2 Experimental Results ... 121
2.7 Multi-segment Acoustic Memory Signal ... 126
2.7.1 Non-exponential Decay Phenomenon of Echo Signals ... 128
2.7.2 Analysis and Discussion ... 133
References ... 137

3 Nonlinear Propagation of Sound Waves in Porous Materials ... 141
3.1 Nonlinear Characteristics of Sound Waves in Viscoelastic Materials Containing Micro-pores ... 142
3.1.1 Theory ... 142
3.1.2 Numerical Calculation ... 152
3.2 Nonlinear Acoustics of Porous Media Considering Dissipation Effects ... 157
3.2.1 Theory ... 157
3.2.2 Numerical Results ... 166
3.2.3 Experimental System and Experimental Results ... 167
References ... 171

4 Non-classical Nonlinear Propagation of Acoustic Waves in Solids with Cracks ... 173
4.1 Classical Nonlinear Effects of Cracks ... 174
4.2 Non-classical Nonlinear Effects of Cracks ... 176
4.2.1 Elastic Hysteresis Effect of Cracks ... 176
4.2.2 Inelastic Hysteresis Effect of Cracks ... 176
4.3 Other Nonlinear Models of Cracks ... 177
4.4 Preisach-Mayergoyz (PM) Model ... 180
4.5 Nonlinear Resonance Spectral Method ... 184
4.5.1 Free at Both Ends ... 184
4.5.2 One End Free, One End Fixed ... 195

4.6 Analysis of High-Order Harmonics of One-Dimensional Non-classical Nonlinearity 214
4.6.1 Nonlinear Stress–Strain Relationship and Its Wave Equation of One-Dimensional Cracked Metal Rod 214
4.6.2 Wave Equation and Its Linear Approximate Solution 215
4.6.3 Harmonic Components Caused by Hysteresis Effect 217
4.6.4 Inversion and Location of Multiple Cracks 223
4.6.5 Discussion on the Validity of Crack Location 229
4.7 Two-Dimensional Non-classical Nonlinear Signal Simulation 230
4.8 Two-Dimensional Non-classical Nonlinear Signal Inversion Imaging 242
4.8.1 Time Reversal Theory 242
4.8.2 Numerical Simulation of Time Reversal 244
References 272

5 Nonlinear Theory and Experimental Study of Sound Waves in Concrete 275
5.1 Nonlinear Theoretical Model of Sound Wave Propagation in Concrete 277
5.2 Experimental Method 288
5.3 Non-classical Nonlinear Analysis of Concrete Under Different Degrees of Damage 289
5.4 Non-classical Nonlinear Changes in Concrete Samples with Different Water Contents 294
5.5 Changes in Non-classical Nonlinear Effects of Concrete Under Sulfate Erosion 297
5.6 Non-classical Nonlinear Research of Concrete with Different Mix Ratios 301
References 311

6 Nonlinear Propagation of Acoustic Waves in Mass-Spring Periodic Structures 313
6.1 Mechanism of Bragg Scattering Type Phonon Crystal Band Gap Formation 314
6.1.1 Linear Band Structure of Mass-Spring System 315
6.1.2 Linear Simulation of the Mass-Spring System 316
6.2 Nonlinear Regulation Based on the Mechanism of Phononic Crystal Bandgap Formation 318
6.2.1 Nonlinear Analysis Based on the Mass-Spring System 318
6.2.2 Nonlinear Simulation of Mass-Spring System 319
6.3 Non-reciprocal Propagation Mechanism 321
6.3.1 Principle of Asymmetric Structure 330
6.3.2 Non-reciprocal Structure Simulation 331

6.4 Propagation of Sound Waves in Nonlinear Metamaterials 333
6.4.1 Formation Mechanism and Characteristics of the Bandgap of Localized Resonance Type Phononic Crystals 333
6.4.2 Nonlinear Research Based on the Band Gap Formation Mechanism of Localized Resonance Type Phononic Crystals 337
References 342

7 Reverse Excitation and Enhancement of Second Harmonics in Metamaterials 345
7.1 Second Harmonic Reverse Excitation Caused by Acoustic Negative Refraction Materials 346
7.2 Quasi-Phase Matching Enhances the Second Harmonic of Dispersion Metamaterials 371
7.2.1 Phase Matching 371
7.2.2 Quasi-Phase Matching 373
7.2.3 Difference Between Phase Matching in Metamaterials and Ordinary Phase Matching 375
References 379

Chapter 1
Fundamentals of Nonlinear Propagation of Sound Waves in Solids

In an ideal fluid, the medium can only produce volumetric deformation, that is, pure compression and expansion deformation. The elasticity of the medium can be characterized by a single volumetric elasticity coefficient. In such a medium, only the alternating process of sparsity and density can occur, which can only propagate longitudinal waves, and the characteristics of this propagation process can be fully described by a scalar (sound pressure). Knowing the sound pressure, we can obtain the particle velocity through the motion equation of the ideal fluid, thereby obtaining some energy relationships of the sound wave. However, the situation in solids is not so simple. In general, solid media not only produce volumetric deformation, but also shear deformation. In addition to volume elasticity, they also have shear elasticity. Therefore, in solids, in addition to propagating compression and expansion longitudinal waves, shear waves can also be propagated. In isotropic solids, the direction of particle vibration of this shear wave is perpendicular to the direction of wave propagation, which is called a transverse wave. In addition, surface waves with amplitudes that decay with surface depth are generated on the free surface of the solid. It can be seen that the propagation of sound waves in solids is much more complicated than in fluids.

The propagation speed of longitudinal waves in solids is $c_l = \sqrt{\frac{\lambda+2\mu}{\rho}}$, and the propagation speed of transverse waves is $c_t = \sqrt{\frac{\mu}{\rho}}$, where λ, μ is the Lame constant, μ is also known as the shear elasticity coefficient, and for liquids $\mu = 0$. When a solid is subjected to external forces, deformation occurs within the body, which is generally described by the physical quantity strain. The relationship between strain and stress in solids is much more complicated than in liquids. In fact, like the fluid dynamics equation, the elastic theory equation is also nonlinear, so when elastic waves propagate in solids, they are also accompanied by nonlinear effects such as waveform distortion. Moreover, because there are not only longitudinal waves but also transverse waves in solids, there is also interaction between sound waves and wave type conversion in solids. The nonlinearity in solids has long been known. The

X. Liu, *Nonlinear Sound Waves in Solids*,
Advances in Acoustics, https://doi.org/10.1007/978-981-96-5253-2_1

asymmetry of the interatomic potential energy curve in crystals (that is, the interaction force is not linearly related to the atomic displacement from the equilibrium position) causes the non-harmonic nature of lattice vibrations, which is the principle of thermal expansion of solids, and also leads to nonlinear propagation of sound waves. The phenomenon that force and displacement are not linearly related is a deviation from Hooke's law, so the greater the deviation from Hooke's law, the greater the thermal expansion. In other words, the non-harmonic nature of lattice vibration, a molecular-level nonlinearity, is manifested macroscopically as elastic nonlinearity, that is, the elastic properties do not follow Hooke's law. Studying the propagation characteristics of finite amplitude sound waves in solids helps to clarify some nonlinear properties of solids. Nonlinear devices can be developed using these nonlinear characteristics.

1.1 Nonlinear Elastic Theory Equation

1.1.1 Deformation

Elastic theory is a mechanics that treats solids as continuous media [1]. Under the influence of forces, solids will undergo different degrees of deformation, that is, they will change their original shape and volume. Observe a point M in the solid, its radius vector $\vec{r}_M$ (its components $x_1 = x, x_2 = y, x_3 = z$), due to deformation, M moves to M', its radius vector $\vec{r}_{M'}$, so the displacement at M is

$$\vec{u}_M = \vec{r}_{M'} - \vec{r}_M \tag{1.1}$$

$\vec{u}_M$ is called the deformation vector, also known as the displacement vector. But u_M cannot be used as the characteristic of solid deformation, because when the solid rotates as a whole or moves as a whole, such displacement can also be obtained. The characteristic of the deformation state should obviously be characterized by the distance between two adjacent points M and N, as shown in Figure 1.1, the distance between MN before deformation is ds

$$ds^2 = |\vec{r}_M - \vec{r}_N|^2 = dx_1^2 + dx_2^2 + dx_3^2 = dx_i^2, \quad (i \text{ denotes the summation}) \tag{1.2}$$

After deformation, the distance of $M'N'$ is ds'

$$ds'^2 = \left|\vec{r}_{M'} - \vec{r}_{N'}\right|^2 = \left|\vec{r}_M + \vec{u}_M - \vec{r}_N - \vec{u}_N\right|^2 = (dx_i + du_i)^2 \tag{1.3}$$

Substitute $\vec{u} = \vec{u}(x_1, x_2, x_3)$, $du_i = \frac{\partial u_i}{\partial x_k} dx_k$ to get

Fig. 1.1 Deformation

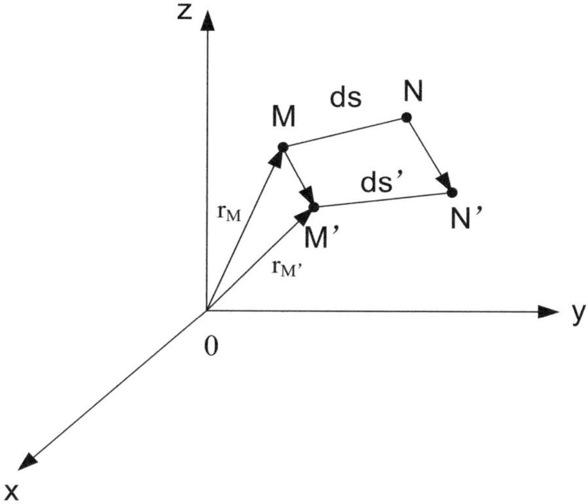

$$ds'^2 = dx_i^2 + 2\frac{\partial u_i}{\partial x_k}dx_i dx_k + \frac{\partial u_i}{\partial x_k}\frac{\partial u_i}{\partial x_l}dx_k dx_l \tag{1.4}$$

$$ds'^2 - ds^2 = 2\frac{\partial u_i}{\partial x_k}dx_i dx_k + \frac{\partial u_i}{\partial x_k}\frac{\partial u_i}{\partial x_l}dx_k dx_l \tag{1.5}$$

Summing the first term on the right side over indices i and k, we get

$$\frac{\partial u_i}{\partial x_k}dx_i dx_k = \frac{\partial u_k}{\partial x_i}dx_k dx_i \tag{1.6}$$

Therefore,

$$2\frac{\partial u_i}{\partial x_k}dx_i dx_k = \left(\frac{\partial u_i}{\partial x_k} + \frac{\partial u_k}{\partial x_i}\right)dx_i dx_k \tag{1.7}$$

For the second term on the right side: swap the positions of i and l

$$\frac{\partial u_i}{\partial x_k}\frac{\partial u_i}{\partial x_l}dx_l dx_k = \frac{\partial u_l}{\partial x_k}\frac{\partial u_l}{\partial x_i}dx_i dx_k \tag{1.8}$$

As per usual rules, we have omitted the summation symbol for vector and tensor indices, and for all twice-occurring indices, we sum over the values 1, 2, 3.

Hence, we get

$$ds'^2 - ds^2 = 2u_{ik}dx_i dx_k \tag{1.9}$$

where

$$u_{ik}=\frac{1}{2}\left(\frac{\partial u_i}{\partial x_k}+\frac{\partial u_k}{\partial x_i}+\frac{\partial u_l}{\partial x_i}\frac{\partial u_l}{\partial x_k}\right) \tag{1.10}$$

u_{ik} is called the deformation tensor, which is symmetric by definition.

In linear acoustics, the third term in Eq. (1.10) is omitted because it is a second-order quantity. In the case of small deformation, the deformation tensor is

$$u_{ik}=\frac{1}{2}\left(\frac{\partial u_i}{\partial x_k}+\frac{\partial u_k}{\partial x_i}\right) \tag{1.11}$$

When discussing finite amplitude sound wave problems, second-order quantities must be considered, so Eq. (1.10) is used (When $i \neq k$, the components of the deformation tensor u_{ik} determine shear deformation).

1.1.2 Elastic Properties of Isotropic Solids

From the knowledge of elasticity, when the deformation of an object is small enough, the elastic potential of the deformed solid is expressed as

$$U=\mu u_{ik}^2+\frac{\lambda}{2}u_{ll}^2,\ \text{due to}\ \lambda=K-\frac{2\mu}{3},$$
$$U=\mu u_{ik}^2+\left(\frac{K}{2}-\frac{\mu}{3}\right)u_{ll}^2 \tag{1.12}$$

K is the compression coefficient.

Like any symmetric tensor, we can use coordinate transformation to move the tensor u_{ik} at each given point to the principal axis. In other words, at any given point, we can choose such a coordinate system (the principal axis of the tensor) that the components of u_{ik} in it only have diagonal elements, i.e., any deformation along the tensor at this point only has expansion and compression. The diagonal elements at this time are

$$\begin{cases} I_1=u_{ll}=u_{11}+u_{22}+u_{33} \\ I_2=\frac{(u_{ll}^2-u_{ik}^2)}{2} \\ I_3=\frac{\left(u_{ik}u_{il}u_{kl}-\frac{3u_{ik}^2u_{ll}}{2}+\frac{u_{ll}^3}{2}\right)}{3} \end{cases} \tag{1.13}$$

The elastic potential energy of the object deformation should be a function of the deformation tensor

$$U=U(I_1,I_2,I_3) \tag{1.14}$$

Expand the series and retain up to third-order quantities

$$U = U(0,0,0) + \frac{\partial U}{\partial I_1}|_0 I_1 + \frac{1}{2}\frac{\partial^2 U}{\partial I_1^2}|_0 I_1^2 + \frac{1}{6}\frac{\partial^3 U}{\partial I_1^3}|_0 I_1^3$$
$$+ \frac{\partial U}{\partial I_2}|_0 I_2 + \frac{1}{2}\frac{\partial^2 U}{\partial I_1 \partial I_2}|_0 I_1 I_2 + \frac{\partial U}{\partial I_3}|_0 I_3 \quad (1.15)$$

We are interested in the increment of elastic potential, $U(0,0,0) = 0$, and the solid is initially in equilibrium

$$\left.\frac{\partial U}{\partial I_1}\right|_0 = 0, \quad \left.\frac{\partial^2 U}{\partial I_1^2}\right|_0 = K + \frac{4}{3}\mu, \quad \left.\frac{\partial U}{\partial I_2}\right|_0 = -2\mu,$$
$$\left.\frac{\partial U}{\partial I_3}\right|_0 = n = A, \quad \left.\frac{\partial^2 U}{\partial I_1 \partial I_2}\right|_0 = -4m = -2A - 2B,$$
$$\left.\frac{\partial^3 U}{\partial I_1^3}\right|_0 = 4m + 2l = 2A + 6B + 2C \quad (1.16)$$

Therefore,

$$U = -2\mu I_2 + \frac{1}{2}\left(K + \frac{4}{3}\mu\right)I_1^2 + AI_3$$
$$+ \frac{1}{2}(-2A - 2B)I_1 I_2 + \frac{1}{6}(2A + 6B + 2C)I_1^3 \quad (1.17)$$

From the above, to describe the nonlinearity of an isotropic solid, in addition to the two linear constants k, μ, three nonlinear constants A, B, C are also needed. Substituting I_1, I_2, I_3 into the above equation, we get

$$U = \mu u_{ik}^2 + \left(\frac{K}{2} - \frac{\mu}{3}\right)u_{ll}^2 + \frac{A}{3}u_{ik}u_{il}u_{kl} + Bu_{ik}^2 u_{ll} + \frac{C}{3}u_{ll}^3 \quad (1.18)$$

where: K, μ are the second-order elastic modulus, linear elastic constants. A, B, C are the third-order elastic modulus, or nonlinear elastic constants.

Substitute the components of the deformation tensor u_{ik} from Eq. (1.10) into Eq. (1.18), and retain up to third-order terms

$$\mu u_{ik}^2 = \frac{\mu}{4}\left(\frac{\partial u_i}{\partial x_k} + \frac{\partial u_k}{\partial x_i} + \frac{\partial u_l}{\partial x_k}\frac{\partial u_l}{\partial x_i}\right)^2$$
$$= \frac{\mu}{4}\left(\frac{\partial u_i}{\partial x_k} + \frac{\partial u_k}{\partial x_i}\right)^2 + \frac{\mu}{2}\left(\frac{\partial u_i}{\partial x_k} + \frac{\partial u_k}{\partial x_i}\right)\frac{\partial u_l}{\partial x_k}\frac{\partial u_l}{\partial x_i}$$
$$+ \text{higher order term}$$
$$= \frac{\mu}{4}\left(\frac{\partial u_i}{\partial x_k} + \frac{\partial u_k}{\partial x_i}\right)^2 + \mu\left(\frac{\partial u_i}{\partial x_k}\right)\frac{\partial u_l}{\partial x_k}\frac{\partial u_l}{\partial x_i}$$
$$+ \text{higher order term} \quad (1.19a)$$

$$\left(\frac{K}{2}-\frac{\mu}{3}\right)u_{ll}^2=\left(\frac{K}{2}-\frac{1}{3}\mu\right)\left[\frac{\partial u_l}{\partial x_l}+\frac{\partial u_l}{\partial x_l}+\frac{\partial u_k}{\partial x_l}\frac{\partial u_k}{\partial x_l}\right]^2\frac{1}{4}$$
$$=\left(\frac{K}{2}-\frac{\mu}{3}\right)\left[\frac{\partial u_l}{\partial x_l}+\frac{1}{2}\left(\frac{\partial u_i}{\partial x_k}\right)^2\right]^2$$
$$=\left(\frac{K}{2}-\frac{\mu}{3}\right)\left(\frac{\partial u_l}{\partial x_l}\right)^2+\left(\frac{K}{2}-\frac{\mu}{3}\right)\left(\frac{\partial u_l}{\partial x_l}\right)\left(\frac{\partial u_i}{\partial x_k}\right)^2$$
$$+\text{ higher order term} \tag{1.19b}$$

$$\frac{A}{3}u_{ik}u_{il}u_{kl}=\frac{A}{3}\frac{1}{8}\left(\frac{\partial u_i}{\partial x_k}+\frac{\partial u_k}{\partial x_i}\right)\left(\frac{\partial u_i}{\partial x_l}+\frac{\partial u_l}{\partial x_i}\right)\left(\frac{\partial u_k}{\partial x_l}+\frac{\partial u_l}{\partial x_k}\right)$$
$$=\frac{A}{24}\left[6\frac{\partial u_i}{\partial x_k}\frac{\partial u_l}{\partial x_i}\frac{\partial u_l}{\partial x_k}+2\frac{\partial u_i}{\partial x_k}\frac{\partial u_k}{\partial x_l}\frac{\partial u_l}{\partial x_i}\right] \tag{1.19c}$$

$$Bu_{ik}^2u_{ll}=\frac{B}{4}\left[\frac{\partial u_i}{\partial x_k}+\frac{\partial u_k}{\partial x_i}+\frac{\partial u_l}{\partial x_i}\frac{\partial u_l}{\partial x_k}\right]^2\left[\frac{\partial u_l}{\partial x_l}+\frac{1}{2}\left(\frac{\partial u_i}{\partial x_k}\right)\right]^2$$
$$=\frac{B}{4}\left[\frac{\partial u_i}{\partial x_k}+\frac{\partial u_k}{\partial x_i}\right]^2\frac{\partial u_l}{\partial x_l}+\text{ higher order term}$$
$$=\frac{B}{4}\left[\left(\frac{\partial u_i}{\partial x_k}\right)^2\frac{\partial u_l}{\partial x_l}+\left(\frac{\partial u_k}{\partial x_i}\right)^2\frac{\partial u_l}{\partial x_l}\right]+\frac{B}{2}\left(\frac{\partial u_i}{\partial x_k}\frac{\partial u_k}{\partial x_i}\frac{\partial u_l}{\partial x_k}\right)$$
$$=\frac{B}{2}\left(\frac{\partial u_i}{\partial x_k}\right)^2\frac{\partial u_l}{\partial x_l}+\frac{B}{2}\left(\frac{\partial u_i}{\partial x_k}\right)\left(\frac{\partial u_k}{\partial x_i}\right)\left(\frac{\partial u_l}{\partial x_l}\right) \tag{1.19d}$$

$$\frac{C}{3}u_{ll}^3=\frac{C}{3}\left(\frac{\partial u_l}{\partial x_l}\right)^3 \tag{1.19e}$$

Therefore,

$$U=\frac{\mu}{4}\left(\frac{\partial u_i}{\partial x_k}+\frac{\partial u_k}{\partial x_i}\right)^2+\left(\frac{K}{2}-\frac{\mu}{3}\right)\left(\frac{\partial u_l}{\partial x_l}\right)^2$$
$$+\left(\mu+\frac{A}{4}\right)\frac{\partial u_i}{\partial x_k}\frac{\partial u_l}{\partial x_i}\frac{\partial u_l}{\partial x_k}+\left(\frac{B+K}{2}-\frac{\mu}{3}\right)\frac{\partial u_l}{\partial x_l}\left(\frac{\partial u_l}{\partial x_k}\right)^2$$
$$+\frac{A}{12}\frac{\partial u_i}{\partial x_k}\frac{\partial u_k}{\partial x_l}\frac{\partial u_l}{\partial x_i}+\frac{B}{2}\frac{\partial u_i}{\partial x_k}\frac{\partial u_k}{\partial x_i}\frac{\partial u_l}{\partial x_l}+\frac{C}{3}\left(\frac{\partial u_l}{\partial x_l}\right)^3$$
$$=U_1+U_2+U_3+U_4+U_5+U_6+U_7$$
$$+\text{ (High - order term ignored).} \tag{1.19}$$

1.1.3 Nonlinear Motion Equations

To derive the isotropic motion equations, we consider the infinitesimal change in internal energy per unit volume.

$$\delta U = \frac{\partial U}{\partial\left(\frac{\partial u_i}{\partial x_k}\right)}\delta\left(\frac{\partial u_i}{\partial x_k}\right) = \sigma_{ik}\delta\left(\frac{\partial u_i}{\partial x_k}\right) \tag{1.20}$$

where:

$$\sigma_{ik} = \frac{\partial U}{\partial\left(\frac{\partial u_i}{\partial x_k}\right)} \tag{1.21}$$

σ_{ik} represents the stress tensor. Equation (1.21) represents the change in internal energy caused by the work done by the stress tensor. Substituting Eq. (1.19) into Eq. (1.20) and retaining up to the second order quantity,

$$\frac{\partial U_1}{\partial\left(\frac{\partial u_i}{\partial x_k}\right)} = \frac{\mu}{2}\left(\frac{\partial u_i}{\partial x_k} + \frac{\partial u_k}{\partial x_i}\right)\left(1 + \frac{\partial\left(\frac{\partial u_k}{\partial x_i}\right)}{\partial\left(\frac{\partial u_i}{\partial x_k}\right)}\right) = \mu\left(\frac{\partial u_i}{\partial x_k} + \frac{\partial u_k}{\partial x_i}\right) \tag{1.22a}$$

$$\frac{\partial U_2}{\partial\left(\frac{\partial u_i}{\partial x_k}\right)} = \left(\frac{K}{2} - \frac{\mu}{3}\right)2\left(\frac{\partial u_l}{\partial x_k}\right)\frac{\partial\left(\frac{\partial u_l}{\partial x_l}\right)}{\partial\left(\frac{\partial u_i}{\partial x_k}\right)}) = 2\left(\frac{K}{2} - \frac{\mu}{3}\right)\frac{\partial u_l}{\partial x_k}\delta_{ik} \tag{1.22b}$$

$$\begin{aligned}
\frac{\partial U_3}{\partial\left(\frac{\partial u_i}{\partial x_k}\right)} &= \left(\mu + \frac{A}{4}\right)\left[\left(\frac{\partial u_l}{\partial x_i}\frac{\partial u_l}{\partial x_k} + \frac{\partial u_i}{\partial x_k}\frac{\partial u_l}{\partial x_i}\frac{\partial\left(\frac{\partial u_l}{\partial x_k}\right)}{\partial\left(\frac{\partial u_i}{\partial x_k}\right)}\right)\right. \\
&\quad \left. + \left(\frac{\partial u_i}{\partial x_k}\frac{\partial u_l}{\partial x_k}\frac{\partial\left(\frac{\partial u_l}{\partial x_i}\right)}{\partial\left(\frac{\partial u_i}{\partial x_k}\right)}\right)\right] = \left(\mu + \frac{A}{4}\right) \\
&\quad \left[\left(\frac{\partial u_l}{\partial x_i}\frac{\partial u_l}{\partial x_k} + \frac{\partial u_l}{\partial x_k}\frac{\partial u_i}{\partial x_l} + \frac{\partial\left(\frac{\partial u_i}{\partial x_k}\right)}{\partial\left(\frac{\partial u_i}{\partial x_k}\right)}\right) + \left(\frac{\partial u_k}{\partial x_l}\frac{\partial u_i}{\partial x_l}\frac{\partial\left(\frac{\partial u_i}{\partial x_k}\right)}{\partial\left(\frac{\partial u_i}{\partial x_k}\right)}\right)\right] \\
&= \left(\mu + \frac{A}{4}\right)\left[\frac{\partial u_l}{\partial x_i}\frac{\partial u_l}{\partial x_k} + \frac{\partial u_l}{\partial x_k}\frac{\partial u_i}{\partial x_l} + \frac{\partial u_k}{\partial x_l}\frac{\partial u_i}{\partial x_l}\right]
\end{aligned} \tag{1.22c}$$

$$\frac{\partial U_4}{\partial\left(\frac{\partial u_i}{\partial x_k}\right)} = \left(\frac{B+K}{2} - \frac{\mu}{3}\right)$$

$$\left[\left(\left(\frac{\partial u_n}{\partial x_m}\right)^2 \frac{\partial\left(\frac{\partial u_l}{\partial x_l}\right)}{\partial\left(\frac{\partial u_i}{\partial x_k}\right)}\right) + \left(2\left(\frac{\partial u_l}{\partial x_l}\right)\left(\frac{\partial u_i}{\partial x_k}\right)\frac{\partial\left(\frac{\partial u_i}{\partial x_k}\right)}{\partial\left(\frac{\partial u_i}{\partial x_k}\right)}\right)\right]$$

$$= \left(\frac{B+K}{2} - \frac{\mu}{3}\right)\left[\left(\frac{\partial u_n}{\partial x_m}\right)^2 \delta_{ik} 2\left(\frac{\partial u_l}{\partial x_l}\right)\left(\frac{\partial u_i}{\partial x_k}\right)\right] \tag{1.22d}$$

$$\frac{\partial U_5}{\partial\left(\frac{\partial u_i}{\partial x_k}\right)} = \frac{A}{12}\left[\left(\frac{\partial u_k}{\partial x_l}\frac{\partial u_l}{\partial x_i} + \left(\frac{\partial u_i}{\partial x_k}\right)\left(\frac{\partial u_k}{\partial x_l}\right)\frac{\partial\left(\frac{\partial u_l}{\partial x_k}\right)}{\partial\left(\frac{\partial u_i}{\partial x_k}\right)}\right)\right.$$

$$\left. + \left(\left(\frac{\partial u_i}{\partial x_k}\right)\left(\frac{\partial u_l}{\partial x_i}\right)\frac{\partial\left(\frac{\partial u_k}{\partial x_l}\right)}{\partial\left(\frac{\partial u_i}{\partial x_k}\right)}\right)\right]$$

$$= \frac{A}{12}\left[\frac{\partial u_k}{\partial x_l}\frac{\partial u_l}{\partial x_i} + \left(\frac{\partial u_l}{\partial x_i}\right)\left(\frac{\partial u_k}{\partial x_l}\right)\frac{\partial\left(\frac{\partial u_i}{\partial x_k}\right)}{\partial\left(\frac{\partial u_i}{\partial x_k}\right)}\right)$$

$$+ \left(\left(\frac{\partial u_l}{\partial x_i}\right)\left(\frac{\partial u_k}{\partial x_l}\right)\frac{\partial\left(\frac{\partial u_i}{\partial x_k}\right)}{\partial\left(\frac{\partial u_i}{\partial x_k}\right)}\right)\Bigg] = \frac{A}{4}\frac{\partial u_k}{\partial x_l}\frac{\partial u_l}{\partial x_i} \tag{1.22e}$$

$$\frac{\partial U_6}{\partial\left(\frac{\partial u_i}{\partial x_k}\right)} = \frac{B}{2}\left[\left(\frac{\partial u_k}{\partial x_i}\frac{\partial u_l}{\partial x_l} + \frac{\partial u_i}{\partial x_k}\frac{\partial u_l}{\partial x_l}\frac{\partial\left(\frac{\partial u_k}{\partial x_i}\right)}{\partial\left(\frac{\partial u_i}{\partial x_k}\right)}\right)\right.$$

$$\left. + \left(\left(\frac{\partial u_i}{\partial x_k}\right)\left(\frac{\partial u_k}{\partial x_i}\right)\frac{\partial\left(\frac{\partial u_l}{\partial x_l}\right)}{\partial\left(\frac{\partial u_i}{\partial x_k}\right)}\right)\right]$$

$$= \frac{B}{2}\left[2\frac{\partial u_k}{\partial x_i}\frac{\partial u_l}{\partial x_l} + \frac{\partial u_l}{\partial x_m}\frac{\partial u_m}{\partial x_l}\delta_{ik}\right] \tag{1.22f}$$

$$\frac{\partial U_7}{\partial\left(\frac{\partial u_i}{\partial x_k}\right)} = C\left(\frac{\partial u_l}{\partial x_l}\right)^2 \delta_{ik} \tag{1.22g}$$

Therefore,

$$\sigma_{ik} = \mu\left(\frac{\partial u_i}{\partial x_k} + \frac{\partial u_k}{\partial x_i}\right) + \left(K - \frac{2}{3}\mu\right)\frac{\partial u_l}{\partial x_l}\delta_{ik}$$

$$+ \left(\frac{K}{2} - \frac{\mu}{3} + \frac{B}{2}\right)\left[\left(\frac{\partial u_l}{\partial x_m}\right)^2 \delta_{ik} + 2\frac{\partial u_i}{\partial x_k}\frac{\partial u_l}{\partial x_l}\right]$$

$$+\frac{A}{4}\frac{\partial u_k}{\partial x_l}\frac{\partial u_l}{\partial x_i}+\frac{B}{2}\left[\frac{\partial u_l}{\partial x_m}\frac{\partial u_m}{\partial x_l}\delta_{ik}+2\frac{\partial u_k}{\partial x_i}\frac{\partial u_l}{\partial x_l}\right]$$
$$+C\left(\frac{\partial u_l}{\partial x_l}\right)^2\delta_{ik} \tag{1.22}$$

On the other hand, we know that the motion equation can be written as

$$\rho_0\frac{\partial^2 u_i}{\partial t^2}=\frac{\partial \sigma_{ik}}{\partial x_k} \tag{1.23}$$

where ρ_0 is the density of the undeformed object.

$\frac{\partial \sigma_{ik}}{\partial x_k}$ represents the total force acting on the unit body in the i direction.

$$\frac{\partial(\sigma_{ik})_1}{\partial x_k}=\mu\frac{\partial^2 u_i}{\partial x_k^2}+\mu\frac{\partial^2 u_k}{\partial x_k\partial x_i} \tag{1.24a}$$

$$\frac{\partial(\sigma_{ik})_2}{\partial x_k}=\left(K-\frac{2}{3}\mu\right)\frac{\partial^2 u_l}{\partial x_i\partial x_l} \tag{1.24b}$$

$$\frac{\partial(\sigma_{ik})_3}{\partial x_k}=\left(\mu+\frac{A}{4}\right)\left[\frac{\partial^2 u_l}{\partial x_k^2}\frac{\partial u_l}{\partial x_i}+\frac{\partial^2 u_k}{\partial x_k^2}\frac{\partial u_i}{\partial x_l}+2\frac{\partial^2 u_l}{\partial x_k\partial x_k}\frac{\partial u_l}{\partial x_k}\right]$$
$$+\left(\mu+\frac{A}{4}\right)\left[\frac{\partial^2 u_l}{\partial x_k\partial x_i}\frac{\partial u_l}{\partial x_k}+\frac{\partial^2 u_k}{\partial x_k\partial x_l}\frac{\partial u_i}{\partial x_l}\right] \tag{1.24c}$$

$$\frac{\partial(\sigma_{ik})_4}{\partial x_k}=\left(K-\frac{2}{3}\mu+B\right)\left[\frac{\partial^2 u_l}{\partial x_k\partial x_i}\frac{\partial u_l}{\partial x_k}+\frac{\partial^2 u_k}{\partial x_k\partial x_l}\frac{\partial u_i}{\partial x_l}\right]$$
$$+\left(K-\frac{2}{3}\mu+B\right)\frac{\partial^2 u_i}{\partial x_k^2}\frac{\partial u_l}{\partial x_l} \tag{1.24d}$$

$$\frac{\partial(\sigma_{ik})_5}{\partial x_k}=\frac{A}{4}\left[\frac{\partial^2 u_k}{\partial x_l\partial x_k}\frac{\partial u_l}{\partial x_i}+\frac{\partial^2 u_i}{\partial x_i\partial x_k}\frac{\partial u_k}{\partial x_l}\right] \tag{1.24e}$$

$$\frac{\partial(\sigma_{ik})_6}{\partial x_k}=B\left[\frac{\partial^2 u_l}{\partial x_l\partial x_k}\frac{\partial u_k}{\partial x_l}+\frac{\partial^2 u_k}{\partial x_l\partial x_k}\frac{\partial u_l}{\partial x_i}\right]$$
$$+B\frac{\partial^2 u_k}{\partial x_k\partial x_i}\frac{\partial u_l}{\partial x_i} \tag{1.24f}$$

$$\frac{\partial(\sigma_{ik})_7}{\partial x_k}=2C\left(\frac{\partial u_l}{\partial x_l}\right)\frac{\partial^2 u_k}{\partial x_i\partial x_k} \tag{1.24g}$$

Substituting Eqs. (1.24a–1.24g) into Eq. (1.23) yields the following motion equation

$$\rho_0 \frac{\partial^2 u_i}{\partial t^2} - \mu \frac{\partial^2 u_i}{\partial x_k^2} - \left(K + \frac{\mu}{3}\right) \frac{\partial^2 u_l}{\partial x_l \partial x_i} = F_i \tag{1.25}$$

where:

$$\begin{aligned} F_i = {} & \left(\mu + \frac{A}{4}\right)\left(\frac{\partial^2 u_l}{\partial x_k^2}\frac{\partial u_l}{\partial x_i} + \frac{\partial^2 u_l}{\partial x_k^2}\frac{\partial u_i}{\partial x_l} + 2\frac{\partial^2 u_i}{\partial x_l \partial x_k}\frac{\partial u_l}{\partial x_k}\right) \\ & + \left(K + \frac{\mu}{3} + \frac{A}{4} + B\right)\left(\frac{\partial^2 u_l}{\partial x_i \partial x_k}\frac{\partial u_l}{\partial x_k} + \frac{\partial^2 u_k}{\partial x_k \partial x_l}\frac{\partial u_i}{\partial x_l}\right) \\ & + \left(\frac{A}{4} + B\right)\left(\frac{\partial^2 u_k}{\partial x_l \partial x_k}\frac{\partial u_l}{\partial x_i} + \frac{\partial^2 u_l}{\partial x_i \partial x_k}\frac{\partial u_k}{\partial x_l}\right) \\ & + \left(K - \frac{2}{3}\mu + B\right)\frac{\partial^2 u_i}{\partial x_k^2}\frac{\partial u_l}{\partial x_l} + (B + 2C)\frac{\partial^2 u_k}{\partial x_k \partial x_i}\frac{\partial u_l}{\partial x_l} \end{aligned}$$

F_i is the i-th component of the body force. Equations (1.24a–1.24g) is the basic equation of the "five constant" nonlinear elasticity theory (the so-called five constants refer to k, μ, A, B, C).

Formally, the source of nonlinearity is due to geometric nonlinearity and the nonlinearity of the generalized Hooke's law. Geometric nonlinearity refers to the second term in the deformation tensor represented by Eq. (1.10), which is purely due to the change in geometric conditions after deformation and does not depend on the physical properties of the deformed object. The nonlinearity of the generalized Hooke's law (Eq. 1.22), also known as physical nonlinearity, is caused by the inclusion of third-order elastic constants in the elastic energy, so it is directly related to the nonlinear elastic constants of the specific solid. In other words, physical nonlinearity is determined by the third-order elastic moduli A, B, C.

1.2 Nonlinear Propagation of Elastic Waves in Solids

When the first-order approximation is only a longitudinal wave, then

$$\vec{u}(x, t) = u_x(x, t)\vec{i}$$

Equations (1.24a–1.24g) becomes

$$\begin{aligned} & \rho_0 \frac{\partial^2 u_x}{\partial t^2} - \mu \frac{\partial^2 u_x}{\partial x^2} - \left(K + \frac{\mu}{3}\right)\frac{\partial^2 u_x}{\partial x^2} \\ & = \left[3\left(K + \frac{4}{3}\mu\right) + 2A + 6B + 2C\right]\frac{\partial^2 u_x}{\partial x^2}\frac{\partial u_x}{\partial x} \end{aligned}$$

$$\text{ream } \alpha = K + \frac{4}{3}\mu, \quad \beta = 3\alpha + 2A + 6B + 2C \tag{1.26}$$

namely,

$$\rho_0 \frac{\partial^2 u_x}{\partial t^2} - \frac{\partial^2 u_x}{\partial x^2} = \beta \frac{\partial^2 u_x}{\partial x^2} \frac{\partial u_x}{\partial x} \tag{1.27}$$

Let the solution be

$u_x = u_x^{(1)} + u_x^{(2)}$ Substituting, we get

First-order approximation equation

$$\rho_0 \frac{\partial^2 u_x^{(1)}}{\partial x^2} - \alpha \frac{\partial^2 u_x^{(1)}}{\partial t^2} = 0, c_l = \sqrt{\alpha/\rho_0} = \sqrt{\left(k + \frac{4}{3}\mu\right)/\rho_0} \tag{1.28}$$

Second-order approximation equation

$$\rho_0 \frac{\partial^2 u_x^{(2)}}{\partial x^2} - \alpha \frac{\partial^2 u_x^{(2)}}{\partial t^2} = \beta \frac{\partial^2 u_x^{(1)}}{\partial x^2} \frac{\partial u_x^{(1)}}{\partial x} \tag{1.29}$$

Solve this equation

When the boundary conditions are: $u_x^{(1)}(0, \mathrm{t}) = u_0(1 - \cos\ \omega \mathrm{t})$

The first-order approximate solution is

$$u_x^1(x, t) = u_0[1 - \cos(\omega t - k_l x)] \tag{1.30}$$

$$v_x^1(\mathrm{x,t}) = u_0\omega \sin(\omega t - k_l x) = v_0 \sin(\omega t - k_l x) \tag{1.31}$$

The solution to Eq. (1.26) is

$$u_x^{(2)} = -\frac{\beta}{\alpha} \frac{k_l^2 x u_0^2}{8} [1 - \cos 2(\omega t - k_l x)] \tag{1.32}$$

where: $k_l = \frac{\omega}{c_l}$

Particle velocity

$$v^{(2)}(x, t) = -\frac{\beta}{4\alpha} \frac{\omega x v_0^2}{c_l^2} \sin 2(\omega t - k_l x) = v_{20} \sin 2(\omega t - k_l x) \tag{1.33}$$

$$v_{20} = -\frac{\beta}{4\alpha}\frac{\omega x p_{10}^2}{\rho_0^2 c_l^4}$$

When there is a longitudinal wave at x = 0, the amplitude of the secondary harmonic (longitudinal wave) generated at distance x, and the corresponding sound pressure are

$$p_{20} = -\frac{\beta}{4\alpha}\frac{\omega x p_{10}^2}{\rho_0 c_l^3} \tag{1.34}$$

$$p_{20} = N\frac{\omega x p_{10}^2}{2\rho_0 c_l^3} \text{ here } N = -\frac{\beta}{2\alpha} = -\left(\frac{3}{2} + \frac{A+3B+C}{\alpha}\right) \tag{1.35}$$

$$\text{Or: } p_{20} = \frac{(n_l+1)\omega x p_{10}^2}{4\rho_0 c_l^3} \text{ here } n_l = -\frac{\beta}{\alpha} - 1 = -4 - \frac{2(A+3B+C)}{\alpha} \tag{1.36}$$

This η_l corresponds to the n used previously in liquids and gases.

For most solids (A + 3B + C) < 0, η_l is approximately 3–14, which is of the same order of magnitude as the n of the liquid, but because the speed of sound in solids is greater than in liquids, the secondary harmonic of the longitudinal wave in solids is at least an order of magnitude smaller than in liquids.

Equation (1.36) can also be written as

$$\frac{A+3B+C}{\alpha} = -\left(\frac{2\rho_0 c_l^3}{\omega x p_{10}}\frac{p_{20}}{p_{10}} + \frac{3}{2}\right) \tag{1.37}$$

If we can determine $p_{20,}p_{10}$ and other quantities in the experiment, then we can find out $\frac{A+3B+C}{\alpha}$, that is, the third-order elastic modulus A + 3B + C, from Eq. (1.35) [2–6].

1.3 Interaction of Elastic Waves in Solids

$$\rho_0\frac{\partial^2 S_i}{\partial t^2} - \mu\frac{\partial^2 S_i}{\partial x_k^2} - \left(K + \frac{\mu}{3}\right)\frac{\partial^2 S_l}{\partial x_l \partial x_i} = F_i \tag{1.38}$$

$$\text{Assuming : } \vec{S} = \vec{S}^{(1)} + \vec{S}^{(s)} \tag{1.39}$$

$\vec{S}^{(1)}$ is the solution to Eq. (1.38) when the right side is zero, $\vec{S}^{(s)}$ is the perturbation solution.

Since we are interested in the interaction of two waves, we assume that $\vec{S}^{(1)}$ is the superposition of two simple harmonic waves

$$\vec{S}^{(1)} = \vec{A}_0\cos\left(\omega_1 t - \vec{k_1}\cdot\vec{r}\right) + \vec{B}_0\cos\left(\omega_2 t - \vec{k_2}\cdot\vec{r}\right) \tag{1.40}$$

$\vec{A}_0$, $\vec{B}_0$ are amplitude vector $\vec{k}_1$, $\vec{k}_2$ are the wave vector,

For shear waves, the amplitude vector is perpendicular to the wave vector $\vec{k}, \omega = c_t k,\ c_t = \sqrt{\frac{\mu}{\rho_0}}$

For longitudinal waves, the amplitude vector is parallel to the wave vector

$$\vec{k}, \omega = c_l k, c_l = \sqrt{\frac{K + \frac{4}{3}\mu}{\rho_0}} \tag{1.41}$$

Substituting $\vec{S}^{(1)}$ into the wave equation, we get

$$\rho_0 \frac{\partial^2 S_i^{(s)}}{\partial t^2} - \mu \frac{\partial^2 S_i^{(s)}}{\partial x_k^2} - \left(K + \frac{\mu}{3}\right) \frac{\partial^2 S_l^{(s)}}{\partial x_l \partial x_i} = p_i \tag{1.42}$$

Since $\vec{S}^{(1)}$ satisfies the homogeneous solution, so $\vec{S}^{(1)}$ Disappearing from the left side of the equation, the vector $\vec{p}$ is the sum of the products of two simple harmonic waves, where some terms represent the interaction of the simple harmonic waves.

Where

$$\begin{aligned} \vec{p} = \vec{I^+} \sin\left[(\omega_1 + \omega_2)\, t - \left(\vec{k_1} + \vec{k_2}\right) \cdot \vec{r}\right] \\ + \vec{I^-} \sin\left[(\omega_1 - \omega_2)\, t - \left(\vec{k_1} - \vec{k_2}\right) \cdot \vec{r}\right] \end{aligned} \tag{1.43}$$

Because

$$\vec{u} = \vec{A}_0 \cos(\omega_1 t - \vec{k}_1 \cdot \vec{r}) + \vec{B}_0 \cos(\omega_2 t - \vec{k}_2 \cdot \vec{r}) = \vec{A}_0 \cos(a) + \vec{B}_0 \cos(b)$$

Here: a $= \omega_1 t - \vec{k}_1 \cdot \vec{r}, b = \omega_2 t - \vec{k}_2 \cdot \vec{r}$

$$\begin{aligned} \nabla^2 \vec{u} &= \vec{A}_0 k_1^2 \cos(a) - \vec{B}_0 k_2^2 \cos(b) \\ \frac{\partial \vec{u}}{\partial x_i} &= \vec{A}_0 \frac{\partial}{\partial x_i}\left(\vec{k}_1 \cdot \vec{r}\right) \sin(a) + \vec{B}_0 \frac{\partial}{\partial x_i}\left(\vec{k}_2 \cdot \vec{r}\right) \sin(b) \\ &= \vec{A}_0 k_1^i \sin(a) + \vec{B}_0 k_2^i \sin(b) \\ \vec{\nabla} u_i &= A_0^i \vec{k}_1 \sin(a) + B_0^i \vec{k}_2 \sin(b) \\ \nabla \cdot \vec{u} &= \left(\vec{A}_0 \cdot \vec{k}_1\right) \sin(a) + \left(\vec{B}_0 \cdot \vec{k}_2\right) \sin(b) \end{aligned} \tag{1}$$

$$\left(\mu + \frac{A}{4}\right)\left(\frac{\partial^2 u_l}{\partial x_k^2} \frac{\partial u_l}{\partial x_i} + \frac{\partial^2 u_k}{\partial x_k^2} \frac{\partial u_i}{\partial x_l} + 2 \frac{\partial^2 u_i}{\partial x_l \partial x_k} \frac{\partial u_l}{\partial x_k}\right)$$

$$= \left(\mu + \frac{A}{4}\right)\left(\nabla^2 u_l \frac{\partial u_l}{\partial x_i} + \nabla^2 u_l \frac{\partial u_i}{\partial x_l} + 2\frac{\partial^2 u_i}{\partial x_l \partial x_k}\frac{\partial u_l}{\partial x_k}\right)$$

$$= \left(\mu + \frac{A}{4}\right)\left(\nabla^2 u \cdot \frac{\partial \vec{u}}{\partial x_i} + \nabla^2 u \cdot \nabla u_i + 2\frac{\partial}{\partial x_k}\vec{\nabla} u_i \cdot \frac{\partial \vec{u}}{\partial x_k}\right)$$

$$= -\left(\mu + \frac{A}{4}\right)\{\left(\vec{A}_0 \cdot \vec{B}_0\right)\left(\overrightarrow{k_2} \cdot \overrightarrow{k_2}\right)\overrightarrow{k_1}\sin(a)\cos(b)$$

$$+ \left(\vec{A}_0 \cdot \vec{B}_0\right)\left(\overrightarrow{k_1} \cdot \overrightarrow{k_1}\right)\overrightarrow{k_2}\sin(b)\cos(a)$$

$$+ \left(\vec{B}_0 \cdot \overrightarrow{k_1}\right)\left(\overrightarrow{k_2} \cdot \overrightarrow{k_2}\right)\vec{A}_0\sin(a)\cos(b) + \left(\vec{A}_0 \cdot \overrightarrow{k_2}\right)\left(\overrightarrow{k_1} \cdot \overrightarrow{k_1}\right)$$

$$\vec{B}_0\sin(b)\cos(a) + 2\left(\vec{A}_0 \cdot \overrightarrow{k_2}\right)\left(\overrightarrow{k_1} \cdot \overrightarrow{k_2}\right)\vec{B}_0\sin(a)\cos(b)$$

$$+ 2\left(\vec{B}_0 \cdot \overrightarrow{k_1}\right)\left(\overrightarrow{k_1} \cdot \overrightarrow{k_2}\right)\vec{A}_0\sin(b)\cos(a)\}$$

$$= -\frac{1}{2}\left(\mu + \frac{A}{4}\right)\{\left(\vec{A}_0 \cdot \vec{B}_0\right)\left(\overrightarrow{k_2} \cdot \overrightarrow{k_2}\right)\overrightarrow{k_1} \pm \left(\vec{A}_0 \cdot \vec{B}_0\right)\left(\overrightarrow{k_1} \cdot \overrightarrow{k_1}\right)\overrightarrow{k_1}$$

$$+ \left(\vec{B}_0 \cdot \overrightarrow{k_1}\right)\left(\overrightarrow{k_2} \cdot \overrightarrow{k_2}\right)\vec{A}_0 \pm \left(\vec{A}_0 \cdot \overrightarrow{k_2}\right)\left(\overrightarrow{k_1} \cdot \overrightarrow{k_1}\right)\vec{B}_0$$

$$\pm 2\left(\vec{A}_0 \cdot \overrightarrow{k_2}\right)\left(\overrightarrow{k_1} \cdot \overrightarrow{k_2}\right)\vec{B}_0 \pm 2\left(\vec{B}_0 \cdot \overrightarrow{k_1}\right)\left(\overrightarrow{k_1} \cdot \overrightarrow{k_2}\right)\vec{A}_0\}$$

$$\times \sin[(\omega_1 \pm \omega_2)t - \left(\vec{k}_1 \pm \vec{k}_2\right) \cdot \vec{r}] \tag{2}$$

$$\left(K + \frac{\mu}{3} + \frac{A}{4} + B\right)\left(\frac{\partial^2 u_l}{\partial x_i \partial x_k}\frac{\partial u_l}{\partial x_k} + \frac{\partial^2 u_k}{\partial x_l \partial x_k}\frac{\partial u_i}{\partial x_l}\right)$$

$$= \left(K + \frac{\mu}{3} + \frac{A}{4} + B\right)\left(\frac{\partial}{\partial x_k}\left(\frac{\partial \vec{u}}{\partial x_i}\right) \cdot \frac{\partial \vec{u}}{\partial x_k} + \frac{\partial}{\partial x_k}\left(\vec{\nabla} u_k\right) \cdot\right.$$

$$\left.\left(\vec{\nabla} u_k\right)\right] = -\left(K + \frac{\mu}{3} + \frac{A}{4} + B\right)$$

$$\{\left(\overrightarrow{A_0} \cdot \overrightarrow{B_0}\right)\left(\overrightarrow{k_1} \cdot \overrightarrow{k_2}\right)\overrightarrow{k_2}\sin(a)\cos(b) + \left(\overrightarrow{A_0} \cdot \overrightarrow{B_0}\right)\left(\overrightarrow{k_1} \cdot \overrightarrow{k_2}\right)$$

$$\overrightarrow{k_1}\sin(b)\cos(a) + \left(\overrightarrow{B_0} \cdot \overrightarrow{k_2}\right)\left(\overrightarrow{k_1} \cdot \overrightarrow{k_2}\right)\overrightarrow{A_0}\sin(a)\cos(b)\}$$

$$= -\frac{1}{2}\left(K + \frac{\mu}{3} + \frac{A}{4} + B\right)\{\left(\overrightarrow{A_0} \cdot \overrightarrow{B_0}\right)\left(\overrightarrow{k_1} \cdot \overrightarrow{k_2}\right)\overrightarrow{k_2}$$

$$\pm \left(\overrightarrow{A_0} \cdot \overrightarrow{B_0}\right)\left(\overrightarrow{k_1} \cdot \overrightarrow{k_2}\right)\overrightarrow{k_1} + \left(\overrightarrow{B_0} \cdot \overrightarrow{k_2}\right)\left(\overrightarrow{k_1} \cdot \overrightarrow{k_2}\right)\overrightarrow{A_0}$$

$$\pm \left(\overrightarrow{A_0} \cdot \overrightarrow{k_1}\right)\left(\overrightarrow{k_1} \cdot \overrightarrow{k_2}\right)\overrightarrow{B_0}\} \times \sin[(\omega_1 \pm \omega_2)t - \left(\overrightarrow{k_1} \pm \overrightarrow{k_2}\right) \cdot \vec{r}\,] \tag{3}$$

$$\left(K - \frac{2}{3}\mu + B\right)\left\{\frac{\partial^2 u_i}{\partial x_k \partial x_k}\frac{\partial u_l}{\partial x_l}\right\}$$

$$= \left(K - \frac{2}{3}\mu + B\right)\{\nabla^2 u_i)(\nabla \cdot \vec{u})\}$$

$$
\begin{aligned}
&= -\frac{1}{2}\left(K - \frac{2}{3}\mu + B\right) \\
&\left\{\left(\vec{A}_0 \cdot \vec{k}_0\right)\left(\vec{k_2} \cdot \vec{k_1}\right)\vec{B}_0 + \left(\vec{B}_0 \cdot \vec{k_2}\right)\left(\vec{k_2} \cdot \vec{k_1}\right)\vec{A}_0\right\} \\
&\times \sin[(\omega_1 \pm \omega_2)t - \left(\vec{k}_1 \pm \vec{k}_2\right) \cdot \vec{r}]
\end{aligned} \tag{4}
$$

$$
\begin{aligned}
&(B + 2C)\left\{\frac{\partial^2 u_k}{\partial x_i \partial x_k}\frac{\partial u_l}{\partial x_l}\right\} = (B + 2C)\{\nabla \cdot \vec{u})(\nabla \cdot \vec{u})\} \\
&= -\frac{1}{2}(B + 2C)\left\{\left(\vec{A}_0 \cdot \vec{k}_1\right)\left(\vec{B_0} \cdot \vec{k_2}\right)\vec{k}_2\right. \\
&\left.\pm \left(\vec{A}_0 \cdot \vec{k_1}\right)\left(\vec{B_0} \cdot \vec{k_2}\right)\vec{k}_1\right\} \times \sin\left[(\omega_1 \pm \omega_2)t - \left(\vec{k}_1 \pm \vec{k}_2\right) \cdot \vec{r}\right]
\end{aligned} \tag{5}
$$

$$
\begin{aligned}
&\left(\frac{A}{4} + B\right)\left(\frac{\partial^2 u_k}{\partial x_l \partial x_k}\frac{\partial u_l}{\partial x_i} + \frac{\partial^2 u_l}{\partial x_i \partial x_k}\frac{\partial u_k}{\partial x_l}\right) \\
&= \left(\frac{A}{4} + B\right)\left\{\frac{\partial}{\partial x_k}\left(\vec{\nabla} u_k\right) \cdot \frac{\partial \vec{u}}{\partial x_i} + \frac{\partial}{\partial x_k}(\nabla \cdot \vec{u}) \cdot \vec{\nabla} u_k\right\} \\
&= -\frac{1}{2}\left(\frac{1}{4}A + B\right)\{\left(\vec{A}_0 \cdot \vec{k_2}\right)\left(\vec{B}_0 \cdot \vec{k_2}\right)\vec{k_1} \\
&\pm \left(\vec{A}_0 \cdot \vec{k_1}\right)\left(\vec{B}_0 \cdot \vec{k_1}\right)\vec{k_2} + \left(\vec{A}_0 \cdot \vec{k_2}\right)\left(\vec{B}_0 \cdot \vec{k_1}\right)\vec{k_2} \\
&\pm \left(\vec{A}_0 \cdot \vec{k_2}\right)\left(\vec{B}_0 \cdot \vec{k_1}\right)\vec{k_1}\} \times \sin[(\omega_1 \pm \omega_2)t - \left(\vec{k}_1 \pm \vec{k}_2\right) \cdot \vec{r}]
\end{aligned}
$$

Therefore

$$
\begin{aligned}
\vec{I^{\pm}} = &-\frac{1}{2}\left(\mu + \frac{A}{4}\right)\left\{\left(\vec{A_0} \cdot \vec{B_0}\right)\left(\vec{k_2} \cdot \vec{k_2}\right)\vec{k_1} \pm \left(\vec{A_0} \cdot \vec{B_0}\right)\left(\vec{k_1} \cdot \vec{k_1}\right)\vec{k_2}\right. \\
&+ \left(\vec{B_0} \cdot \vec{k_1}\right)\left(\vec{k_2} \cdot \vec{k_2}\right)\vec{A_0} \pm \left(\vec{A_0} \cdot \vec{k_2}\right)\left(\vec{k_1} \cdot \vec{k_1}\right)\vec{B_0} \\
&\left.+ 2\left(\vec{A_0} \cdot \vec{k_2}\right)\left(\vec{k_1} \cdot \vec{k_2}\right)\vec{B_0} \pm 2\left(\vec{B_0} \cdot \vec{k_1}\right)\left(\vec{k_1} \cdot \vec{k_2}\right)\vec{A_0}\right\} \\
&- \frac{1}{2}\left(K + \frac{\mu}{3} + \frac{A}{4} + B\right)\left\{\left(\vec{A_0} \cdot \vec{B_0}\right)\left(\vec{k_1} \cdot \vec{k_2}\right)\vec{k_2}\right. \\
&\left.\pm \left(\vec{A_0} \cdot \vec{B_0}\right)\left(\vec{k_1} \cdot \vec{k_2}\right)\vec{k_1} + \left(\vec{B_0} \cdot \vec{k_2}\right)\left(\vec{k_1} \cdot \vec{k_2}\right)\vec{A_0} \pm \left(\vec{A_0} \cdot \vec{k_1}\right)\left(\vec{k_1} \cdot \vec{k_2}\right)\vec{B_0}\right\} \\
&- \frac{1}{2}\left(K - \frac{2}{3}\mu + B\right)\left\{\left(\vec{A_0} \cdot \vec{k_1}\right)\left(\vec{k_2} \cdot \vec{k_1}\right)\vec{B_0} \pm \left(\vec{B_0} \cdot \vec{k_2}\right)\left(\vec{k_1} \cdot \vec{k_2}\right)\vec{A_0}\right\} \\
&- \frac{1}{2}\left(\frac{A}{4} + B\right)\{\left(\vec{A_0} \cdot \vec{k_2}\right)\left(\vec{B_0} \cdot \vec{k_2}\right)\vec{k_1} \pm \left(\vec{A_0} \cdot \vec{k_1}\right)\left(\vec{B_0} \cdot \vec{k_1}\right)\vec{k_2} \\
&+ \left(\vec{A_0} \cdot \vec{k_2}\right)\left(\vec{B_0} \cdot \vec{k_1}\right)\vec{k_2} \pm \left(\vec{A_0} \cdot \vec{k_2}\right)\left(\vec{B_0} \cdot \vec{k_1}\right)\vec{k_1}\} - \frac{1}{2}(B + 2C) \\
&\left\{\left(\vec{A_0} \cdot \vec{k_1}\right)\left(\vec{B_0} \cdot \vec{k_2}\right)\vec{k_2} \pm \left(\vec{A_0} \cdot \vec{k_1}\right)\left(\vec{B_0} \cdot \vec{k_2}\right)\vec{k_1}\right\}
\end{aligned} \tag{1.44}
$$

Rewrite (1.42) as

$$\rho_0 \frac{\partial^2 S_i^{(s)}}{\partial t^2} - \mu \frac{\partial^2 S_i^{(s)}}{\partial x_k^2} - \left(K + \frac{4}{3}\mu\right) \frac{\partial^2 S_l^{(s)}}{\partial x_l \partial x_i} + \mu \frac{\partial^2 S_l^{(s)}}{\partial x_l \partial x_i} = p_i \tag{1.45}$$

Rewritten to vector mode

$$\because \frac{\partial^2 S_i}{\partial x_k^2} = \nabla^2 S_i, \frac{\partial^2 S_l^{(s)}}{\partial x_l \partial x_i} = \frac{\partial}{\partial x_i}\left(\frac{\partial S_l}{\partial x_l}\right) = \frac{\partial}{\partial x_i}\left(\nabla \cdot \vec{S}\right)$$

$$\therefore \frac{\partial^2 \vec{S}}{\partial t^2} - c_t^2 \nabla^2 \overline{S} - c_l^2 \nabla\left(\nabla \cdot \vec{S}\right) + c_t^2 \nabla\left(\nabla \cdot \vec{S}\right) = 4\pi \vec{q}, \quad 4\pi \vec{q} = \frac{\overline{p}}{\rho_0}$$

namely $$\frac{\partial^2 \vec{S}}{\partial t^2} - c_l^2 \nabla\left(\nabla \cdot \vec{S}\right) + c_t^2 \nabla \times \left(\nabla \times \vec{S}\right) = 4\pi \vec{q} \tag{1.46}$$

Scattered wave

$$\vec{S}(r,t) = \frac{\overrightarrow{I^+} \cdot \hat{r}}{4\pi c_l^2 \rho_0} \frac{\hat{r}}{r}$$

$$\int_V \sin\left[\left(\frac{\omega_1 + \omega_2}{c_l}\hat{r} - \overrightarrow{k_1} - \overrightarrow{k_2}\right) \cdot \overrightarrow{r'} - (\omega_1 + \omega_2)\left(\frac{r}{c_l} - t\right)\right] dV$$

$$+ \frac{\overrightarrow{I^-} \cdot \hat{r}}{4\pi c_l^2 \rho_0} \frac{\hat{r}}{r} \int_V \sin\left[\left(\frac{\omega_1 - \omega_2}{c_l}\hat{r} - \overrightarrow{k_1} + \overrightarrow{k_2}\right) \cdot \overrightarrow{r'} - (\omega_1 - \omega_2)\left(\frac{r}{c_l} - t\right)\right] dV$$

$$+ \frac{\overrightarrow{I^+} - \left(\hat{r} \cdot \overrightarrow{I^+}\right)}{4\pi c_t^2 \rho_0} \frac{\hat{r}}{r} \int_V \sin\left[\left(\frac{\omega_1 + \omega_2}{c_t}\hat{r} - \overrightarrow{k_1} - \overrightarrow{k_2}\right) \cdot \overrightarrow{r'} - (\omega_1 + \omega_2)\left(\frac{r}{c_t} - t\right)\right] dV$$

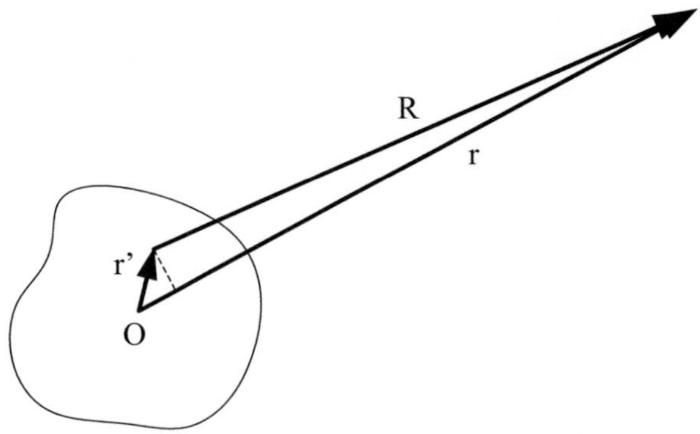

Fig. 1.2 The relationship between the position of the sound source and the observer

$$+\frac{\overrightarrow{I^-}-\left(\hat{r}\cdot\overrightarrow{I^-}\right)\hat{r}}{4\pi c_t^2\rho_0}\frac{\hat{r}}{r}\int\limits_V \sin\left[\left(\frac{\omega_1-\omega_2}{c_t}\hat{r}-\overrightarrow{k_1}+\overrightarrow{k_2}\right)\cdot\overrightarrow{r'}-(\omega_1-\omega_2)\left(\frac{r}{c_t}-t\right)\right]dV \tag{1.47}$$

$$\hat{r}=\frac{\overrightarrow{r'}}{r}\ \text{is a unit vector in the}\ \vec{r}\text{direction}$$

$$\text{far field}\ R\approx r-\hat{r}\cdot\overrightarrow{r'}$$

(See Fig. 1.2)

resonance conditions:

$$\begin{aligned}
&\frac{\omega_1+\omega_2}{c_l}\hat{r}_s-\left(\overrightarrow{k_1}+\overrightarrow{k_2}\right)=0 \quad\Rightarrow\quad (c_l,\omega_1+\omega_2)\\
&\frac{\omega_1-\omega_2}{c_l}\hat{r}_s-\left(\overrightarrow{k_1}-\overrightarrow{k_2}\right)=0 \quad\Rightarrow\quad (c_l,\omega_1-\omega_2)\\
&\frac{\omega_1+\omega_2}{c_t}\hat{r}_s-\left(\overrightarrow{k_1}+\overrightarrow{k_2}\right)=0 \quad\Rightarrow\quad (c_t,\omega_1+\omega_2)\\
&\frac{\omega_1-\omega_2}{c_t}\hat{r}_s-\left(\overrightarrow{k_1}-\overrightarrow{k_2}\right)=0\Rightarrow(c_t,\omega_1-\omega_2)
\end{aligned} \tag{1.48}$$

Taking the first term of (1.46) as an example, discuss the resonance condition (Fig. 1.3)

$$\left(\frac{\omega_1+\omega_2}{c_l}\right)^2=k_1^2+k_2^2+2\overrightarrow{k_1}\cdot\overrightarrow{k_2}=k_1^2+k_2^2+2k_1k_2\cos\varphi \tag{1.49}$$

There are three situations of interaction between two intersecting sound waves:

$$\begin{aligned}
&L+L\rightarrow L\\
&T+T\rightarrow L\\
&L+T\rightarrow L
\end{aligned}$$

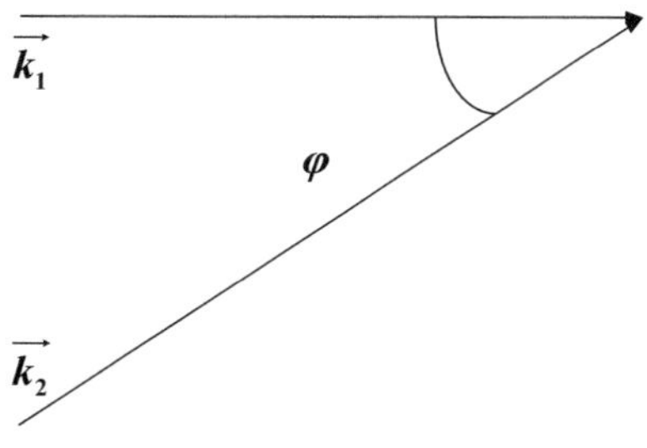

Fig. 1.3 $\vec{k}_1$, $\vec{k}_2$ is the initial incident wave, with an angle of φ

$L + L \to L$

(1) $\left(\frac{\omega_1 + \omega_2}{c_l}\right)^2 = \left(\frac{\omega_1}{c_l}\right)^2 + \left(\frac{\omega_2}{c_l}\right)^2 + 2\left(\frac{\omega_1}{c_l}\right)\left(\frac{\omega_2}{c_l}\right)\cos\varphi \quad \Rightarrow \quad \varphi = 0$

$T + T \to L$

$$\left(\frac{\omega_1 + \omega_2}{c_l}\right)^2 = \left(\frac{\omega_1}{c_t}\right)^2 + \left(\frac{\omega_2}{c_t}\right)^2 + 2\left(\frac{\omega_1}{c_t}\right)\left(\frac{\omega_2}{c_t}\right)\cos\varphi$$

(2) ream $c = \frac{c_t}{c_l}, \quad a = \frac{\omega_2}{\omega_1},$

obtain $\cos\varphi = c^2 + \frac{1}{2a}(c^2 - 1)\left(a^2 + 1\right)$

$\because |\cos\varphi| \le 1 \quad \therefore \frac{1-c}{1+c} < a < \frac{1+c}{1-c}$

(3) $T + L \to L$

$$\left(\frac{\omega_1 + \omega_2}{c_l}\right)^2 = \left(\frac{\omega_1}{c_t}\right)^2 + \left(\frac{\omega_2}{c_l}\right)^2 + 2\left(\frac{\omega_1}{c_t}\right)\left(\frac{\omega_2}{c_l}\right)\cos\varphi \quad \text{ream} c = \frac{c_t}{c_l}, \quad a = \frac{\omega_2}{\omega_1},$$

$$\text{obtain } \cos\varphi = c + \frac{1}{2c}a\left(c^2 - 1\right) \because |\cos\varphi| \le 1 \therefore 0 < a < \frac{2c}{1-c} \tag{1.50}$$

For $\frac{\omega_2}{\omega_1}$ in this range, you can choose the angle between the initial wave vectors $\vec{k}_1$ and $\vec{k}_2$ to obtain a scattered wave with a wave vector direction of $\vec{k}_2$.

$$\text{scattered wave} \quad \vec{S}(\vec{r}, t) = \frac{\overrightarrow{I^+} \cdot \hat{r}_s \, \hat{r}_s}{4\pi c_l^2 \rho_0} \frac{V}{r} \sin(\omega_1 + \omega_2)\left(t - \frac{r}{c_l}\right)$$

$$\text{scattering wave amplitude} \quad S_A = \frac{\overrightarrow{I^+} \cdot \hat{r}_s \, \hat{r}_s}{4\pi c_l^2 \rho_0} \frac{V}{r}$$

for instance $T + T \to L$

For two initial transverse waves

$\because \vec{A}_0 \cdot \overrightarrow{k_1} = \vec{B}_0 \cdot \overrightarrow{k_2} = 0$

$$\therefore \overrightarrow{I^+} = -\frac{1}{2}\left(\mu + \frac{1}{4}A\right)$$
$$\left\{\begin{array}{l}\left(\vec{A}_0 \cdot \vec{B}_0\right)\left(k_2^2 \overrightarrow{k_1} + k_1^2 \overrightarrow{k_2}\right) + \left(\vec{B}_0 \cdot \overrightarrow{k_1}\right)\left(k_2^2 + 2\overrightarrow{k_1} \cdot \overrightarrow{k_2}\right)\vec{A}_0 + \\ \left(\vec{A}_0 \cdot \overrightarrow{k_2}\right)\left(k_1^2 + 2\overrightarrow{k_1} \cdot \overrightarrow{k_2}\right)\vec{B}_0\end{array}\right\}$$
$$-\frac{1}{2}\left(K + \frac{1}{3}\mu + \frac{1}{4}A + B\right)\left\{\left(\vec{A}_0 \cdot \vec{B}_0\right)\left(\overrightarrow{k_1} \cdot \overrightarrow{k_2}\right)\left(\overrightarrow{k_1} + \overrightarrow{k_2}\right)\right\}$$

$$-\frac{1}{2}\left(\frac{1}{4}A+B\right)\left\{\left(\vec{A}_0\cdot\overrightarrow{k_2}\right)\left(\vec{B}_0\cdot\overrightarrow{k_1}\right)\left(\overrightarrow{k_1}+\overrightarrow{k_2}\right)\right\} \tag{1.51}$$

(1) If a sequence of initial waves is perpendicular to the $\vec{k}_1\ \vec{k}_2$ plane $\vec{B}_0 \perp \left(\overrightarrow{k_1}, \overrightarrow{k_2}\right)$ plane,

Another set of initial waves is located in the $\vec{k}_1\vec{k}_2$ plane, $\vec{A}_0 // \left(\overrightarrow{k_1}, \overrightarrow{k_2}\right)$plane,

namely $\vec{A}_0\cdot\vec{B}_0=\vec{B}_0\cdot\overrightarrow{k_1}=0$, $\overrightarrow{I^+}=-\frac{1}{2}\left(\mu+\frac{1}{4}A\right)\left(\overrightarrow{A_0}\cdot\overrightarrow{k_2}\right)\left(k_1^2+2\overrightarrow{k_1}\cdot\right.$

$\left.\overrightarrow{k_2}\right)\vec{B}_0$, The direction of $\hat{r}_s$ is $\overrightarrow{k_1}+\overrightarrow{k_2}$, $\overrightarrow{I^+}\cdot\hat{r}_s=0$, $S_A=0$

S_A has no scattered waves. (1.52)

(2) If the two initial waves are perpendicular to the $\vec{k}_1\ \vec{k}_2$ plane,

$$\vec{A}_0\cdot\overrightarrow{k_2}=\vec{B}_0\cdot\overrightarrow{k_1}=0$$

$$S_A=\frac{A_0B_0\omega_1^3V}{16\pi\rho_0 r}\frac{1}{c_t^4c_l}\left\{\begin{array}{l}-\left(\mu+\frac{1}{4}A\right)\left[\left(a^3+1\right)\left(c^2-1\right)+a(a+1)\left(c^2+1\right)\right]\\ -\left(K+\frac{1}{3}\mu+\frac{1}{4}A+B\right)\left[\begin{array}{l}c^2\left(3c^2-1\right)\left(a+a^2\right)\\ +c^2\left(c^2-1\right)\left(a^3+1\right)\end{array}\right]\end{array}\right\}$$

where: $\frac{\omega_2}{\omega_1}$ is the frequency ratio, $c=\frac{c_t}{c_l}$ is the sound velocity ratio. (1.53)

Taking the third term of (1.48) as an example, discuss the resonance condition Taking $T+T\rightarrow T$ as an example:

resonance conditions

$$\left(\frac{\omega_1+\omega_2}{c_t}\right)^2=\left(\frac{\omega}{c_t}\right)^2+\left(\frac{\omega}{c_t}\right)^2+2\left(\frac{\omega}{c_t}\right)\left(\frac{\omega}{c_t}\right)\cos\phi$$

$$\varphi=0$$

amplitude

$$S_A=\frac{\vec{I}^+-\left(\hat{r}_s\times\vec{I}^+\right)\hat{r}_s}{4\pi c_t^2\rho_0 r}V$$

(1) if $\vec{A}_0 // \vec{k}_1, \vec{k}_2$ plane

$\vec{B}_0 \perp (\vec{k}_1, \vec{k}_2)$ plane

$$\vec{A}_0\cdot\vec{k}_1=\vec{B}_0\cdot\vec{k}_0=0,\ \vec{A}_0\cdot\vec{B}_0=\vec{A}_0\cdot\vec{k}_2=0$$

$$\vec{I}^{+} = 0$$

$$S_A = 0$$

(2) if $\vec{A}_0$, $\vec{B}_0$ lie on $\vec{k}_1$, $\vec{k}_2$ plane

$$\vec{A}_0 \cdot \vec{k}_1 = \vec{B}_0 \cdot \vec{k}_0 = 0, \vec{A}_0 \cdot \vec{k}_2 = \vec{B}_2 \cdot \vec{k}_1 = 0$$

$\vec{I}^{+}$ and $\hat{r}_s$, that is $\vec{k}_1$, $\vec{k}_2$ are in the same direction.

$$S_A = 0 \tag{1.54}$$

The relationship and constraints between scattered waves and initial waves are shown in Table 1.1.

Table 1.1 Relationship and constraint conditions between scattered waves and initial waves

Initial wave	Scattered wave	Frequency	$\cos\varphi$	Frequency constraint condition	The behavior of initial shear wave	Scattering wave amplitude
T+T	L	$\omega_1 + \omega_2$	$c^2 + \frac{1}{2a}(c^2 - 1)(a^2 + 1)$	$\frac{1-c}{1+c} < a < \frac{1+c}{1-c}$	//// //⊥ ⊥⊥	S_{A1} 0 S_{A2}
L+T	L	$\omega_1 + \omega_2$	$c + \frac{1}{2c}a(c^2 - 1)$	$0 < a < \frac{2c}{1-c}$	// ⊥	S_{A3} 0
L+L	T	$\omega_1 - \omega_2$	$\frac{1}{c^2} + \frac{1}{2ac^2}(c^2 - 1)(a^2 + 1)$	$\frac{1-c}{1+c} < a < \frac{1+c}{1-c}$		S_{A4}
L+T	L	$\omega_1 - \omega_2$	$c + \frac{1}{2c}a(1 - c^2)$	$0 < a < \frac{2c}{1-c}$	// ⊥	S_{A5} 0
L+T	T	$\omega_1 - \omega_2$	$\frac{1}{c} + \frac{1}{2ac}(c^2 - 1)$	$\frac{1-c}{2} < a < \frac{1+c}{2}$	// ⊥	S_{A6} S_{A7}

1.4 Nonlinear Acoustics at Solid Interfaces

1.4.1 Mechanism of High Order Harmonic Generation

When ultrasonic waves hit a non-ideal interface, they generate higher harmonics. The generation of higher harmonics is related to the nonlinearity of the elastic behavior of the material, which is manifested as the nonlinearity of the relationship between stress and strain in the material. This stress-strain nonlinearity can be described by the nonlinear Hooke's law.

$$\sigma = E_0\varepsilon\left(1 + \beta\varepsilon + \gamma\varepsilon^2 + \cdots\right) \tag{1.55}$$

where δ represents stress, ε represents strain, E_0 is the elastic modulus, β and γ represent the second and third order nonlinear elastic coefficients respectively. However, the nonlinearity generated by the non-ideal interface is different from the nonlinearity generated by the material's constitutive properties. The high harmonic response phenomenon generated by the non-ideal interface is called acoustic contact nonlinearity [7–12]. Its physical mechanism is that the defects in the non-ideal interface are stretched or squeezed by longitudinal sound waves, causing continuous vibration of the defects on the non-ideal surface. The contact stiffness of the defects on the non-ideal surface is higher during the compression phase of the wave than during the stretching phase, so the defects close during wave compression, and the wave can continue to propagate after passing through the defects, while the defects open during wave stretching, and the wave propagation is blocked. This repeated collision between the defect contact surfaces generates the acoustic contact nonlinearity effect, as shown in Fig. 1.4.

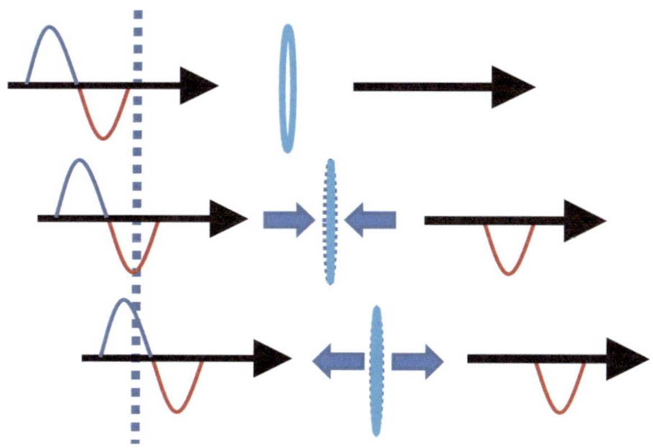

Fig. 1.4 Schematic diagram of the principle of acoustic contact nonlinearity

1.4.2 Theoretical Model

As shown in Fig. 1.5, the adhesive interface is located at x = 0, T_0 is the adhesion force of the adhesive interface, $X_+(t)$ and $X_-(t)$ are the displacements of the right and left interfaces of the adhesive layer respectively, $P_+(t)$ and $P_-(t)$ are the stresses on the two interfaces respectively.

In the absence of incident waves, the initial condition is

$$\begin{cases} X_+(t) = X_-(t) = 0 \\ P_+(t) = P_-(t) = -T_0 \end{cases} \tag{1.55}$$

That is to say, the two interfaces are located at x = 0, and are only subjected to the static stress T_0.

A column of sound longitudinal waves propagates in the sample as shown in Fig. 1.5, where $f(x-ct)$ is the incident longitudinal wave, $g(x-ct)$ and $h(x-ct)$ are the emitted and transmitted longitudinal waves respectively. Since the adhesive layer is very thin, for simplicity, we assume that there are no waves propagating in the adhesive layer. The incident wave can be expressed as

$$f(x-ct) = -A\cos(kx - \omega t + \varphi_0) \tag{1.57}$$

where: $k = 2\pi/(cT)$, $\omega = 2\pi/T$, c is the longitudinal wave speed of the solid sample, A is the amplitude, φ_0 is the initial phase, T is the time period.

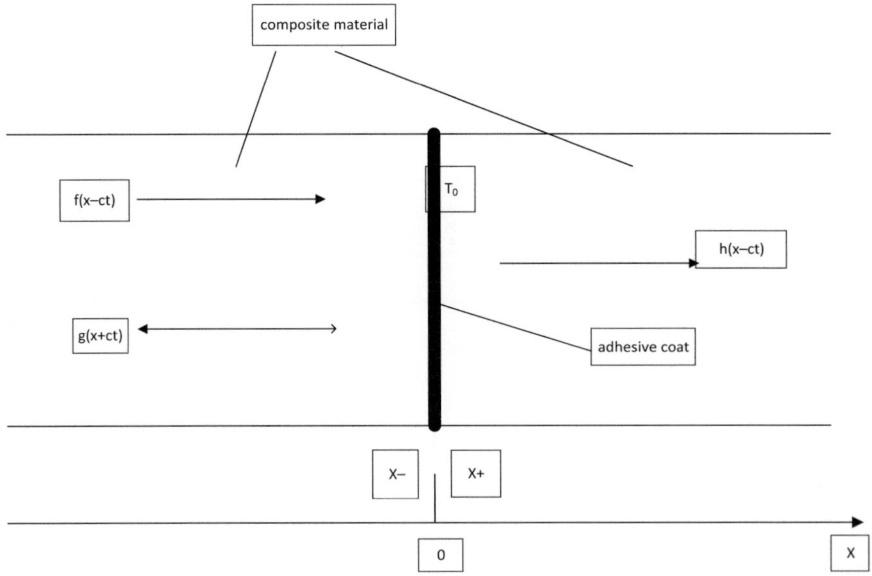

Fig. 1.5 Model of interface in solid

Therefore, the displacements and stresses of the two interfaces are

$$\begin{cases} X_+(t) = h(-ct) \\ X_-(t) = f(-ct) + g(ct) \end{cases} \tag{1.58}$$

$$\begin{cases} P_+(t) = C\frac{\partial h(x-ct)}{\partial x} - T_0 = -\rho c\frac{\partial X_+(t)}{\partial t} - T_0 \\ P_-(t) = C\frac{\partial (f(x-ct)+g(x+ct))}{\partial x} - T_0 = -F + \rho c\frac{\partial X_-(t)}{\partial t} - T_0 \end{cases} \tag{1.59}$$

where F is the driving force generated by the incident sound wave on the left interface, which can be expressed as

$$F(t) = -2C\frac{\partial f(x-ct)}{\partial x} = F_0 \sin(\omega t) \tag{1.60}$$

where: $F_0 = 2\rho c\omega A$, ρ is the density of the solid material, $C = \rho c^2$ is the elastic constant of the solid material.

Under the initial conditions, the bonding layer is closed, and the two interfaces are in contact under the action of adhesion force. In the first half of the period ($0 < t < T/2$), the left interface moves in the +x direction under the drive of the incident wave, the bonding layer is closed, and the two interfaces move synchronously in the +x direction, so the boundary condition is

$$\begin{cases} X_+(t) = X_-(t) \\ P_+(t) = P_-(t) \end{cases} \tag{1.61}$$

Therefore

$$X_+(t) = X_-(t) = f(-ct) \tag{1.62}$$

In the latter half of the cycle ($T/2 < t < T$), the left interface moves in the -x direction under the drive of the incident wave. If the amplitude of the incident wave is large enough, a crack will be opened in the adhesive layer. When the adhesive layer is opened, $X_+(t) > X_-(t)$, the boundary conditions are: $P_+(t) = P_-(t) = 0$

From Eq. (1.59), we can get

$$\begin{cases} \rho c\frac{\partial X_+(t)}{\partial t} = -T_0 \\ \rho c\frac{\partial X_-(t)}{\partial t} = F(t) + T_0 \end{cases} \tag{1.63}$$

Assume that at $t = t_1$, the adhesive layer changes from closed to open, from Eq. (1.63) we can get

$$\rho c(X_+(t) - X_-(t)) = -\int_{t_1}^{t_1+\Delta t} (F(t) + 2T_0)dt \tag{1.64}$$

From Eq. (1.64), it can be seen that only when $F(t) + 2T_0 < 0$ is, the adhesive layer is opened, ($X_+(t) - X_-(t) > 0$), contact acoustic nonlinearity is generated.

Let $\Delta t \to 0, X_+(t) = X_-(t)$, we can get t_1

$$\sin \omega t_1 = -\gamma, t_1 = \frac{(\pi + \sin^{-1} \gamma)}{\omega} \tag{1.65}$$

where: $\gamma = T_0/\rho c\omega A = 2T_0/F_0$, characterizes the relationship between the adhesion force T_0 and the maximum value of the acoustic driving force F.

When $t < t_1$, the adhesive layer is closed, the two interfaces move synchronously, and the movement can be obtained from Eq. (1.53). When $t > t_1$, the adhesive layer is opened, the two interfaces move asynchronously. According to method (1.63), the right interface moves in the -x direction, and the direction of the left interface's displacement changes with the incident wave. It starts moving in the -x direction, reaches the maximum amplitude, and then moves in the +x direction. Assume $t = t_2$, $X_+(t) = X_-(t)$, the two interfaces collide and move synchronously, so the movement of the right interface in one cycle can be represented as:

$$h(x - ct) = \begin{cases} -A\cos[k(x - ct_1) - A\gamma\omega(t - t_1), t_1 < t < t_2 \\ -A\cos[k(x - ct)], t_2 \le t < t_1 + T \end{cases} \tag{1.66}$$

where t_1 is the moment when the adhesive layer changes from closed to open, t_2 is the moment when the adhesive layer recovers from open to closed. The time interval $(t_2 - t_1)$ can be obtained from the previous discussion. From Eq. (1.65), it can be seen that the value of γ should satisfy $0 < \gamma < 1$. When $\gamma > 1(T_0 > F_0/2)$, t_1 has no solution, that is, the adhesive layer cannot be opened, and contact acoustic nonlinearity will not be generated, so the incident wave must have sufficient amplitude.

From the previous description, it can be seen that $h(x - ct)$ has fundamental and harmonic components. FFT can be performed on $h(x - ct)$ to obtain its fundamental amplitude A_1 and second harmonic amplitude A_2.

Therefore, for the solid interface model in Fig. 1.5, the expressions for the incident wave, reflected wave, and transmitted wave are:

$$\begin{cases} f(x - ct) = -A\cos(kx - \omega t), \\ g(x + ct) = -(1/2)Y(t + x/c), \\ h(x - ct) = f(x - ct) + 1/2Y(t - x/c), \end{cases} \tag{1.67}$$

$$Y = 2A[\cos \omega t + \cos \omega t_1 - \eta\omega(t - t_1)] \tag{1.68}$$

where Y is the gap (non-adhesive interface) state caused by wave propagation, t_1 is the time when the gap starts to move, η is a parameter representing the width of the gap. When $\eta \le 0$, the width of the gap is greater than the amplitude of the vibration, and the sound wave cannot pass through the gap; when $\eta \ge 1$, $Y = 0$; when η

is between 0 and 1, the interface vibration may exhibit nonlinearity, and second harmonics are easily generated.

For interfaces with artificial defects (micro-cracks or incomplete bonding), under the action of ultrasound, the transmitted wave in the defect-free area is $h_1(x-ct)$

$$h_1(x-ct)=f(x-ct), \tag{1.69}$$

In the defective area, when the defect cracks, the sound wave cannot pass through the area where the defect exists, but the sound wave can pass through other parts smoothly; when the defect is closed, the sound wave can pass through all interface areas smoothly. Therefore, the transmitted wave in the presence of defects $h_2(x-ct)$ is expressed as

$$h_2(x-ct)=\begin{cases}-A\cos[k(x-ct_1)]-A\gamma\omega(t-t_1), t_1<t<t_2\\ -A\cos[k(x-ct)], t_2\le t<t_1+T\end{cases} \tag{1.70}$$

Therefore, the total transmitted wave is

$$h(x-ct)=\begin{cases}\int_{t_1}^{t}\sqrt{(1-R)\omega^2A^2\sin^2[k(x-ct)]+RA^2\gamma^2\omega^2}dt, t_1<t<t_2\\ -A\cos[k(x-ct)], t_2\le t<t_1+T\end{cases} \tag{1.71}$$

1.5 Nonlinear Lamb Waves in Plates

1.5.1 Nonlinear Lamb Waves

Lamb waves are defined as elastic waves generated when plane strain propagates in a uniform and isotropic plate, which are formed by the coupling of transverse waves and longitudinal waves through reflection on the upper and lower surfaces. According to the motion of the particles relative to the center plane of the plate, the propagation of Lamb waves in the plate shell is divided into two basic modes: symmetric and antisymmetric. The symmetric mode (S) includes S_0, S_1, S_2, …, S_n modes, and the antisymmetric mode (A) includes A_0, A_1, A_2, …, A_n modes. Ultrasonic waves usually propagate in the form of Lamb waves in plate materials. Lamb waves are a special form of stress wave formed by the coupling of longitudinal waves and transverse waves in sound-guided waves (such as metal thin plates) with a thickness and excitation sound wave wavelength of the same order of magnitude. They will produce different propagation modes under different thicknesses and excitation frequencies. Deng Mingxi, Lima, and Hamilton independently proposed the mode expansion analysis method using guided wave excitation [13–31], analyzing the second harmonic generation effect of Lamb waves. The basic point is that under the second-order perturbation approximation, the double-frequency body force and surface driving stress tensor accompanying the propagation of the fundamental Lamb wave, their function is to excite a series of double-frequency Lamb wave modes in the plate, which add up to form the second harmonic sound field of the fundamental Lamb wave; when the phase velocity of the fundamental Lamb wave is equal to the phase velocity of a double-frequency Lamb wave mode, the double-frequency Lamb wave mode accumulates with the propagation distance, at this time it is equivalent to the second harmonic with cumulative effect obtained based on the interface nonlinear sound reflection theory.

1.5.2 Using the Mode Expansion Analysis Method [32]

The motion equation of nonlinear elastic waves in the plate is:

$$(\lambda+\mu)\nabla(\nabla\cdot u)-\mu\nabla\times(\nabla\times u)+f=\rho_0\frac{\partial^2 u}{\partial t^2} \tag{1.72}$$

$$\text{The boundary condition } L \text{ is: } \left(S^L-\bar{S}\right)\cdot n_y=0 \tag{1.73}$$

Here u is the particle displacement, λ, μ are the Lame constants, ρ_0 is the density, y is the thickness direction, n_y is perpendicular to the surface L unit vector, S^L is the linear part of the Piola-Kirchhoff stress tensor, $\bar{S}$ and f are the nonlinear parts.

Considering the first-order nonlinearity, the expression of energy is as follows

$$\begin{aligned}E=&\frac{1}{4}\mu\left(u_{i,j}+u_{j,i}\right)^2+\left(\frac{1}{2}K-\frac{1}{3}u\right)\left(u_{i,i}\right)^2\\&+\left(\mu+\frac{1}{4}A\right)\left(u_{i,j}u_{k,i}u_{k,j}\right)+\left(\frac{1}{2}K-\frac{1}{3}u+\frac{1}{2}B\right)\left[u_{i,i}\left(u_{j,k}\right)^2\right]\\&+\frac{1}{12}A\left(u_{i,j}u_{j,k}u_{k,i}\right)+\frac{1}{2}B\left(u_{j,k}u_{k,j}u_{i,i}\right)+\frac{1}{3}C\left(u_{i,i}\right)^3+\cdots\end{aligned} \tag{1.74}$$

where A, B, and C are third-order nonlinear elastic constants, μ and K are the shear and bulk modulus, respectively, here

$$\varepsilon_{ij}=\frac{1}{2}\left(u_{i,j}+u_{j,i}\right) \tag{1.75}$$

$$S_{ij}=\frac{\partial E}{\partial\left(u_{i,j}\right)};f_i=S_{ij} \tag{1.76}$$

The solution in the plate can be written as a linear superposition of existing waveguide modes

$$v(y, z, t) = \frac{1}{2}\sum_{m=1}^{\infty} A_m(z)v_m(z)e^{-i\omega t} \tag{1.77}$$

$$S(y, z, t) = \frac{1}{2}\sum_{m=1}^{\infty} A_m(z)S_m(y) \cdot n_z e^{-i\omega t} \tag{1.78}$$

Here: $v = \frac{\partial u}{\partial t}$, v_m is the particle velocity of the m-th mode, S_m is the strain tensor of the m-th mode, A_m is the amplitude of the m-th mode.

A_m is determined by the following equation

$$4P_{mm}\left(\frac{d}{d} - ik_n^*\right)A_m(z) = \left(f_n^{surf} + f_n^{vol}\right)e^{ikz}, m = 1, 2, ... \tag{1.79}$$

Here

$$P_{mn} = -\frac{1}{8}\int_{-h}^{h} \left(v_n^* \cdot S_m + v_m \cdot S_n^*\right) \cdot n_z d\Omega, \tag{1.80}$$

where

$$f_n^{surf} = -\frac{1}{2}v_n^* S \cdot n_y|_{y=-h}^{y=h}, \tag{1.81}$$

$$f_n^{vol} = \frac{1}{2}\int_{-h}^{h} f \cdot v_n^* dy \tag{1.82}$$

k_n is the wave number of the wave that is not perpendicular to the k_m mode.

Use perturbation method to solve the equation

$$u = u^1 + u^2 \tag{1.83}$$

u^2 is a perturbation caused by nonlinearity, which is very small compared to u^1, u^1 is the solution of the following linear equation

$$(\lambda + 2\mu)\nabla\left(\nabla \cdot u^1\right) - \mu\nabla \times \left(\nabla \times u^1\right) - \rho_0\frac{\partial^2 u}{\partial t^2} = 0 \tag{1.84}$$

$$\text{On the boundary L} : S^L\left(u^1\right) \cdot n_y = 0 \tag{1.85}$$

u^2 satisfies the following equation

$$(\lambda + 2\mu)\nabla\left(\nabla \cdot u^2\right) - \mu\nabla \times \left(\nabla \times u^2\right) - \rho_0 \frac{\partial^2 u^2}{\partial t^2} = -f^1 \tag{1.86}$$

$$\text{On the boundary L} : S^L\left(u^2\right) \cdot n_y = -S^1 \tag{1.87}$$

S^1 and f^1 are the surface traction and body force obtained from u^1.

S^1 and f^1 are obtained by substituting the linear solution into Eq. (1.71). If the frequency of the main excitation mode of the waveguide is ω, the energy Eq. (1.69) is first-order nonlinear, S^1 and f^1 will contain second harmonic 2ω. If the energy equation contains second-order nonlinearity, S^1 and f^1 will contain third harmonic 3ω.

The solution of Eq. (1.79) is

$$A_m(z) = \underset{m}{\overline{A}}(z)e^{i(2\kappa z)} - \underset{m}{\overline{A}}(0)e^{i\kappa_n^* z} \tag{1.88}$$

Here

$$\underset{m}{\overline{A}}(z) = i\frac{\left(f_n^{vol} + f_n^{surf}\right)}{4P_{mn}\left[\kappa_n^* - 2\kappa\right]};\ \kappa_n^* \neq 2\kappa \tag{1.89}$$

$$\underset{m}{\overline{A}}(z) = \frac{\left(f_n^{vol} + f_n^{surf}\right)}{4P_{mn}};\ \kappa_n^* = 2\kappa \tag{1.90}$$

Here A_m is the mode amplitude at 2ω, κ is the wave vector of the main wave.

For a solid plate in Fig. 1.6, when the frequency is f and the order is l, the Lamb wave propagates along the Oz axis. Due to geometric nonlinearity and the bulk elasticity nonlinearity of the solid, there will be a second harmonic driving stress tensor $P_l^{(2f)}$ and a second harmonic driving body force $F_l^{(2f)}$ on the surface and inside the solid plate, respectively. Their function is to excite a series of second harmonic Lamb wave modes in the solid plate. These second harmonic Lamb waves superimpose to form the second harmonic acoustic field $U^{(2f)}$ of the lth order fundamental Lamb wave, that is,

$$U^{(2f)} = \sum a_n(z)U_n^{(2f)}(y) \tag{1.91}$$

where

$U^{(2f)}$ represents the acoustic field function of the nth order second harmonic Lamb wave mode, $a_n(z)$ represents the expansion coefficient of the nth order second harmonic Lamb wave mode, its formal solution is as follows

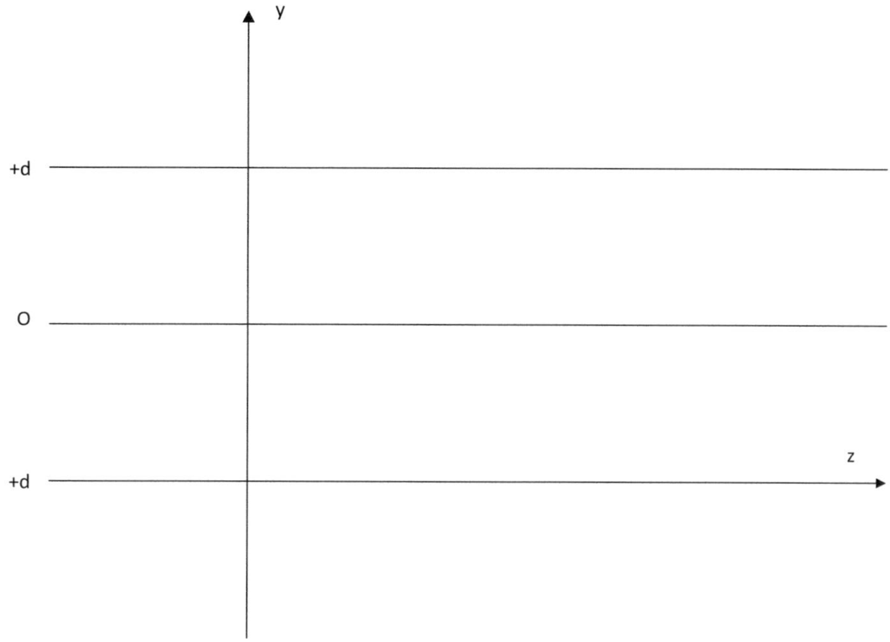

Fig. 1.6 Solid plate and calculation coordinate system

$$a_n(z) = \frac{f_{n,l}^{(V)}(z) + f_{n,l}^{(S)}(z)}{4P_{nn} \times 2\pi f}\left[\frac{c_n^{(2f)} c_l^{(f)}}{c_l^{(f)} - c_n^{(2f)}}\right]\sin\left[2\pi f \frac{c_n^{(2f)} - c_l^{(f)}}{c_n^{(2f)} c_l^{(f)}} z\right], \tag{1.92}$$

where P_{nn} is the average power flow (per unit width along the Ox axis) of the nth order second harmonic Lamb wave mode in the propagation direction, $c_l^{(f)}$ and $c_n^{(2f)}$ represent the phase velocities of the lth order fundamental Lamb wave and the nth order second harmonic Lamb wave mode, respectively. When the lth order fundamental Lamb wave mode with frequency f propagates in the solid, $P_l^{(2f)}$ and $F_l^{(2f)}$ are coupled with the acoustic field of the nth order second harmonic Lamb wave, resulting in $f_{n,l}^{(S)}(z)$ and $f_{n,l}^{(V)}(z)$ in Eq. (1.83), which are called the surface driving source and body driving source of the nth order second harmonic Lamb wave mode, respectively. $f_{n,l}^{(S)}(z)$ and $f_{n,l}^{(V)}(z)$ are related to the density of the plate, the second and third order elastic constants, and the acoustic field function of the nth order second harmonic Lamb wave, and are proportional to the square of the amplitude of the lth order fundamental Lamb wave.

1.6 Nonlinear Sound Propagation in Non-uniform Media

1.6.1 Theory

The form of the Westervelt equation under the three-dimensional Cartesian coordinate system [33–35]:

$$\rho\nabla\cdot\left(\frac{1}{\rho}\nabla p\right)-\frac{1}{c^2}\frac{\partial^2 p}{\partial t^2}+\frac{\delta}{c^4}\frac{\partial^3 p}{\partial t^3}+\frac{\beta}{\rho c^4}\frac{\partial^2 p^2}{\partial t^2}=0 \tag{1.93}$$

where p is the sound pressure, δ is the diffusion factor of the medium, σ is the nonlinearity coefficient of the medium, σ is the density of the medium, c is the speed of sound in the medium. Let $f=\frac{p}{\sqrt{\rho}}$, the nonlinear wave equation is obtained:

$$\begin{aligned}\nabla^2 f-\frac{1}{c_0^2}\frac{\partial^2 f}{\partial t^2}&=\sqrt{\rho}\nabla^2\left(\frac{1}{\sqrt{\rho}}\right)f\\&+\frac{1}{c_0^2}\left(\frac{c_0^2}{c^2}-1\right)\frac{\partial^2 f}{\partial t^2}-\frac{\delta}{c^4}\frac{\partial^3 f}{\partial t^3}-\frac{\beta}{\sqrt{\rho}c^4}\frac{\partial^2 f^2}{\partial t^2}\end{aligned} \tag{1.94}$$

Let: $w=f+v$, $v=\left(\frac{c_0^2}{c^2}-1\right)f$, Eq. (1.87) can be simplified to:

$$-\frac{1}{c_0}\frac{\partial^2 w}{\partial t^2}=\nabla^2 v-\nabla^2 w+\frac{q-h-d}{c_0^2} \tag{1.95}$$

$$\begin{aligned}q&=c_0^2\sqrt{\rho}\nabla^2\frac{1}{\sqrt{\rho}}f\\h&=c_0^2\frac{\beta}{\sqrt{\rho}c^4}\frac{\partial^2 f^2}{\partial t^2}\\d&=c_0^2\frac{\delta}{c^4}\frac{\partial^3 f}{\partial t^3}\end{aligned} \tag{1.96}$$

c_0 is the speed of sound in the background medium. Fourier transform (1.88) in space to get the k-space equation:

$$\frac{\partial^2 W}{\partial t^2}=c_0^2k^2(V-W)-(Q-H-D) \tag{1.97}$$

where W, Q, H, D are the Fourier transforms of w, q, h, d respectively.

The non-standard central difference method is used to discretize (1.90):

$$W(t+\Delta t)=2W(t)-W(t-\Delta t)4\sin^2\left(\frac{c_0k\Delta t}{2}\right)$$

$$\left[V - W - \frac{1}{c_0^2 k^2}(Q - H - D)\right] \tag{1.98}$$

The stability of Eq. (1.90) requires $\sin\left(\frac{CFLc_0\pi}{2c_{max}\frac{c_0}{c_{max}}}\right)$, for $c(r) < c_0$, it always satisfies the stability condition.

1.6.2 Calculation Results and Discussion

1. One-dimensional medium

(1) **One-dimensional homogeneous medium**

According to the above method, we first calculate the propagation of sound waves along the x-axis in a one-dimensional homogeneous medium. A Gaussian beam is used as the incident sound wave: $p_i = p_0 \sin(\omega_0 \tau)e^{-\tau^2/(2\alpha^2)}$, where $\tau = t - (x - x_0)/c_0$ is the relaxation time, $\alpha = 10^{-5}$, $p_0 = 20$ kPa, $\omega_0 = 0.2\,\text{MHz} \times 2\pi$. The acoustic parameters of the homogeneous medium are: $c_0 = 1500$ m/s, $\rho = 1000\,\text{kg/m}^3$, $\beta_0 = 3.5$. The sound pressure is observed at a distance of 22.5 cm from the sound source.

As can be seen from Fig. 1.7, due to the influence of nonlinearity and attenuation, the waveform will be distorted, making the peak sound pressure reach 24 kPa, while the valley value is −17.5 kPa, and the energy is converted from the fundamental wave to the higher harmonics. The spectrum from the fundamental wave to the fifth harmonic can be clearly displayed in the figure.

(2) **One-dimensional non-homogeneous medium**

To simulate the characteristics of a non-homogeneous medium, we introduce a Gaussian correlation function: $Cor_\phi(x) = exp(-x^2/l^2)$, $N(f_x) = \sqrt{F(Cor_\phi(x))}$ is.

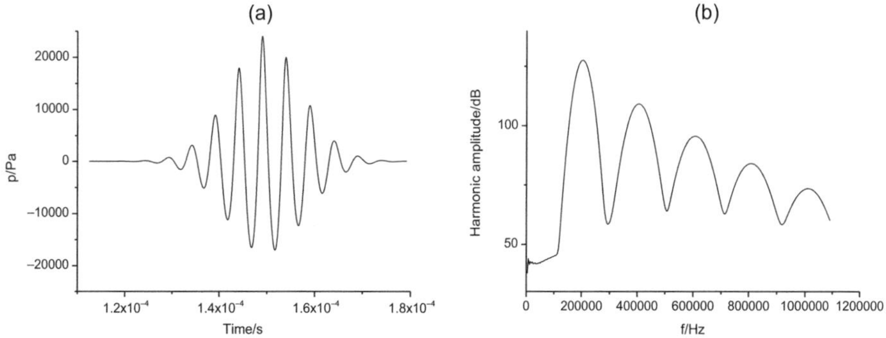

Fig. 1.7 **a** Waveform at x = 22.5 cm in a homogeneous medium; **b** Spectrum at x = 22.5 cm in a homogeneous medium

The Fourier change of $Cor_{\phi}(x)$, l is the correlation length and generates a random function with zero mean and standard deviation of σ $s_i(x)$ $(i = 1, 2, 3...)$. The Fourier transform of $s_i(x)$ is performed: $S_i(f_x) = F\{s_i(x)\}$, $\phi_i(f_x) = N(f_x)S_i(f_x)$, This generates the phase change function: $\varphi_i = F^{-1}\{\phi_i(f_x)\}$, which is used to simulate the changes in sound speed c, density ρ, and nonlinear coefficient β in a non-homogeneous medium: $c = c_0(1 + \varphi_1(x))$, $\rho = \rho_0(1 + \varphi_2(x))$, $\beta = \beta_0(1 + \varphi_3(x))$. The sound pressure is also observed at a distance of 22.5 cm from the sound source.

A. Influence of correlation length

The larger the l, the stronger the non-homogeneity of the medium. Fix $\sigma = 0.15$, I increases from 2 to 14 mm. As can be seen from Fig. 1.8, l increase will lead to an increase in p^-, a decrease in p^+, energy conversion from low-order harmonics to high-order harmonics, where the amplitudes of the fourth and fifth harmonics can increase by 5.5 and 8 dB respectively.

B. Impact of Standard Deviation

The larger the standard deviation σ, the greater the degree of change in the medium parameters. With a fixed correlation length $l = 8$ mm, σ increases from 0.03 to 0.27. Figure 1.9 shows that p^+ will decrease by 32%, and p^- fluctuates between 90 and 120%. The fundamental wave remains basically unchanged, and compared with the uniform situation, the amplitude differences of the second, third, and fourth harmonics vary between 3 and 0 dB, −4.8 and 0.8 dB, 0 and 4.0 dB, and −0.5 and 4.7 dB respectively.

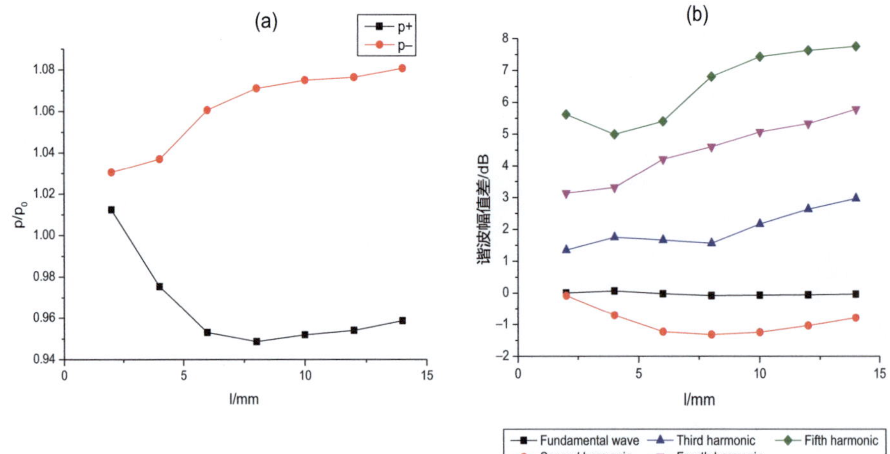

Fig. 1.8 The impact of l on the peak and valley values (**a**) and amplitude of each harmonic (**b**)

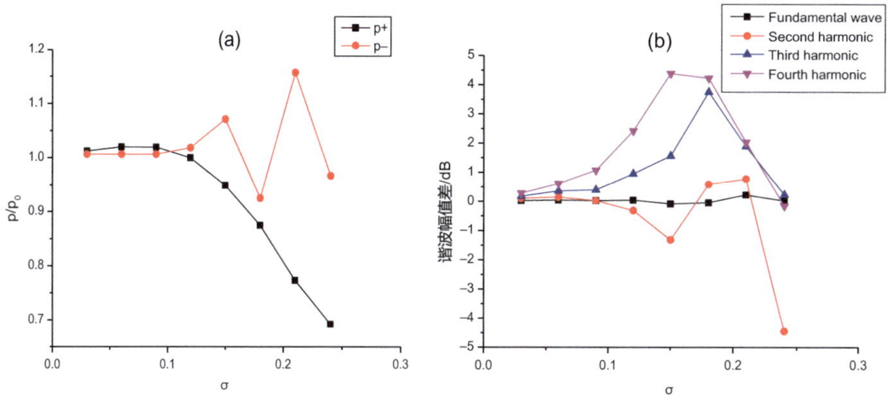

Fig. 1.9 The impact of σ on the peak and valley values (**a**) and amplitude of each harmonic (**b**)

C. Inserting a Uniform Layer

With $\sigma = 0.15$ fixed, $l = 8$ mm, a uniform layer with a thickness of d is inserted 1 cm away from the sound source in the non-uniform medium, and the total thickness of the medium layer is 22.5 cm. Figure 1.10 shows that as the thickness of the inserted uniform layer increases, p^+ and p^- move from 95 and 109% respectively towards 1. When the thickness of the uniform layer accounts for more than 70% of the total thickness of the medium layer, the attenuation change of the fundamental and second harmonic amplitudes is almost 0 dB.

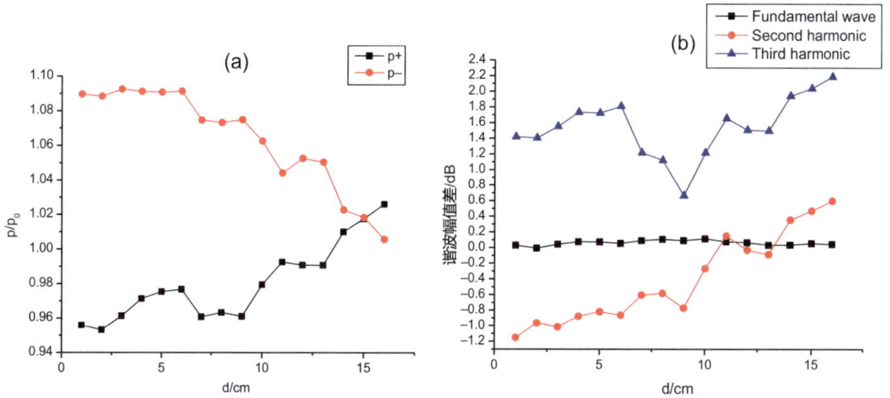

Fig. 1.10 The impact of a inserted uniform layer with a thickness of d on the peak and valley values (**a**) and amplitude of each harmonic (**b**)

D. Inserting a Non-uniform Layer

A non-uniform layer with a thickness of d, $l = 8$ mm, and $\sigma = 0.15$ is inserted 1 cm away from the sound source in the uniform medium, and the total thickness of the medium is 22.5 cm. Figure 1.11 shows that when the inserted thickness is less than 27% of the total thickness, the attenuation changes of the fundamental, second, and third harmonic amplitudes are all within 1 dB, and the thickness has a greater impact on the higher harmonics.

2. **Two-dimensional Medium**

(1) **Two-dimensional Uniform Medium**

Using the above method, we calculate the situation of sound wave propagation along the x-axis in a two-dimensional uniform medium. A Gaussian sound beam is used as the incident sound wave, with a source radius of 1 cm: $p_i = p_0 \sin(\omega_0 \tau) e^{-\tau^2/(2\alpha^2)}$, where $\tau = t - (x - x_0)/c_0$ is the relaxation time, $\alpha = 10^{-5}$, $p_0 = 20\,\text{kPa}$, $\omega_0 = 1\,\text{MHz} \times 2\pi$. The acoustic parameters of the uniform medium are: $c_0 = 1500\,\text{m/s}$, $\rho = 1000\,\text{kg/m}^3$, $\beta_0 = 4.0$. The sound pressure is observed at a distance of 20.3 mm from the sound source on the x-axis.

As can be seen from Fig. 1.12, due to the effects of nonlinearity and attenuation, the waveform will be distorted, allowing the peak sound pressure to reach 17.2 kPa, while the valley value is −16.2 kPa, and energy is transferred from the fundamental wave to the higher harmonics. The spectrum from the fundamental wave to the fifth harmonic can be clearly displayed in the figure.

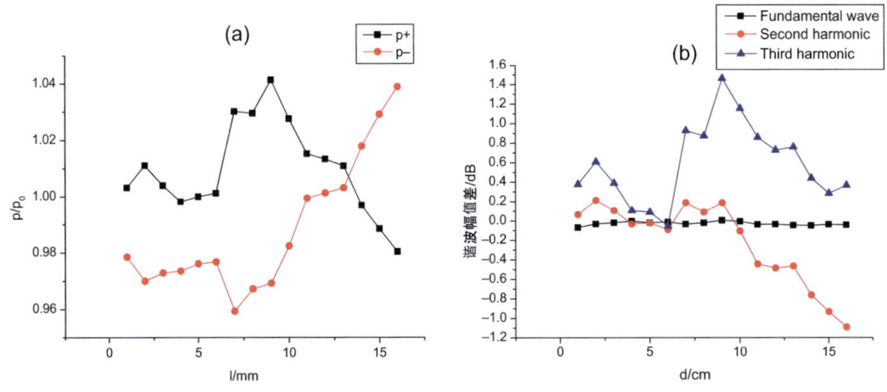

Fig. 1.11 The impact of a inserted non-uniform layer with a thickness of d on the peak and valley values (**a**) and amplitude of each harmonic (**b**)

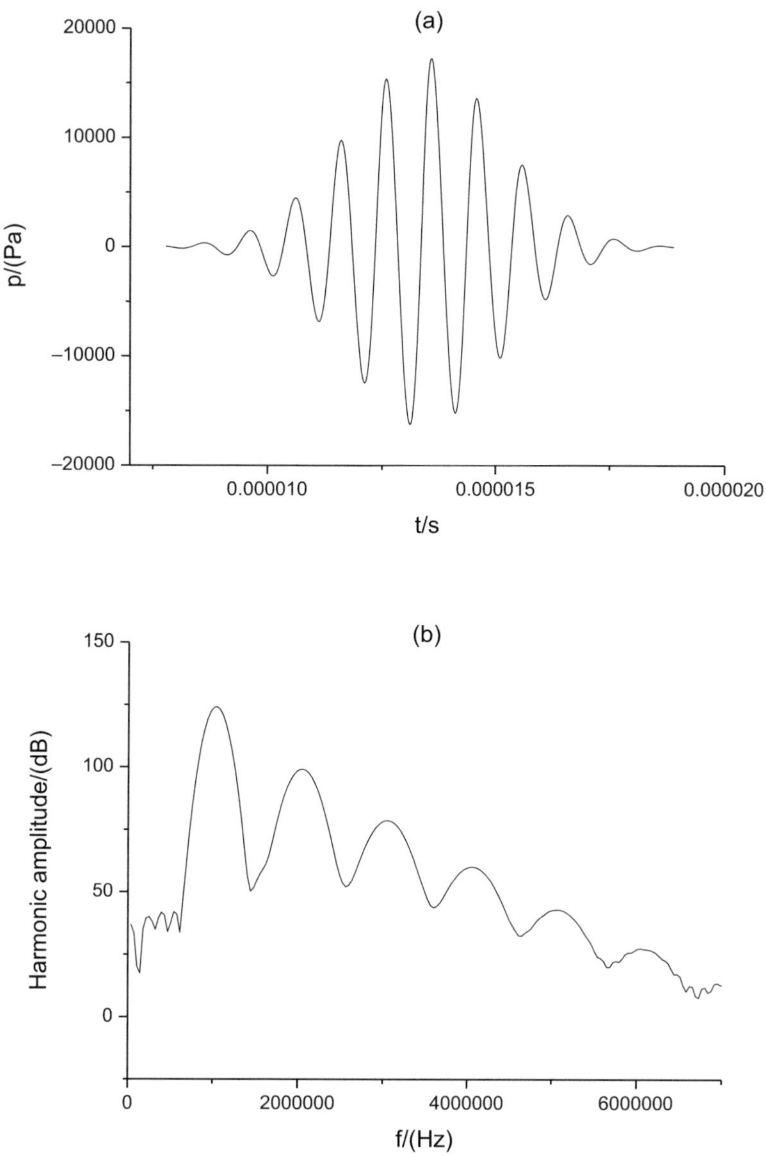

Fig. 1.12 **a** The waveform at x = 20.3 mm, y = 0 cm in the uniform medium; **b** The spectrum at x = 20.3 mm, y = 0 cm in the uniform medium

(2) **Two-dimensional Non-uniform Medium**

To simulate the characteristics of a non-uniform medium, a Gaussian correlation function is introduced: $Cor_\phi(x, y) = \exp(-(x^2 + y^2)/l^2)$, $N(f_x, f_y) = \sqrt{F(Cor_\phi(x, y))}$ is the Fourier transform of $Cor_\phi(x, y)$, l is the related length, and it generates a random function with a mean of zero and a standard deviation of σ $s_i(x, y)$ $(i = 1, 2, 3...)$. The two-dimensional Fourier transform of $s_i(x, y)$ is: $S_i(f_x, f_y) = F\{s_i(x, y)\}$, $\phi_i(f_x, f_y) = N(f_x, f_y)S_i(f_x, f_y)$, which generates the phase change function: $\varphi_i = F^{-1}\{\phi_i(f_x, f_y)\}$, used to simulate the changes in c, ρ, β in non-uniform media: $c = c_0(1 + \varphi_1(x, y))$, $\rho = \rho_0(1 + \varphi_2(x, y))$, $\beta = \beta_0(1 + \varphi_3(x, y))$.

Figure 1.13 is the waveform diagram at times $t_1 = 5T_0$, $t_2 = 10T_0$, $t_3 = 15T_0$. From the figure, it can be seen that the non-uniformity of the medium causes significant fluctuations in the sound wave, with energy dispersed over a larger range in the propagation direction.

A. The impact of related length

The larger l is, the stronger the non-uniformity of the medium. Fix $\sigma = 0.25$, l increases from 2 to 15 mm. Unlike the one-dimensional case where "increasing l leads to an increase in p^-, a decrease in p^+, and energy transfer from low-order harmonics to high-order harmonics", in two-dimensional media, the sound beam will diverge along the y-axis, so increasing l will cause the amplitudes of p^+, p^-, and all harmonics to decrease simultaneously. When $l < 7$ mm, the differences in amplitude of the fundamental, second, and third harmonics are all within – 1 dB, and the ratios of p^+, p^- to those in a uniform medium are all within 95%. When l reaches 15 mm, p^+, p^- are 57 and 60% of those in a uniform medium, respectively, and the fundamental, second, and third harmonics are attenuated by 2.5, 4.6, and 7 dB, respectively (Fig. 1.14).

B. The impact of standard deviation

The standard deviation σ, The larger it is, the greater the degree of change in the dielectric parameters. Fix the related length $l = 8$ mm, σ increases from 0.03 to 0.42. In a one-dimensional medium, at this time p^+ decreases by 32%, p^- fluctuates between 90 and 120%. In a two-dimensional medium, due to the divergence of the sound beam, p^+, p^- will decrease simultaneously, reaching 78 and 80% respectively. The fundamental wave remains basically unchanged, and the amplitudes of the second and third harmonics have attenuated by 6.3 and 7.2 dB respectively. $\sigma < 0.18$ almost ignores the non-uniformity of the medium, $\sigma < 0.24$ when, p^+, p^- the ratio is all within 95%, the difference in amplitude of the fundamental wave and the second harmonic is less than — 0.3 dB, and the difference in amplitude of the third harmonic is kept within – 1 dB (Fig. 1.15)

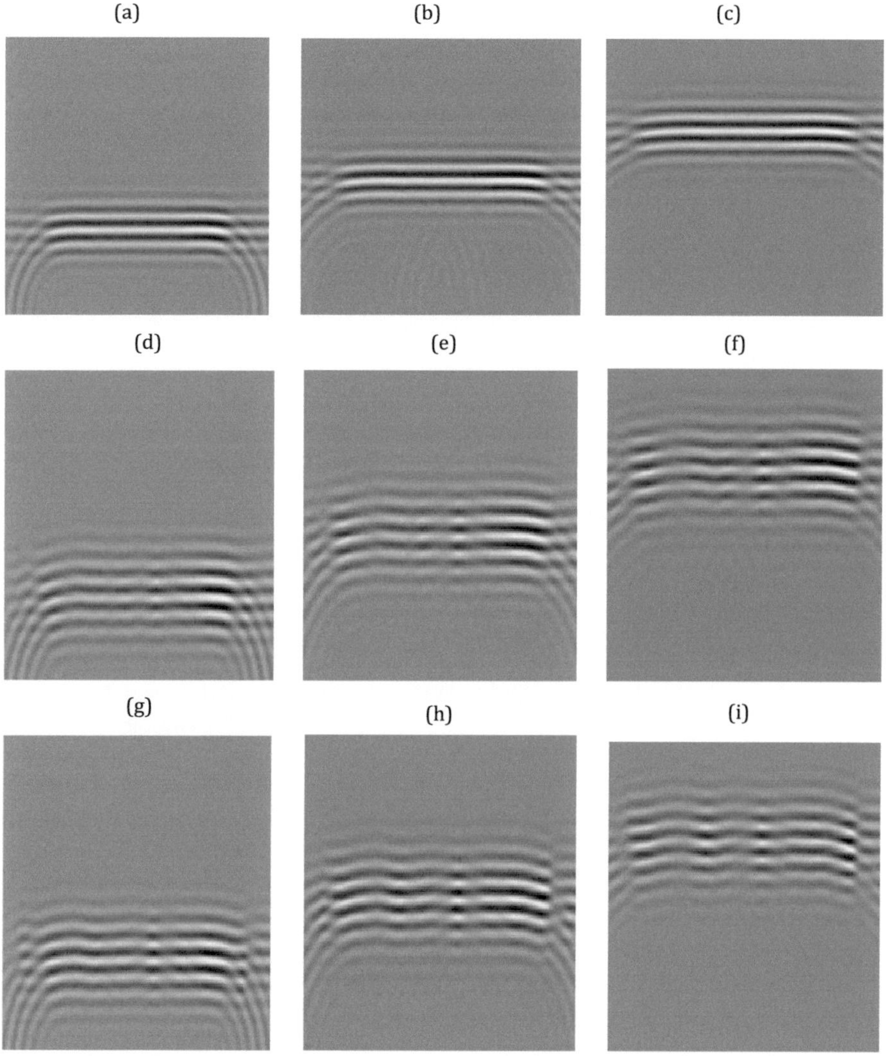

Fig. 1.13 The waveform at $t_1 = 5T_0$, $t_2 = 10T_0$, $t_3 = 15T_0$, the area is 3 cm * 3 cm, T_0 is the signal period. **a–c** are the waveform diagrams at times t_1, t_2, and t_3 in a uniform medium; **d–f** are the waveform diagrams at times t_1, t_2, and t_3 in a non-uniform medium with $l = 12$ mm, $\sigma = 0.25$; **g–i** are the waveform diagrams at times t_1, t_2, and t_3 in a non-uniform medium with $l = 8$ mm, $\sigma = 0.39$

C. Inserting a uniform layer

Fix $\sigma = 0.27$, $l = 10$ mm, insert a uniform layer with a thickness of d at a distance of 2 mm from the sound source in the non-uniform medium, and the total thickness of the medium layer is 20.3 mm. Figure 1.16 shows that as the thickness of the inserted uniform layer increases, p^+, p^- all gradually increase. When $d > 10$ mm, p^+, p^- reach

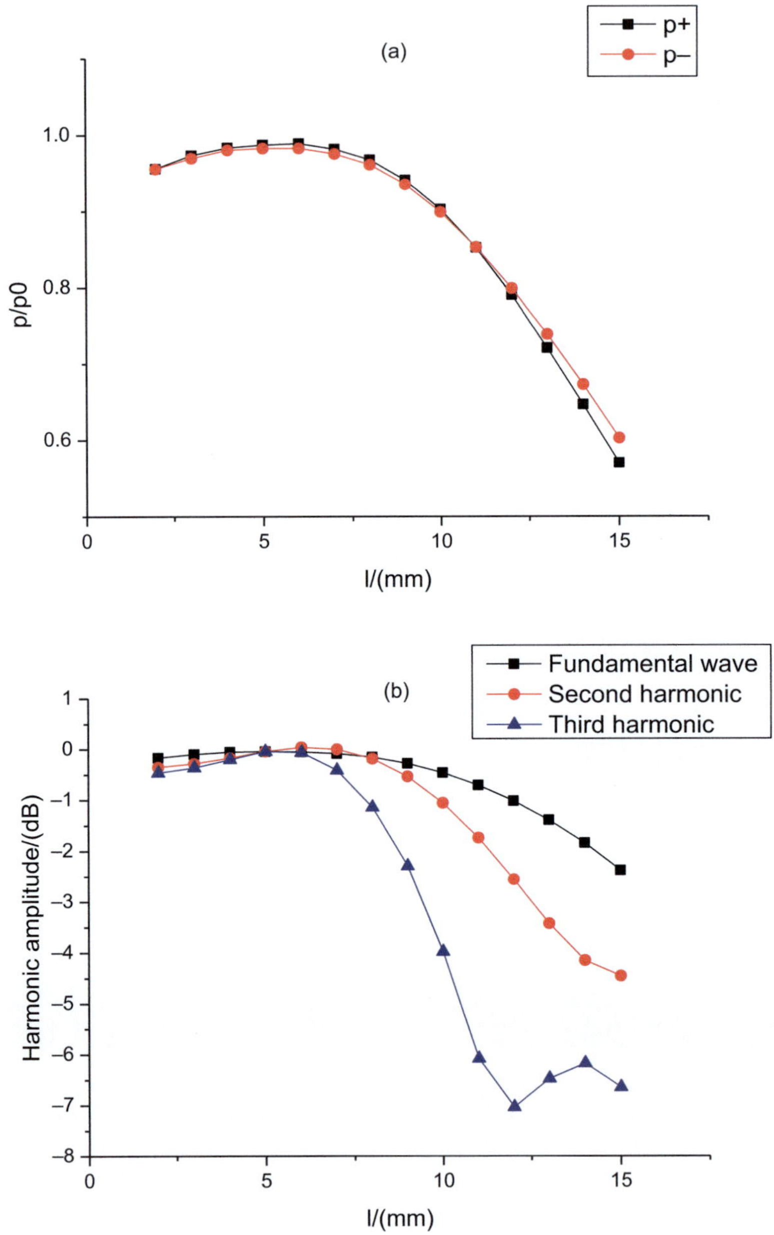

Fig. 1.14 The impact of l on the peak and valley values (**a**) and amplitude of each harmonic (**b**)

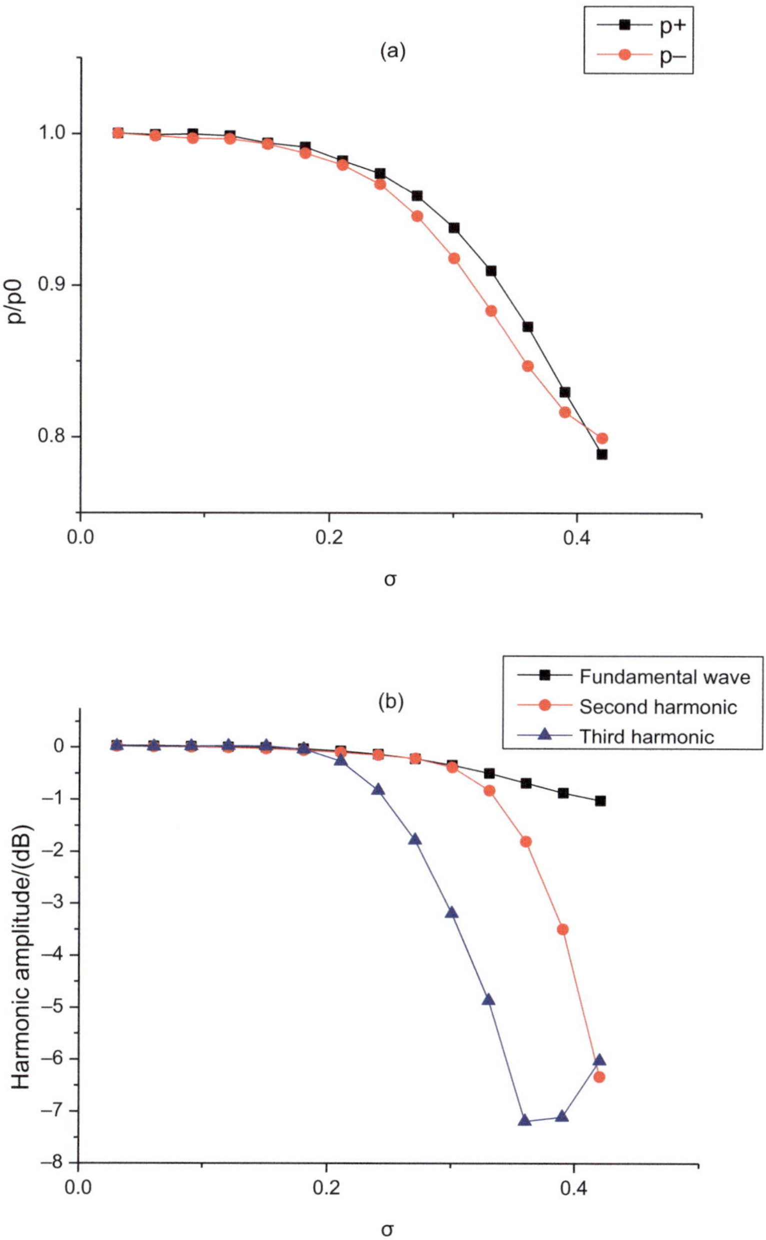

Fig. 1.15 The impact of σ on the peak and valley values (**a**) and amplitude of each harmonic (**b**)

92 and 90% respectively, and the attenuation of the fundamental wave, second, and third harmonics are within – 0.5, – 0.2, and – 1.8 dB respectively. When d > 12 mm, that is, the thickness of the uniform layer accounts for more than 60% of the total thickness of the medium layer, the non-uniformity of the medium can basically be ignored, which is different from the 70% in the one-dimensional case. This shows that due to the diffusion of sound waves on the y-axis, the actual impact of the non-uniformity of the medium on the sound field in the two-dimensional medium is smaller.

D. Inserting a non-uniform layer

Insert a non-uniform layer with a thickness of d, $l = 10$ mm, $\sigma = 0.27$ at a distance of 2 mm from the sound source in the uniform medium, and the total thickness of the medium is 20.3 mm. Figure 1.17 shows that when d < 9 mm, that is, the inserted thickness is less than 44.3% of the total thickness, the fundamental wave hardly attenuates, and the attenuation of the second and third harmonics is within 1.1 dB. The thickness of the inserted layer has a greater impact on the higher harmonics.

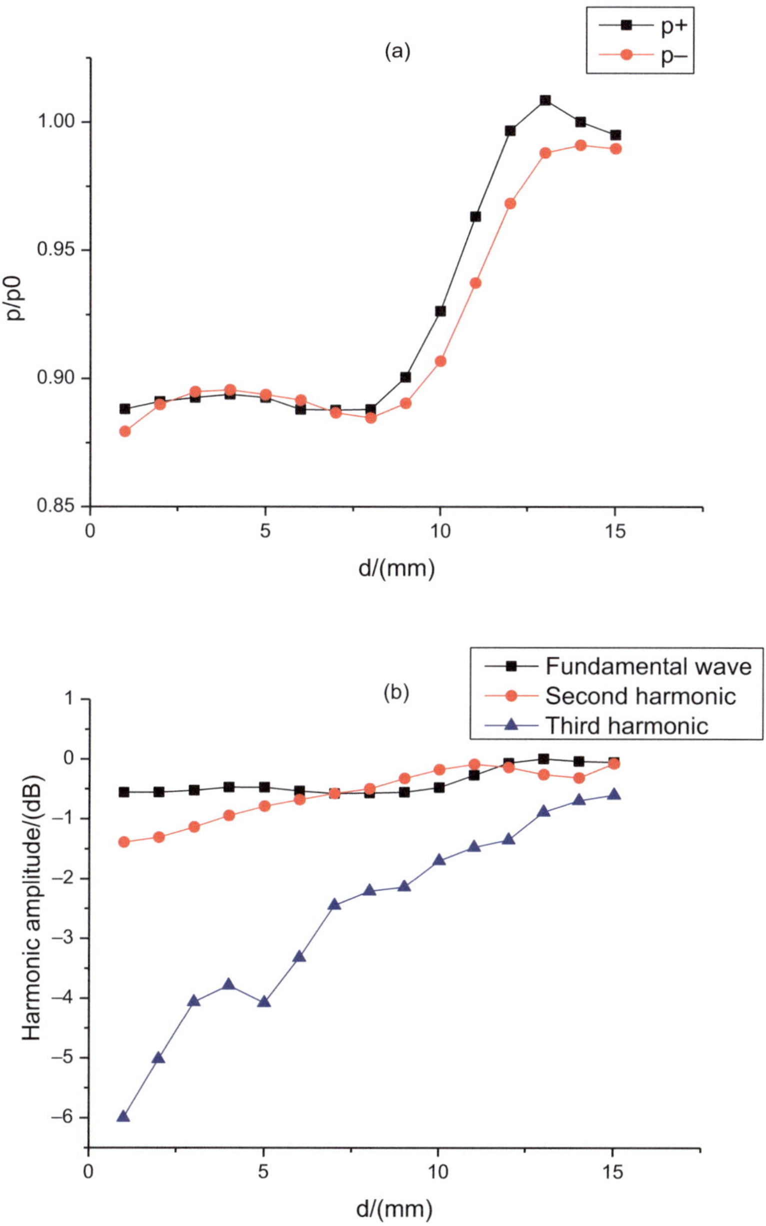

Fig. 1.16 The impact of a inserted uniform layer with a thickness d on the peak and valley values (**a**) and amplitude of each harmonic (**b**)

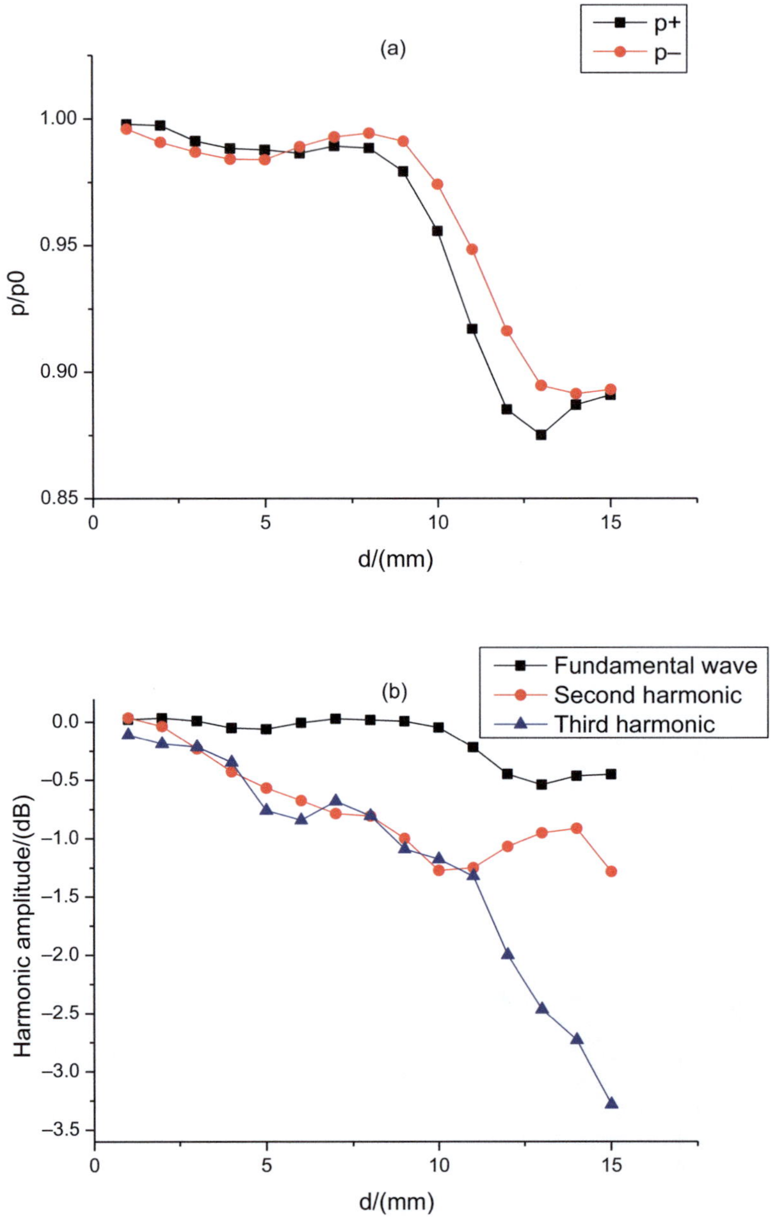

Fig. 1.17 The impact of a inserted non-uniform layer with a thickness d on the peak and valley values (**a**) and amplitude of each harmonic (**b**)

References

1. Landau LD, Lifshitz EM. Theory of elasticity. 3rd ed. New York: Pergamon Press; 1986.
2. Guitong Y. Nonlinear waves in solids. Adv Mechan. 1982;01:94–104.
3. Gonghuan Du, Zheming Z, Xiufen G. Fundamentals of acoustics. Nanjing: Nanjing University Press; 2001.
4. Hamilton MF, Blackstock DT. Nonlinear acoustics: theory and applications. New York: Academic Press Inc.; 1997.
5. Zuwen Q. Nonlinear acoustics. Beijing: Science Press; 1992.
6. Zuwen Q. Nonlinear acoustics. 2nd ed. Beijing: Science Press; 2009.
7. Chen J, Zhang D, Mao Y. Research on the evaluation of the bonding state of the solid bonding interface by the contact sound nonlinearity in the reflected wave. In: 2010 national acoustics conference, Harbin; 2010.
8. Chen J, Zhang D, Mao Y. Contact acoustic nonlinearity phenomenon of sound wave propagation in composite materials. In: Third national symposium on piezoelectric and acoustic theory and device technology, Nanjing; 2008.
9. Jianjun C, De Z. Research on contact acoustic nonlinearity of asymmetric adhesive force interface. Acoust Technol. 2014;33(2):23–6.
10. Wu WQ, Ni YF, Qiu G, et al. Investigation of contact acoustic nonlinearity at solid. In: IEEE ultrasonic symposium; 1998. p. 1175–79.
11. Chen JJ, Zhang D, Mao YW, et al. A unique method to describe the bonding strength in a bounded solid-solid interface by contact acoustic nonlinearity. Chin Phys Lett. 2009;26:014302.
12. Chen J, Mao Y, Zhang D, et al. Research on the adhesion force of the solid adhesive interface by the contact acoustic nonlinearity method. In: 2006 harmonious development of Western China acoustic academic conference, Dujiangyan; 2006.
13. Deng MX, Pei JF. Assessment of accumulated fatigue damage in solid plates using nonlinear Lamb wave approach. Appl Phys Lett. 2007;90(12):121902.
14. Deng MX, Xiang YX, Liu LB. Time-domain analysis and experimental examination of cumulative second-harmonic generation by primary Lamb wave propagation. J Appl Phys. 2011;109:113525.
15. Deng MX, Xiang YX. Analysis of second-harmonic generation by primary ultrasonic guided wave propagation in a piezoelectric plate. Ultrasonics. 2015;61:121–5.
16. Deng MX, Wang P, Lv XF. Experimental observation of cumulative second harmonic generation of Lamb-wave propagation in an elastic plate. J Phys D: Appl Phys. 2005;38:344–53.
17. Deng MX. Cumulative second-harmonic generation of Lamb mode propagation in a solid plate. J Appl Phys. 1999;85:3051–8.
18. Deng MX. Analysis of second-harmonic generation of Lamb modes using a modal analysis approach. J Appl Phys. 2003;94:4153–9.
19. Deng MX, Yang J. Characterization of elastic anisotropy of a solid plate using nonlinear Lamb wave approach. J Sound Vibr. 2007;308:201–11.
20. Deng M, Price DC, Scott DA. Experimental observation of Lamb wave nonlinear effects. Acoustics. 2005;30:37–46.
21. Mingxi D, Junfeng P. Nonlinear ultrasonic Lamb wave method for nondestructive evaluation of fatigue damage in solid plates. Acoustics. 2008;33:360–9.
22. Mingxi D, Yanxun X, Junfeng P, et al. Time-domain measurement method of ultrasonic Lamb wave second harmonic based on group velocity mismatch. Acoustics. 2012;37:621–8.
23. Mingxi D. Nonlinear research of Lamb waves. Acoustics. 1996;21:429–36.
24. Deng M. Nonlinear research of Lamb waves (II). Acoustics. 1997;22:182–7.
25. Mingxi D. Mode expansion analysis of second harmonic generation of Lamb waves in layered structures. Acoustics. 2005;30:132–42.
26. Mingxi D. A nonlinear ultrasonic Lamb wave method for determining the properties of the adhesive layer in composite plates. Acoustics. 2005;30:542–51.
27. Mingxi D. Experimental observation of Lamb wave nonlinear effects (II). Acoustics. 2006;31:1–7.

28. Liu Yaolu Hu, Ning DM, et al. Nonlinear Lamb waves in plate-shell structures. Adv Mechan. 2017;47:503–33.
29. Li WB, Deng MX, Xiang YX. Review on second-harmonic generation of ultrasonic guided waves. Chin Phys B. 2017;26(11):114302.
30. Mingxi D. Nonlinear Lamb waves in solid plates. Beijing: Science Press; 2006.
31. de Lima WJ, Hamilton MF. Finite-amplitude waves in isotropic elastic plates. J Sound Vibr. 2003;265:819–39.
32. Srivastava A, Lanza di Scalea F. On the existence of antisymmetric or symmetric Lamb waves at nonlinear higher harmonics. J Sound Vibr. 2009;323:932–43.
33. Dagrau F, Renier M, Marchiano R, et al. Acoustic shock wave propagation in a heterogeneous medium: a numerical simulation beyond the parabolic approximation. J Acoust Soc Am. 2011;130(1):20–32.
34. Jing Y, Wang TR, Clement GT. A k-space method for moderately nonlinear wave propagation. IEEE Trans Ultrason Ferroelectr Freq Control. 2012;59(8):1664–73.
35. Xue J, Shilei L, Xiaozhou L, et al. Study on the acoustic field characteristics in one-dimensional non-uniform media. Acoust Technol. 2013;32(4):25–8.

Chapter 2
Nonlinear Effects and Acoustic Memory Phenomena of Sound Waves in Crystals and Ceramic Materials

With the development of science and technology and industry, people's requirements for the characteristics of crystals and ceramic materials are getting higher and higher, and the requirements for materials are also getting higher and higher. People are no longer satisfied with the characteristics of cubic crystal systems, and are also interested in the characteristics of various other crystal systems. At the same time, people are no longer satisfied with the inherent characteristics of single-component materials. Multi-component crystals can integrate the advantages of each component material, and the fusion of each component material can obtain some new characteristics and receive a lot of research. Studying the physical characteristics of crystals and guiding the corresponding design has always been a rapidly developing field with application prospects. This chapter studies the nonlinear acoustic propagation characteristics and acoustic memory phenomena of sound waves in various crystal systems and ceramic materials.

2.1 Nonlinear Interactions of Sound Waves in Crystals

2.1.1 Theory

The interaction of elastic waves in isotropic solids has been studied deeply in theory and experiment [1–4]. Domanski et al. studied the propagation and interaction of weak non-elastic waves in cubic crystals [5]. Researchers theoretically and experimentally found that finite amplitude transverse waves do not deform when propagating in uniform isotropic solids, that is, they do not generate second harmonics [6]. However, in anisotropic solids, the propagation of transverse and longitudinal waves is different from that in isotropic solids. In anisotropic solids, transverse waves can generate second harmonics, and two collinear transverse waves can also interact [7]. Researchers have studied the interaction of collinear transverse sound waves

X. Liu, *Nonlinear Sound Waves in Solids*,
Advances in Acoustics, https://doi.org/10.1007/978-981-96-5253-2_2

in cubic crystals, and experiments have proved the interaction of transverse sound waves in cubic crystals [8]. This section derives the wave equation of the interaction of transverse and longitudinal waves propagating in different directions in different crystals through the elastic energy expression. Through the wave equation, we obtain the interaction coefficient of plane waves propagating in different crystals, including cubic, trigonal, tetragonal, and hexagonal crystals. This section also obtains the interaction coefficient of waves in general anisotropic solids through the wave equation, and emphasizes the nonlinear terms, and obtains the analytical solution of the interacting sound waves and the analytical solution of the second-order nonlinear sound wave interaction coefficient. The results show that transverse and longitudinal waves will deform and couple during propagation, and two waves of different frequencies f_1 and f_2 will generate sum frequency waves and difference frequency waves ($f_1 \pm f_2$), these results indicate that the inherent nonlinear characteristics in anisotropic solids can generate interacting elastic sound waves.

For any crystal, the wave equation of sound waves is [8]

$$
\begin{aligned}
\rho_0 u_{1,tt} - C'_{11}u_{1,xx} - C'_{16}u_{2,xx} - C'_{15}u_{3,xx} &= f\left(u_{1,x}, u_{2,x}, u_{3,x}\right)\\
\rho_0 u_{2,tt} - C'_{16}u_{1,xx} - C'_{66}u_{2,xx} - C'_{56}u_{3,xx} &= g\left(u_{1,x}, u_{2,x}, u_{3,x}\right)\\
\rho_0 u_{3,tt} - C'_{15}u_{1,xx} - C'_{56}u_{2,xx} - C'_{55}u_{3,xx} &= h\left(u_{1,x}, u_{2,x}, u_{3,x}\right)
\end{aligned} \tag{2.1}
$$

where

$$
\begin{aligned}
f(\alpha, \beta, \gamma) &= \frac{\partial}{\partial x}\left\{\begin{array}{c} \frac{1}{2}(3C'_{11} + C'_{111})\alpha^2 + \frac{1}{2}(C'_{11} + C'_{166})\beta^2 + \frac{1}{2}(C'_{11} + C'_{155})\gamma^2 \\ +(C'_{16} + C'_{116})\alpha\beta + (C'_{15} + C'_{115})\alpha\gamma + C'_{156}\beta\gamma \end{array}\right\}\\
g(\alpha, \beta, \gamma) &= \frac{\partial}{\partial x}\left\{\begin{array}{c} \frac{1}{2}(C'_{16} + C'_{116})\alpha^2 + \frac{1}{2}(3C'_{16} + C'_{666})\beta^2 + \frac{1}{2}(C'_{16} + C'_{556})\gamma^2 \\ +(C'_{11} + C'_{166})\alpha\beta + C'_{156}\alpha\gamma + (C'_{15} + C'_{556})\beta\gamma \end{array}\right\}\\
h(\alpha, \beta, \gamma) &= \frac{\partial}{\partial x}\left\{\begin{array}{c} \frac{1}{2}(C'_{15} + C'_{115})\alpha^2 + \frac{1}{2}(C'_{15} + C'_{556})\beta^2 + \frac{1}{2}(3C'_{15} + C'_{555})\gamma^2 \\ +C'_{156}\alpha\beta + (C'_{11} + C'_{155})\alpha\gamma + (C'_{16} + C'_{556})\beta\gamma \end{array}\right\}
\end{aligned} \tag{2.2}
$$

ρ is the density of the solid, u_i is the displacement of the sound wave, C'_{ij} and C'_{ijk} are the equivalent second and third order elastic constants in different coordinate systems, respectively.

This section uses the following three different coordinate systems.

Coordinate axis 1: [100], [010], [001]

Coordinate axis 2: [110], [110], [001]

Coordinate axis 3: [111], $[11\bar{2}]$, $[\overline{11}0]$

2.1.1.1 The Case of Two Longitudinal Waves

Solve the equation using perturbation method.

Let $u_1 = u_1^{[I]} + u_1^{[II]}$, $u_2 = 0$, $u_3 = 0$.

The equation becomes:

$$\rho_0 u_{1,tt}^{[II]} - C'_{11} u_{1,xx}^{[II]} = f\left(u_{1,x}^{[I]}, 0, 0\right) = \frac{\partial}{\partial x}\left\{\frac{1}{2}(3C'_{11} + C'_{111})(u_{1,x}^{[I]})^2\right\}$$
$$= \left(3C'_{11} + C'_{111}\right) * u_{1,x}^{[I]} * u_{1,xx}^{[I]} \tag{2.3}$$

Let $u_1^{[I]} = acos\Omega_1 + bcos\Omega_2$

Where $\Omega_i = \omega_i t - k_i x + \theta_i$, $k_i^2 C'_{11} = \rho\omega_i^2$, $k_1 k_2 > 0$, a, b are the amplitudes of the two waves respectively.

Therefore, Eq. (2.1) becomes

$$\rho_0 u_{1,tt}^{[II]} - C'_{11} u_{1,xx}^{[II]} = f\left(u_{1,x}^{[I]}, 0, 0\right) = \left(3C'_{11} + C'_{111}\right) * (ak_1 \sin\Omega_1 + bk_2 \sin\Omega_2) *$$
$$\left(-ak_1^2 \cos\Omega_1 - bk_2^2 \cos\Omega_2\right)$$
$$= -\left(3C'_{11} + C'_{111}\right)\left(\frac{1}{2}a^2 k_1^3 \sin(2\Omega_1) + \frac{1}{2}b^2 k_2^3 \sin(2\Omega_2)\right.$$
$$\left. + abk_1 k_2 (k_1 \cos\Omega_1 \sin\Omega_2 + k_2 \cos\Omega_2 \sin\Omega_1)\right)$$
$$= -\left(3C'_{11} + C'_{111}\right)\left(\frac{1}{2}a^2 k_1^3 \sin(2\Omega_1)\right.$$
$$\left. + \frac{1}{2}b^2 k_2^3 \sin(2\Omega_2) + \frac{1}{2}abk_1 k_2 ((k_1 + k_2)\sin(\Omega_1 + \Omega_2) - (k_1 - k_2)\sin(\Omega_1 - \Omega_2))\right) \tag{2.4}$$

Let

$$u_1^{[II]} = B_1 x \sin(2\Omega_1) + B_2 x \cos(2\Omega_2) + C_1 x \sin(2\Omega_2) + C_2 x \cos(2\Omega_2)$$
$$+ D_1 x \sin(\Omega_1 + \Omega_2) + D_2 x \cos(\Omega_1 + \Omega_2)$$
$$+ E_1 x \sin(\Omega_1 - \Omega_2) + E_2 x \cos(\Omega_1 - \Omega_2) \tag{2.5}$$

where: $B_i, C_i, D_i, E_i (i = 1, 2)$ are constants, so we have

$$u_{tt}^{[II]} = -4\omega_1^2 B_1 x \sin(2\Omega_1) + 4\omega_1^2 B_2 x \cos(2\Omega_1)$$
$$+ 4\omega_2^2 C_1 x \sin(2\Omega_2) + 4\omega_2^2 C_2 x \cos(2\Omega_2)$$
$$+ (\omega_1 + \omega_2)^2 D_1 x \sin(\Omega_1 + \Omega_2) + (\omega_1 + \omega_2)^2 D_2 x \cos(\Omega_1 + \Omega_2)$$
$$+ (\omega_1 - \omega_2)^2 E_1 x \sin(\Omega_1 - \Omega_2)\big) + (\omega_1 - \omega_2)^2 E_2 x \cos(\Omega_1 - \Omega_2)\big)$$

$$
\begin{aligned}
u_{xx}^{[II]} &= B_1\left(-4k_1\cos(2\Omega_1) - 4k_1^2 x\sin(2\Omega_1)\right) + B_2(4k_1\sin(2\Omega_1)\\
&-4k_1^2 x\cos(2\Omega_1)\big) + C_1\left(-4k_2\cos(2\Omega_2) - 4k_2^2 x\sin(2\Omega_2)\right) + C_2(4k_2\sin(2\Omega_2)\\
&-4k_2^2 x\cos(2\Omega_2)\big) + D_1\left(-2(k_1+k_2)\cos(\Omega_1+\Omega_2) - (k_1+k_2)^2 x\sin(\Omega_1+\Omega_2)\right)\\
&+D_2\left(2(k_1+k_2)\sin(\Omega_1+\Omega_2) - (k_1+k_2)^2 x\cos(\Omega_1+\Omega_2)\right) + E_1(-2(k_1\\
&-k_2)\cos(\Omega_1-\Omega_2) - (k_1-k_2)^2 x\sin(\Omega_1-\Omega_2)\big) + E_2(2(k_1-k_2)\sin(\Omega_1-\Omega_2)\\
&-(k_1-k_2)^2 x\cos(\Omega_1-\Omega_2)\big) = \sin(2\Omega_1)\left[-4k_1^2 xB_1 + 4k_1B_2\right] + \cos(2\Omega_1)[-4k_1B_1\\
&-4k_1^2 xB_2\big] + \sin(2\Omega_2)\left[-4k_2^2 xC_1 + 4k_2C_2\right] + \cos(2\Omega_2)\left[-4k_2C_1 - 4k_2^2 xC_2\right] + \sin(\Omega_1\\
&+\Omega_2)\left[-(k_1+k_2)^2 xD_1 + 2(k_1+k_2)D_2\right] + \cos(\Omega_1+\Omega_2)[-2(k_1+k_2)D_1 - (k_1\\
&+k_2)^2 xD_2\big] + \sin(\Omega_1-\Omega_2)\left[-(k_1-k_2)^2 xE_1 + 2(k_1-k_2)E_2\right] + \cos(\Omega_1\\
&-\Omega_2)\left[-2(k_1-k_2)E_1 - (k_1-k_2)^2 xE_2\right]
\end{aligned}
\tag{2.6}
$$

Substituting $u_{tt}^{[II]}$, $u_{xx}^{[II]}$ and setting the coefficients on both sides equal, we get

$$
\begin{aligned}
&-4\rho_0\omega_1^2 B_1 x - C'_{11}\left(-4k_1^2 xB_1 + 4k_1B_2\right) = -\left(3C'_{11} + C'_{111}\right)\left(\frac{1}{2}a^2k_1^3\right)\\
&-4\rho_0\omega_1^2 B_2 x - C'_{11}\left(-4k_1B_1 - 4k_1^2 xB_2\right) = 0 - 4\rho_0\omega_2^2 C_1 x - C'_{11}\left(-4k_2^2 xC_1 + 4k_2C_2\right)\\
&= -\left(3C'_{11} + C'_{111}\right)\left(\frac{1}{2}b^2k_2^3\right) - 4\rho_0\omega_2^2 C_2 x - C'_{11}\left(-4k_2C_1 - 4k_2^2 xC_2\right)\\
&= 0 - \rho_0(\omega_1+\omega_2)^2 D_1 x - C'_{11}\left(-(k_1+k_2)^2 xD_1 + 2(k_1+k_2)D_2\right)\\
&= -\left(3C'_{11} + C'_{111}\right)\frac{1}{2}abk_1k_2(k_1+k_2) - \rho_0(\omega_1+\omega_2)^2 D_2 x\\
&-C'_{11}\left(2(k_1+k_2)D_1 - (k_1+k_2)^2 xD_2\right) = 0 - \rho_0(\omega_1-\omega_2)^2 E_1 x\\
&-C'_{11}\left(-(k_1-k_2)^2 xE_1 + 2(k_1-k_2)E_2\right) = \left(3C'_{11} + C'_{111}\right)\frac{1}{2}abk_1k_2(k_1-k_2)\\
&-\rho_0(\omega_1-\omega_2)^2 E_2 x - C'_{11}\left(2(k_1-k_2)E_1 - (k_1-k_2)^2 xE_2\right) = 0
\end{aligned}
\tag{2.7}
$$

Solving this, we get

$$
\begin{aligned}
B_1 &= 0, \quad B_2 = \frac{3\left(C'_{11} + C'_{111}\right)a^2k_1^2}{8C'_{11}}\\
C_1 &= 0, \quad C_2 = \frac{3\left(C'_{11} + C'_{111}\right)b^2k_2^2}{8C'_{11}}\\
D_1 &= 0, \quad D_2 = \frac{3\left(C'_{11} + C'_{111}\right)abk_1k_2}{4C'_{11}}
\end{aligned}
$$

$$E_1 = 0, \quad E_2 = -\frac{3(C'_{11} + C'_{111})abk_1k_2}{4C'_{11}} \tag{2.8}$$

Therefore, the final sound wave displacement is:

$$\begin{aligned} u = u_1^{[I]} + u_1^{[II]} &= a\cos\Omega_1 + b\cos\Omega_2 + \frac{3(C'_{11} + C'_{11})a^2k_1^2}{8C'_{11}} x\cos(2\Omega_2) \\ &+ \frac{3(C'_{11} + C'_{11})b^2k_2^2}{8C'_{11}} x\cos(2\Omega_2) + \frac{3(C'_{11} + C'_{111})abk_1k_2}{4C'_{11}} x\cos(\Omega_1 + \Omega_2) \\ &- \frac{3(C'_{11} + C'_{111})abk_1k_2}{4C'_{11}} \times \cos(\Omega_1 - \Omega_2)\Omega_i = \omega_i t - k_i x + \theta_i, \end{aligned} \tag{2.9}$$

2.1.1.2 The Case of Two Transverse Waves

If two transverse waves propagate in the medium, then $u_1 = 0$.

The equation becomes

$$\begin{aligned} \rho_0 u_{1,tt}^{[II]} - C'_{11} u_{1,xx}^{[II]} &= \frac{\partial}{\partial x}\left\{\frac{1}{2}(C'_{11} + C'_{155})[(u_{2,x}^{[I]})^2 + \left(u_{3,x}^{[I]})^2\right]\right\} \\ \rho_0 u_{2,tt}^{[II]} - C'_{55} u_{2,xx}^{[II]} &= \frac{\partial}{\partial x}\left\{\frac{1}{2}C'_{666}[(u_{2,x}^{[I]})^2 - \left(u_{3,x}^{[I]})^2\right] - C'_{555} u_{2,x}^{[I]} u_{3,x}^{[I]}\right\} \\ \rho_0 u_{3,tt}^{[II]} - C'_{55} u_{3,xx}^{[II]} &= \frac{\partial}{\partial x}\left\{-\frac{1}{2}C'_{555}[(u_{2,x}^{[I]})^2 - \left(u_{3,x}^{[I]})^2\right] - C'_{666} u_{2,x}^{[I]} u_{3,x}^{[I]}\right\} \end{aligned} \tag{2.10}$$

Let:

$$\begin{pmatrix} u_1^{[I]} \\ u_2^{[I]} \\ u_3^{[I]} \end{pmatrix} = \begin{pmatrix} 0 \\ a\cos\phi_1 \\ a\sin\phi_1 \end{pmatrix} \cos\Omega_1 + \begin{pmatrix} 0 \\ b\cos\phi_2 \\ b\sin\phi_2 \end{pmatrix} \cos\Omega_2$$

where

$$\Omega_i = \omega_i t - k_i x + \theta_i, \; k_i^2 C'_{55} = \rho\omega_i^2, \quad k_1k_2 > 0 \tag{2.11}$$

ϕ_1, ϕ_2 is the angle between the two waves in the Y–Z plane and the Y-axis.

Then the wave equation becomes

$$\rho_0 u_{1,tt}^{[II]} - C'_{11} u_{1,xx}^{[II]} = \frac{\partial}{\partial x}\left\{\frac{1}{2}(C'_{11} + C'_{155})[(u_{2,x}^{[I]})^2 + \left(u_{3,x}^{[I]})^2\right]\right\}$$

$$\begin{aligned} \rho_0 u_{2,tt}^{[II]} - C'_{55} u_{2,xx}^{[II]} &= \frac{\partial}{\partial x}\left\{\frac{1}{2}C'_{666}\left[\left(u_{2,x}^{[I]}\right)^2 - \left(u_{3,x}^{[I]}\right)^2\right] - C'_{555} u_{2,x}^{[I]} u_{3,x}^{[I]}\right\} \\ &= C'_{666}\left(u_{2,x}^{[I]} u_{2,xx}^{[I]} - u_{3,x}^{[I]} u_{3,xx}^{[I]}\right) - C'_{555}\left(u_{2,xx}^{[I]} u_{3,x}^{[I]} + u_{2,x}^{[I]} u_{3,xx}^{[I]}\right) \end{aligned}$$

$$
\begin{aligned}
&= \frac{1}{2}C'_{666}abk_1k_2\{\cos(\phi_1+\phi_2)((k_1+\ k_2)\sin(\Omega_1+\Omega_2)-(k_1-k_2)\sin(\Omega_1-\Omega_2))\}\\
&\quad -\frac{1}{2}C'_{555}(abk_1k_2\sin(\phi_1+\phi_2)((k_1+\ k_2)\sin(\Omega_1+\Omega_2)-(k_1-k_2)\sin(\Omega_1-\Omega_2)))\\
&= \sin(\Omega_1+\Omega_2)abk_1k_2(k_1+k_2)\frac{1}{2}\left(C'_{666}\cos(\phi_1+\phi_2)-C'_{555}\sin(\phi_1+\phi_2)\right)\\
&\quad -\sin(\Omega_1-\Omega_2)abk_1k_2(k_1-k_2)\frac{1}{2}\left(C'_{666}\cos(\phi_1+\phi_2)-C'_{555}\sin(\phi_1+\phi_2)\right)\\
&\rho_0u^{[II]}_{3,tt}-C'_{55}u^{[II]}_{3,xx}=\frac{\partial}{\partial x}\left\{-\frac{1}{2}C'_{555}\left[\left(u^{[I]}_{2,x}\right)^2-\left(u^{[I]}_{3,x}\right)^2\right]-C'_{666}u^{[I]}_{2,x}u^{[I]}_{3,x}\right\}\\
&= C'_{666}\left(u^{[I]}_{2,x}u^{[I]}_{2,xx}-u^{[I]}_{3,x}u^{[I]}_{3,xx}\right)-C'_{555}\left(u^{[I]}_{2,xx}u^{[I]}_{3,x}+u^{[I]}_{2,x}u^{[I]}_{3,xx}\right)\\
&= -\frac{1}{2}C'_{555}abk_1k_2\{\cos(\phi_1+\phi_2)((k_1+k_2)\sin(\Omega_1+\ \Omega_2)-(k_1-k_2)\sin(\Omega_1-\Omega_2))\}\\
&\quad -\frac{1}{2}C'_{666}(abk_1k_2\sin(\phi_1+\phi_2)((k_1+\ k_2)\sin(\Omega_1+\Omega_2)-(k_1-k_2)\sin(\Omega_1-\Omega_2)))\\
&= \sin(\Omega_1+\Omega_2)abk_1k_2(k_1+k_2)\frac{1}{2}\left(-C'_{555}\cos(\phi_1+\phi_2)-C'_{666}\sin(\phi_1+\phi_2)\right)\\
&\quad -\sin(\Omega_1-\Omega_2)abk_1k_2(k_1-k_2)-\frac{1}{2}\left(C'_{555}\cos(\phi_1+\phi_2)-C'_{666}\sin(\phi_1+\phi_2)\right)
\end{aligned}
\tag{2.12}
$$

If only considering the sum-frequency and difference-frequency waves

$$
\begin{aligned}
u^{[II]}_2 &= A_1x\sin(\Omega_1+\Omega_2)+A_2x\cos(\Omega_1+\Omega_2)\\
&\quad +B_1x\sin(\Omega_1-\Omega_2)+B_2x\cos(\Omega_1-\Omega_2)\\
u^{[II]}_3 &= D_1x\sin(\Omega_1+\Omega_2)+D_2x\cos(\Omega_1+\Omega_2)\\
&\quad +E_1x\sin(\Omega_1-\Omega_2)+E_2x\cos(\Omega_1-\Omega_2)
\end{aligned}
$$

$$
\begin{aligned}
u^{[II]}_{2,tt} &= -(\omega_1+\omega_2)^2A_1x\sin(\Omega_1+\Omega_2)-(\omega_1+\omega_2)^2A_2x\cos(\Omega_1+\Omega_2)\\
&\quad -(\omega_1-\omega_2)^2B_1x\sin(\Omega_1-\Omega_2)-(\omega_1-\omega_2)^2B_2x\cos(\Omega_1-\Omega_2)
\end{aligned}
$$

$$
\begin{aligned}
u^{[II]}_{3,tt} &= -(\omega_1+\omega_2)^2D_1x\sin(\Omega_1+\Omega_2)-(\omega_1+\omega_2)^2D_2x\cos(\Omega_1+\Omega_2)\\
&\quad -\ (\omega_1-\omega_2)^2E_1x\sin(\Omega_1-\Omega_2)-(\omega_1-\omega_2)^2E_2x\cos(\Omega_1-\Omega_2)
\end{aligned}
$$

$$
\begin{aligned}
u^{[II]}_{2,xx} &= \sin(\Omega_1+\Omega_2)\left[-(k_1+k_2)^2xA_1+2(k_1+k_2)A_2\right]\\
&\quad +\cos(\Omega_1+\Omega_2)[-2(k_1+\ k_2)A_1-(k_1+k_2)^2xA_2\big]\\
&\quad +\sin(\Omega_1-\Omega_2)\left[-(k_1-k_2)^2xV_1+2(k_1-k_2)V_2\right]\\
&\quad +\cos(\Omega_1-\Omega_2)\left[-2(k_1-k_2)V_1-(k_1-k_2)^2xV_2\right]
\end{aligned}
$$

$$
u^{[II]}_{3,xx} = \sin(\Omega_1+\Omega_2)\left[-(k_1+k_2)^2xD_1+2(k_1+k_2)D_2\right]
$$

$$+\cos(\Omega_1+\Omega_2)[-2(k_1+k_2)D_1-(k_1+k_2)^2xD_2]$$
$$+\sin(\Omega_1-\Omega_2)\left[-(k_1-k_2)^2xE_1+2(k_1-k_2)E_2\right]$$
$$+\cos(\Omega_1-\Omega_2)\left[-2(k_1-k_2)E_1-(k_1-k_2)^2xE_2\right] \quad (2.13)$$

Substituting into the wave equation, we get

$$
\begin{aligned}
A_1 &= B_1 = D_1 = E_1 = 0;\\
A_2 &= \frac{abk_1k_2}{4C'_{55}}\left[C'_{666}\cos(\phi_1+\phi_2)-C'_{555}\sin(\phi_1+\phi_2)\right]\\
B_2 &= -\frac{abk_1k_2}{4C'_{55}}\left[C'_{666}\cos(\phi_1+\phi_2)-C'_{555}\sin(\phi_1+\phi_2)\right]\\
D_2 &= -\frac{abk_1k_2}{4C'_{55}}\left[-C'_{555}\cos(\phi_1+\phi_2)-C'_{666}\sin(\phi_1+\phi_2)\right]\\
E_2 &= \frac{abk_1k_2}{4C'_{55}}\left[-C'_{555}\cos(\phi_1+\phi_2)-C'_{666}\sin(\phi_1+\phi_2)\right]
\end{aligned} \quad (2.14)
$$

Therefore, the sound wave displacement (considering only the sum-frequency and difference-frequency waves) is:

$$
\begin{aligned}
u =& \frac{abk_1k_2}{4C'_{55}}\left[C'_{666}\cos(\phi_1+\phi_2)-C'_{555}\sin(\phi_1+\phi_2)\right]x\cos(\Omega_1+\Omega_2)\\
&-\frac{abk_1k_2}{4C'_{55}}\left[C'_{666}\cos(\phi_1+\phi_2)-C'_{555}\sin(\phi_1+\phi_2)\right]x\cos(\Omega_1-\Omega_2)\\
&-\frac{abk_1k_2}{4C'_{55}}\left[-C'_{555}\cos(\phi_1+\phi_2)-C'_{666}\sin(\phi_1+\phi_2)\right]x\cos(\Omega_1+\Omega_2)\\
&+\frac{abk_1k_2}{4C'_{55}}\left[-C'_{555}\cos(\phi_1+\phi_2)-C'_{666}\sin(\phi_1+\phi_2)\right]x\cos(\Omega_1-\Omega_2)
\end{aligned} \quad (2.15)
$$

2.1.1.3 All Crystal Systems

For all crystal systems, the key to finding the interaction solution of sound waves in them is to find the equivalent second-order and third-order elastic constants in different coordinate systems.

Trigonal crystals (3-point group, 32-point group, 3 m-point group)

Tetragonal crystals (4-point group, 422, 4 mm-point group, 42 m-point group)

Hexagonal crystals (6,622, 6 mm, 6 m^2-point group)

Cubic crystals (23-point group, 43 m-point group)

Solution process

Step 1: Obtain the independent second-order, third-order elastic constants and non-zero non-independent second-order, third-order elastic constants of different structure crystals in the original coordinate system through literature, and write out the elastic energy expression through them.

Step 2: Obtain the transformation relationship of elastic energy density under the working coordinate system through coordinate transformation.

Step 3: According to the principle of conservation of elastic energy, substitute the elastic energy density into the expression of elastic energy under the original coordinate system to obtain the expression of elastic energy under the working coordinate system.

Step 4: Derive the elastic energy density through the expression of elastic energy under the working coordinate system to obtain the stress matrix.

Step 5: Substitute the stress matrix into the fluctuation equation, omit the part above the second order, and integrate and merge to obtain the fluctuation equation and equivalent elastic constant under the working coordinate system.

The case of cubic crystal propagating along the [100] direction:

Take the cubic crystal (432 point group) as an example:

If the propagation direction is the a-axis, the displacement component can be expressed as

$$u = u(a,t), v = v(a,t), w = w(a,t)$$
$$x = a + u, y = b + v, z = c + w$$

$$\text{Jacobian matrix:} J = \begin{pmatrix} 1 + u_a & 0\ 0 \\ v_a & 1\ 0 \\ w_a & 0\ 1 \end{pmatrix} \tag{2.16}$$

$$\text{Strain component}: \eta = \frac{1}{2}(J * J - I)$$

$$\begin{aligned} \eta_{11} &= u_a + \frac{1}{2}\left(u_a^2 + v_a^2 + w_a^2\right) \\ \eta_{12} &= \eta_{21} = \frac{1}{2}v_a \\ \eta_{13} &= \eta_{31} = \frac{1}{2}w_a \end{aligned} \tag{2.17}$$

Elastic energy:

$$\phi = \frac{1}{2}C_{11}\eta_{11}^2 + C_{44}\left(\eta_{12}^2 + \eta_{21}^2 + \eta_{31}^2 + \eta_{13}^2\right)$$

$$+\frac{1}{6}C_{111}\eta_{11}^3 + C_{166}\left[\eta_{11}\left(\eta_{12}^2 + \eta_{21}^2 + \eta_{31}^2 + \eta_{13}^2\right)\right] \tag{2.18}$$

Substitute into the stress tensor: $T_{ik} = J_{il}\frac{\partial\phi}{\partial\eta_{kl}}$ and the fluctuation equation: (ignoring the piezoelectric effect)

$$\begin{aligned}
\rho_0\frac{\partial^2 u}{\partial t^2} &= \frac{\partial T_{11}}{\partial a} + \frac{\partial T_{12}}{\partial b} + \frac{\partial T_{13}}{\partial c} \\
\rho_0\frac{\partial^2 v}{\partial t^2} &= \frac{\partial T_{21}}{\partial a} + \frac{\partial T_{22}}{\partial b} + \frac{\partial T_{23}}{\partial c} \\
\rho_0\frac{\partial^2 w}{\partial t^2} &= \frac{\partial T_{31}}{\partial a} + \frac{\partial T_{32}}{\partial b} + \frac{\partial T_{33}}{\partial c}
\end{aligned} \tag{2.19}$$

Then obtain the set of fluctuation equations:

$$\begin{aligned}
\frac{\rho_0}{C_{11}}u_{tt} - u_{aa} &= \frac{3C_{11} + C_{111}}{C_{11}}u_a u_{aa} + \left(1 + \frac{C_{166}}{C_{11}}\right)[v_a v_{aa} + w_a w_{aa}] \\
\frac{1}{C_t^2}v_{tt} - v_{aa} &= \frac{C_{11} + C_{166}}{C_{44}}[u_a v_{aa} + v_a u_{aa}] \\
\frac{1}{C_t^2}w_{tt} - w_{aa} &= \frac{C_{11} + C_{166}}{C_{44}}[u_a w_{aa} + w_a u_{aa}]
\end{aligned} \tag{2.20}$$

Cubic crystal propagating along the [110] direction:
The coordinate transformation matrix is:

$$\begin{pmatrix} a' \\ b' \\ c' \end{pmatrix} = \frac{1}{\sqrt{2}}\begin{pmatrix} 1 & 1 & 0 \\ -1 & 1 & 0 \\ 0 & 0 & \sqrt{2} \end{pmatrix}\begin{pmatrix} a \\ b \\ c \end{pmatrix} \tag{2.21}$$

Therefore, the displacement in the new coordinates is:

$$\begin{aligned}
&u' = u'(a', t), v' = v'(a', t), w' = w'(a', t) \\
&\eta_{11}' = u_{a'}' + \frac{1}{2}\left(u_{a'}'^2 + v_{a'}'^2 + w_{a'}'^2\right) \\
&\eta_{12}' = \eta_{21}' = \frac{1}{2}v_{a''} \\
&\eta_{13'} = \eta_{31}' = \frac{1}{2}w_{a''}
\end{aligned} \tag{2.22}$$

Transformation formula:

$$\eta' = \alpha^*\eta\alpha \tag{2.23}$$

Hence we have:

$$\eta_{11} = \frac{1}{2}\left(\eta'_{11} - 2\eta'_{21}\right), \eta_{12} = \frac{1}{2}\eta'_{11}, \eta_{13} = \frac{1}{\sqrt{2}}\eta'_{13},$$
$$\eta_{21} = \frac{1}{2}\eta'_{11}, \eta_{22} = \frac{1}{2}\left(\eta'_{11} + 2\eta'_{12}\right), \eta_{23} = \frac{1}{\sqrt{2}}\eta'_{13},$$
$$\eta_{31} = \frac{1}{\sqrt{2}}\eta'_{13}, \eta_{32} = \frac{1}{\sqrt{2}}\eta'_{13}, \eta_{33} = 0 \tag{2.24}$$

The elastic energy after the rotation axis transformation is:

$$\begin{aligned}
\Phi(\eta') = \frac{1}{48}\Big(&6\left(-\eta'_{11} + \eta'_{12} + \eta'_{21}\right)^2 c_{11} + \left(\eta'_{11} - \eta'_{12} - \eta'_{21}\right)^3 c_{111} \\
&+3\left(-\eta'_{11} + b_{12} + \eta'_{21}\right)^2\left(\eta'_{11} + \eta'_{12} + \eta'_{21}\right)c_{112} \\
&+ 12\left(\eta'_{11} - \eta'_{12} - \eta'_{21}\right)\left(\eta'_{11} + \eta'_{12} + \eta'_{21}\right)c_{12} \\
&+ 3\left(\eta'_{11} - \eta'_{12} - \eta'_{21}\right)\left(\eta'_{11} + \eta'_{12} + \eta'_{21}\right)^2 c_{122} \\
&+ 6\left(\eta'_{11} + \eta'_{12} + \eta'_{21}\right)^2 c_{22} + 12\left(\eta'_{11} - \eta'_{12} - \eta'_{21}\right)\left(b13^2 + b31^2\right)c_{144} \\
&+ 12\left(\eta'_{11} - \eta'_{12} - \eta'_{21}\right)\left(\eta'^2_{13} + \eta'^2_{31}\right)c_{155} + \left(\eta'_{11} + b\eta'_{12} + \eta'_{21}\right)^3 c_{222} \\
&+ 12\left(\eta'_{11} + \eta'_{12} + \eta'_{21}\right)\left(\eta'^2_{13} + \eta'^2_{31}\right)c_{244} + 12\left(\eta'^2_{11} + \left(\eta'_{12} - \eta'_{21}\right)^2\right) \\
&\left(\eta'_{11} - \eta'_{12} - \eta'_{21}\right)c_{166} + 24\left(\eta'^2_{13} + \eta'^2_{31}\right)c_{44} + 12\left(\eta'_{11} + \eta'_{12} + \eta'_{21}\right) \\
&\left(\eta'^2_{13} + \eta'^2_{31}\right)c_{255} + 12\left(\eta'^2_{11} + \left(\eta'_{12} - \eta'_{21}\right)^2\right)\left(\eta'_{11} + \eta'_{12} + \eta'_{21}\right)c_{266} \\
&+ 24\left(\eta'^2_{13} + \eta'^2_{31}\right)c_{55} + 24\eta'_{11}\left(\eta'_{13} + \eta'_{31}\right)^2 c_{456} \\
&+24\left(\eta'^2_{11} + \left(\eta'_{12} - \eta'_{21}\right)^2\right)c_{66}\Big)
\end{aligned} \tag{2.25}$$

Take the derivative of $\Phi(\eta')$ with respect to η'_{ij}, then $\eta'_{12} = \eta'_{21}, \eta'_{13} = \eta'_{31}$ substitute into

$$T_{ik} = J_{il}\frac{\partial\Phi(\eta')}{\partial\eta'_{kl}} \tag{2.26}$$

Substitute into the set of fluctuation equations to get:

$$\begin{aligned}
&\frac{1}{C_l^2}u_{tt} - u_{aa} = \beta u_a u_{aa} + \gamma_1 v_a v_{aa} + \gamma_2 w_a w_{aa} \\
&\frac{1}{C_{t1}^2}v_{tt} - v_{aa} = \gamma_1\left[u_a v_{aa} + v_a u_{aa}\right] \\
&\frac{1}{C_{t2}^2}w_{tt} - w_{aa} = \gamma_2\left[u_a w_{aa} + w_a u_{aa}\right]
\end{aligned} \tag{2.27}$$

According to the above-mentioned solution steps, the equivalent second and third order elastic constants of triclinic, cubic, tetragonal, and hexagonal crystals in three coordinate systems are obtained, so that the analytical solutions of wave interactions in different crystals in various coordinate systems can be obtained.

Cubic Crystal

	[100]	[110]	[111]
C'_{11}	c_{11}	$\frac{1}{2}(c_{11}+c_{12}+2c_{44})$	$\frac{1}{3}(c_{11}+2c_{12}+4c_{44})$
C'_{15}	0	0	0
C'_{16}	0	0	0
C'_{55}	c_{44}	c_{44}	$\frac{1}{3}(c_{11}-c_{12}+c_{44})$
C'_{56}	0	0	0
C'_{66}	c_{44}	$\frac{1}{2}(c_{11}-c_{12})$	$\frac{1}{3}(c_{11}-c_{12}+c_{44})$

23 point groups	[100]	[110]	[111]
C'_{111}	c_{111}	$\frac{1}{8}\begin{pmatrix}2c_{111}+3c_{112}+3c_{113}\\+12c_{155}+12c_{166}\end{pmatrix}$	$\frac{1}{9}\begin{pmatrix}c_{111}+3c_{112}+3c_{113}+2c_{123}\\+12c_{144}+12c_{155}+12c_{166}+16c_{456}\end{pmatrix}$
C'_{115}	0	0	0
C'_{116}	0	$-\frac{1}{8}(c_{112}-c_{113}-4c_{155}+4c_{166})$	0
C'_{155}	c_{155}	$\frac{1}{4}(2c_{144}+c_{155}+c_{166}+4c_{456})$	$\frac{1}{9}\begin{pmatrix}c_{111}-c_{123}-3c_{144}\\+3c_{155}+3c_{166}-2c_{456}\end{pmatrix}$
C'_{156}	0	0	0
C'_{166}	c_{166}	$\frac{1}{8}(2c_{111}-c_{112}-c_{113})$	$\frac{1}{9}\begin{pmatrix}c_{111}-c_{123}-3c_{144}\\+3c_{155}+3c_{166}-2c_{456}\end{pmatrix}$
C'_{555}	0	0	$\frac{1}{2\sqrt{6}}(c_{112}-c_{113}-c_{155}+c_{166})$
C'_{556}	0	$\frac{1}{4}(-c_{155}+c_{166})$	$\frac{1}{18\sqrt{2}}\begin{pmatrix}2c_{111}-3c_{112}-3c_{113}+4c_{123}\\+6c_{144}-3c_{155}-3c_{166}-4c_{456}\end{pmatrix}$
C'_{566}	0	0	$-\frac{1}{2\sqrt{6}}(c_{112}-c_{113}-c_{155}+c_{166})$
C'_{666}	0	$\frac{3}{8}(c_{112}-c_{113})$	$-\frac{1}{18\sqrt{2}}\begin{pmatrix}2c_{111}-3c_{112}-3c_{113}+4c_{123}\\+6c_{144}-3c_{155}-3c_{166}-4c_{456}\end{pmatrix}$

432 and 43 m point groups	[100]	[110]	[111]
C'_{111}	c_{111}	$\frac{1}{4}(c_{111}+3c_{112}+12c_{155})$	$\frac{1}{9}\begin{pmatrix} c_{111}+6c_{112}+2c_{123} \\ +12c_{144}+24c_{155}+16c_{456} \end{pmatrix}$
C'_{115}	0	0	0
C'_{116}	0	0	0
C'_{155}	c_{111}	$\frac{1}{2}(c_{144}+c_{155}+2c_{456})$	$\frac{1}{9}(c_{111}-c_{123}-3c_{144}+6c_{155}-2c_{456})$
C'_{156}	0	0	0
C'_{166}	c_{111}	$\frac{1}{4}(c_{111}-c_{112})$	$\frac{1}{9}(c_{111}-c_{123}-3c_{144}+6c_{155}-2c_{456})$
C'_{555}	0	0	0
C'_{556}	0	0	$\frac{1}{9\sqrt{2}}\begin{pmatrix} c_{111}-3c_{112}+2c_{123} \\ +3c_{144}-3c_{155}-2c_{456} \end{pmatrix}$
C'_{566}	0	0	0
C'_{666}	0	0	$-\frac{1}{9\sqrt{2}}\begin{pmatrix} c_{111}-3c_{112}+2c_{123} \\ +3c_{144}-3c_{155}-2c_{456} \end{pmatrix}$

Trigonal Crystals

32 point groups and 3 m point groups	[100]	[110]	[111]
C'_{11}	c_{11}	c_{11}	$\frac{1}{9}(4c_{11}+4c_{13}+8c_{14}+c_{33}+8c_{44})$
C'_{15}	0	$\frac{c_{14}}{\sqrt{2}}$	$-\sqrt{\frac{2}{3}}c_{14}$
C'_{16}	0	0	$\frac{1}{9}\sqrt{2}(2c_{11}-c_{13}+c_{14}-c_{33}-2c_{44})$
C'_{55}	c_{44}	c_{44}	$\frac{1}{3}(c_{11}-c_{12}-2c_{14}+c_{44})$
C'_{56}	c_{14}	$-\frac{c_{14}}{\sqrt{2}}$	0
C'_{66}	$\frac{1}{2}(c_{11}-c_{12})$	$\frac{1}{2}(c_{11}-c_{12})$	$\frac{1}{9}\begin{pmatrix} 2c_{11}-4c_{13}-2c_{14} \\ +2c_{33}+c_{44} \end{pmatrix}$
32 point groups and 3 m point groups	[100]	[110]	[111]
C'_{111}	c_{111}	$\frac{1}{2}(c_{111}+c_{222})$	$\frac{1}{27}\begin{pmatrix} 4c_{111}+12c_{113}+24c_{114}+48c_{124} \\ +6c_{133}+24c_{134}+48c_{155}+4c_{222} \\ +c_{333}+24c_{344}-16c_{444} \end{pmatrix}$

(continued)

(continued)

32 point groups and 3 m point groups	[100]	[110]	[111]
C'_{115}	0	$\frac{(c_{114}+2c_{124})}{\sqrt{2}}$	$\frac{2}{9}\sqrt{\frac{2}{3}}\left(\begin{array}{c} c_{111}-3c_{114}-6c_{124} \\ -3c_{134}-c_{222}+2c_{444} \end{array}\right)$
C'_{116}	0	$\frac{1}{2}(c_{111}-c_{222})$	$\frac{\sqrt{2}}{27}\left(\begin{array}{c} 2c_{111}+6c_{114}+12c_{124} \\ -3c_{133}-6c_{134}+2c_{222} \\ -c_{333}-12c_{344}+4c_{444} \end{array}\right)$
C'_{155}	c_{155}	c_{155}	$\frac{1}{9}\left(\begin{array}{c} -c_{112}+c_{113}-2c_{114}-c_{123} \\ -2c_{124}-2c_{134}-2c_{144}+4c_{155} \\ +c_{222}+c_{344}+2c_{444} \end{array}\right)$
C'_{156}	$\frac{1}{2}(c_{114}+3c_{124})$	$-\frac{(c_{114}+3c_{124})}{2\sqrt{2}}$	$\frac{1}{9\sqrt{3}}\left(\begin{array}{c} 2c_{111}-3c_{114}-3c_{124} \\ +6c_{134}-2c_{222}-2c_{444} \end{array}\right)$
C'_{166}	$\frac{1}{4}\left(\begin{array}{c} -2c_{111}-c_{112} \\ +3c_{222} \end{array}\right)$	$\frac{1}{4}(-c_{112}+c_{222})$	$\frac{1}{27}\left(\begin{array}{c} 2c_{111}-6c_{113}-6c_{134}-6c_{155} \\ +2c_{222}+2c_{333}+9c_{344}-2c_{444} \end{array}\right)$
C'_{555}	0	$\frac{-c_{444}}{\sqrt{2}}$	$-\frac{1}{3\sqrt{6}}(2c_{111}+6c_{124}-2c_{222}+c_{444})$
C'_{556}	0	0	$\frac{1}{9\sqrt{2}}\left(\begin{array}{c} -c_{112}-2c_{113}-2c_{114}+2c_{123} \\ +4c_{124}+4c_{134}+4c_{144}-2c_{155} \\ +c_{222}-2c_{344}-c_{444} \end{array}\right)$
C'_{566}	0	$-\frac{c_{124}}{\sqrt{2}}$	$\frac{1}{9\sqrt{6}}(2c_{111}+6c_{124}-2c_{222}+c_{444})$
C'_{666}			$\frac{1}{27\sqrt{2}}\left(\begin{array}{c} 2c_{111}-12c_{113}-6c_{114}-12c_{124} \\ +12c_{133}+12c_{134}+6c_{155} \\ +2c_{222}-4c_{333}-6c_{344}+c_{444} \end{array}\right)$
3 point groups	[100]	[110]	[111]
C'_{11}	c_{11}	c_{11}	$\frac{1}{9}(4c_{11}+4c_{13}+8c_{14}+8c_{25}+c_{33}+8c_{44})$
C'_{15}	$-c_{25}$	$\frac{(c_{14}+c_{25})}{\sqrt{2}}$	$-\sqrt{\frac{2}{3}}(c_{14}-c_{25})$
C'_{16}	0	0	$\frac{1}{9}\sqrt{2}(2c_{11}-c_{13}+c_{14}+c_{25}-c_{33}-2c_{44})$
C'_{55}	c_{44}	c_{44}	$\frac{1}{3}(c_{11}-c_{12}-2c_{14}-2c_{25}+c_{44})$
C'_{56}	c_{14}	$\frac{(-c_{14}+c_{25})}{\sqrt{2}}$	0
C'_{66}	$\frac{1}{2}(c_{11}-c_{12})$	$\frac{c_{11}-c_{12}}{2}$	$\frac{1}{9}(2c_{11}-4c_{13}-2c_{14}-2c_{25}+2c_{33}+c_{44})$
3 point groups	[100]	[110]	[111]

(continued)

(continued)

32 point groups and 3 m point groups	[100]	[110]	[111]
C'_{111}	c_{111}	$\frac{1}{2}(c_{111}-2c_{116}+c_{222})$	$\frac{1}{27}\begin{pmatrix} 4c_{111}+12c_{113}+24c_{114}-24c_{115} \\ -8c_{116}+48c_{124}+6c_{133}+24c_{134} \\ -24c_{135}+48c_{155}+4c_{222}+c_{333} \\ +24c_{344}-16c_{444}+16c_{445} \end{pmatrix}$
C'_{115}	c_{115}	$\frac{1}{\sqrt{2}}(c_{114}-c_{115}+2c_{124})$	$\frac{1}{9}\sqrt{\frac{2}{3}}\begin{pmatrix} 2c_{111}-6c_{114}-6c_{115} \\ -12c_{124}-6c_{134}-6c_{135} \\ -2c_{222}+4c_{444}+4c_{445} \end{pmatrix}$
C'_{116}	c_{116}	$\frac{1}{2}(c_{111}-c_{222})$	$\frac{\sqrt{2}}{27}\begin{pmatrix} 2c_{111}+6c_{114}-6c_{115}-4c_{116} \\ +12c_{124}-3c_{133}-6c_{134} \\ +6c_{135}+2c_{222}-c_{333} \\ -12c_{344}+4c_{444}-4c_{445} \end{pmatrix}$
C'_{155}	c_{155}	c_{155}	$\frac{1}{9}\begin{pmatrix} -c_{112}+c_{113}-2c_{114}+2c_{115} \\ +4c_{116}-c_{123}-2c_{124}+2c_{125} \\ -2c_{134}+2c_{135}-2c_{144}+4c_{155} \\ +c_{222}+c_{344}+2c_{444}-2c_{445} \end{pmatrix}$
C'_{156}	$\frac{1}{2}(c_{114}+3c_{124})$	$-\frac{1}{2\sqrt{2}}\begin{pmatrix} c_{114}+c_{115} \\ +3c_{124}-c_{125} \end{pmatrix}$	$\frac{1}{9\sqrt{3}}\begin{pmatrix} 2c_{111}-3c_{114}-3c_{115}-3c_{124} \\ -3c_{125}+6c_{134}+6c_{135}+6c_{145} \\ -2c_{222}-2c_{444}-2c_{445} \end{pmatrix}$
C'_{166}	$\frac{1}{4}\begin{pmatrix} -2c_{111} \\ -c_{112}+3c_{222} \end{pmatrix}$	$\frac{1}{4}(-c_{112}+4c_{116}+c_{222})$	$\frac{1}{27}\begin{pmatrix} 2c_{111}-6c_{113}-4c_{116}-6c_{134} \\ +6c_{135}-6c_{155}+2c_{222}+2c_{333} \\ +9c_{344}-2c_{444}+2c_{445} \end{pmatrix}$
C'_{555}	$-c_{445}$	$\frac{(-c_{444}+c_{445})}{\sqrt{2}}$	$-\frac{1}{3\sqrt{6}}\begin{pmatrix} 2c_{111}+6c_{124}-6c_{125} \\ +18c_{14}-6c_{145}-2c_{222} \\ -18c_{25}+c_{444}+c_{445} \end{pmatrix}$
C'_{556}	$-c_{145}$	$-c_{145}$	$\frac{1}{9\sqrt{2}}\begin{pmatrix} -c_{112}-2c_{113}-2c_{114}+2c_{115} \\ +4c_{116}+2c_{123}+4c_{124}+8c_{125} \\ +4c_{134}-4c_{135}+4c_{144}-2c_{155} \\ +c_{222}-2c_{344}-c_{444}+c_{445} \end{pmatrix}$
C'_{566}	c_{125}	$-\frac{(c_{124}+c_{125})}{\sqrt{2}}$	$\frac{1}{9\sqrt{6}}\begin{pmatrix} 2c_{111}+6c_{124}-6c_{125} \\ -6c_{145}-2c_{222}+c_{444}+c_{445} \end{pmatrix}$

(continued)

(continued)

32 point groups and 3 m point groups	[100]	[110]	[111]
C'_{666}	$-c_{116}$	$-\frac{1}{2}(c_{111}-c_{222})$	$\frac{\sqrt{2}}{54}\begin{pmatrix} 2c_{111}-12c_{113}-6c_{114}+6c_{115} \\ -4c_{116}-12c_{124}+12c_{133}+12c_{134} \\ -12c_{135}+6c_{155}+2c_{222}-c_{445} \\ -4c_{333}-6c_{344}+c_{444} \end{pmatrix}$

Tetrahedral Crystal

4 point groups	[100]	[110]	[111]
C'_{11}	c_{11}	$\frac{1}{2}(c_{11}+c_{12}+2c_{66})$	$\frac{1}{9}(2c_{11}+2c_{12}+4c_{13}+c_{33}+8c_{44}+4c_{66})$
C'_{15}	0	0	$-\frac{1}{3}\sqrt{\frac{2}{3}}c_{16}$
C'_{16}	$\frac{1}{2}c_{16}$	$-\frac{1}{2}c_{16}$	$\frac{1}{9\sqrt{2}}\begin{pmatrix} 2c_{11}+2c_{12}-2c_{13} \\ -2c_{33}-4c_{44}+4c_{66} \end{pmatrix}$
C'_{55}	c_{44}	c_{44}	$\frac{1}{3}(c_{11}-c_{12}+c_{44})$
C'_{56}	0	0	$-\frac{c_{16}}{3\sqrt{3}}$
C'_{66}	c_{66}	$\frac{1}{2}(c_{11}-c_{12})$	$\frac{1}{9}(c_{11}+c_{12}-4c_{13}+2c_{33}+c_{44}+2c_{66})$
4 point groups	[100]	[110]	[111]
C'_{111}	c_{111}	$\frac{1}{4}(c_{111}+3c_{112}+12c_{166})$	$\frac{1}{27}\begin{pmatrix} 2c_{111}+6c_{112}+6c_{113}+6c_{123} \\ +6c_{133}+24c_{144}+24c_{155} \\ +24c_{166}+c_{333}+24c_{344}+12c_{366} \\ +24c_{446}+48c_{456}-24c_{466} \end{pmatrix}$
C'_{115}	0		$-\frac{1}{9}\sqrt{\frac{2}{3}}\begin{pmatrix} 4c_{116}+4c_{136}+8c_{145} \\ -4c_{446}-4c_{466} \end{pmatrix}$
C'_{116}	c_{116}	$-c_{116}$	$\frac{\sqrt{2}}{27}\begin{pmatrix} c_{111}+3c_{112}-3c_{133} \\ +12c_{166}-c_{333}-12c_{344} \end{pmatrix}$
C'_{155}	c_{155}	$\frac{1}{2}\begin{pmatrix} c_{144}+c_{155}+c_{446} \\ +2c_{456}-c_{466} \end{pmatrix}$	$\frac{1}{9}\begin{pmatrix} c_{111}-c_{112}+c_{113}-c_{123}-3c_{144} \\ +5c_{155}+c_{344}+c_{446}-2c_{456}-c_{466} \end{pmatrix}$
C'_{156}	0	0	$-\frac{1}{9\sqrt{3}}\begin{pmatrix} 4c_{116}-2c_{136}-4c_{145} \\ -c_{446}-c_{466} \end{pmatrix}$

(continued)

(continued)

4 point groups	[100]	[110]	[111]
C'_{166}	c_{166}	$\frac{1}{4}(c_{111}-c_{112})$	$\frac{1}{27}\begin{pmatrix} c_{111}+3c_{112}-3c_{113}-3c_{123} \\ -3c_{144}-3c_{155}+12c_{166} \\ +2c_{333}+9c_{344}-6c_{366} \\ -3c_{446}-6c_{456}+3c_{466} \end{pmatrix}$
C'_{555}	0	0	$\frac{2}{\sqrt{6}}c_{145}$
C'_{556}	$-c_{466}$	$-c_{145}$	$\frac{\sqrt{2}}{18}\begin{pmatrix} c_{111}-c_{112}-2c_{113}+2c_{123}+3c_{144} \\ -c_{155}-2c_{344}+c_{446}-2c_{456}-c_{466} \end{pmatrix}$
C'_{566}	0	0	$-\frac{1}{9}\sqrt{\frac{2}{3}}\begin{pmatrix} 2c_{116}-4c_{136}+c_{145} \\ +c_{446}+c_{466} \end{pmatrix}$
C'_{666}	0	0	$\frac{1}{27\sqrt{2}}\begin{pmatrix} c_{111}+3c_{112}-6c_{113}-6c_{123} \\ +12c_{133}+3c_{144}+3c_{155} \\ +12c_{166}-4c_{333}-6c_{344} \\ -12c_{366}+3c_{446}+6c_{456}-3c_{466} \end{pmatrix}$
422 groups	[100]	[110]	[111]
C'_{11}	c_{11}	$\frac{1}{2}(c_{11}+c_{12}+2c_{66})$	$\frac{1}{9}(2c_{11}+2c_{12}+4c_{13}+c_{33}+8c_{44}+4c_{66})$
C'_{15}	0	0	0
C'_{16}	0	0	$\frac{\sqrt{2}}{9}(c_{11}+c_{12}-c_{13}-c_{33}-2c_{44}+2c_{66})$
C'_{55}	c_{44}	c_{44}	$\frac{1}{3}(c_{11}-c_{12}+c_{44})$
C'_{56}	0	0	0
C'_{66}	c_{66}	$\frac{1}{2}(c_{11}-c_{12})$	$\frac{1}{9}(c_{11}+c_{12}-4c_{13}+2c_{33}+c_{44}+2c_{66})$
422 groups	[100]	[110]	[111]
C'_{111}	c_{111}	$\frac{1}{4}(c_{111}+3c_{112}+12c_{166})$	$\frac{1}{27}\begin{pmatrix} 2c_{111}+6c_{112}+6c_{113}+6c_{123} \\ +6c_{133}+24c_{144}+24c_{155}+12c_{166} \\ +c_{333}+24c_{344}+12c_{366}+48c_{456} \end{pmatrix}$
C'_{115}	0	0	$\frac{2}{9}\sqrt{\frac{2}{3}}c_{166}$
C'_{116}	0	0	$\frac{\sqrt{2}}{27}\begin{pmatrix} c_{111}+3c_{112}-3c_{133} \\ +6c_{166}-c_{333}-12c_{344} \end{pmatrix}$
C'_{155}	c_{155}	$\frac{1}{2}(c_{144}+c_{155}+2c_{456})$	$\frac{1}{9}\begin{pmatrix} c_{111}-c_{112}+c_{113}-c_{123} \\ -3c_{144}+5c_{155}+c_{344}-2c_{456} \end{pmatrix}$

(continued)

(continued)

4 point groups	[100]	[110]	[111]
C'_{156}	0	0	$\frac{2c_{166}}{9\sqrt{3}}$
C'_{166}	c_{166}	$\frac{1}{4}(c_{111}-c_{112})$	$\frac{1}{27}\begin{pmatrix} c_{111}+3c_{112}-3c_{113}-3c_{123} \\ -3c_{144}-3c_{155}+6c_{166}+2c_{333} \\ +9c_{344}-6c_{366}-6c_{456} \end{pmatrix}$
C'_{555}	0	0	0
C'_{556}	0	0	$\frac{1}{9\sqrt{2}}\begin{pmatrix} c_{111}-c_{112}-2c_{113}+2c_{123} \\ +3c_{144}-c_{155}-2c_{344}-2c_{456} \end{pmatrix}$
C'_{566}	0	0	$\frac{1}{9}\sqrt{\frac{2}{3}}c_{166}$
C'_{666}	0	0	$\frac{1}{27\sqrt{2}}\begin{pmatrix} c_{111}+3c_{112}-6c_{113}-6c_{123} \\ +12c_{133}+3c_{144}+3c_{155}+6c_{166} \\ -4c_{333}-6c_{344}-12c_{366}+6c_{456} \end{pmatrix}$

Hexagonal Crystals

	[100]	[110]	[111]
C'_{11}	c_{11}	c_{11}	$\frac{1}{9}(4c_{11}+4c_{13}+c_{33}+8c_{44})$
C'_{15}	0	0	0
C'_{16}	0	0	$\frac{1}{9}\sqrt{2}(2c_{11}-c_{13}-c_{33}-2c_{44})$
C'_{55}	c_{44}	c_{44}	$\frac{1}{3}(c_{11}-c_{12}+c_{44})$
C'_{56}	0	0	0
C'_{66}	$\frac{1}{2}(c_{11}-c_{12})$	$\frac{1}{2}(c_{11}-c_{12})$	$\frac{1}{9}(2c_{11}-4c_{13}+2c_{33}+c_{44})$
6 point groups	[100]	[110]	[111]
C'_{111}	c_{111}	$\frac{1}{2}(c_{111}-2c_{116}+c_{222})$	$\frac{1}{27}\begin{pmatrix} 4c_{111}+12c_{113}-8c_{116}+6c_{133} \\ +48c_{155}+4c_{222}+c_{333}+24c_{344} \end{pmatrix}$
C'_{115}	0	0	$\frac{2}{9}\sqrt{\frac{2}{3}}(c_{111}-c_{222})$
C'_{116}	c_{116}	$\frac{1}{2}(c_{111}-c_{222})$	$\frac{\sqrt{2}}{27}\begin{pmatrix} 2c_{111}-4c_{116}-3c_{133} \\ +2c_{222}-c_{333}-12c_{344} \end{pmatrix}$
C'_{155}	c_{155}	c_{155}	$\frac{1}{9}\begin{pmatrix} -c_{112}+c_{113}+4c_{116}-c_{123} \\ -2c_{144}+4c_{155}+c_{222}+c_{344} \end{pmatrix}$
C'_{156}	0	0	$\frac{2}{9\sqrt{3}}(c_{111}+3c_{145}-c_{222})$
C'_{166}	$\frac{1}{4}\begin{pmatrix} -2c_{111}-c_{112} \\ +3c_{222} \end{pmatrix}$	$\frac{1}{4}(-c_{112}+4c_{116}+c_{222})$	$\frac{1}{27}\begin{pmatrix} 2c_{111}-6c_{113}-4c_{116}-6c_{155} \\ +2c_{222}+2c_{333}+9c_{344} \end{pmatrix}$

(continued)

(continued)

	[100]	[110]	[111]
C'_{555}	0	0	$-\frac{1}{3}\sqrt{\frac{2}{3}}(c_{111}-3c_{145}-c_{222})$
C'_{556}	$-c_{145}$	$-c_{145}$	$\frac{1}{9\sqrt{2}}\left(\begin{array}{c}-c_{112}-2c_{113}+4c_{116}+2c_{123}\\+4c_{144}-2c_{155}+c_{222}-2c_{344}\end{array}\right)$
C'_{566}	0	0	$\frac{1}{9}\sqrt{\frac{2}{3}}(c_{111}-3c_{145}-c_{222})$
C'_{666}	$-c_{116}$	$-\frac{1}{2}(c_{111}-c_{222})$	$\frac{\sqrt{2}}{27}\left(\begin{array}{c}c_{111}-6c_{113}-2c_{116}+6c_{133}\\+3c_{155}+c_{222}-2c_{333}-3c_{344}\end{array}\right)$
622 point groups	[100]	[110]	[111]
C'_{111}	c_{111}	$\frac{1}{2}(c_{111}+c_{222})$	$\frac{1}{27}\left(\begin{array}{c}4c_{111}+12c_{113}+6c_{133}+48c_{155}\\+4c_{222}+c_{333}+24c_{344}\end{array}\right)$
C'_{115}	0	0	$\frac{2}{9}\sqrt{\frac{2}{3}}(c_{111}-c_{222})$
C'_{116}	0	$\frac{1}{2}(c_{111}-c_{222})$	$\frac{1}{27}\sqrt{2}\left(\begin{array}{c}2c_{111}-3c_{133}+2c_{222}\\-c_{333}-12c_{344}\end{array}\right)$
C'_{155}	c_{155}	c_{155}	$\frac{1}{9}\left(\begin{array}{c}-c_{112}+c_{113}-c_{123}-2c_{144}\\+4c_{155}+c_{222}+c_{344}\end{array}\right)$
C'_{156}	0	0	$\frac{2}{9\sqrt{3}}(c_{111}-c_{222})$
C'_{166}	$\frac{1}{4}\left(\begin{array}{c}-2c_{111}-c_{112}\\+3c_{222}\end{array}\right)$	$\frac{1}{4}(-c_{112}+c_{222})$	$\frac{1}{27}\left(\begin{array}{c}2c_{111}-6c_{113}-6c_{155}\\+2c_{222}+2c_{333}+9c_{344}\end{array}\right)$
C'_{555}	0	0	$-\frac{1}{3}\sqrt{\frac{2}{3}}(c_{111}-c_{222})$
C'_{556}	0	0	$\frac{1}{9\sqrt{2}}\left(\begin{array}{c}-c_{112}-2c_{113}+2c_{123}+4c_{144}\\-2c_{155}+c_{222}-2c_{344}\end{array}\right)$
C'_{566}	0	0	$\frac{1}{9}\sqrt{\frac{2}{3}}(c_{111}-c_{222})$
C'_{666}	0	$-\frac{1}{2}(c_{111}-c_{222})$	$\frac{\sqrt{2}}{27}\left(\begin{array}{c}c_{111}-6c_{113}+6c_{133}+3c_{155}\\+c_{222}-2c_{333}-3c_{344}\end{array}\right)$

2.2 Interaction of Collinear Acoustic Waves Propagating Along Pure Mode Directions in Crystals

Previous research on the interaction of collinear acoustic waves has been dedicated to acoustic waves propagating along pure mode directions in cubic crystals. In this section, we will calculate that can be extended to all crystal point groups. Here, the nonlinear parameter of nonlinear interaction is defined, and the effective third-order elastic constants of the crystal can be calculated using the method proposed in this section. The results obtained are of great significance for studying the elastic nonlinearity of any symmetric crystal.

It is well known that the interaction of collinear acoustic waves will produce harmonics and sum and difference frequency waves. Nonlinear effects can be used to determine the third-order high-order elastic constants (TOECs) of the material. These high-order constants characterize the non-harmonicity of the crystal lattice and are important for understanding many properties of the crystal, such as thermal transport, temperature and pressure dependence of the second-order elastic constants, high-frequency phonon decay, etc.

There has been a lot of theoretical and experimental research on the interaction and harmonic generation of elastic waves [1–4, 7, 9,10]. The third-order elastic constants of some cubic crystals and their temperature dependence have been measured using ultrasonic second harmonic technology [11, 12]. The anisotropic nonlinear elastic properties of icosahedral quasi-crystals have also been studied [13, 14], and the interaction of shear waves in cubic and other crystals has been observed and calculated [6, 15–19], although such interaction does not exist in homogeneous isotropic crystals. Nevertheless, most research still focuses on the study of longitudinal waves in isotropic solids or cubic crystals such as metals. It is well known that many useful materials developed in recent decades belong to systems other than cubic crystals. For example, domain-engineered PMN-PT single crystals, which have mm^2, 4 mm or 3 m macroscopic symmetry when polarized in their unnatural polarization direction [20], studying the elastic nonlinearity of these new crystals is of great reference value for their high-power applications.

Here, we will extend the theory of collinear acoustic wave interaction in cubic crystal systems to cover all crystal point groups, thereby promoting further research on crystals. The basic equation of second-order nonlinearity is first given. Among the 32 point groups, 20 are piezoelectric bodies, but in the current calculations, only elastic nonlinearity is considered. Here, we study the nonlinear interaction of collinear acoustic waves propagating along the pure mode direction of the crystal, in order to simplify calculations and experiments. We define the nonlinear parameters that characterize the interaction, and some special TOECs or their combinations appear in the nonlinear parameters, so that we can use these nonlinear parameters to obtain effective TOECs. The calculation results of tetragonal crystals (point groups 422, 4 mm, $\bar{4}2$ m, 4/mmm) are given as examples.

2.2.1 Basic Equation

The basic equation of collinear acoustic wave interaction [7, 12], its motion stress equation can be written as [21]

$$\delta_{iM}\rho_0\ddot{u}_M = \frac{\partial P_{Ji}}{\partial a_J}, \delta_{iM}\rho_0\ddot{u}_M = \frac{\partial P_{Ji}}{\partial a_J}, \tag{2.28}$$

Here ρ_0 is the static mass density of the crystal, δ_{iM} is the Kronecker function, which characterizes the transformation from the deformed state to the non-deformed state, P_{Ji} is the Piola–Kirchhoff stress tensor, It is a two-point tensor, representing the force situation of a unit internal plane under non-deformation, which can be expressed as

$$P_{Ji} = \frac{\partial x_i}{\partial a_I}\frac{\partial \varphi}{\partial \eta_{IJ}}. \tag{2.29}$$

In Eq. (2.29), a_I and x_I are the positions of the same material particles before and after deformation, φ It is the strain energy density, which is invariant under symmetric operations and can be expressed as a function of the Lagrangian strain tensor,

$$\varphi = \frac{1}{2}c_{IJKL}\eta_{IJ}\eta_{KL} + \frac{1}{6}c_{IJKLMN}\eta_{IJ}\eta_{KL}\eta_{MN} + \cdots, \tag{2.30}$$

The uppercase subscript here represents the physical quantity in the undeformed state.

The stress of Lagrange can be expressed as:

$$\eta_{KL} = \frac{1}{2}\left[u_{K,L} + u_{L,K} + u_{P,L}u_{P,K}\right]. \tag{2.31}$$

Approximated to second-order nonlinearity, the Piola–Kirchhoff strain tensor can be expressed as:

$$P_{Ji} = \delta_{iI}\left[c_{IJKL}u_{K,L} + \left(\frac{1}{2}c_{IJNL}\delta_{MK} + c_{NJKL}\delta_{IM} + \frac{1}{2}c_{IJKLMN}\right)u_{K,L}u_{M,N}\right]. \tag{2.32}$$

Taking this result to the stress equation, approximated to second-order nonlinearity, the equation of motion displacement can be written as:

$$\tilde{c}_{IJKLMN} \tag{2.33a}$$

$$\begin{aligned} \rho_0\ddot{u}_I &= c_{LJKL}u_{K,LJ} + \tilde{c}_{IJLLMN}\left(u_{K,LJ}u_{M,N} + u_{K,L}u_{M,NJ}\right) \\ \tilde{c}_{IJKLMN} &= \frac{1}{2}c_{IJNL}\delta_{MK} + c_{NJKL}\delta_{IM} + \frac{1}{2}c_{IJKLMN} \end{aligned} \tag{2.33b}$$

$$\tilde{c}_{IJKLMN} = \frac{1}{2}c_{IJNL}\delta_{MK} + c_{NJKL}\delta_{IM} + \frac{1}{2}c_{IJKLMN}$$

Here, the Kronecker function is omitted. (Note: $\tilde{c}_{IJKLMN}$ and c_{IJKLMN} are not the same symmetry)

For the calculation of plane wave propagation, it is usually convenient to choose a coordinate system, that is, the direction of wave propagation is parallel to the axis of the coordinate system. When the a-axis is the direction of wave propagation, the displacement of the particles only depends on the coordinate a in space. In this case, Eqs. (2.33a, 2.33b) can be rewritten as [7, 19]

$$\begin{aligned} \rho_0 \ddot{u}_J - \gamma_{J111} u_{1,11} - \gamma_{J121} u_{2,11} - \gamma_{J131} u_{3,11} &= f_J(u_1, u_2, u_3), \ (J = 1-3) \\ \rho_0 \ddot{u}_J - \gamma_{J111} u_{1,11} - \gamma_{J121} u_{2,11} - \gamma_{J131} u_{3,11} &= f_J(u_1, u_2, u_3), \end{aligned} \tag{2.34}$$

Here the a-axis is represented as the 1-axis, and the comma in the subscript here is the partial derivative of the coordinate system.

$$\begin{aligned} f_J &= g_{J1} u_{1,11} u_{1,1} + g_{J2} u_{2,11} u_{2,1} + g_{J3} u_{3,11} u_{3,1} \\ f_J &= g_{J1} u_{1,11} u_{1,1} + g_{J2} u_{2,11} u_{2,1} + g_{J3} u_{3,11} u_{3,1} + g_{J6}\left(u_{1,11} u_{2,1} + u_{1,1} u_{2,11}\right) \\ &\quad + g_{J5}\left(u_{1,11} u_{3,1} + u_{1,1} u_{3,11}\right) + g_{J4}\left(u_{2,11} u_{3,1} + u_{2,1} u_{3,11}\right) \end{aligned} \tag{2.35}$$

And

$$\begin{aligned} & g_{11} = (3\gamma_{11} + \gamma_{111}),\ g_{22} = (3\gamma_{16} + \gamma_{666}),\ g_{33} = (3\gamma_{15} + \gamma_{555}), \\ & g_{12} = g_{26} = (\gamma_{11} + \gamma_{166}),\ g_{13} = g_{35} = (\gamma_{11} + \gamma_{155}), \\ & g_{14} = g_{25} = g_{36} = \gamma_{156},\ g_{15} = g_{31} = (\gamma_{15} + \gamma_{115}),\ g_{16} = g_{21} = (\gamma_{16} + \gamma_{116}), \\ & g_{23} = g_{34} = (\gamma_{16} + \gamma_{556}),\ \phi = \varphi. \end{aligned}$$

Equation (2.34) replaces c_{IJ} and c_{IJK} using γ_{IJ} and γ_{IJK} to emphasize the constants in the calculation coordinates, using the usual omission notation, the right side of Eq. (2.34) can be considered as a perturbation, the nonlinear equation can be solved by the method of step-by-step approximation: that is, let

$$u_I = \delta u_I^{(I)} + \delta^2 u_I^{(II)} + \cdots \tag{2.36}$$

Here in $\delta \leq 1$, its exponent indicates the order of step-by-step approximation in Eq. (2.36), and the approximate equations of the first and second orders are

$$\rho_0 \begin{bmatrix} \ddot{u}_1^{(I)} \\ \ddot{u}_2^{(I)} \\ \ddot{u}_3^{(I)} \end{bmatrix} - \begin{bmatrix} \gamma_{11} & \gamma_{16} & \gamma_{15} \\ \gamma_{16} & \gamma_{66} & \gamma_{56} \\ \gamma_{15} & \gamma_{56} & \gamma_{55} \end{bmatrix} \begin{bmatrix} u_{1,11}^{(I)} \\ u_{2,11}^{(I)} \\ u_{3,11}^{(I)} \end{bmatrix} = 0, \tag{2.37a}$$

$$\rho_0 \begin{bmatrix} \ddot{u}_1^{(II)} \\ \ddot{u}_2^{(II)} \\ \ddot{u}_3^{(II)} \end{bmatrix} - \begin{bmatrix} \gamma_{11} & \gamma_{16} & \gamma_{15} \\ \gamma_{16} & \gamma_{66} & \gamma_{56} \\ \gamma_{15} & \gamma_{56} & \gamma_{55} \end{bmatrix} \begin{bmatrix} u_{1,11}^{(II)} \\ u_{2,11}^{(II)} \\ u_{3,11}^{(II)} \end{bmatrix}$$

$$= \begin{bmatrix} g_{11} & g_{12} & g_{13} & g_{14} & g_{15} & g_{16} \\ g_{21} & g_{22} & g_{23} & g_{24} & g_{25} & g_{26} \\ g_{31} & g_{32} & g_{33} & g_{34} & g_{35} & g_{36} \end{bmatrix} \begin{bmatrix} h_1 \\ h_2 \\ h_3 \\ h_4 \\ h_5 \\ h_6 \end{bmatrix}, \tag{2.37b}$$

Here g_{IJ} can be given by Eq. (2.35)

$$\begin{aligned} h_J &= u^{(I)}_{J,1} u^{(I)}_{J,11} (J = 1, 2, 3) \\ h_\alpha &= u^{(I)}_{I,1} u^{(I)}_{J,11} + u^{(I)}_{I,11} u^{(I)}_{J,1} \\ &[\alpha = 4(I, J = 2, 3), 5(I, J = 1, 3), 6(I, J = 1, 2)]. \end{aligned} \tag{2.37c}$$

It can be seen that the first-order approximation is a non-perturbative linear equation. For a plane, the equation is equivalent to the Christoffel equation [22], there are three modes: one quasi-longitudinal wave, two quasi-transverse waves, all along the a-axis direction of the crystal. The usual particle displacement u_J includes the contributions of these three waves, that is to say, the three displacement components may be coupled. Obviously, when $\gamma_{15} = \gamma_{16} = 0$, the displacement u_1 will not be coupled with u_2 and u_3. At this time, the direction of wave propagation is the first type of pure mode direction. If the axes 2 and 3 of the coordinate system are chosen to be consistent with the polarization directions of the two pure shear waves, i.e., $\gamma_{56} = 0$ [22]. Then, for crystals in the first type of pure mode direction, which have a certain symmetry, there is $\gamma_{15} = \gamma_{16} = \gamma_{56} = 0$, and for the second type of pure mode direction, the conditions to be met are: [28] $\gamma_{15} \neq 0, \gamma_{16} = \gamma_{56} = 0$ or $\gamma_{16} \neq 0$, $\gamma_{15} = \gamma_{56} = 0$.

In ultrasonic measurements, the first type of pure mode direction or simple pure mode direction is preferred because it simplifies the interpretation of experimental results. Since there is no coupling of particle displacements, when dealing with longitudinal waves $c_{112} = c_{122}$ $c_{113} = c_{223}$, when dealing with shear waves, c_{123} $c_{133} = c_{233}$, or $c_{144} = c_{255}$ $c_{155} = c_{244}$ or $c_{166} = c_{266}$, c_{333}, $u^I_3 \neq 0$. Moreover, it is well known that second harmonic or sum and difference frequency waves (hereafter referred to as mixing waves) are strongest when the interactions are synchronized, that is, when the propagation speed of the waves in the interaction is consistent with the free propagation of the second harmonic or mixing waves. Therefore, the shear waves generated by the nonlinear interaction of longitudinal waves are irrelevant to the actual experiment. Ignoring the non-synchronicity of the interaction, the step-by-step approximation equation for the pure mode direction can be simplified. For pure longitudinal waves

$$\rho_0 \ddot{u}^{(I)}_1 - \gamma_{11} u^{(I)}_{1,11} = 0 \tag{2.38a}$$

$$\rho_0 \ddot{u}^{(II)}_1 - \gamma_{11} u^{(II)}_{1,11} = (3\gamma_{11} + \gamma_{111}) u^{(I)}_{1,1} u^{(I)}_{1,11} \tag{2.39b}$$

Along the pure mode direction, the speeds of the two shear waves may be the same or different. For the latter, the step-by-step approximation equation can be written as

$$\rho_0 \ddot{u}_2^{(I)} - \gamma_{66} u_{2,11}^{(I)} = 0 \tag{2.40a}$$

$$\rho_0 \ddot{u}_2^{(II)} - \gamma_{66} u_{2,11}^{(II)} = \gamma_{666} u_{2,1}^{(I)} u_{2,11}^{(I)} \tag{2.40b}$$

or

$$\rho_0 \ddot{u}_3^{(I)} - \gamma_{55} u_{3,11}^{(I)} = 0 \tag{2.41a}$$

$$\rho_0 \ddot{u}_3^{(II)} - \gamma_{55} u_{3,11}^{(II)} = \gamma_{555} u_{3,1}^{(I)} u_{3,11}^{(I)} \tag{2.41b}$$

For the former situation, usually referred to as the degenerate case, the self-action and interaction between the two shear waves are synchronized. Therefore, the equation is written as

$$\rho_0 \ddot{u}_2^{(I)} - \gamma u_{2,11}^{(I)} = 0, \tag{2.42a}$$

$$\rho_0 \ddot{u}_3^{(I)} - \gamma u_{3,11}^{(I)} = 0, \tag{2.42b}$$

$$\begin{aligned}\rho_0 \ddot{u}_2^{(II)} - \gamma u_{2,11}^{(II)} &= \gamma_{666} u_{2,1}^{(I)} u_{2,11}^{(I)} + \gamma_{556} u_{3,1}^{(I)} u_{3,11}^{(I)} \\ &\quad + \gamma_{566}\left(u_{2,1}^{(I)} u_{3,11}^{(I)} + u_{2,11}^{(I)} u_{3,1}^{(I)}\right),\end{aligned} \tag{2.43a}$$

$$\begin{aligned}\rho_0 \ddot{u}_3^{(II)} - \gamma u_{3,11}^{(II)} &= \gamma_{566} u_{2,1}^{(I)} u_{2,11}^{(I)} + \gamma_{555} u_{3,1}^{(I)} u_{3,11}^{(I)} \\ &\quad + \gamma_{556}\left(u_{2,1}^{(I)} u_{3,11}^{(I)} + u_{2,11}^{(I)} u_{3,1}^{(I)}\right),\end{aligned} \tag{2.43b}$$

Here: $\gamma = \gamma_{55} = \gamma_{66}$.

2.2.2 Generation of Second Harmonic and Mixing Waves

Due to the absence of pure modal directional mode coupling, longitudinal and transverse waves can be solved separately to obtain second-order approximate equations. For example, let the solution of Eq. (2.43a) be

$$u_1^{(I)} = A_1 \sin(\omega_1 t - k_1 a) + A_2 \sin(\omega_2 t - k_2 a). \tag{2.44}$$

Here $k_j^2 = \rho_0 \omega_j^2 / \gamma_{11}$ (j = 1, 2), (2.43b) becomes

$$\rho_0 \ddot{u}_1^{(II)} - \gamma_{11} u_{1,11}^{(II)} = \frac{1}{2}(3\gamma_{11} + \gamma_{111})\{k_1^3 A_1^2 \sin[2(\omega_1 t - k_1 a)] + k_2^3 A_2^2 \sin[2(\omega_2 t - k_2 a)] + k_1 k_2 (k_1 + k_2) A_1 A_2 \sin[(\omega_1 + \omega_2)t - (k_1 + k_2)a] + k_1 k_2 (k_1 - k_2) A_1 A_2 \sin[(\omega_1 - \omega_2)t - (k_1 - k_2)a]\} \tag{2.45}$$

Equation (2.45) is a linear non-homogeneous differential equation. The terms on the right side of the equation represent the self-action and interaction of the fundamental wave, which can be considered as the driving force for the generation of second harmonic and mixing waves. Since formula (2.45) becomes linear, the special solutions of the second harmonic and mixing waves can be represented separately.

For the second harmonic, it is

$$u_1^{(II)} = U_{2\omega_1} \cos[2(\omega_1 t - k_1 a)] \text{ or } u_1^{(II)} = U_{2\omega_2} \cos[2(\omega_2 t - k_2 a)] \tag{2.46a}$$

$$U_{2\omega_1} = \frac{1}{8}\beta_L k_1^2 A_1^2 a \text{ or } U_{2\omega_2} = \frac{1}{8}\beta_L k_2^2 A_2^2 a \tag{2.46b}$$

For sum and difference frequency waves, it is

$$u_1^{(II)} = U_{\omega_1 \pm \omega_2} \cos[(\omega_1 \pm \omega_2)t - (k_1 \pm k_2)a] \tag{2.46c}$$

$$U_{\omega_1 \pm \omega_2} = \frac{1}{4}\beta_L k_1 k_2 A_1 A_2 a \tag{2.46d}$$

Here

$$\beta_L = -\frac{3\gamma_{11} + \gamma_{111}}{\gamma_{11}} \tag{2.47}$$

is called the nonlinear parameter [23]. It can be seen that the amplitude of the second harmonic or mixing wave driven by synchronous interaction will increase with the increase of propagation distance, so if the length of the sample is long enough, the energy will accumulate to a measurable level. Both the second-order elastic constants (SOECs) and TOECs contribute to the definition of the nonlinear parameter. The former is the contribution of induced nonlinearity caused by finite strain, and the latter is the contribution of the material's inherent nonlinearity.

Transverse waves can also obtain similar results. The non-degenerate shear wave has exactly the same solution as the longitudinal wave, but the corresponding nonlinear parameter is:

$$\beta_{S1} = -\frac{\gamma_{555}}{\gamma_{55}} \tag{2.48a}$$

or

$$\beta_{S2} = -\frac{\gamma_{666}}{\gamma_{66}} \tag{2.48b}$$

For degenerate shear waves, the setup in an experiment may be such that transducers emitting at angular frequencies ω_1 and ω_2 are used to simultaneously generate the fundamental shear waves. In this case, the solution of the first-order approximate Eqs. (2.42a, 2.42b) can be written as

$$u_2^{(I)} = A_1 \sin(\omega_1 t - k_1 a) + A_2 \sin(\omega_2 t - k_2 a) \tag{2.49a}$$

$$u_3^{(I)} = B_1 \sin(\omega_1 t - k_1 a) + B_2 \sin(\omega_2 t - k_2 a) \tag{2.49b}$$

Here $k_{1,2} = \omega_{1,2}\sqrt{\rho_0/\gamma}$ is the wave number of the shear fundamental wave, replacing (2.49a, 2.49b) into (2.43a, 2.43b) we can get

$$\begin{aligned}
\rho_0 \ddot{u}_2^{(II)} - \gamma u_{2,11}^{(II)} &= \frac{1}{2}k_1^3\left(\gamma_{666}A_1^2 + \gamma_{556}B_1^2 + 2\gamma_{566}A_1B_1\right)\\
&\sin[2(\omega_1 t - k_1 a)] + \frac{1}{2}k_2^3\left(\gamma_{666}A_2^2 + \gamma_{556}B_2^2 + 2\gamma_{566}A_2B_2\right)\\
&\sin[2(\omega_2 t - k_2 a)] + \frac{1}{2}k_1k_2(k_1 + k_2)\\
&\left[\gamma_{666}A_1A_2 + \gamma_{556}B_1B_2 + \gamma_{566}(A_1B_2 + A_2B_1)\right]\\
&\sin[(\omega_1 + \omega_2)t - (k_1 + k_2)a] + \frac{1}{2}k_1k_2(k_1 - k_2)\\
&\left[\gamma_{666}A_1A_2 + \gamma_{556}B_1B_2 + \gamma_{566}(A_1B_2 + A_2B_1)\right]\\
&\sin[(\omega_1 - \omega_2)t - (k_1 - k_2)a]
\end{aligned} \tag{2.50a}$$

$$\begin{aligned}
\rho_0 \ddot{u}_3^{(II)} - \gamma u_{3,11}^{(II)} &= \frac{1}{2}k_1^3\left(\gamma_{566}A_1^2 + \gamma_{555}B_1^2 + 2\gamma_{556}A_1B_1\right)\\
&\sin[2(\omega_1 t - k_1 a)] + \frac{1}{2}k_2^3\left(\gamma_{566}A_2^2 + \gamma_{556}B_2^2 + 2\gamma_{566}A_2B_2\right)\\
&\sin[2(\omega_2 t - k_2 a)] + \frac{1}{2}k_1k_2(k_1 + k_2)\\
&\left[\gamma_{566}A_1A_2 + \gamma_{555}B_1B_2 + \gamma_{556}(A_1B_2 + A_2B_1)\right]\\
&\sin[(\omega_1 + \omega_2)t - (k_1 + k_2)a] + \frac{1}{2}k_1k_2(k_1 - k_2)\\
&\left[\gamma_{566}A_1A_2 + \gamma_{555}B_1B_2 + \gamma_{556}(A_1B_2 + A_2B_1)\right]\\
&\sin[(\omega_1 - \omega_2)t - (k_1 - k_2)a]
\end{aligned} \tag{2.50b}$$

Assuming that the two shear wave fundamentals have different amplitudes and polarizations, it can be assumed that

$$A_1 = A\cos\phi,\ B_1 = A\sin\phi,\ A_2 = B\cos\psi,\ B_2 = B\sin\psi \tag{2.51}$$

Here φ and ψ are the angles between the polarization direction of the transducer and the displacement u_2, A and B are the amplitudes of the shear waves with frequencies ω_1 and ω_2, then Eqs. (2.50a, 2.50b) can be written as:

$$\begin{aligned}
&\rho_0 \ddot{u}_2^{(II)} - \gamma u_{2,11}^{(II)} = \frac{1}{2}k_1^3\left(\gamma_{666}\cos^2\varphi + \gamma_{556}\sin^2\varphi + \gamma_{566}\sin 2\varphi\right)\\
&A^2\sin[2(\omega_1 t - k_1 a)] + \frac{1}{2}k_2^3\left(\gamma_{666}\cos^2\psi + \gamma_{556}\sin^2\psi + \gamma_{566}\sin 2\psi\right)\\
&B^2\sin[2(\omega_2 t - k_2 a)] + \frac{1}{2}k_1 k_2(k_1 + k_2)\\
&\left[\gamma_{666}\cos\varphi\cos\psi + \gamma_{556}\sin\varphi\sin\psi + \gamma_{566}\sin(\varphi + \psi)\right]\\
&AB\sin[(\omega_1 + \omega_2)t - (k_1 + k_2)a] + \frac{1}{2}k_1 k_2(k_1 - k_2)\\
&\left[\gamma_{666}\cos\varphi\cos\psi + \gamma_{556}\sin\varphi\sin\psi + \gamma_{566}\sin(\varphi + \psi)\right]\\
&AB\sin[(\omega_1 - \omega_2)t - (k_1 - k_2)a]
\end{aligned} \tag{2.52a}$$

$$\begin{aligned}
&\rho_0 \ddot{u}_3^{(II)} - \gamma u_{3,11}^{(II)} = \frac{1}{2}k_1^3\left(\gamma_{566}\cos^2\varphi + \gamma_{555}\sin^2\varphi + \gamma_{556}\sin 2\varphi\right)\\
&A^2\sin[2(\omega_1 t - k_1 a)] + \frac{1}{2}k_2^3\left(\gamma_{566}\cos^2\psi + \gamma_{555}\sin^2\psi + \gamma_{556}\sin 2\psi\right)\\
&B^2\sin[2(\omega_1 t - k_1 a)] + \frac{1}{2}k_1 k_2(k_1 + k_2)\\
&\left[\gamma_{566}\cos\varphi\cos\psi + \gamma_{555}\sin\varphi\sin\psi + \gamma_{556}\sin(\varphi + \psi)\right]\\
&AB\sin[(\omega_1 + \omega_2)t - (k_1 + k_2)a] + \frac{1}{2}k_1 k_2(k_1 - k_2)\\
&\left[\gamma_{566}\cos\varphi\cos\psi + \gamma_{555}\sin\varphi\sin\psi + \gamma_{556}\sin(\varphi + \psi)\right]\\
&AB\sin[(\omega_1 - \omega_2)t - (k_1 - k_2)a]
\end{aligned} \tag{2.52b}$$

It can be seen that Eqs. (2.52a, 2.52b) is similar to Eq. (2.45), the interaction of the two degenerate shear waves will generate the second harmonic and mixed frequency wave of each fundamental wave, these waves can be represented as

$$u_{2\_2\omega_1}^{(II)} = U_{2\_2\omega_1}\cos[2(\omega_1 t - k_1 a)], \tag{2.53a}$$

$$u_{2\_2\omega_2}^{(II)} = U_{2\_2\omega_2}\cos[2(\omega_2 t - k_2 a)], \tag{2.53b}$$

$$u_{2\_\omega_1\pm\omega_2}^{(II)} = U_{2\_\omega_1\pm\omega_2}\cos[(\omega_1 \pm \omega_2)t - (k_1 \pm k_2)a)], \tag{2.53c}$$

$$u_{3\_2\omega_1}^{(II)} = U_{3\_2\omega_1}\cos[2(\omega_1 t - k_1 a)], \tag{2.53d}$$

$$u_{3\_2\omega_2}^{(II)} = U_{3\_2\omega_2}\cos[2(\omega_2 t - k_2 a)] \tag{2.53e}$$

$$u^{(II)}_{3\_\omega_1\pm\omega_2} = U_{3\_\omega_1\pm\omega_2}\cos[(\omega_1 \pm \omega_2)t - (k_1 \pm k_2)a)] \tag{2.53f}$$

Here

$$U_{2\_2\omega_1} = \frac{1}{8}\beta_{S3}k_1^2A^2a, \tag{2.54a}$$

$$U_{2\_2\omega_2} = \frac{1}{8}\beta_{S4}k_2^2B^2a, \tag{2.54b}$$

$$U_{3\_2\omega_1} = \frac{1}{8}\beta_{S5}k_1^2A^2a \tag{2.54c}$$

$$U_{3\_2\omega_2} = \frac{1}{8}\beta_{S6}k_2^2B^2a \tag{2.54d}$$

$$U_{2\_\omega_1\pm\omega_2} = \frac{1}{4}\beta_{S7}k_1k_2ABa, \tag{2.54e}$$

$$U_{3\_\omega_1\pm\omega_2} = \frac{1}{4}\beta_{S8}k_1k_2ABa \tag{2.54f}$$

and

$$\beta_{S3} = -\frac{\gamma_{666}\cos^2\phi + \gamma_{556}\sin^2\phi + \gamma_{566}\sin 2\phi}{\gamma} \tag{2.55a}$$

$$\beta_{S4} = -\frac{\gamma_{666}\cos^2\psi + \gamma_{556}\sin^2\psi + \gamma_{566}\sin 2\psi}{\gamma} \tag{2.55b}$$

$$\beta_{S5} = -\frac{\gamma_{566}\cos^2\phi + \gamma_{555}\sin^2\phi + \gamma_{556}\sin 2\phi}{\gamma} \tag{2.55c}$$

$$\beta_{S6} = -\frac{\gamma_{566}\cos^2\psi + \gamma_{555}\sin^2\psi + \gamma_{556}\sin 2\psi}{\gamma} \tag{2.55d}$$

$$\beta_{S7} = -\frac{\gamma_{666}\cos\phi\cos\psi + \gamma_{556}\sin\phi\sin\psi + \gamma_{566}\sin(\phi+\psi)}{\gamma} \tag{2.55e}$$

$$\beta_{S8} = -\frac{\gamma_{566}\cos\phi\cos\psi + \gamma_{555}\sin\phi\sin\psi + \gamma_{556}\sin(\phi+\psi)}{\gamma} \tag{2.55f}$$

In experiments, a broadband transducer is usually used to generate two shear waves with very small frequency differences, at which time $\phi = \psi$

and

$$\beta_{S3} = \beta_{S4} = \beta_{S7} = -\frac{\gamma_{666}\cos^2\phi + \gamma_{556}\sin^2\phi + \gamma_{566}\sin 2\phi}{\gamma} \tag{2.56a}$$

$$\beta_{S5} = \beta_{S6} = \beta_{S8} = -\frac{\gamma_{566}\cos^2\phi + \gamma_{555}\sin^2\phi + \gamma_{556}\sin 2\phi}{\gamma} \tag{2.56b}$$

Meanwhile, if a receiving transducer is used to detect a second harmonic or mixed wave, the amplitude detected by the receiving transducer can be represented as:

$$U = U_2\cos\theta + U_3\sin\theta, \tag{2.57}$$

Here U_2 and U_3 represent the amplitudes of the second harmonic or mixed wave along the displacement directions of u_2 and u_3, θ is the angle between the polarization direction of the receiving transducer and the displacement u_2.

Usually: $\theta = \varphi$.

From Eqs. (2.48a, 2.48b) and (2.54a–2.54f), it can be seen that only the intrinsic nonlinearity of the material contributes to the second-order nonlinearity effect of pure shear waves.

2.2.3 *Effective Second-Order and Third-Order Elastic Constants of Crystals*

The above discussion shows that the third-order elastic constants used to determine the second-order nonlinear interaction of collinear acoustic waves along the pure mode direction of crystals by TOECs (third-order elastic constants) are γ_{111}, γ_{555}, γ_{666}, γ_{556} and γ_{566}. These constants are quoted into the calculation coordinates, which may be different from those in the constitutive coordinate system. They can be obtained by the following tensor transformation to the constants defined in the constitutive coordinates,

$$\gamma_{IJKLMN} = \alpha_{IP}\alpha_{JQ}\alpha_{KR}\alpha_{LS}\alpha_{MT}\alpha_{NU}c_{PQRSTU}, \tag{2.58}$$

Here α_{AB} are the elements of the coordinate transformation matrix, that is, the direction cosines of the rotation axis A relative to the original coordinate B axis. Therefore, γ_{IJKLMN} may just be a combination of a single c_{IJKLMN} or c_{IJKLMN}, so it is called the effective TOEC third-order elastic constant.

The pure mode directions of various crystals have been studied by Brugger [22]. The relationship between the calculation coordinate system and the constitutive coordinate system is α_{AB}, can be found in literature [22], where the computational coordinates of the a-axis (or 1-axis) are always parallel to the propagation of the wave, that is, the displacement direction of the pure mode longitudinal wave. The other two axes are parallel to the displacement directions of the two pure mode shear waves, as mentioned above.

The third-order elastic constant is a sixth-order tensor, with a total of 729 elements, which means there are 729 combinations in Eq. (2.58). However, due to the symmetry of the third-order elastic constants, the maximum number of independent TOECs is 56. The 56 independent TOECs can be grouped by possible ways of permuting position indices, as shown in Table 2.1.

Now, for simplicity, we define the following coefficients, where the superscript letters x, y, z of $A_{ijklmn}^{x,y,z}$ correspond to the subscripts of the related third-order elastic constants cxyz:

$$
\begin{aligned}
A_{ijklmn}^{(ppp)} &= a_p b_p c_p, \text{ for } c_{111}, c_{222}, c_{333} \\
A_{ijklmn}^{(ppq)} &= a_p b_p c_q + a_p b_q c_p + a_q b_p c_p c_{112}, c_{113}, c_{122}, c_{133}, c_{223}, c_{233} \\
A_{ijklmn}^{(ppQ)} &= a_p b_p c_Q + a_p b_Q c_p + a_Q b_p c_p c_{114}, c_{115}, c_{116}, c_{224}, c_{225}, c_{226}, c_{334}, c_{335}, c_{336} \\
A_{ijklmn}^{(pqQ)} &= a_p\left(b_q c_Q + b_Q c_q\right) + a_q\left(b_p c_Q + b_Q c_p\right) + a_Q\left(b_p c_q + b_q c_p\right) \\
&\quad c_{124}, c_{125}, c_{126} c_{134}, c_{135}, c_{136}, c_{234}, c_{235}, c_{236} \\
A_{ijklmn}^{(pQQ)} &= a_p b_Q c_Q + a_Q b_p c_Q + a_Q b_Q c_p c_{144}, c_{155}, c_{166}, c_{244}, c_{255}, c_{266} c_{344}, c_{355}, c_{366}, \\
A_{ijklmn}^{(pQR)} &= a_p\left(b_Q c_R + b_R c_Q\right) + b_p\left(a_Q c_R + a_R c_Q\right) + c_p\left(a_Q b_R + a_R b_Q\right) \\
&\quad c_{145, c_{146}}, c_{156}, c_{245}, c_{246}, c_{256}, c_{345}, c_{346}, c_{356} \\
A_{ijklmn}^{(QQR)} &= a_Q b_Q c_R + a_Q b_R c_Q + a_R b_Q c_Q c_{445}, c_{446}, c_{455}, c_{466}, c_{556}, c_{566} \\
A_{ijklmn}^{(QQQ)} &= a_Q b_Q c_Q \text{ for } c_{444}, c_{555}, c_{666} \\
A_{ijklmn}^{(123)} &= a_1(b_2 c_3 + b_3 c_2) + a_2(b_1 c_3 + b_3 c_1) + a_3(b_1 c_2 + b_2 c_1) c_{123} \\
A_{ijklmn}^{(456)} &\to a_4(b_5 c_6 + b_6 c_5) + a_5(b_4 c_6 + b_6 c_4) + a_6(b_5 c_4 + b_4 c_5) c_{456}
\end{aligned} \tag{2.59}
$$

Here:

$$a_p = \alpha_{ip}\alpha_{jp}, b_p = \alpha_{kp}\alpha_{lp}, c_p = \alpha_{mp}\alpha_{np}$$

Table 2.1 56 independent TOECs (M represents the possible number of permutations of position indices)

M	TOECs
1	$c_{111}, c_{222}, c_{333}$
3	$c_{112}, c_{113}, c_{122}, c_{133}, c_{223}, c_{233}$
6	$c_{114}, c_{115}, c_{116}, c_{224}, c_{225}, c_{226}, c_{334}, c_{335}, c_{336}, c_{123}$
8	$c_{444}, c_{555}, c_{666}$
12	$c_{124}, c_{125}, c_{126}, c_{134}, c_{135}, c_{136}, c_{234}, c_{235}, c_{236}, c_{144}, c_{155}, c_{166}, c_{344}, c_{355}, c_{366}, c_{254}, c_{255}, c_{266}$
24	$c_{145}, c_{146}, c_{156}, c_{245}, c_{246}, c_{256}, c_{345}, c_{346}, c_{356}, c_{445}, c_{446}, c_{455}, c_{466}, c_{556}, c_{566}$
48	c_{456}

$$
\begin{aligned}
a_Q &= \alpha_{ip}\alpha_{jq} + \alpha_{iq}\alpha_{jp} \quad b_Q = \alpha_{kp}\alpha_{lq} + \alpha_{kq}\alpha_{lp} c_Q \\
&= \alpha_{mp}\alpha_{nq} + \alpha_{mq}\alpha_{np} \\
p, q &= 1, 2, 3, Q, R = \begin{cases} 4\ p, q = 2, 3 \\ 5\ p, q = 3, 1 \\ 6\ p, q = 1, 2 \end{cases}
\end{aligned} \tag{2.60}
$$

Each $A^{()}_{ijklmn}$ contains the sum of the M times accumulation of the product of $\alpha_{is}\alpha_{jt}\alpha_{ku}\alpha_{lv}\alpha_{mw}\alpha_{nx}$. Then, the sum of formula (2.58) can be divided into N groups, where N equals the number of independent TOECs determined by the crystal symmetry.

For example: The independent and non-zero third-order elastic constants of the tetragonal point groups 422, 4 mm, $\overline{4}2$ m and 4/mmm are 12 and 20, respectively,

$$
\begin{aligned}
c_{111} &= c_{222}, c_{112} = c_{122}, c_{113} = c_{223}, c_{123}, c_{133} \\
&= c_{233}, c_{144} = c_{255}, c_{155} = c_{244}, c_{166} \\
&= c_{266}, c_{333}, c_{344} = c_{355}, c_{366}, c_{456}.
\end{aligned}
$$

The third-order elastic constants under any rotated coordinate system can be written as

$$
\begin{aligned}
\gamma_{ijklmn} = {} & \left(A^{(111)}_{ijklmn} + A^{(222)}_{ijklmn}\right)c_{111} + \left(A^{(112)}_{ijklmn} + A^{(122)}_{ijklmn}\right)c_{112} \\
& + \left(A^{(113)}_{ijklmn} + A^{(223)}_{ijklmn}\right)c_{113} + A^{(123)}_{ijklmn}c_{123} + \left(A^{(133)}_{ijklmn} + A^{(233)}_{ijklmn}\right)c_{133} \\
& + \left(A^{(144)}_{ijklmn} + A^{(255)}_{ijklmn}\right)c_{144} + \left(A^{(155)}_{ijklmn} + A^{(244)}_{ijklmn}\right)c_{155} \\
& + \left(A^{(166)}_{ijklmn} + A^{(266)}_{ijklmn}\right)c_{166} + A^{(333)}_{ijklmn}c_{333} + \left(A^{(344)}_{ijklmn} + A^{(355)}_{ijklmn}\right)c_{344} \\
& + A^{(366)}_{ijklmn}c_{366} + A^{(456)}_{ijklmn}c_{456}
\end{aligned} \tag{2.61}
$$

Obviously, the calculation of formula (2.61) can be implemented by a computer. The effective TOECs of the tetragonal point groups 422, 4 mm, $\overline{4}2$ m and 4/mmm in the pure mode direction are shown in Table 2.2. The symbols of the pure mode direction used here are defined in reference [22]. For other point groups, formulas similar to formula (2.61) can be easily obtained.

Similarly, the effective second-order elastic constants can be written as:

$$
\gamma_{ijkl} = \alpha_{ip}\alpha_{jq}\alpha_{rk}\alpha_{sl}c_{pqrs} \tag{2.62}
$$

They can be obtained by the method used to calculate the effective TOECs, or by the M-matrix listed by Auld [24]. The tetragonal point groups 422, 4 mm, $\overline{4}2$ m and the effective SOECs for 4/mmm calculations are also shown in Table 2.2.

In this section, we studied the interaction of collinear elastic waves propagating along the pure mode direction of the crystal. We provided the nonlinear parameters

Table 2.2 Effective SOECs and TOECs of tetragonal crystals (tetragonal crystals 422, 4 mm, $\bar{4}2$ m and 4/mmm)

	TIα	TIβ[a]	TIπ[a]	TIγ	TIκ
γ_{11}	c_{33}	$\frac{1}{2}c_{11}\cos^4\mu$ $+\frac{1}{2}c_{12}\cos^4\mu$ $+c_{66}\cos^4\mu$ $+2c_{13}\cos^2\mu\sin^2\mu$ $+4c_{44}\cos^2\mu\sin^2\mu$ $+c_{33}\sin^4\mu$	$c_{11}\cos^4\theta$ $+2c_{13}\cos^2\theta\sin^2\theta$ $+4c_{44}\cos^2\theta\sin^2\theta$ $+c_{33}\sin^4\theta$	$\frac{1}{2}c_{11}$ $+\frac{1}{2}c_{12}+c_{66}$	c_{11}
γ_{55}	c_{44}	$\frac{1}{2}c_{11}\cos^2\mu\sin^2\mu$ $+\frac{1}{2}c_{12}\cos^2\mu\sin^2\mu$ $-2c_{13}\cos^2\mu\sin^2\mu$ $+c_{33}\cos^2\mu\sin^2\mu$ $+c_{66}\cos^2\mu\sin^2\mu$ $+c44\cos^2 2\mu$	$c_{11}\cos^2\theta\sin^2\theta$ $-2c_{13}\cos^2\theta\sin^2\theta$ $+c_{33}\cos^2\theta\sin^2\theta$ $+c_{44}\cos^2 2\theta$	c_{44}	c_{44}
γ_{66}	c_{44}	$\frac{1}{2}c_{11}\cos^2\mu$ $-\frac{1}{2}c_{12}\cos^2\mu$ $+c_{44}\sin^2\mu$	$c_{44}\sin^2\theta$ $+c_{66}\cos^2\theta$	$-\frac{1}{2}c_{12}+$ $\frac{1}{2}c_{11}$	c_{66}

(continued)

Table 2.2 (continued)

	TIα	TIβ[a]	TIπ[a]	TIγ	TIκ
γ_{15}	0	$-\frac{1}{2}c_{11}\cos^3\mu\sin\mu$ $-\frac{1}{2}c_{12}\cos^3\mu\sin\mu$ $-c_{66}\cos^3\mu\sin\mu$ $+c_{33}\cos\mu\sin^3\mu$ $+2c_{44}\cos\mu\sin\mu\cos2\mu$ $+c_{13}\cos\mu\sin\mu\cos2\mu$	$-c_{11}\cos^3\theta\sin\theta$ $+c_{33}\cos\theta\sin^3\theta$ $+2c_{44}\cos\theta\sin\theta\cos2\theta$ $+c_{13}\cos\theta\sin\theta\cos2\theta$	0	0
γ_{16}	0	0	0	0	0
γ_{56}	0	0	0	0	0

(continued)

Table 2.2 (continued)

	TIα	TIβ[a]	TIπ[a]	TIγ	TIκ
γ_{111}	c_{333}	$\frac{1}{4}c_{111}\cos^6\mu$ $+\frac{3}{4}c_{112}\cos^6\mu$ $+3c_{166}\cos^6\mu$ $+\frac{3}{2}c_{113}\cos^4\mu\sin^2\mu$ $+\frac{3}{2}c_{123}\cos^4\mu\sin^2\mu$ $+6c_{144}\cos^4\mu\sin^2\mu$ $+6c_{155}\cos^4\mu\sin^2\mu$ $+3c_{366}\cos^4\mu\sin^2\mu$ $+12c_{456}\cos^4\mu\sin^2\mu$ $+3c_{133}\cos^2\mu\sin^4\mu$ $+12c_{344}\cos^2\mu\sin^4\mu$ $+c_{333}\sin^6\mu$	$c_{111}\cos^6\theta$ $+3c_{113}\cos^4\theta\sin^2\theta$ $+12c_{155}\cos^4\theta\sin^2\theta$ $+3c_{133}\cos^2\theta\sin^4\theta$ $+12c_{344}\cos^2\theta\sin^4\theta$ $+c_{333}\sin^6\theta$	$\frac{1}{4}C_{111}$ $+\frac{3}{4}C_{112}$ $+3c_{166}$	c_{111}

(continued)

Table 2.2 (continued)

	TIα	TIβ[a]	TIπ[a]	TIγ	TIκ
γ_{555}	0	$-\frac{1}{4}c_{111}\cos^3\mu\,\sin^3\mu$ $-\frac{3}{4}\,c_{112}\cos^3\mu\,\sin^3\mu$ $+\frac{3}{2}c_{113}\cos^3\mu\,\sin^3\mu$ $+\frac{3}{2}\,c_{123}\cos^3\mu\,\sin^3\mu$ $-3\,c_{133}\cos^3\mu\,\sin^3\mu$ $-3c_{166}\cos^3\mu\,\sin^3\mu$ $+c_{333}\cos^3\mu\,\sin^3\mu$ $+3c_{366}\,\cos^3\mu\,\sin^3\mu$ $-\frac{3}{2}c_{144}\,\cos\mu\,\sin\mu\,\cos^2 2\mu$ $-\frac{3}{2}\,c_{155}\cos\mu\,\sin\mu\,\cos^2 2\mu$ $+3c_{344}\cos\mu\,\sin\mu\,\cos^2 2\mu$ $-3\,c_{456}\cos\mu\,\sin\mu\,\cos^2 2\mu$	$-c_{111}\cos^3\theta\,\sin^3\theta$ $+3\,c_{113}\cos^3\theta\,\sin^3\theta$ $-3\,c_{133}\cos^3\theta\,\sin^3\theta$ $+c_{333}\cos^3\theta\,\sin^3\theta$ $-3c_{155}\,\cos\theta\,\sin\theta\,\cos^2 2\theta$ $+3\,c_{344}\,\cos\theta\,\sin\theta\,\cos^2 2\theta$	0	0
γ_{666}	0	0	0	0	0

(continued)

Table 2.2 (continued)

	TIα	TIβ[a]	TIπ[a]	TIγ	TIκ
γ_{556}	0	0	0	0	0
γ_{566}	0	$-\frac{1}{4}c_{111}\cos^3\mu\,\sin\mu$ $+\frac{1}{4}c_{112}\cos^3\mu\,\sin\mu$ $+\frac{1}{2}c_{113}\cos^3\mu\,\sin\mu$ $-\frac{1}{2}c_{123}\cos^3\mu\,\sin\mu$ $+c_{344}\cos\mu\,\sin^3\mu$ $+c_{456}\cos\mu\,\sin^3\mu$ $+c_{144}\left(-\frac{1}{2}\cos\mu\,\sin^3\mu-\cos\mu\,\sin\mu\,\cos 2\mu\right)$ $+c_{155}\left(-\frac{1}{2}\cos\mu\,\sin^3\mu+\cos\mu\,\sin\mu\,\cos 2\mu\right)$	$-c_{166}\cos^3\theta\,\sin\theta$ $+c_{366}\cos^3\theta\,\sin\theta$ $-c_{144}\cos\theta\,\sin^3\theta$ $+c_{344}\cos\theta\,\sin^3\theta$ $+2\,c_{456}\cos\theta\,\sin\theta\,\cos 2\theta$	0	0

[a] This is the second pure mode direction, the particle displacement along the 2-axis shear wave is a pure mode, and the other shear wave is coupled with the longitudinal wave

describing the interaction. The effective third-order elastic parameters contained in the parameters are c_{111}, c_{555} and c_{666} (when the shear wave is non-degenerate) or c_{111}, c_{555}, c_{666}, c_{556} and c_{566} (when the shear wave is degenerate). We proposed a computer method for calculating the third-order effective elastic parameters of the crystal. The results show that the calculation method for the cubic crystal given in this paper is the same as the results given in reference [19], thus proving the effectiveness of this method.

It is observed that γ_{111} will never disappear. Therefore, the secondary nonlinear interaction of the longitudinal wave always exists, regardless of the direction of wave propagation. By Eq. (2.58)

$$\gamma_{111} = \gamma_{111111} = \alpha_{1p}\alpha_{1q}\alpha_{1r}\alpha_{1s}\alpha_{1t}\alpha_{1u}c_{pqrstu} = \alpha_{11}^{6}c_{111} + \alpha_{12}^{6}c_{222} + \alpha_{13}^{6}c_{333} + \cdots \tag{2.63}$$

It is well known that for any crystal (including isotropic solids), c_{111}, c_{222} and c_{333} are not zero. Among them, at least one of α_{11}, α_{12} and α_{13} is non-zero, therefore, γ_{111} is never zero, at least equal to one of c_{111}, c_{222} and c_{333}. But in fact, for γ_{555} and γ_{666} the situation is different.

$$\begin{aligned} \gamma_{555} &= \gamma_{131313} = \alpha_{1p}\alpha_{3q}\alpha_{1r}\alpha_{3s}\alpha_{1t}\alpha_{3u}c_{porstu}, \ \gamma_{666} = \gamma_{121212} \\ &= \alpha_{1p}\alpha_{2q}\alpha_{1r}\alpha_{2s}\alpha_{1t}\alpha_{2u}c_{pqrstu} \end{aligned} \tag{2.64}$$

They contain the product of transformation matrix elements in different rows. If one of the α_{ij} is zero, then the product $\alpha_{1p}\alpha_{3q}\alpha_{1r}\alpha_{3s}\alpha_{1t}\alpha_{3u} = 0$, Furthermore, due to symmetry, some crystals c_{555} and c_{666} are zero in themselves. Therefore, γ_{555} and γ_{666} have a very high probability of being zero. Our calculations show that most pure modal directions $c\gamma_{555}$ are zero, and in all calculated cases γ_{666} are zero. Many pure modal directions still prohibit the secondary nonlinear interaction of shear waves.

Our research results propose a general method for calculating effective TOECs [25], which can be used for any symmetric system and is very helpful for experimental research on the nonlinear elastic properties of crystals.

2.3 Study of Pure Mode Axes of PMN-PT Crystals with Different PT Contents

The lead-based composite perovskite structure relaxor ferroelectric single crystal $(1-x)Pb(Mg_{1/3}Nb_{2/3})O_3 - xPbTiO_3$ (abbreviated as PMN-xPT) is a solid solution composed of relaxor ferroelectric $Pb(Mg_{1/3}Nb_{2/3})O_3$ (PMN) and normal ferroelectric $PbTiO_3$ (PT). Its successful growth is considered a revolutionary breakthrough in the ferroelectric field in the past 50 years. Single crystals with compositions in the MPB (morphotropic phase boundary) range have excellent piezoelectric constants, electromechanical coupling coefficients, and electrostrictive properties [26]. Its piezoelectric constant d_{33} can reach 1500–2000 pC/N, the electromechanical coupling coefficient k_{33} can reach 0.92–0.94, and the maximum electrostrictive strain can reach 1.7% [27], which has far exceeded the currently used piezoelectric materials. It is a new material with a major breakthrough in the ferroelectric field in the past 50 years. Such excellent piezoelectric and electrostrictive performance is expected to greatly improve the performance of piezoelectric devices in medical ultrasound imaging, underwater sonar, non-destructive testing, solid-state large displacement piezoelectric actuators, and make PMN-PT single crystals important in ultrasound transduction and electromechanical conversion. Applications, with broad application prospects.

The Curie point of PMN is around − 17 °C, and it is a pseudocubic (trigonal) structure at room temperature. In the PMN-PT system, as the PT content increases, the system gradually transitions from a trigonal phase to a tetragonal phase, and there is a morphotropic phase boundary (MPB) at the PT content of 32–35 mol%. Materials near this phase boundary have better dielectric and piezoelectric properties [28–30]. The pure mode of PMN-PT crystals with different PT contents is also different.

Because the crystal is elastic, when a mechanical disturbance occurs in a part of the crystal, this disturbance will propagate in the crystal, forming elastic waves. The propagation of elastic waves in crystals has some properties closely related to the symmetry and anisotropy of the crystal. In the problem of elastic wave propagation, the direction of energy speed and phase speed is of great significance, this direction is the pure mode axis of the crystal.

2.3.1 *Pure Mode of PMN-PT Crystals*

Given the wave vector of a specific axis $\vec{k}$, substitute its unit vector $\vec{k}^0 = \vec{k}/|k|$ into the Christoffel equation, the phase velocity of the elastic wave $\vec{v}$ can be solved, and then use the slowness surface equation $F(k_1/\omega, k_2/\omega, k_3/\omega)$ to solve the energy propagation speed of the elastic wave $\vec{v}_e$, according to $\vec{v}_e$ and $\vec{v}$ the direction relationship determines whether this axis is a pure mode axis. When the direction of energy speed is consistent with the direction of the wave vector (i.e., $\vec{v}_e // \vec{v}$), this direction is generally referred to as the pure mode axis [31]. If calculations are performed directly, it would be very cumbersome and the amount of computation would be large. We can use the slowness surface and make judgments directly based on the shape of the surface.

The slowness surface represents the plane wave equation of the phase velocity of the elastic wave propagation, that is, the Christoffel equation expression [31, 32] is:

$$\left(\Gamma_{il} - \rho v^2 \delta_{il}\right) u_{l0} = 0, \ (i, l = 1, 2, 3) \tag{2.65}$$

where $\Gamma_{il} = c_{ijkl} l_j l_k$ is defined as the Christoffel tensor. To have a non-zero solution, it is required that

$$Q(\upsilon, l_1, l_2, l_3) = \left|\Gamma_{il} - \rho \upsilon^2 \delta_{il}\right| = 0 \tag{2.66}$$

Rewritten as a surface equation:

$$F(k_1/\omega, k_2/\omega, k_3/\omega) = 0 \tag{2.67}$$

This surface is called the slowness surface. It can be seen as a surface in a three-dimensional variable space (i.e., $\xi_i = k_i/\omega(i = 1, 2, 3)$). The length of the radial vector at each point on the surface is $(\sum_i \xi_i^2)^{1/2} = \left|\vec{k}\right|/\omega = 1/\upsilon$. The reciprocal of the velocity $1/\upsilon$ is generally defined as slowness. Given any wave vector direction (l_1, l_2, l_3), solving the equation gives the phase velocity of the elastic wave in this direction $\vec{\upsilon}$. There are three positive solutions for each direction, so the slowness surface is a three-layer surface. It describes the speed of elastic wave propagation in all directions in the crystal as a whole.

2.3.2 Energy Velocity of Elastic Waves Represented by the Slowness Surface

In reality, there is no existence of monochromatic waves. Any actual wave is composed of many monochromatic waves with different frequencies and propagation directions. This wave composed of superposition is called a wave packet or wave group. However, when the frequency and direction of the elastic wave propagating in the medium do not differ much, and the phase velocity also does not differ much, that is, there is no significant dispersion phenomenon, and the loss in the medium can also be ignored, it can be considered that the propagation speed of the center or maximum value of the wave packet, that is, the overall propagation speed of the wave group, is the speed of energy propagation [31]. It can be proved that the normal direction of any point on the slowness surface is the direction of energy propagation, as shown in Fig. 2.1. $\overrightarrow{OA}$ is the direction of phase velocity, $\overrightarrow{OB}$ is the direction of group velocity.

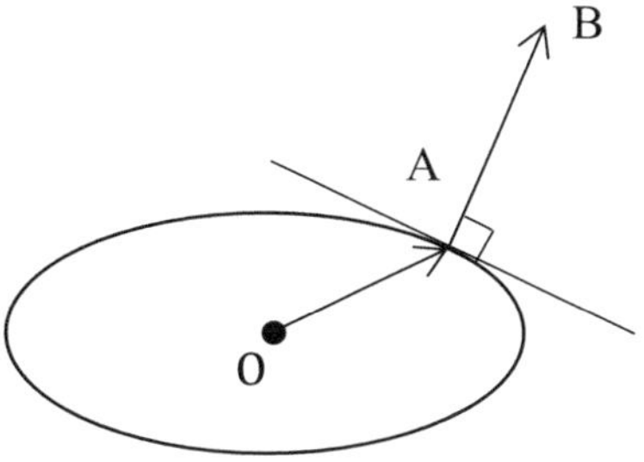

Fig. 2.1 Relationship between slowness surface and energy velocity

2.3.3 Specific Analysis of the Propagation Speed and Pure Mode of PMN-PT Crystal

According to Table 2.3 about the density and compliance constants of PMN-PT crystal, the slowness surfaces of PMN-PT crystal on different planes are drawn, as shown in Figs. 2.2, 2.3, and 2.4.

As shown in the figure, the slowness curve in Fig. 2.2a is symmetrical about the x-axis, and the slowness curve intersects the x-axis at three points, the normal direction of these three points is the same as the x-axis. In Fig. 2.2b, the slowness curve is also symmetrical about the x-axis, and the normal direction of the three points where the slowness curve intersects the x-axis is also the same as the x-axis. Therefore, the x-axis is a pure mode axis. Similarly, the y-axis and z-axis are also pure mode axes. From Fig. 2.2a, it can be seen that in the direction of the angle with the x-axis at 45°, 135°, 225°, and 315°, the direction of the tangent of the curve is perpendicular to the radial vector, that is, the direction of the phase velocity is parallel to the direction of the energy velocity, that is, these directions are all pure mode axes.

From Figs. 2.2 and 2.3, it can be seen that due to the similar structures of PMN-0.30PT and PMN-0.33PT crystals, the shapes of the slowness surfaces on various planes of these two crystals are basically the same. When elastic waves propagate in these two crystals, the longitudinal waves propagate slowest in the coordinate axis direction and fastest in the 45°, 135°, 225°, and 315° directions. As the propagation

Table 2.3 Density and compliance constants of PMN-PT crystal (10^{-12})

	PMN-42%PT	PMN-33%PT	PMN-30%PT
$\rho(\mathrm{kg/m^3})$	8010	8060	8040
s_{11}^E	9.43	69.0	52.0
s_{12}^E	− 1.68	− 11.1	− 18.9
s_{13}^E	− 6.13	− 55.7	− 31.1
s_{33}^E	19.21	119.6	67.7
s_{44}^E	35.09	14.5	14.0
s_{66}^E	12.5	15.2	15.2

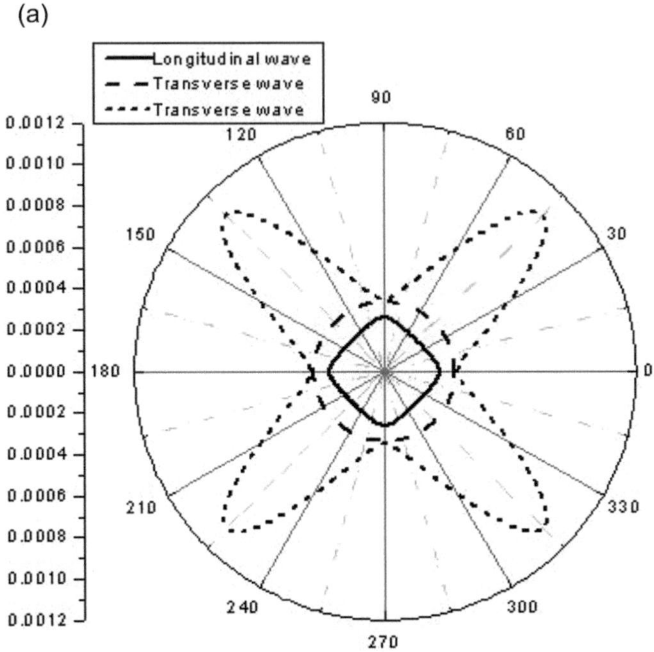

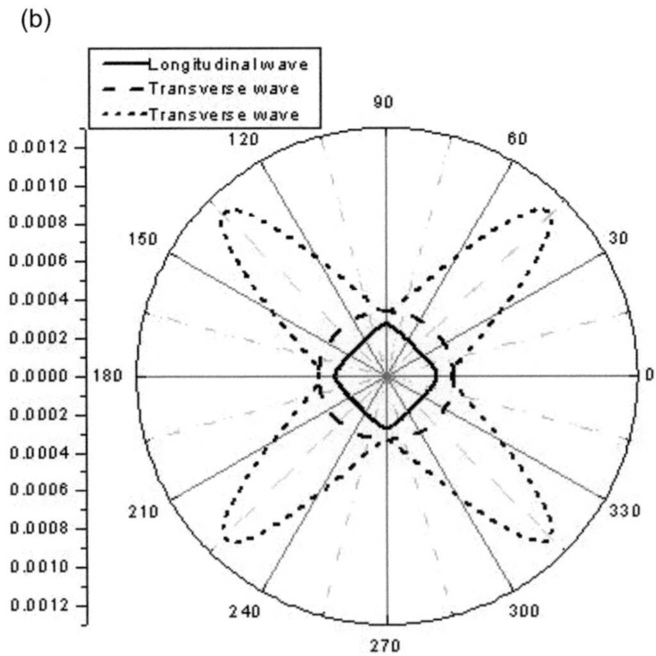

Fig. 2.2 **a** Slowness curve of PMN-0.30PT in the X–Y plane, **b** slowness curve of PMN-0.30PT in the X–Z plane, **c** slowness curve of PMN-0.30PT in the 110–001 plane

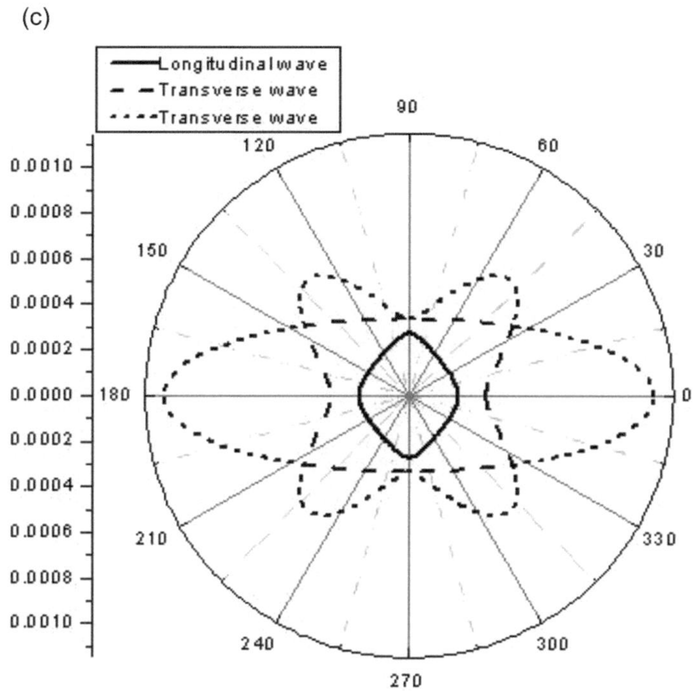

Fig. 2.2 (continued)

direction changes, the speed of one shear wave changes very little and can be approximated as constant, while the speed of the other shear wave changes significantly, being fastest when propagating along the coordinate axis and slowest when propagating in the 45°, 135°, 225°, and 315° directions. Figure 2.2c also show that when elastic waves propagate in the [110–001] plane, the longitudinal wave propagates slowest along the z-axis and fastest along the [110] direction. As the propagation direction changes, the speeds of both shear waves change significantly.

Compared with Figs. 2.2 and 2.3, Fig. 2.4 clearly shows that the shapes of the slowness surfaces when elastic waves propagate in PMN-PT crystals with different PT contents are very different, indicating the structural differences of the crystals. It can also be seen that the number of pure mode axes is significantly different. In Fig. 2.4a, the slowness curves are symmetric about the x-axis, and the slowness curves intersect the x-axis at three points, the normal directions of these three points are in the same direction as the x-axis. In Fig. 2.4b, the slowness curve is also symmetric about the x-axis, and the normal directions of the three points where the slowness curve intersects the x-axis are also in the same direction as the x-axis. Therefore, the x-axis is a pure mode axis. Similarly, the y-axis and z-axis are also pure mode axes. From Fig. 2.4a, it can be seen that in the direction at an angle of 45°, 135°, 225°, and 315° with the x-axis, the direction of the tangent of the curve is perpendicular to the radial vector, i.e., the direction of the phase velocity is parallel to the direction of the

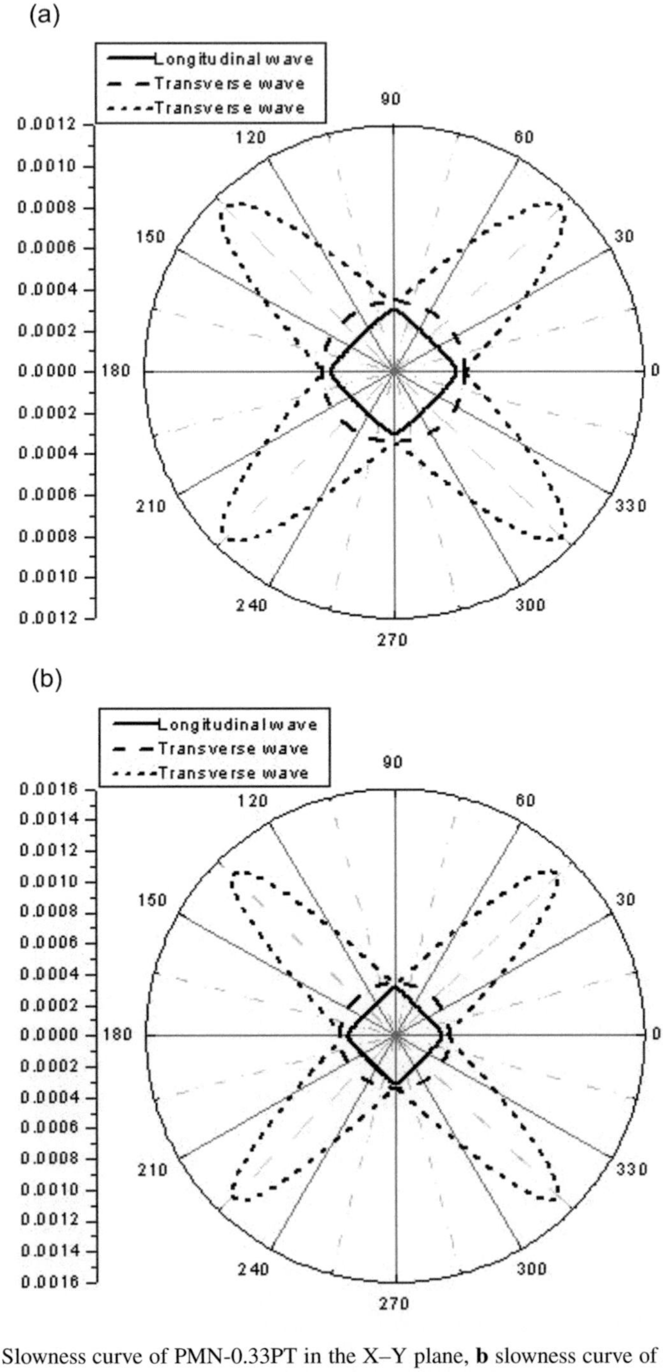

Fig. 2.3 **a** Slowness curve of PMN-0.33PT in the X–Y plane, **b** slowness curve of PMN-0.33PT in the X–Z plane, **c** slowness curve of PMN-0.33PT in the 110–001 plane

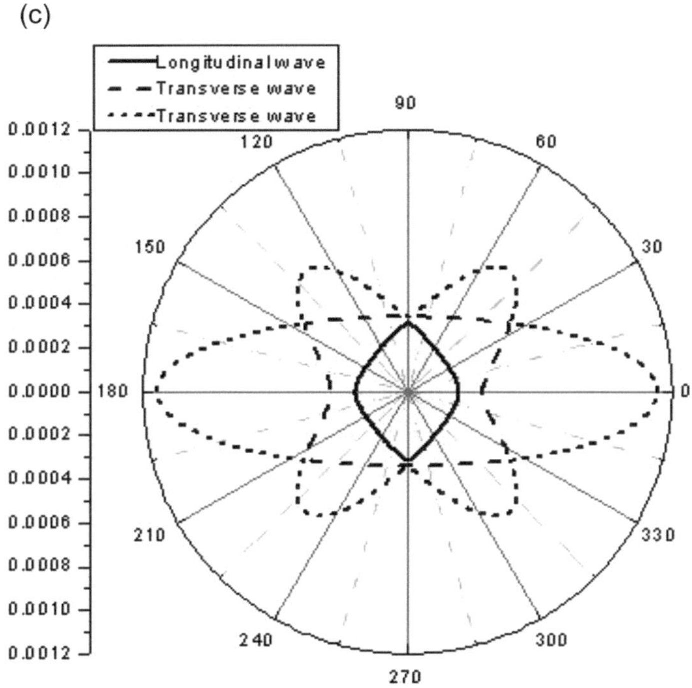

Fig. 2.3 (continued)

energy velocity, i.e., these directions are all pure mode axes. However, in Fig. 2.4b, c, the number of pure mode axes is significantly reduced, and the direction at an angle of 45° with the coordinate axis is no longer a pure mode direction. This is because when the PT content is 42%, the anisotropy of PMN-PT is significantly weakened compared to when the PT content is 33%, so the number of pure mode directions is reduced.

2.4 Study on the Acoustic Characteristics of Lithium Tantalate and PIN-PMN-PT Crystals

Ferroelectric lithium niobate and lithium tantalate crystals are "universal" materials in the field of functional materials. They have good mechanical and physical properties and low cost, and as nonlinear optical crystals, electro-optic crystals, piezoelectric crystals, acousto-optic crystals, and birefringent crystals, they have been widely used in the IT industry centered on optical technology.

(a)

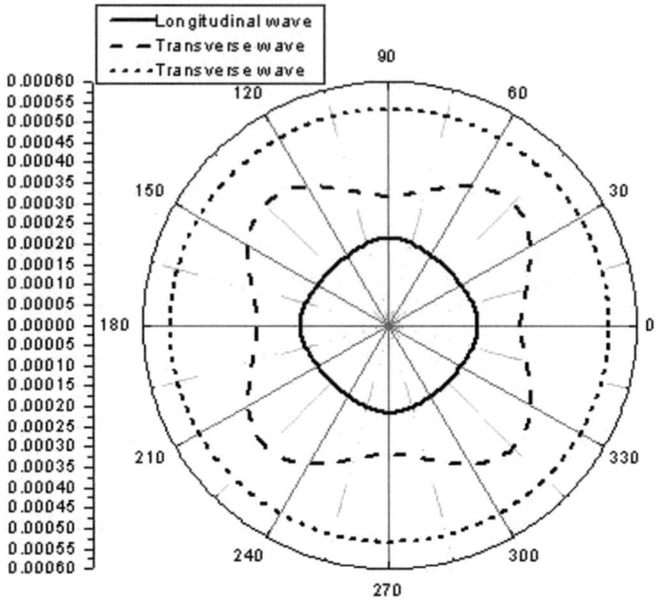

(b)

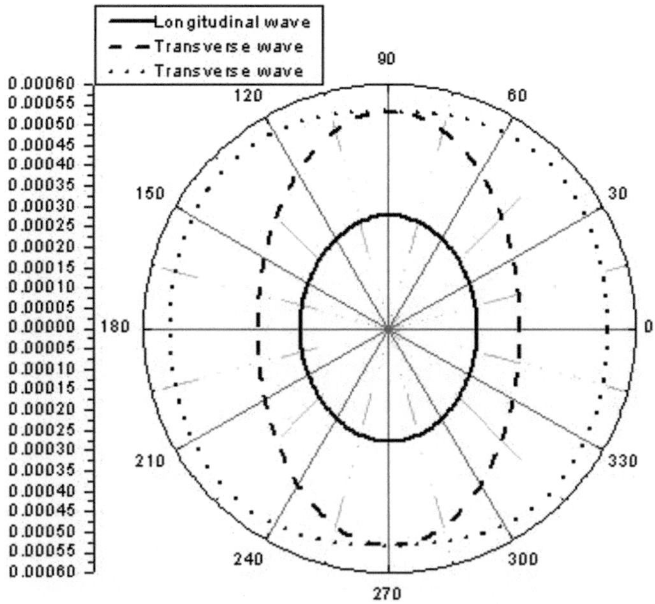

Fig. 2.4 **a** Slowness curve of PMN-0.42PT in the X–Y plane, **b** slowness curve of PMN-0.42PT in the Y–Z plane, **c** slowness curve of PMN-0.42PT in the 110–001 plane

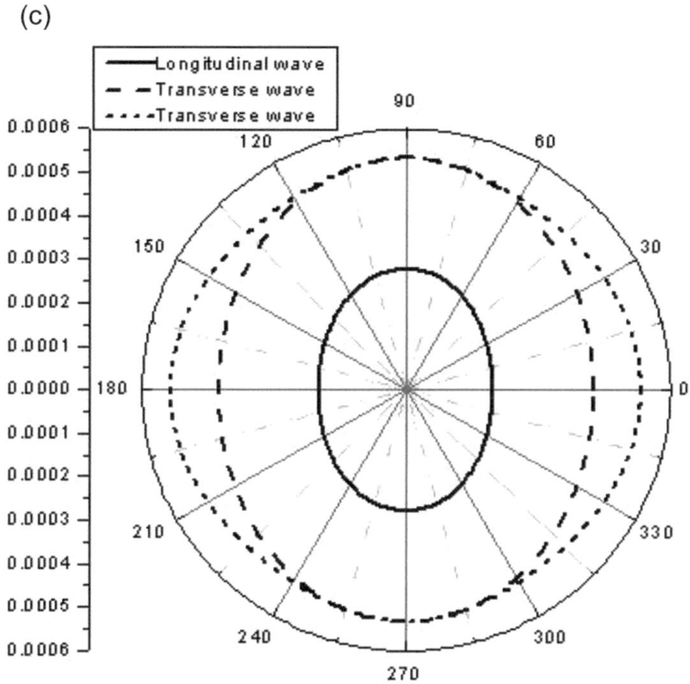

Fig. 2.4 (continued)

Lithium tantalate ($LiTaO_3$) crystal is an important multifunctional crystal material, which has excellent piezoelectric, ferroelectric, acousto-optic, and electro-optic effects, thus becoming a basic functional material in the fields of surface acoustic wave (SAW) devices, optical communication, lasers, and optoelectronics. Polished lithium tantalate wafers are widely used in the manufacture of resonators, filters, transducers, and other electronic communication devices, especially for the manufacture of high-frequency surface acoustic wave devices due to their good electromechanical coupling and temperature coefficient, and are used in mobile phones, walkie-talkies, satellite communications, aerospace, and many other high-end communication fields [33].

$xPb(In_{1/2}\ Nb_{1/2})O_3-(1-x-y)0.45Pb(Mg_{1/3}\ Nb_{2/3})O_3-yPbTiO_3$, i.e., lead indium niobate-lead magnesium niobate-lead titanate (PIN-PMN-PT) crystals, have received great attention in recent years. PZT ceramics have always been the main material for ultrasonic transducers, but PZT has shown large acoustic attenuation and sound speed divergence effects due to acoustic scattering at grain boundaries [40]. Therefore, the application range of PZT is very small above 25 MHz.

High-frequency broadband transducers have attracted great attention in medical imaging in recent years because they have higher lateral and axial resolution [35, 36]. So far, the characteristics of single crystals with ultra-high piezoelectric performance are not as good as those of multi-domain crystals, mainly for two reasons. First, for practical applications, ultra-high piezoelectric phenomena only occur in multi-domain crystals polarized in the [001] and [011] directions. Second, single-domain crystals are not as stable as multi-domain crystals.

Recently, it has been proven that relaxor ferroelectric single crystals $(1-x)Pb(Mg_{1/3}\ Nb_{2/3})O_3-xPbTiO_3$ (PMN-PT) have superior electromechanical performance [37, 38]. However, the phase transition temperature of PMN-PT from rhombohedral to tetragonal is relatively low (~ 75–95 °C). Moreover, the coercive electric field *Ec* (~ 2.5 kV/cm) of PMN-PT is too low for high-intensity applications, and its performance will decrease when integrated into devices. Therefore, in recent years, crystals with similar performance to PMN-PT, high piezoelectric coupling and electromechanical coupling coefficients, but also with higher coercive electric fields and phase transition temperatures, have been a hot topic of research.

The ternary crystal $xPb(In_{1/2}\ Nb_{1/2})O_3-(1-x-y)Pb(Mg_{1/3}\ Nb_{2/3})O_3-yPbTiO_3$ (PIN-PMN-PT) has such characteristics. Its electromechanical coupling performance is similar to that of PMN-PT crystals, but its phase transition temperature is about 30° higher than that of PMN-PT crystals, and its coercive electric field has also increased by 2.5 times [39–42]. Tests have shown that PIN-PMN-PT has better thermal stability than PMN-PT [43].

There have been studies on the theory and experiment of limited amplitude ultrasound propagation in crystals. Domanski studied the propagation and nonlinear interaction of ultrasound in cubic crystals [17], Du studied the pure mode and nonlinear propagation in crystals [44], Jiang and others studied the degenerate field in piezoelectric bodies and its impact on nonlinear parameters [45]. Albert and others studied the method of measuring the third-order elastic constants of cubic crystals by ultrasonic pulse [46]. Philip studied the third-order elastic parameters of silicon and germanium at temperatures from 300 to 3 °K [11], and measured the nonlinearity of $CsCdF_3$ and $KZnF_3$ crystals [12]. Jiang also studied the generation of the second harmonic of shear waves in crystals [15], and studied the excitation of the third harmonic of transverse waves in crystals and ceramics [16]. The nonlinear term indicates that the transverse wave is not a pure transverse wave and is accompanied by a longitudinal wave mode. On the other hand, the longitudinal wave still maintains the form of a pure longitudinal wave, but is accompanied by a second harmonic. Therefore, nonlinear phenomena will occur in the propagation of longitudinal waves in the pure mode direction. Existing literature has studied the linear characteristics of lithium niobate crystals and found abnormal sound speed (sound attenuation) changes at 75 °C [47]. However, there have been no reports on the study of the nonlinear acoustic characteristics of lithium tantalate crystals. We measured the changes in the acoustic parameters of lithium tantalate crystals with temperature, analyzed the results, and used the second harmonic generation technology to explain the nonlinear changes of sound waves in lithium tantalate. This chapter studies the nonlinear characteristics of such materials, based on the second harmonic generation technology, deeply explores the propagation speed of sound waves in lithium tantalate crystals, the amplitude of the fundamental wave, the amplitude of the second harmonic with temperature changes, and thus obtains the relationship between the nonlinearity of lithium tantalate and the crystal structure.

At the same time, the experimental method compares the changes in sound speed, sound attenuation, first and second harmonics, and nonlinear coefficients of [001] polarized and [111] polarized PIN-PMN-PT crystals in the range of 0–110 °C. The acoustic characteristics of [001] polarized PIN-PMN-PT crystals basically change monotonically, while the acoustic characteristics of [111] polarized PIN-PMN-PT crystals have inflection points, which are completely different from [001] polarized crystals.

2.4.1 Experimental Method

2.4.1.1 Transducer Sensitivity Calibration

The most commonly used methods for absolute measurement of sound wave amplitude are mainly [17]: (1) Using a capacitive transducer; (2) Using a laser probe; (3) Using a piezoelectric transducer. The first two are non-contact methods, the propagation of sound waves is not disturbed, but they have higher requirements for the sample surface and measurement environment. The third method is a contact method, the coupling quality between the transducer and the sample may affect the measurement. But from a practical point of view, this method is more feasible. Because the transducer band is narrow, the contact transducer has a high sensitivity. When measuring the absolute amplitude of sound waves with a piezoelectric transducer, in principle, the sensitivity of the transducer must be absolutely corrected. The correction experiment framework is shown in Fig. 2.5.

The signal generator sends sound waves to the sample through the transducer and receives the returned sound waves. The incident current I_{in}, incident voltage V_{in} and the output current I_{out}, output voltage V_{out} returned by the sample after one reflection are obtained. The data is collected by the computer and the amplitude spectrum is obtained through Fourier analysis. Therefore, the response function of the transducer

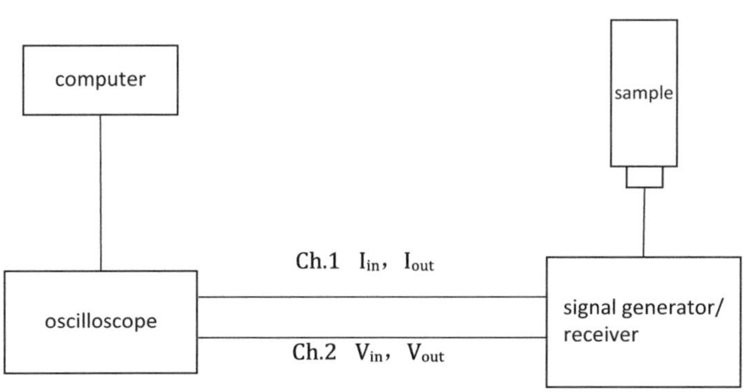

Fig. 2.5 Transducer sensitivity correction framework diagram

is obtained:

$$|H(\omega)| = \sqrt{\frac{\left|I_{in}(\omega)\left(\frac{V_{out}(\omega)}{I_{out}(\omega)}\right) + V_{in}(\omega)\right|}{2\omega^2 \rho \upsilon a |I_{out}(\omega)|}} \tag{2.68}$$

where, ρ is the density of the sample, υ is the speed of sound propagation in the crystal, a is the area of the transducer.

The absolute amplitude of the sound wave can be obtained by the product of the response function and the output current:

$$|A_0(\omega_0)| = |H(\omega_0)| * |I_{out}(\omega_0)| \tag{2.69}$$

$$|A_1(2\omega_0)| = |H(2\omega_0)| * |I_{out}(2\omega_0)| \tag{2.70}$$

2.4.1.2 Experimental System

The experimental device is shown in Fig. 2.6. The modulated signal generated by the function generator is amplified by the power amplifier (center frequency is 2.5 MHz, the pulse repetition frequency is 100 Hz, the pulse width is 4 μs) to excite lead zirconate titanate (PZT), generating an ultrasonic longitudinal wave with a center frequency of 2.5 MHz. After the ultrasonic wave propagates in the crystal, it is received at the other end of the sample with a PZT with a center frequency of 4.8 MHz. The received signal is received and stored by the digital storage oscilloscope. The sound axis offset angle of the transmitting and receiving transducers is less than 2°. First, the transmitting and receiving transducers and the test block are fixed on the experimental bracket to ensure that the axis of the test block and the axis of the transmitting and receiving transducers are kept in a line, to ensure that the relative position of the sample and the transmitting and receiving transducers does not change during the experiment, and to ensure that the pressure between the test block and the transmitting and receiving transducers does not change. Vaseline is used as the ultrasonic coupling agent. The pulse generator/receiver is added at the receiving end to ensure impedance matching.

In the experiment, the aluminum block sample is a cylinder with a height of 3.5 cm and a base diameter of 3 cm. The lithium tantalate sample is a cylinder with a height of 3.5 cm and a base diameter of 3.1 cm. The PIN-PMN-PT experiment uses a sample of 0.28PIN-0.42PMN-0.30PT, which are polarized in the [001] direction and the [111] direction. The crystal size in the [001] polarization direction is 10.50 mm*10.325 mm*11.224 mm, The crystal size in the [111] polarization direction is 11.53 mm*11.035 mm*10.52 mm.

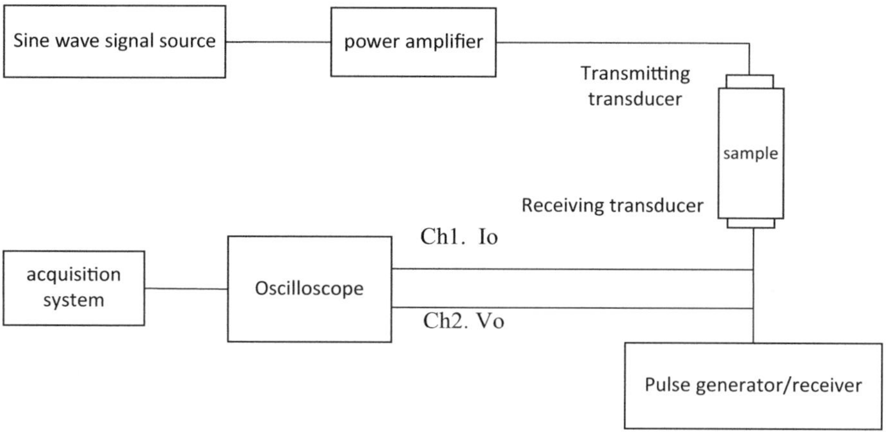

Fig. 2.6 Experimental device diagram

2.4.1.3 Experimental Results

Experimental Results of Aluminum Block

Figures 2.7 and 2.9 are the voltage and current received by the oscilloscope at room temperature of 20 °C for the aluminum block. The first, second, and third echoes of the aluminum block can be clearly seen, and the current measured by the ammeter matches the voltage in the same proportion. The second and third harmonic components of the system can be clearly seen from Fig. 2.8.

In order to eliminate system linear and nonlinear errors, first use the aluminum block for nonlinear experiments, the second harmonic changes with temperature as shown in Fig. 2.10.

From Fig. 2.10, which shows the change in the second harmonic of the aluminum block with temperature, it can be seen that between 0 and 110 °C, the change in the amplitude of the second harmonic of the aluminum block is not significant. It can be considered that within the allowable error range, when the temperature exceeds 110 °C, the amplitude of the second harmonic drops sharply, indicating that the PZT transducer can no longer work normally. Therefore, the measurement temperature of PIN-PMN-PT is limited to within 110 °C.

Experimental Results of Lithium Tantalate

Lithium tantalate, like lithium niobate, belongs to the 3 m point group in the ferroelectric phase structure at room temperature, and to the $\overline{3}$ m point group in the paraelectric phase structure above the Curie point, belonging to the trigonal crystal system.

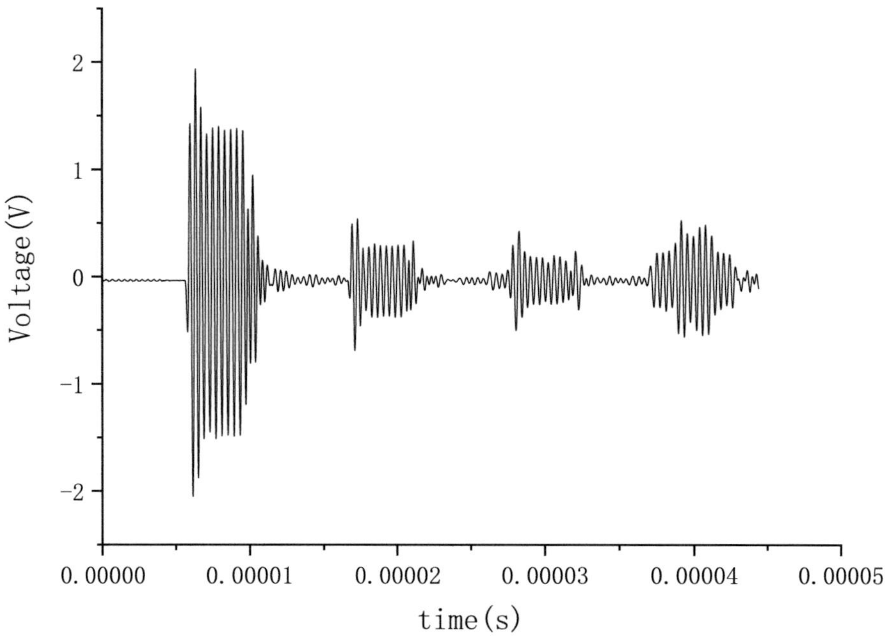

Fig. 2.7 Voltage in aluminum material

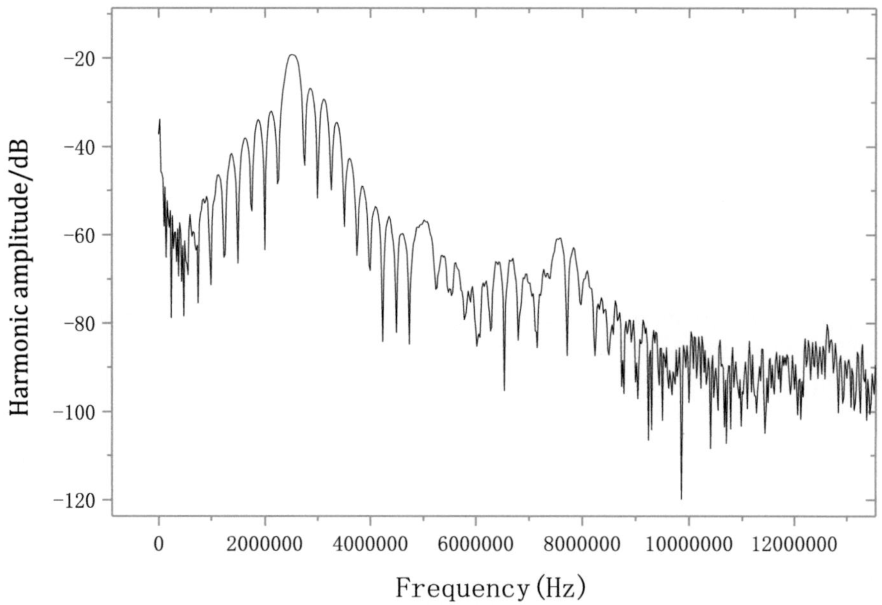

Fig. 2.8 Spectrum of direct wave voltage in aluminum material

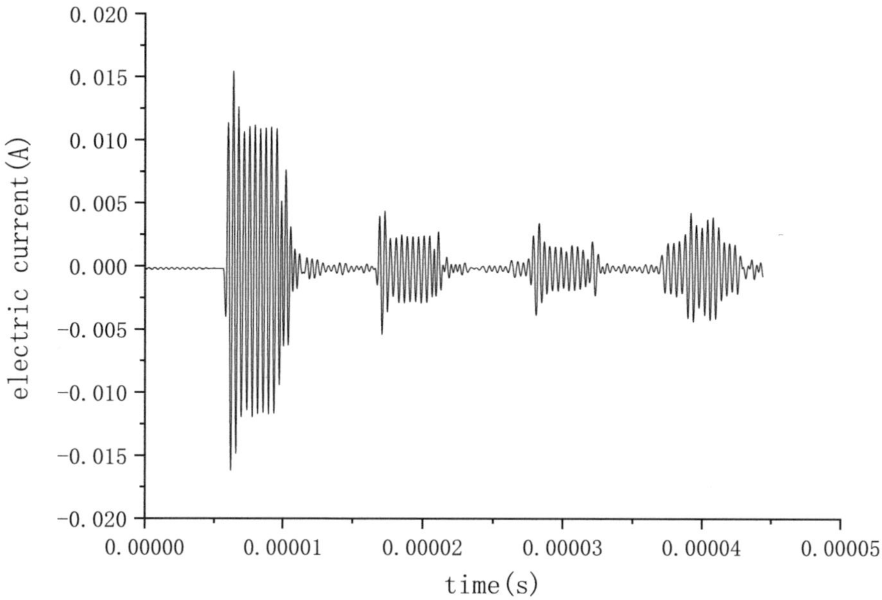

Fig. 2.9 Electric current in aluminum material

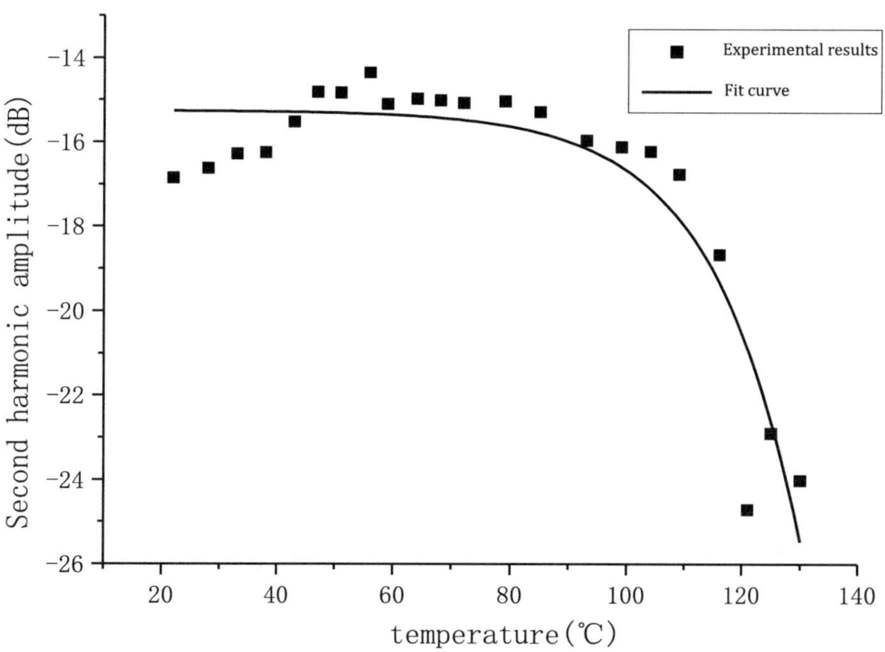

Fig. 2.10 Second harmonic in aluminum material changes with temperature

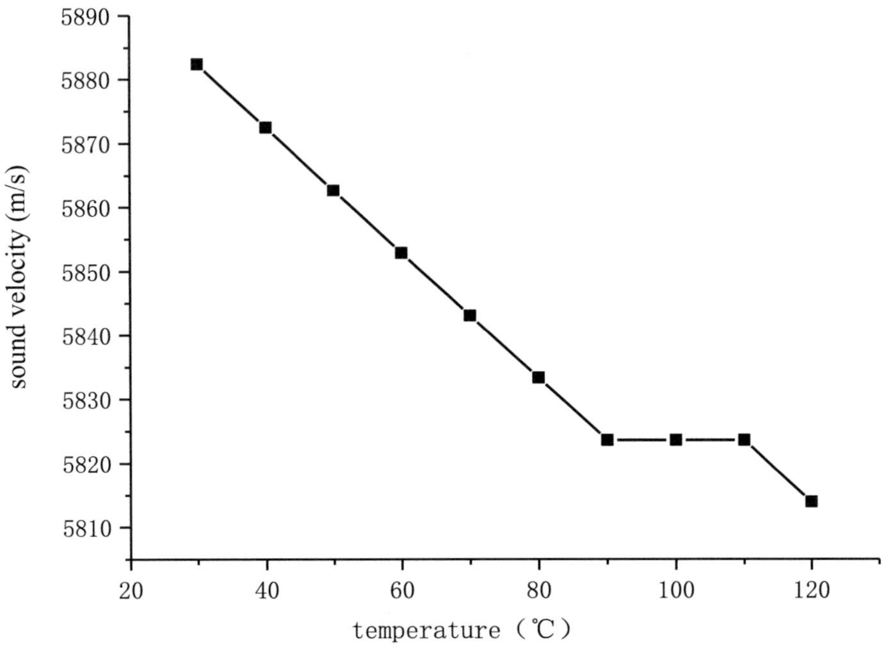

Fig. 2.11 Sound speed in lithium tantalate changes with temperature

Sound Speed

Between 30 and 120 °C, the propagation speed of sound waves in lithium tantalate changes with temperature as shown in Fig. 2.11.

As can be seen from Fig. 2.11, as the temperature increases, the change in the speed of sound in lithium tantalate is not drastic, and it gradually decreases. If the sound speed at 30 °C is taken as the standard, within the range of 120 °C, the change in sound speed is at most only 1.2%.

Fundamental and Second Harmonics

Through the measurement of sound speed and formulas (2.69) and (2.70), the changes in the amplitudes of the fundamental and second harmonics with temperature are obtained.

Figure 2.12 shows the relationship between the amplitude of the fundamental wave in lithium tantalate and temperature. Figure 2.13 shows the relationship between the amplitude of the second harmonic in lithium tantalate and temperature. The solid line is the fitting curve of the data. As can be seen from the figure, as the temperature increases, both the amplitude of the fundamental wave and the second harmonic show a downward trend.

As shown in Fig. 2.14, the ratio of the amplitude of the second harmonic A_2 to the amplitude of the fundamental wave A_1 shows that as the temperature increases, the

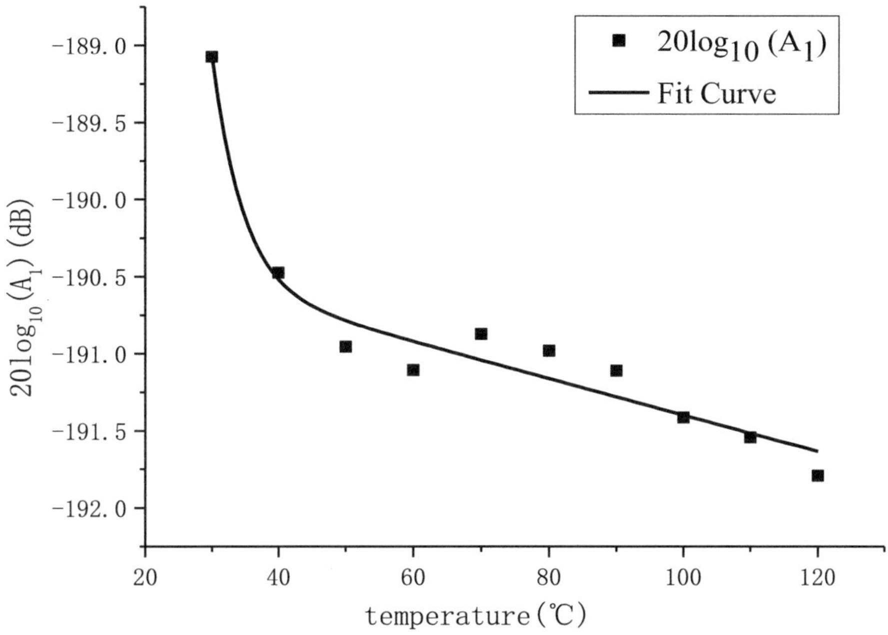

Fig. 2.12 Relationship between the amplitude of the fundamental wave in lithium tantalate and temperature

degree of nonlinearity of the lithium tantalate crystal weakens. Since the temperature characteristics of the lithium tantalate crystal are relatively good, it is more stable compared to the lithium niobate crystal. As the temperature increases, the lithium tantalate crystal shows a weakening of its nonlinear characteristics, indicating that the nonlinear relationship between stress and strain in the structure of the lithium tantalate crystal weakens, making it less likely to generate nonlinear harmonics, and therefore more suitable for linear working ranges.

PIN-PMN-PT Experimental Results

Figures 2.15 and 2.17 are the voltage and current signals received by the oscilloscope for the [001] polarized PIN-PMN-PT crystal at 21 °C. Figure 2.16 is the spectrum signal obtained from the Fourier analysis of the direct wave voltage.

Figures 2.18 and 2.20 are the voltage and current signals received by the oscilloscope for the [111] polarized PIN-PMN-PT crystal at 23 °C. Figure 2.19 is the spectrum signal obtained from the Fourier analysis of the direct wave voltage.

Sound Speed and Sound Attenuation Changes From Fig. 2.21 As can be seen from Fig. 2.22, the speed of sound in [001] polarized crystals decreases with the increase in temperature, but the change is not significant, only 3.57%. The speed of sound in [111] polarized crystals also shows a decreasing trend, but the change is

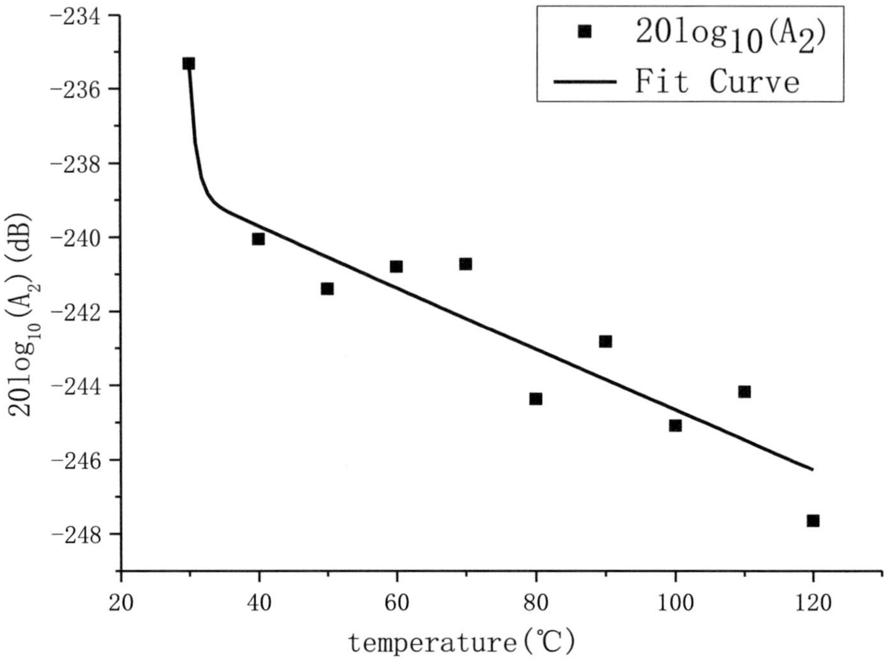

Fig. 2.13 Relationship between the amplitude of the second harmonic in lithium tantalate and temperature

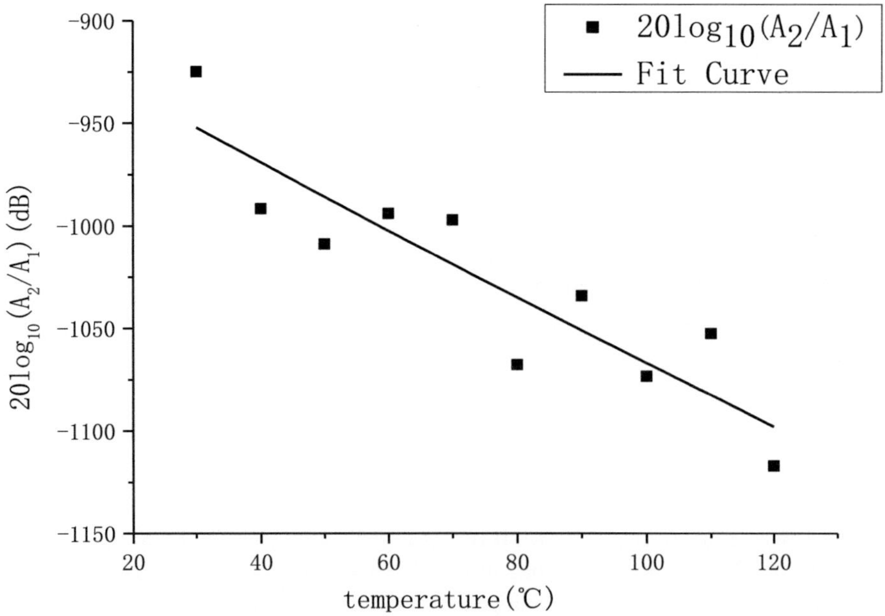

Fig. 2.14 Relationship of A_2/A_1 with temperature

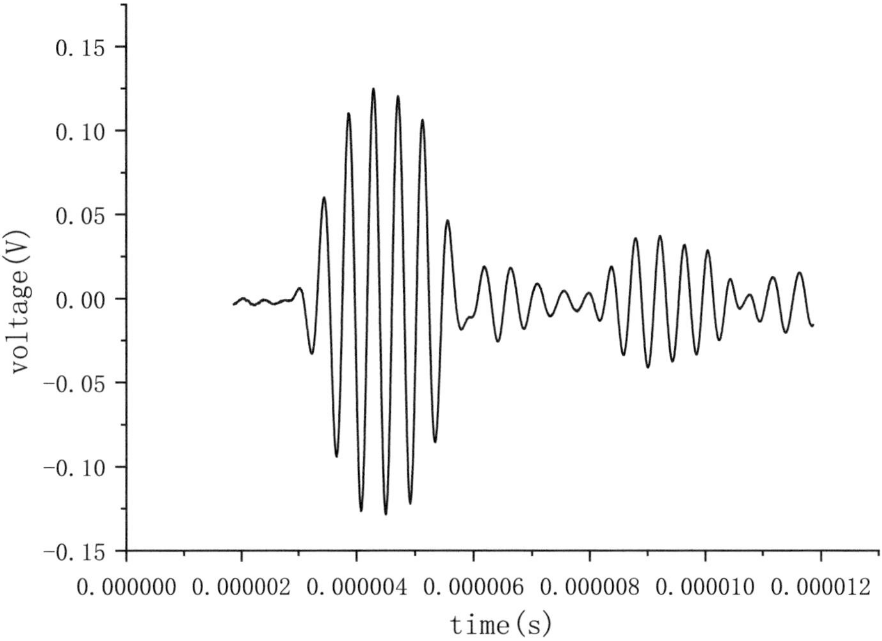

Fig. 2.15 Voltage of [001] polarized PIN-PMN-PT crystal

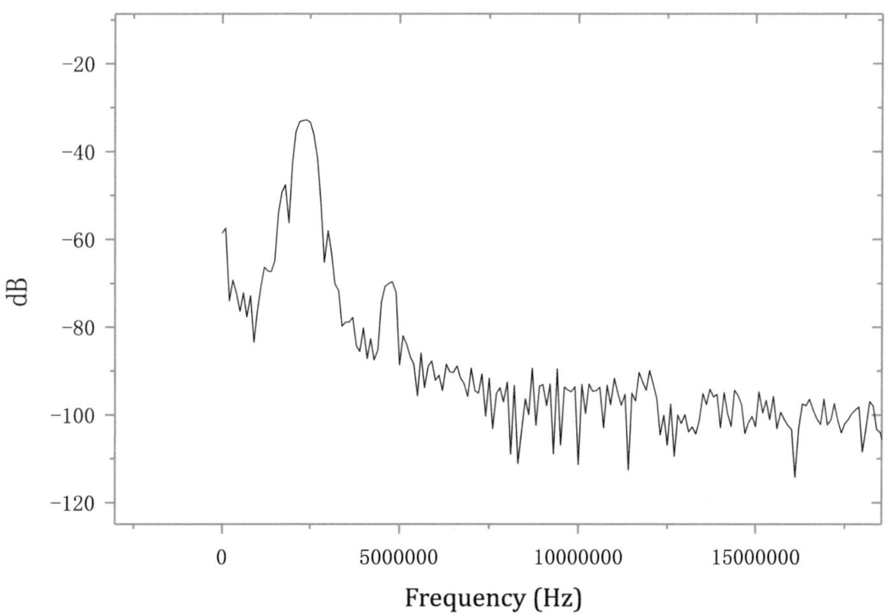

Fig. 2.16 Spectrum of direct wave of [001] polarized PIN-PMN-PT crystal

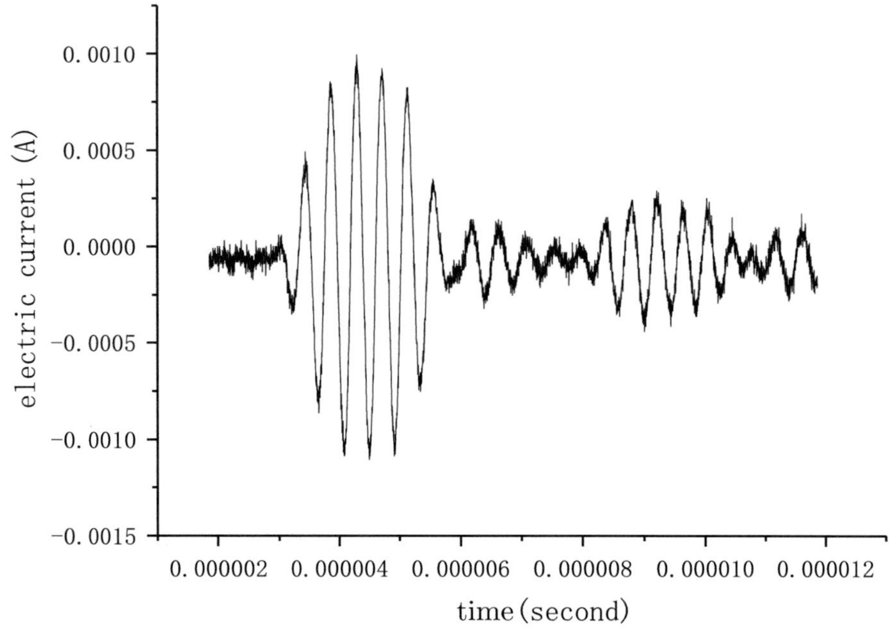

Fig. 2.17 Electric current of [001] polarized PIN-PMN-PT crystal

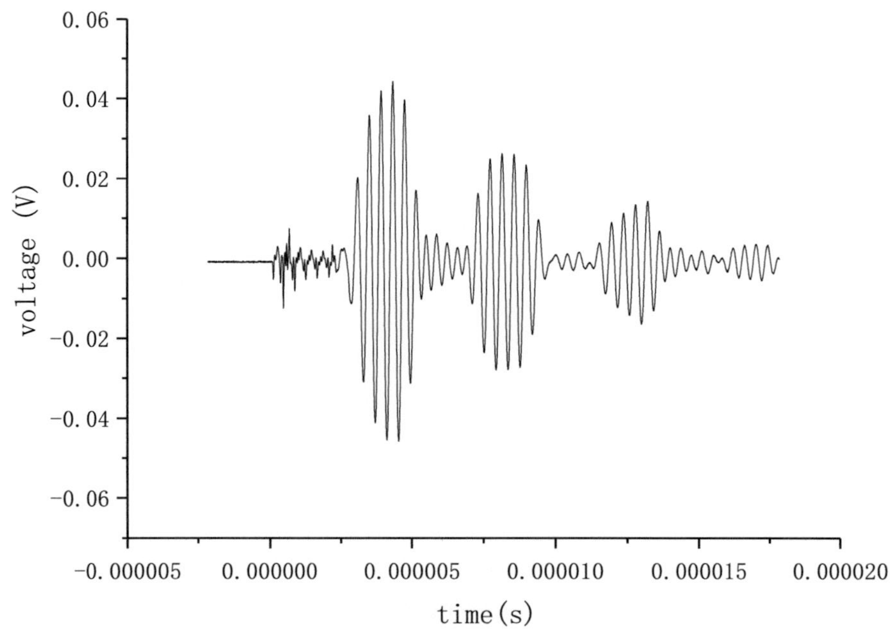

Fig. 2.18 Voltage of [111] polarized PIN-PMN-PT crystal

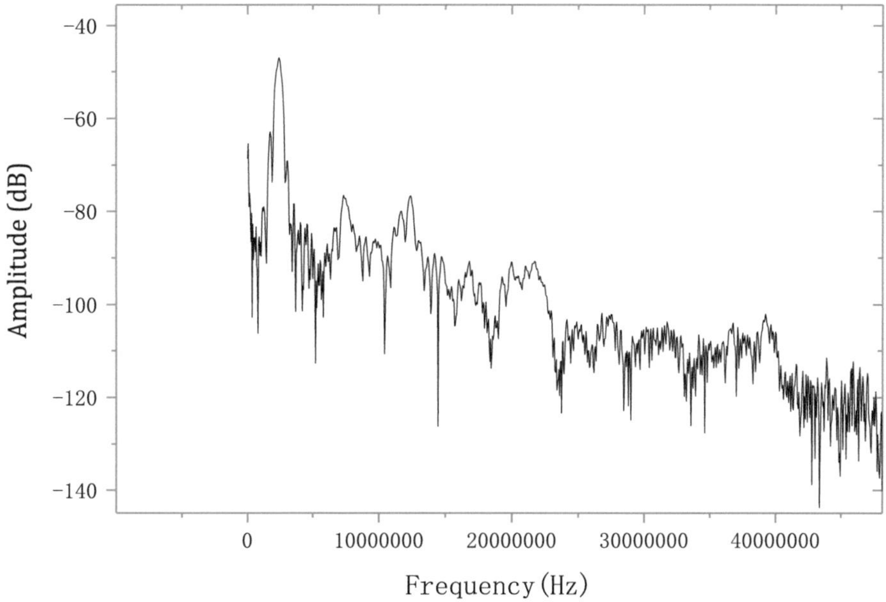

Fig. 2.19 Spectrum of direct wave voltage of [111] polarized PIN-PMN-PT crystal

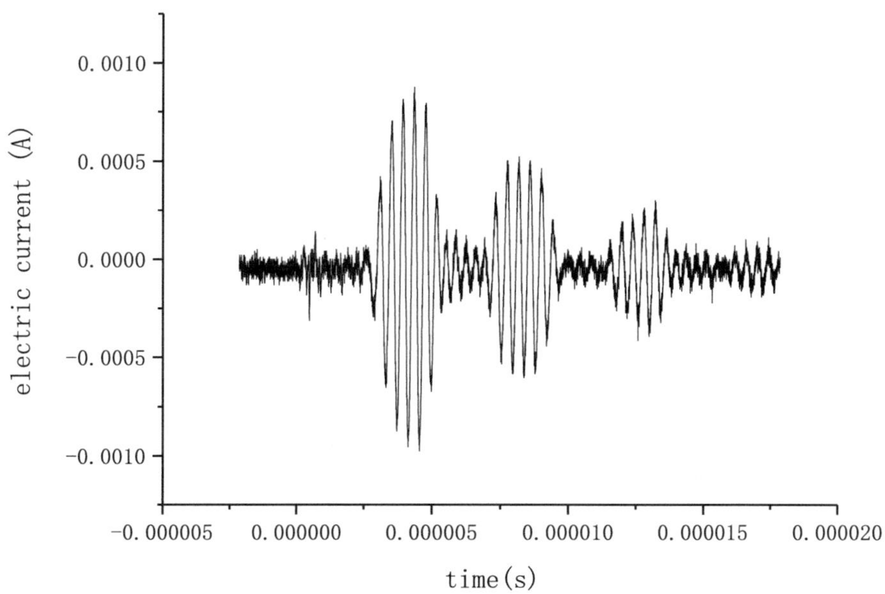

Fig. 2.20 Electric current of [111] polarized PIN-PMN-PT crystal

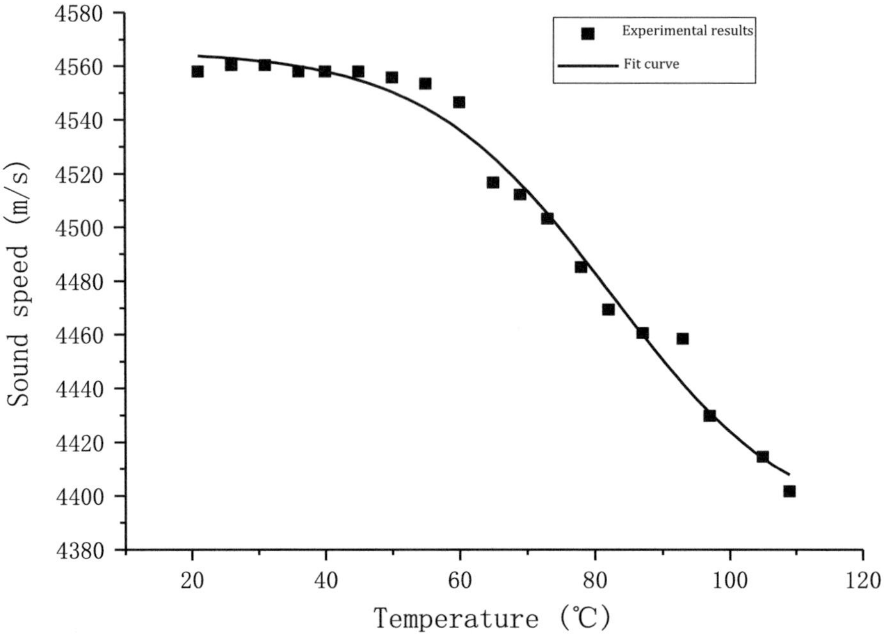

Fig. 2.21 Sound speed change with temperature for [001] polarized PIN-PMN-PT crystal

even smaller, only 1.31%. Especially within 90 °C, it is only 0.6%, which can be considered that the speed of sound has basically not changed. Numerically, the speed of sound in [111] polarized crystals is higher than that in [001] polarized crystals, about 8.8% higher at 20 °C.

From Figs. 2.23 and 2.24, it can be seen that the sound attenuation of [001] polarized crystals decreases first and then increases with the increase in temperature, decreasing within 0–90 °C, and increasing from 90 to 110 °C. The sound attenuation of [111] polarized crystals shows an oscillating trend, with a maximum value around 68 °C. Numerically, the sound attenuation is between 0.15 and 0.55 dB/cm, and the difference is not significant.

Harmonics and Nonlinear Coefficients Change with Temperature From Figs. 2.25 and 2.26, it can be seen that the first and second harmonics of [001] polarized crystals decrease with the increase in temperature, and from Figs. 2.27 and 2.28, it can be seen that the fundamental amplitude of [111] polarized crystals first increases and then decreases, and the second harmonic amplitude also first increases and then decreases, also having a maximum value near 68 °C.

For these two types of crystals, the nonlinear coefficient $\beta = \frac{8}{3}\frac{1}{k^2 a}\frac{A_2}{A_1^2}$, where k is the wave number, $k = \frac{2\pi f}{c}$, a is the length of the sample, A_1, A_2 are the fundamental and second harmonic amplitudes respectively.

From Figs. 2.29 and 2.30, it can be clearly seen that there are significant differences in the changes of the nonlinear coefficients of crystals with different polarization

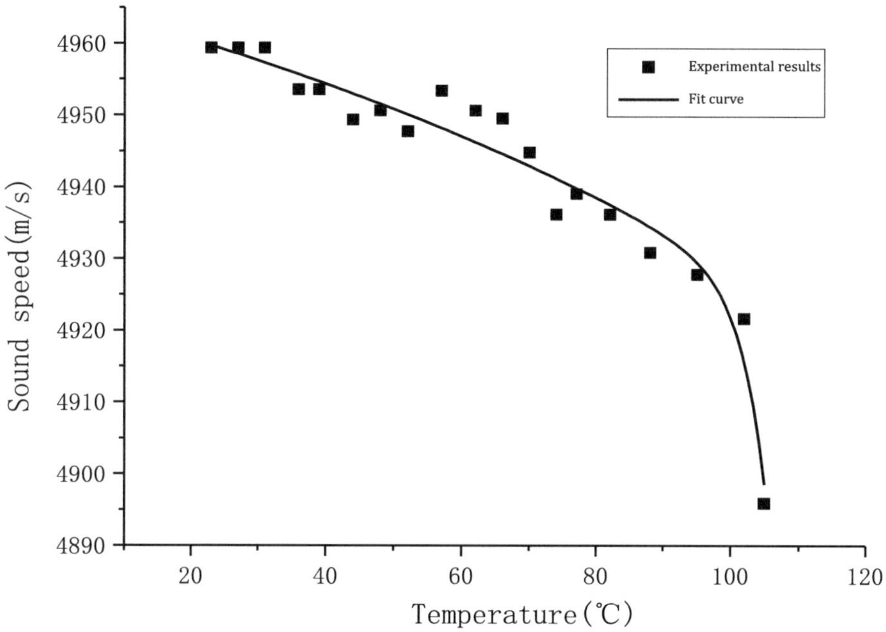

Fig. 2.22 Sound speed changes with temperature for [111] polarized PIN-PMN-PT crystal

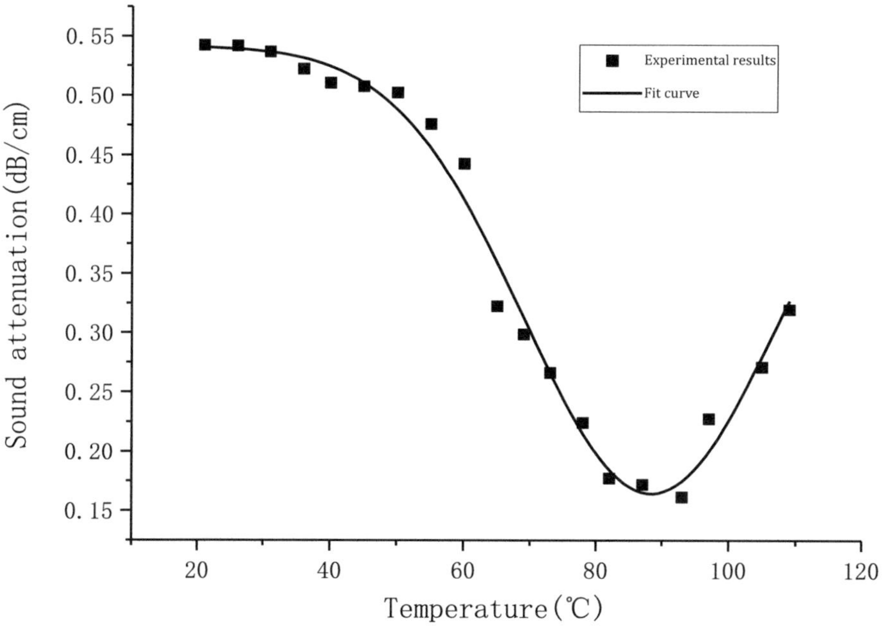

Fig. 2.23 Sound attenuation changes with temperature for [001] polarized PIN-PMM-PT crystal

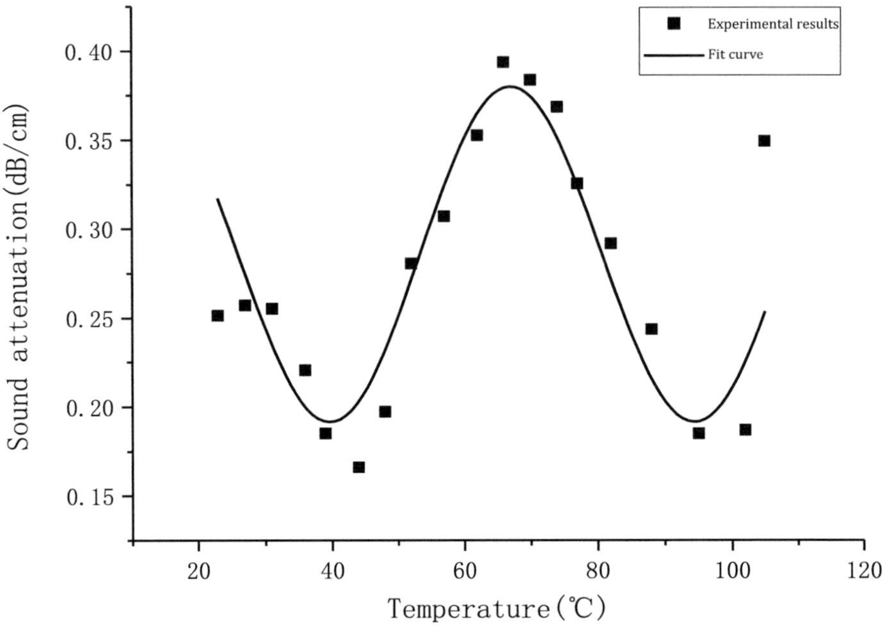

Fig. 2.24 Sound attenuation change with temperature for [111] polarized PIN-PMM-PT crystal

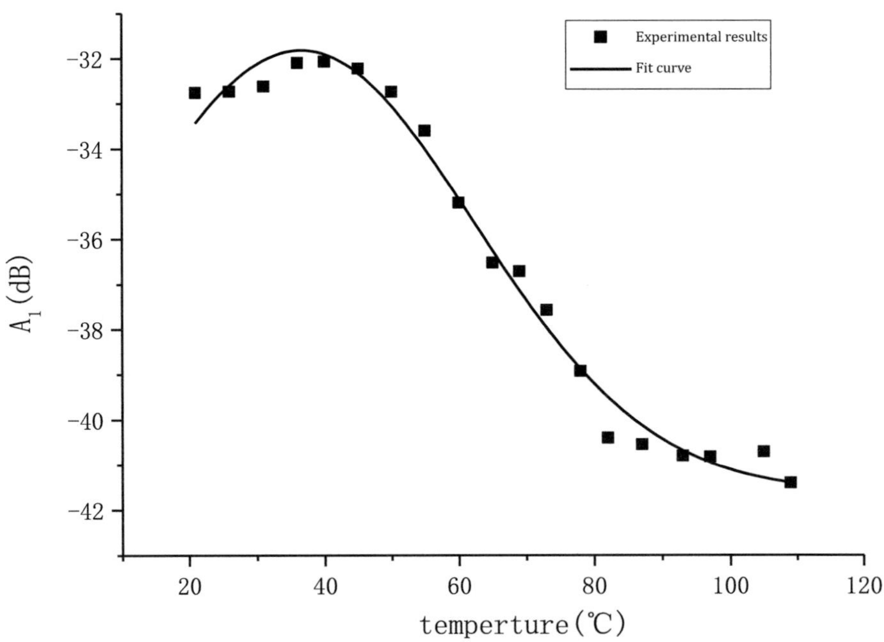

Fig. 2.25 A_1 changes with temperature for [001] polarized PIN-PMM-PT crystal

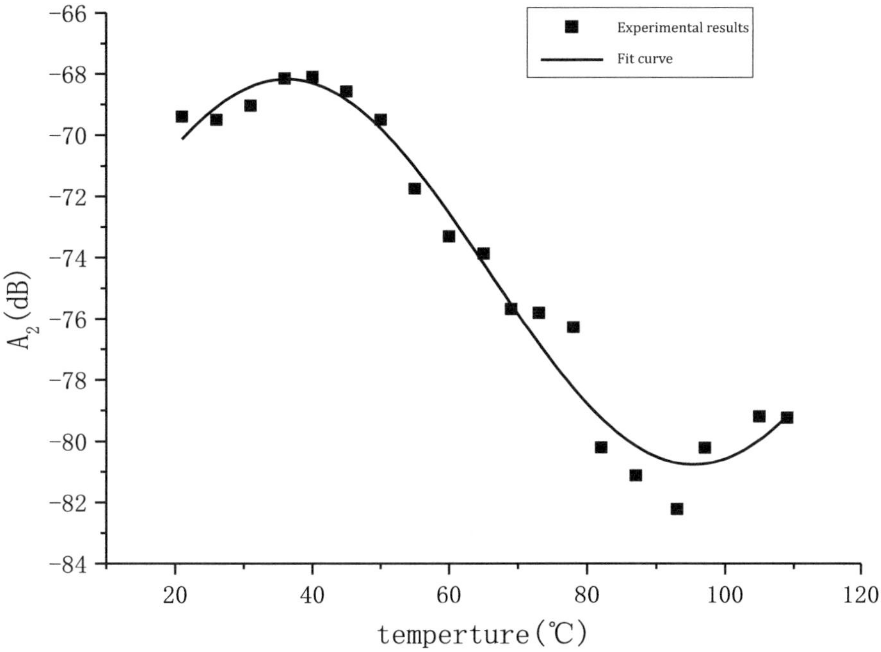

Fig. 2.26 A_2 changes with temperature for [001] polarized PIN-PMM-PT crystal

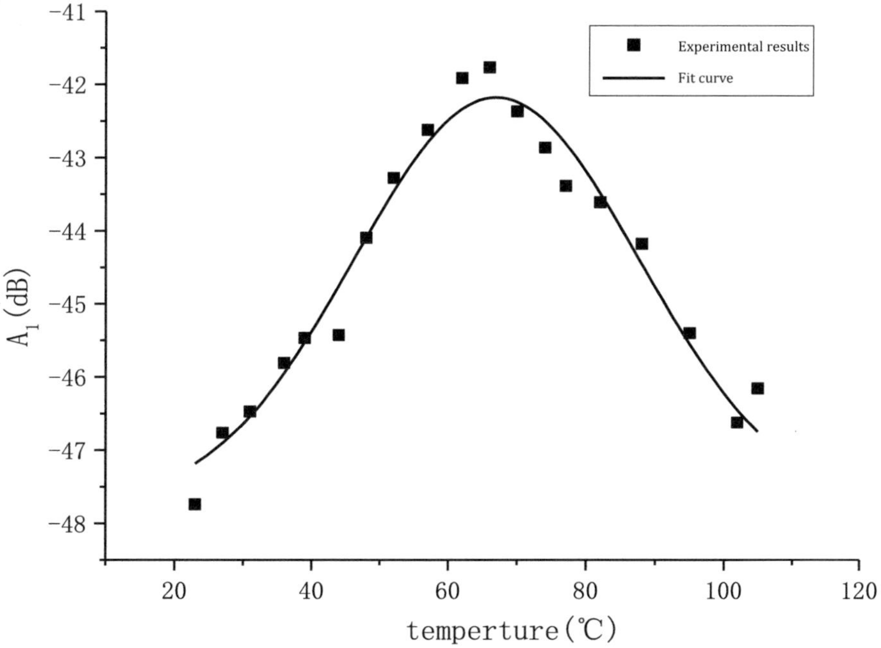

Fig. 2.27 A_1 changes with temperature for [111] polarized PIN-PMM-PT crystal

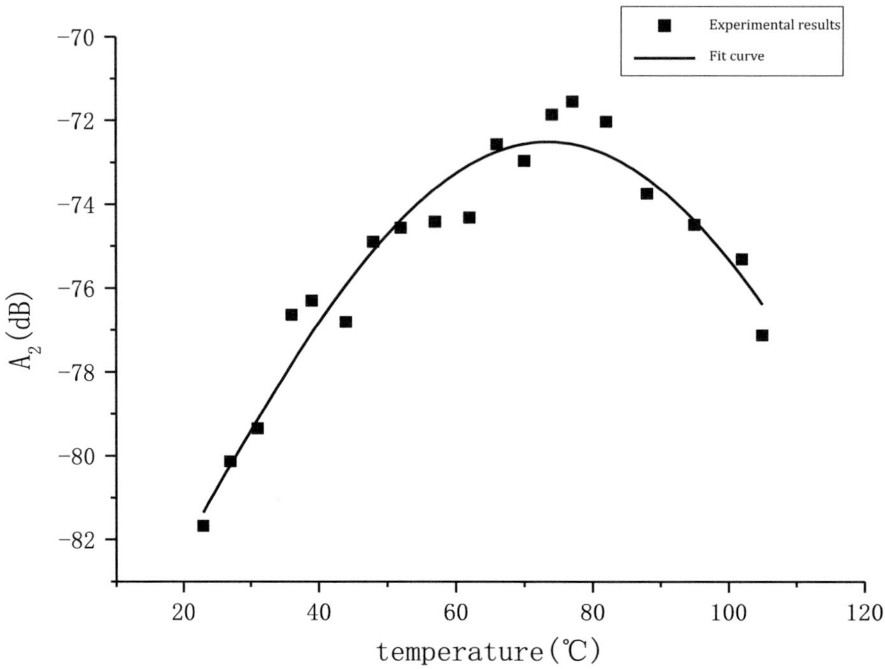

Fig. 2.28 A_2 changes with temperature for [111] polarized PIN-PMM-PT crystal

directions with temperature. The nonlinear coefficient of [001] polarized crystals increases with the increase in temperature, showing a monotonic trend, while the [111] direction polarized samples decrease first and then increase with the increase in temperature, with a minimum value between 60 and 70 °C. Numerically, the β of [001] polarized crystals is between [0.5, 7], while the β of [111] polarized crystals is between [1.75, 5], the range of change is smaller than that of [001] polarized crystals. This indirectly indicates that different polarization directions will lead to different crystal structures, thereby changing some physical properties of the crystal, such as sound speed, sound attenuation, second harmonic, and nonlinear coefficient.

2.5 Ceramic Phase Transition and Nonlinear Effects

Piezoelectric ceramics are widely used in electronics, optics, heat, acoustics and other fields, becoming an important functional material in defense industry, civilian industry and daily life, and one of the main research directions of current high technology. At present, the most widely used piezoelectric ceramics are lead zirconate titanate (PZT) and its ternary or quaternary ceramics. In these ceramics, there is a coexistence area of cubic and tetragonal phases, that is, the morphotropic phase

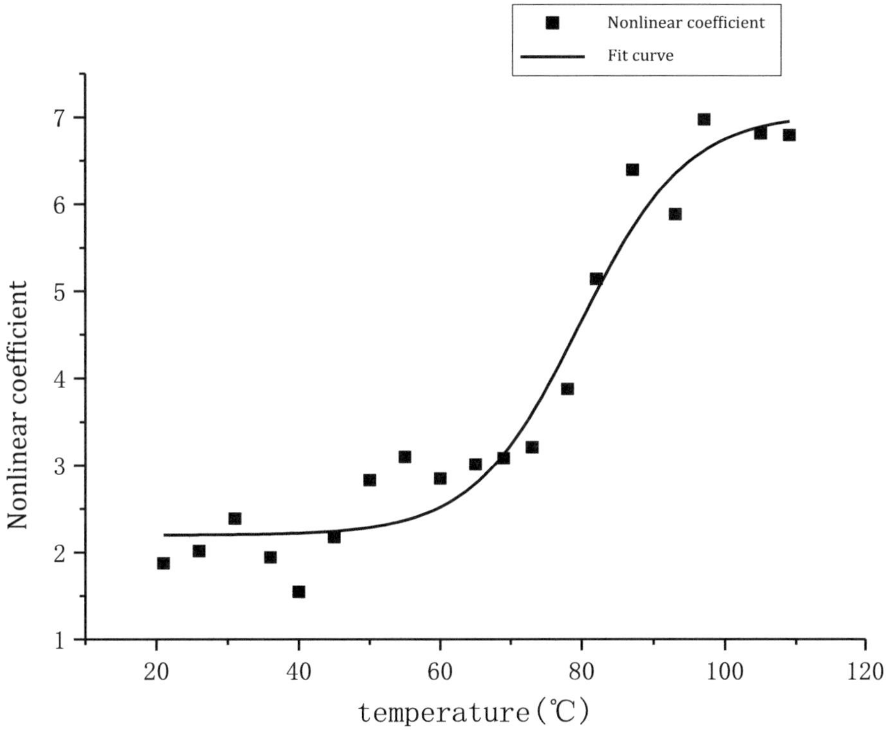

Fig. 2.29 Nonlinear coefficient changes with temperature for [001] polarized PIN-PMM-PT crystal

boundary (MPB), which is the area where strong ferroelectric piezoelectricity occurs. The composition of practical piezoelectric ferroelectric materials often lies near the MPB.

The concept of morphotropic phase boundary (MPB) was first proposed by Jaffe in the early 70s when studying PZT [48].

Theoretically, when the free energy of the three-phase and four-phase are equal, the two phases coexist. The boundary corresponding to this point is called the quasi-homotypic phase boundary. If the composition deviates from this point, the phase with lower free energy can stably exist. In fact, many factors such as changes in the preparation process of materials can cause energy fluctuations, forming a two-phase coexistence in the vicinity of the MPB. Therefore, the quasi-homotypic phase boundary refers to a finite region, not a clear boundary. Lead zirconate titanate $Pb(Zr_x, Ti_{1-x})O_3$ (PZT) piezoelectric ceramics and their derivatives are widely used in ultrasonic transducers and other electromechanical devices due to their high piezoelectric coefficients. Lead zirconate titanate is a solid solution composed of $PbTiO_3$ and $PbZrO_3$, abbreviated as PZT. This material has good performance and has been widely used in pyroelectric infrared detectors and pyroelectric imaging devices. Its main advantages are [49]: 1. Most of the excellent pyroelectric materials are single crystals, and the spontaneous polarization rate of the corresponding polycrystalline

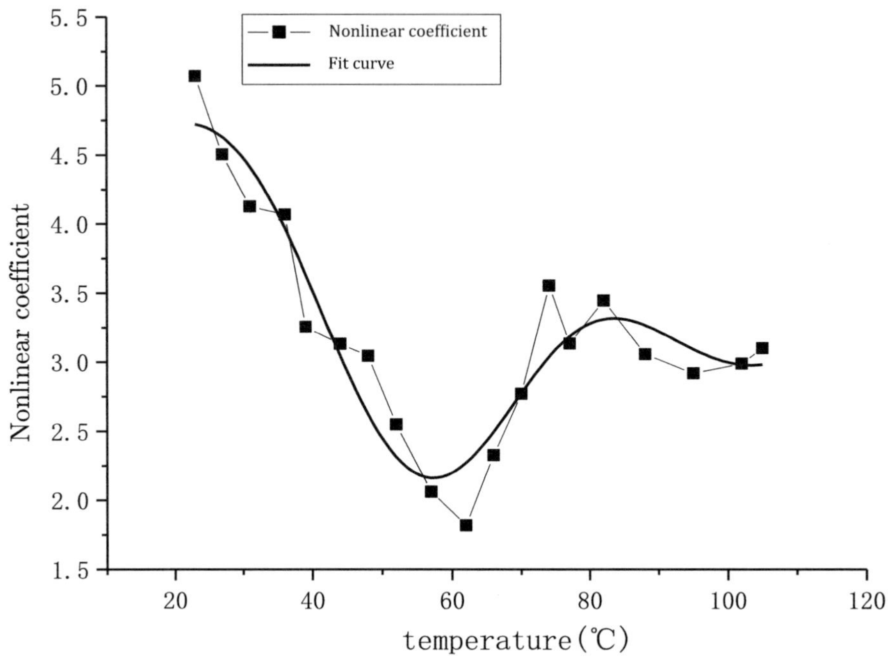

Fig. 2.30 Nonlinear coefficient changes with temperature for [111] polarized PIN-PMM-PT crystal

materials cannot reach half of the corresponding single crystal materials. In PZT, the spontaneous polarization rate of PZT ferroelectric polycrystalline materials that transition from the cubic phase paraelectric phase to the tetragonal phase or rhombohedral phase ferroelectric phase can reach 86.6 and 91.2% of the single crystal materials; 2. In PZT, a large amount of other ions are allowed to be doped and form a solid solution system with PZT, which can improve performance through doping; 3. PZT ceramics are easy to prepare, have a high Curie point, stable performance, and good mechanical strength. When PZT is near the phase boundary of the tetragonal phase and the rhombohedral phase, its coupling coefficient and dielectric constant are the highest among piezoelectric ceramics. This is because near the phase boundary, it is easier to reorient during polarization. Its electromechanical coupling coefficient k_{33} can reach 0.6, and the piezoelectric constant d_{33} can reach 200×10^{-12} C/N. There are two types of commercially available PZT ceramics, one is hard doping, forming oxygen vacancies (anions), and the other is soft doping, causing metal (cation) vacancies. They are classified as hard piezoelectric ceramics (such as PZT-4, PZT-8, which are doped with Mn^{2+}, 3^+ and Fe^{2+}, 3^+ ions respectively) and soft piezoelectric ceramics (PZT-5A, usually doped with Nb^{5+} ions). Soft PZT, due to the addition of high-valence ions such as Nb^{5+}, forms a certain amount of positive ion vacancies in the lattice, which makes the domain wall in the grain easy to move, resulting in a decrease in the coercive field, making the polarization of the ceramic easier, and therefore usually has a higher piezoelectric coefficient. However, the internal friction caused by the movement of the domain wall will cause a large attenuation, resulting in a decrease in the mechanical quality factor and electrical quality factor. However, for hard PZT, the anion vacancies cause the unit cell to shrink, inhibit the movement of the domain wall, reduce the ion diffusion rate, increase the coercive field, make polarization difficult, thereby reducing the internal loss of the material, but this is at the expense of reducing the piezoelectric coefficient. Hard PZT ceramics are usually used in high-power situations, while soft PZT is used in piezoelectric sensors or low-power transmitters. In high-power applications, nonlinear effects cannot be ignored [50]. The nonlinear electrical response and electromechanical response of PZT ceramics under different excitation levels under an external AC electric field have been measured [51]. The nonlinear dielectric response of soft and hard PZT was studied experimentally, which is below the coercive field level, and the effect of defect disorder on the nonlinear dielectric response and hysteresis effect was discussed. Defect disorder is related to dielectric nonlinearity, which can lead to the generation of odd harmonics [52]. The simulation results are based on the nonlinear constitutive relationship, and the results are consistent with the discovered inelastic behavior of soft and hard PZT ceramics [53]. The nonlinear dielectric properties of $Pb(Zn_{1/3}Nb_{2/3})O_3$–$xPbTiO_3$ single crystals have been reported, and the second and third responses of soft and hard PZT ceramics have also been detected in the crystal [54]. On the other hand, Ken Yamada. The study of nonlinear vibrations in piezoelectric materials. For the vibration of a uniform plate, odd harmonics are obvious. However, after heat treatment, the piezoelectric coefficient is no longer uniform, which triggers the second harmonic [55]. This section studies the relationship between the nonlinear behavior of PZT ceramics after polarization and time. The higher harmonics of the piezoelectric resonator come from the geometric constraints and boundary conditions of the resonator, and it is assumed that the material is linear. In our research, the higher harmonics are generated by the nonlinearity of the material during wave propagation. The amplitude of nonlinearity increases with the propagation distance. There is no resonance in harmonic generation technology.

Traditional piezoelectric ceramics have lead volatilization during the preparation process, which not only deviates the stoichiometric ratio of ceramics, but also causes environmental pollution. The development of non-lead-based piezoelectric ferroelectric ceramics is a topic of great practical significance. At present, research on NBT ceramics has become a hot spot in the field of lead-free piezoelectric ceramics research.

Sodium bismuth titanate $Na_{0.5}$ $Bi_{0.5}$ TiO_3 (abbreviated as NBT) is an A-site composite perovskite structure (ABO_3) compound discovered by Smolenskii et al. in 1960 [56], It is considered to be the most promising lead-free piezoelectric ceramic to replace lead-based piezoelectric materials. The Curie temperature of NBT is 320 °C, it is a trigonal structure at room temperature, with a large residual polarization strength $Pr = 38.0\ \mu C/cm^2$ and a high coercive electric field strength $Ec = 7.3$ kV/mm, showing strong ferroelectricity.

Because NBT has a high coercive electric field, the polarization of pure NBT ceramics is difficult, and its piezoelectric performance cannot be fully demonstrated.

Therefore, in order to reduce the coercive electric field and improve its piezoelectric activity, doping replacement is often used. NBT-based lead-free ceramics is one of the hot research areas of lead-free piezoelectric materials. Due to the composite substitution structure of multiple ions at the A-site in NBT ceramics, it has special ferroelectric phase transitions and relaxation characteristics. In order to improve the piezoelectric performance of NBT-based ceramics, most methods adopt various ion doping replacements, which will inevitably change the ferroelectric characteristics and relaxation phase transition characteristics of NBT. The change in the temperature of the ferroelectric-antiferroelectric phase transition will cause a change in the depolarization temperature of the piezoelectric ceramics, which will affect the use of the piezoelectric ceramics. Therefore, this phenomenon must be given enough attention. How to improve the piezoelectric performance while maintaining a sufficiently high phase transition temperature still needs further research.

Barium zirconate titanate (BZT) is also an important lead-free piezoelectric ceramic, which is characterized by a high piezoelectric constant, but the Curie temperature is very low, about 100 °C, which limits its application. $Ba(Zr_y\ Ti_{10y})O_3$ is a solid solution of barium titanate ($BaTiO_3$) and barium zirconate ($BaZrO_3$), with three trigonal, orthorhombic, tetragonal and cubic crystal forms and corresponding three phase transition temperatures. Its phase transition corresponds to the Curie temperature is a function of the Zr component. When $y > 0.15$, the three phase transitions of BZT ceramics shrink [57]; and when $y \sim 0.2$, only one phase exists [58], the phase transition temperature is close to room temperature. Below this temperature, the monoclinic phase is stable, and above this temperature, the cubic phase is stable [59]. When $y > 0.25$, BZT exhibits relaxor behavior [60].

NBT-BZT system ceramics is a solid solution formed by NBT and BZT phases. The addition of BZT can improve the piezoelectric performance of NBT ceramics, and the addition of NBT can increase the Curie temperature of BZT and reduce its sintering temperature, thereby obtaining lead-free piezoelectric ceramics with good piezoelectric performance and meeting the use requirements. In NBT-BZT ceramics, there is a quasi-homotypic phase boundary that transitions from the trigonal phase to the tetragonal phase. The material near the phase boundary has excellent piezoelectric and dielectric properties. Therefore, the study of the NBT-BZT system has high practical significance.

2.5.1 High-Order Nonlinear Changes Caused by Polarization in PZT

2.5.1.1 Experimental System

The hard PZT, PZT-4 and PZT-8, as well as soft PZT, PZT-5A, are used in this section.

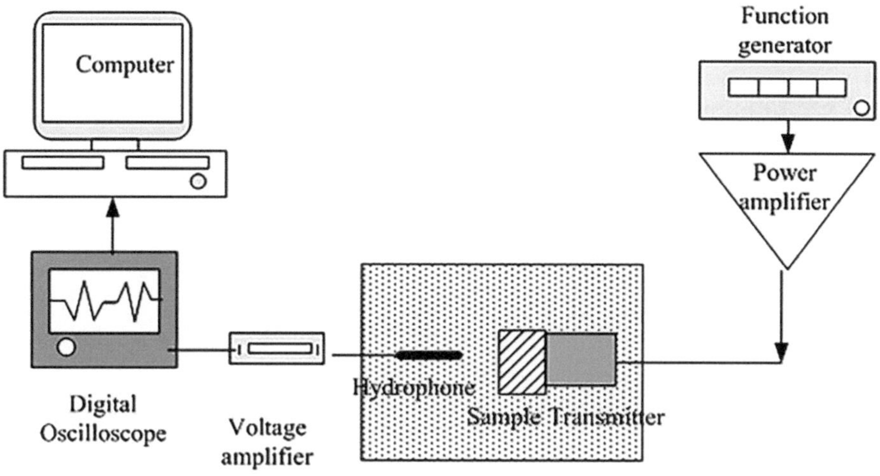

Fig. 2.31 Experimental setup diagram

The experimental setup is shown in Fig. 2.31, where the burst signal is input from the signal generator to the power amplifier. It is then output to the transmitting transducer with a center frequency of 1.2 MHz. After the signal passes through the sample, it is detected by a hydrophone (PVDF, frequency 1–10 MHz, aperture 1 mm) and transmitted to the digital oscilloscope, then collected by the computer. The dimensions of the samples PZT-4, PZT-8, PZT-5 are 3.02 × 2.01 × 1.74, 2.94 × 2.14 × 1.83, 3.22 × 2.26 × 1.97 cm^3 respectively. The samples are immersed in silicone oil and polarized for 20 min at 140 °C under a 4 kV/mm electric field. The direction of wave propagation is the same as the polarization direction.

2.5.1.2 Experimental Results

In the absence of a sample, the corresponding waveform and spectrum are received by the hydrophone, as shown in Fig. 2.32. In this case, the second and third harmonics of the transducer are very small and can be ignored.

After the sound wave passes through the PZT-4 sample before polarization, the waveform and its spectrum are as shown in Fig. 2.33, and the waveform and its spectrum after the PZT-4 sample is polarized for one hour are as shown in Fig. 2.34. In both cases, due to the nonlinearity of the ceramic sample, the second and third harmonics can be clearly observed.

Table 2.4 summarizes the relationship between the harmonics and time when the wave propagates in PZT ceramics. In the case before polarization, the amplitude of the third harmonic is smaller than that of the second harmonic, but after polarization, the amplitude of the third harmonic is larger than that of the second harmonic. The

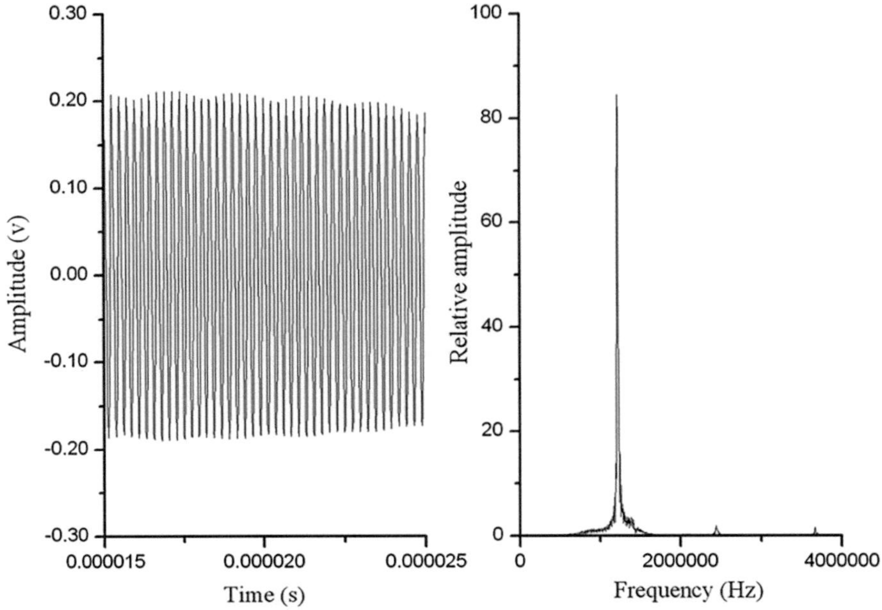

Fig. 2.32 Received waveform and its spectrum in the absence of a sample

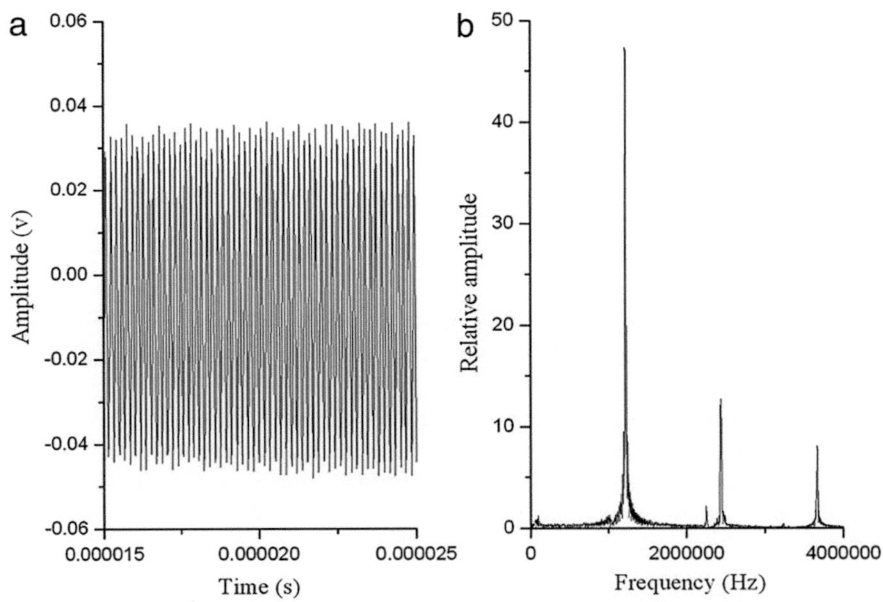

Fig. 2.33 Received waveform and its spectrum after the sound wave passes through the unpolarized PZT-4 sample

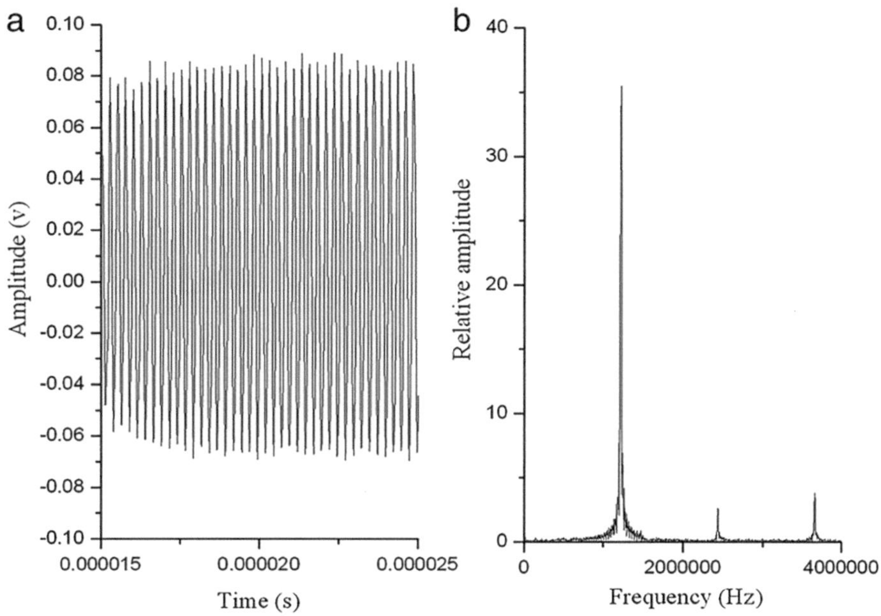

Fig. 2.34 Received waveform and its spectrum after the sound wave passes through the PZT-4 sample polarized for 1 h

Table 2.4 Ratio of the amplitude of the second and third harmonics to the fundamental amplitude after different polarization times for each sample

	Unpolarized		Polarized for 1 h		Polarized for 1 day		Polarized for 2 days		Polarized for 3 days	
	p_2/p_1	p_3/p_1	p_2/p_1	p_3/p_1	p_2/p_1	p_3/p_1	p_2/p_1	p_3/p_1	p_2/p_1	p_3/p_1
PZT-4	0.268	0.171	0.074	0.107	0.080	0.121	0.082	0.123	0.084	0.125
PZT-8	0.212	0.175	0.085	0.121	0.093	0.129	0.095	0.133	0.092	0.134
PZT-5A	0.201	0.076	0.065	0.126	0.070	0.136	0.073	0.139	0.072	0.141

longer the polarization time, the more stable the nonlinearity of PZT, which saturates after 2 days of polarization and reaches a stable value.

2.5.1.3 Analysis and Conclusion

Before polarization, ceramic materials are considered isotropic, and the amplitude of the second harmonic should be greater than the third harmonic, which is a characteristic of ordinary solid nonlinearity. As is well known, elasticity can be described by the expansion of elastic strain. For ordinary solids, the linear part dominates, and the higher the order of strain, the smaller the contribution of the high-order part. If there are micro-cracks in the solid, the nonlinear behavior will change because the impact of micro-cracks on high-order nonlinearity is higher than that on low-order nonlinearity, which we call non-classical nonlinearity. After polarization, ferroelectric materials exhibit strong anisotropy due to non-zero polarization. Due to the transformation of domains during polarization, mechanical damage may occur, so the classical nonlinearity of the material will also become non-classical nonlinearity, and there is also a hysteresis relationship between stress and strain. Due to non-classical nonlinearity, the growth rate of the third harmonic amplitude is greater than the second harmonic, resulting in a higher third harmonic amplitude than the second harmonic. This figure is consistent with the results in reference [50]. This phenomenon is more pronounced in soft PZT (see Table 2.4), because the transformation of domains in soft PZT is easier than in hard PZT.

This section studies the nonlinear changes of soft PZT and hard PZT ceramics before and after polarization. The results show that before polarization, the nonlinearity is classical nonlinearity, and the amplitude of the second harmonic is greater than the third harmonic. After polarization, the nonlinearity becomes non-classical nonlinearity, and the amplitude of the third harmonic is greater than the second harmonic. This phenomenon is more pronounced in soft PZT than in hard PZT. The change of domains leads to the generation of micro-cracks, and produces this interesting nonlinear phenomenon. Nonlinearity will also change over time, but it tends to stabilize after 2 days of polarization.

Relationship Between NBT and BZT Ceramic Phase Transition and Sound Speed

2.5.1.4 Experimental System

The transducer sensitivity calibration is shown in Fig. 2.5, and the experimental framework is shown in Fig. 2.6. The experimental samples are NBT, BZT-7, and BZT-15. The size of NBT is 11.435 × 11.845 × 21.75 mm^3, the size of BZT-7 is 23.582 × 11.7 × 9.23 mm^3, its composition is $BaZr_{0.07}\ Ti_{0.93}\ O_3$, the size of BZT-15 is 10.37 × 8.70 × 24.63 mm^3, its composition is $BaZr_{0.15}\ Ti_{0.85}\ O_3$.

2.5.1.5 Experimental Results

The dielectric constant K of NBT, BZT-7, and BZT-15 changes with temperature as shown in Figs. 2.35, 2.36, and 2.37. The relationship between the sound speed of NBT ceramics and temperature is shown in Fig. 2.38. During the temperature rise from 20 to 90 °C, the sound speed of NBT drops from 5220 to 5140 m/s, a change of only 1.53%, which can be considered as no change.

However, the relationship between the sound speed of BZT-7 and BZT-15 ceramics and temperature is completely different, as shown in Figs. 2.39 and 2.40.

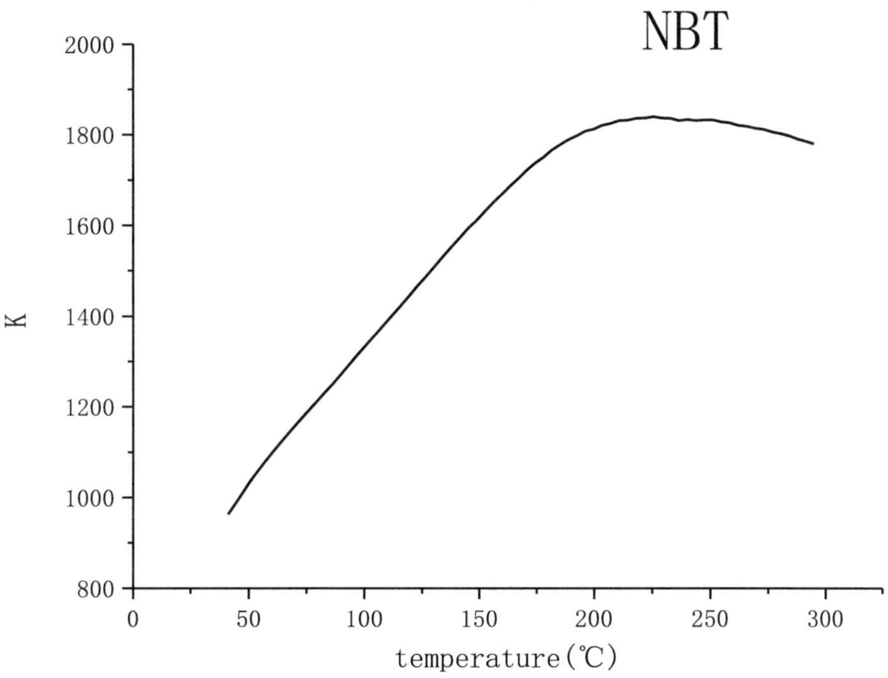

Fig. 2.35 Dielectric constant K of NBT ceramics changes with temperature

Figure 2.39 shows that around 70 °C, the sound speed of BZT-7 changes significantly, with a change of up to 12.5%, and the phase transition temperature of BZT-7 is 70 °C. Figure 2.40 shows that around 60 °C, the sound speed of BZT-15 ceramics also changes significantly, which is more obvious than BZT-7, with a change of up to 34.6%, and the phase transition temperature of BZT-15 is 58 °C.

From the existing data, it is known that the phase transition temperature of NBT is relatively high, above 200 °C. In the range of 0–90 °C, the speed of sound of NBT hardly changes, which is consistent with the fact that the phase transition temperature of NBT is high.

2.6 Acoustic Memory Phenomenon in Crystals

Lithium niobate ($LiNbO_3$, abbreviated as LN) has always been a hotspot in basic science and applied research due to its many excellent physical and chemical properties. It has excellent piezoelectric, nonlinear optical, and electro-optical properties, and is widely used in surface acoustic wave, electro-optic, acousto-optic, storage and other devices [61–63]. To the polarization direction of the crystal, the acoustic memory phenomenon of lithium niobate crystals has different amplitude-frequency

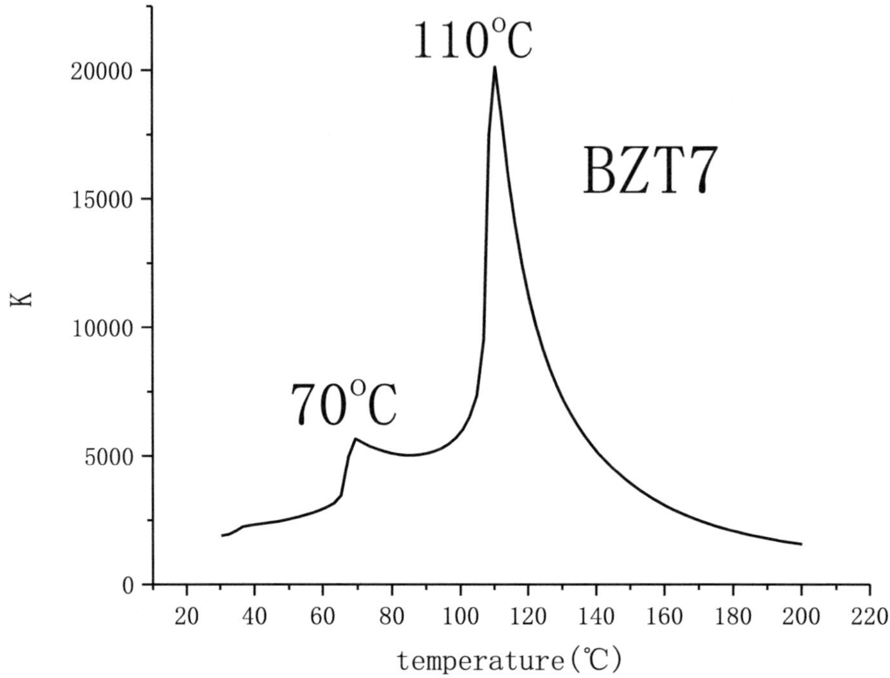

Fig. 2.36 Dielectric constant K of BZT-7 ceramics changes with temperature

characteristics from the common ferroelectric memory phenomenon [64–66], and is also different from the common photoelectric memory phenomenon [66, 67]. In 2002, Breazeale [67, 68] and others discovered the acoustic memory phenomenon of lithium niobate, that is, when lithium niobate is excited by ultrasonic waves, a part of the sound energy is stored in the lithium niobate crystal, and is emitted after a period of time (about 40–70 μs), this signal is the acoustic memory signal, and this phenomenon is the acoustic memory phenomenon. However, the ultrasonic frequency band of their research is 16–30 MHz.

The acoustic memory phenomenon is a new physical phenomenon. In lithium niobate crystals, we observed the acoustic memory phenomenon in the frequency range of 2.5–10 MHz. The acoustic memory phenomenon is determined by the dielectric properties of the ferroelectric medium. The experiment found that as the temperature rises, the amplitude of the acoustic memory signal increases, and vice versa. In addition, we found the amplitude hysteresis phenomenon of the direct wave signal's first harmonic and second harmonic, the amplitude hysteresis phenomenon of the acoustic memory signal, and the nonlinear frequency offset. The above three characteristics and the non-exponential decay relationship between echo and distance all prove the non-classical nonlinear acoustics characteristics of lithium niobate crystals. When the excitation signal is large enough, or each modulation pulse in the excitation signal contains enough cycles, we first discovered multiple memory signals. Based

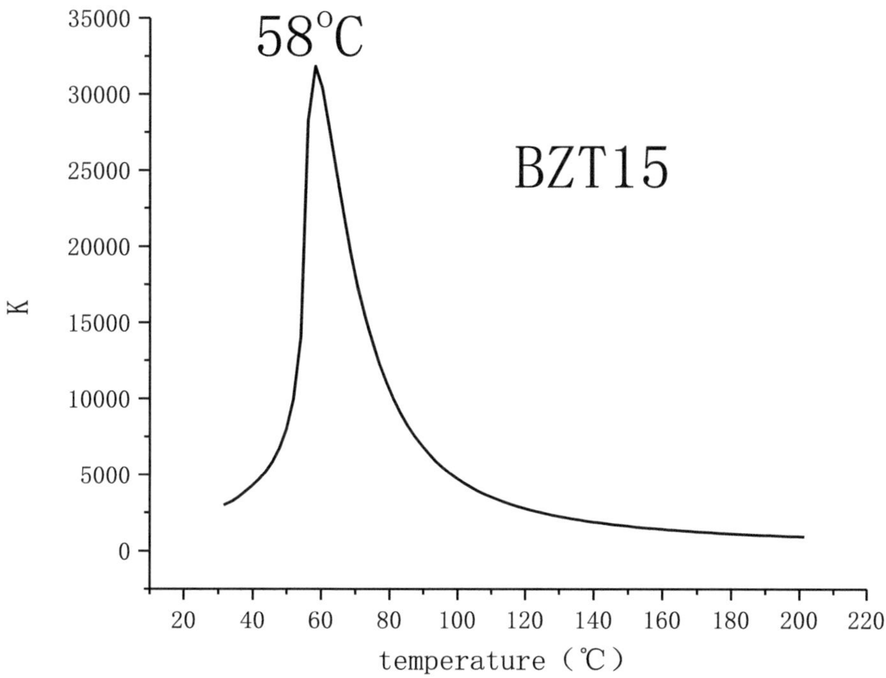

Fig. 2.37 Dielectric constant K of BZT-15 ceramics changes with temperature

on the changes in the crystal structure after ultrasonic excitation and non-classical nonlinear theory, we qualitatively explained the above experimental phenomena [69].

2.6.1 Experimental Method

The experimental device is shown in Fig. 2.41. The modulation signal generated by the function generator is amplified by the power amplifier (center frequency is 2.5 MHz, pulse repetition frequency is 100 Hz, pulse width is 4 μs) to excite the lead zirconate titanate (PZT) transmitting transducer (Shantou Ultrasound Co., Ltd., aperture 28 mm), generating a longitudinal ultrasonic wave with a center frequency of 2.5 MHz. After the ultrasonic wave propagates in the lithium niobate crystal, the signal is received at the other end of the sample with the same PZT transducer. And the received signal is received and stored by the digital storage oscilloscope. The sound axis offset angle of the transmitting and receiving transducers is less than 2^0.

First, fix the transmitting and receiving transducers and the lithium niobate test block on the experimental bracket to ensure that the axis of the lithium niobate test block and the axis of the transmitting and receiving transducers are kept in a straight line, to ensure that the relative position of the lithium niobate sample and the

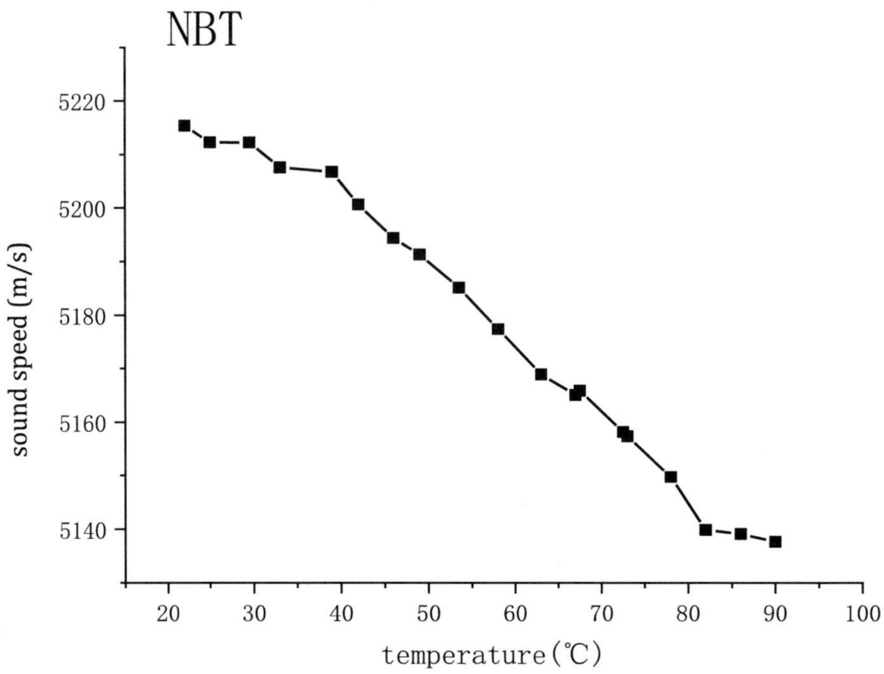

Fig. 2.38 Relationship between the sound speed of NBT ceramics and temperature

transmitting and receiving transducers does not change during the experiment, and to ensure that the pressure between the lithium niobate test block and the transmitting and receiving transducers does not change. Vaseline is used as the ultrasonic coupling agent. Two lithium niobate samples are used as experimental samples. The first lithium niobate crystal is cylindrical, with a radius of 2.0 cm and a length of 2.5 cm, and the polarization direction is the axial direction. The second lithium niobate sample is rectangular in shape, with lengths in the X, Y, and Z directions of 3.3, 3.4, and 3.4 cm, respectively, and the polarization direction is along the Z direction.

The acoustic contact surface of the experimental sample should be defect-free and smooth. The surfaces of the two experimental samples should be parallel and regular to avoid wall reflection and possible waveform conversion. In non-parallel samples, after several echo signals, a signal similar to the memory signal will appear, but this is not a memory signal, but is caused by wall reflection and waveform conversion of the sample. In experimental samples with uneven density, uneven sound propagation speed may also produce a shape similar to the acoustic memory signal. However, these are not acoustic memory signals.

We first observed acoustic memory signals in cylindrical lithium niobate samples. At this time, the amplitude of the excitation signal from the signal generator was 200 mv, the frequency was 2.5 MHz, each pulse signal contained 8 cycles, the repetition frequency of the pulse was 100 Hz, and the experimental temperature was 20 °C. Under the same experimental conditions, we measured the waveform of the

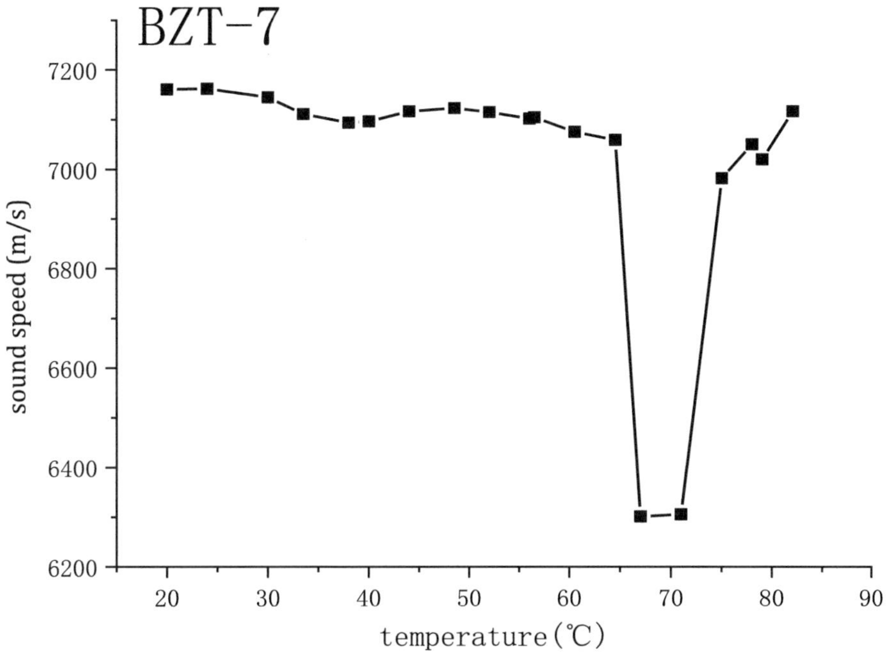

Fig. 2.39 Relationship between the sound speed of BZT-7 ceramics and temperature

received signal in an aluminum block and found no memory signal, indicating that the memory signal observed in lithium niobate was not due to errors in the experimental system. To illustrate the relationship between the amplitude of the memory signal and the polarization direction of the crystal, we compared the amplitudes of the memory signals in the polarized and non-polarized directions in cubic crystals.

At room temperature, when the excitation voltage gradually changes from 100 to 1000 mv, and then decreases to 100 mv. By comparing the changes in the amplitude of the memory signal during the process of increasing and decreasing the excitation voltage, we obtained the hysteresis curve of the amplitude of the acoustic memory signal. We also obtained the hysteresis curves of the fundamental and second harmonic amplitudes of the direct wave. Since the transmission and reception performance of the transducer will change with the change in excitation voltage, it is necessary to correct the directly measured experimental results based on the changes in transducer performance.

When the excitation voltage is 200 mv, when the number of cycles of each excitation pulse signal increases from 4 to 64, we found multiple segments of memory signals. To avoid waveform superposition between the direct wave and the reflected wave, the limit value of the number of cycles is 67 based on the sample thickness. After each change in the number of cycles, it takes ten minutes for the transmission and reception performance of the transducer to fully stabilize.

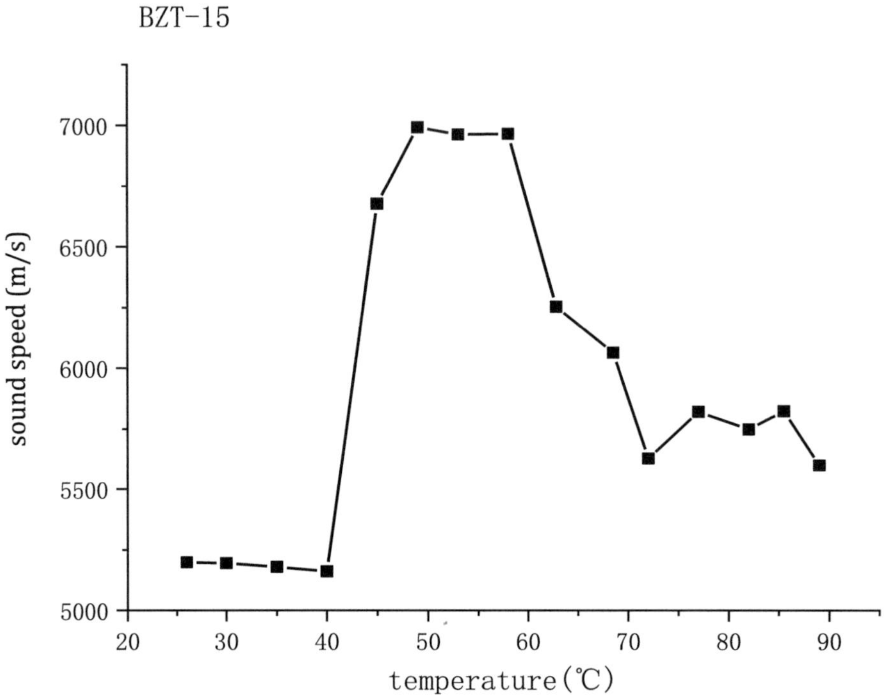

Fig. 2.40 Relationship between the sound speed of BZT-15 ceramics and temperature

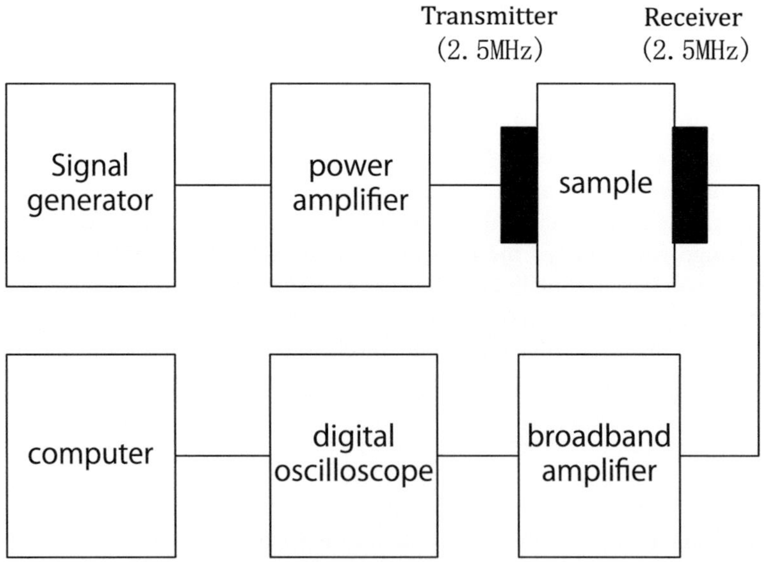

Fig. 2.41 Schematic diagram of the experimental system

An ultrasonic wave is emitted using a PZT ultrasonic transducer with a center frequency of 2.5 MHz. A 2.5 MHz ultrasonic transducer is used as the receiving transducer to receive the signal, and a digital filter with a passband of 2–3 MHz is used to extract the first harmonic. A 5 MHz ultrasonic transducer is used as the receiving transducer, and a digital filter with a passband of 4–6 MHz is used to extract the second harmonic. By gradually changing the amplitude of the excitation voltage, we obtained the hysteresis phenomenon of the fundamental and second harmonic ultrasonic amplitudes of the direct wave.

In the experiment of measuring the relationship between the memory signal and temperature, an oven is used as the heat source, and it is naturally cooled. Since Vaseline melts at temperatures above 40^0, a high-temperature coupling agent is used instead of Vaseline as the experimental coupling agent.

In order to make the temperature distribution inside the test block uniform, the heating and cooling process is very slow. During the experiment, it takes about 15 min for each temperature increase of 5 °C and about 20 min for each decrease of 5 °C. A pair of temperature probes are placed against the top and bottom surfaces of the lithium niobate test block, and their average value is taken as the actual temperature inside the lithium niobate test block. To avoid experimental errors caused by repeatedly fixing the test block, the experimental bracket, along with the transducer and lithium niobate test block already fixed on the bracket, are placed in the oven for heating and cooling. The operating temperature range of the ultrasonic transducer is limited, so the temperature cannot be too high during the experiment, and the temperature range of this experiment is 5–55 °C. Within the operating temperature range of the transducer, the transmission and reception characteristics will also be affected by the temperature, so it is necessary to measure the temperature amplitude characteristics of the experimental system and correct the measurement results.

At excitation voltages of 100 mv and 1 v, the amplitudes of the direct wave and multiple echoes were compared, and the non-exponential decay phenomenon of the lithium niobate crystal was obtained.

2.6.2 Experimental Results

2.6.2.1 Memory Signal

In the cylindrical lithium niobate crystal, the acoustic memory signal at 2.5 MHz is shown in Fig. 2.42. After seven echoes, the echo signal decays to the noise level. After a period of time (10–20 μs), an envelope signal appears, which is the acoustic memory signal. The acoustic memory signal is generated by releasing the energy stored in the crystal again. When an aluminum block is used as the test sample, the waveforms of the direct wave and multiple echoes are shown in Fig. 2.43. The echo signal decays exponentially to a very small level. No acoustic memory signal is generated. This indicates that the acoustic memory signal is not generated by errors in the experimental system, but is determined by the characteristics of lithium niobate

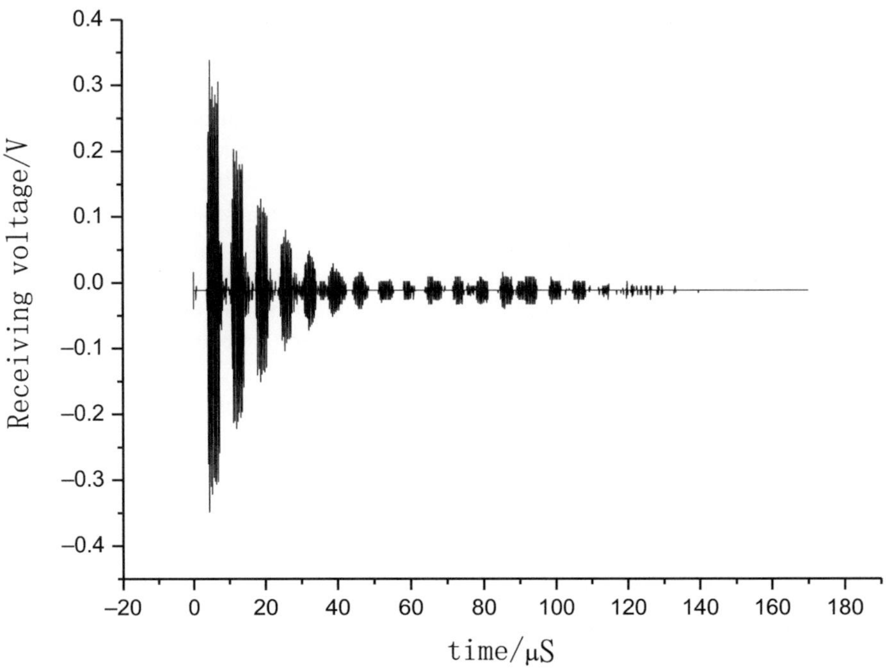

Fig. 2.42 Acoustic memory phenomenon in cylindrical lithium niobate sample

itself. Figure 2.44 analyzes the frequency of the acoustic memory signal, and we found that the frequency of the acoustic memory is 2.44 kHz. This indicates that there is a frequency offset of 60 kHz in the memory signal. This is determined by the non-classical nonlinearity of the lithium niobate crystal, which further indicates that the acoustic memory signal is not caused by multiple reflections and waveform conversions of ultrasonic waves in the sample.

We also observed memory signals when the excitation signal was 5 MHz, as shown in Fig. 2.45a. When a 10 MHz ultrasonic transducer is used to receive the signal, the observed memory signal is as shown in Fig. 2.45b.

2.6.2.2 Relationship Between Acoustic Memory Signal and Polarization Direction

We studied the relationship between the acoustic memory signal and the crystal polarization direction in a rectangular lithium niobate sample. The polarization direction of this crystal is in the **Z** direction. As shown in Fig. 2.6a, the acoustic memory signal in the **Z** direction is larger than that in the non-polarized **Y** direction, as shown in Fig. 2.46b. This indicates that the acoustic memory signal is related to the polarization direction.

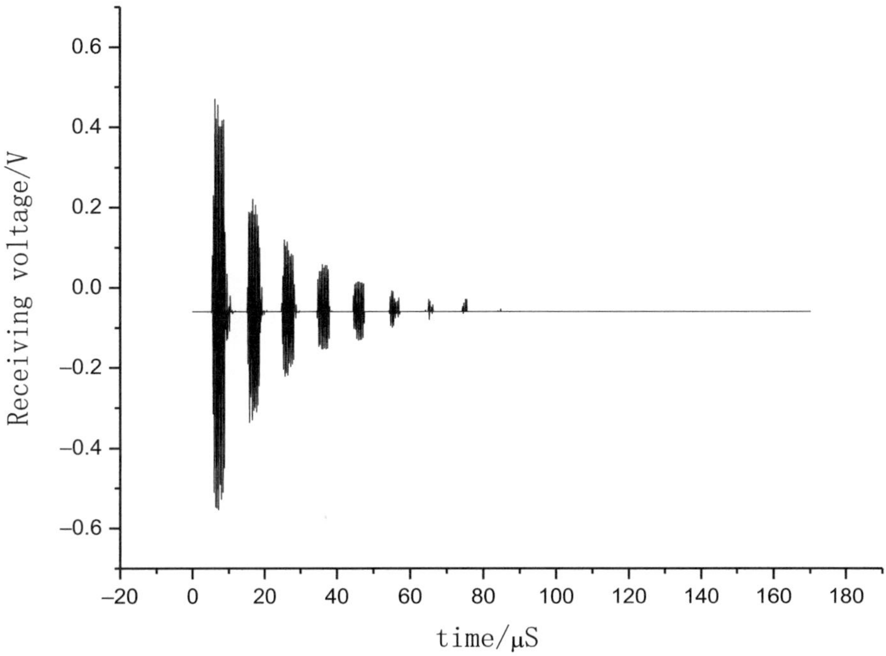

Fig. 2.43 Each echo in aluminum material

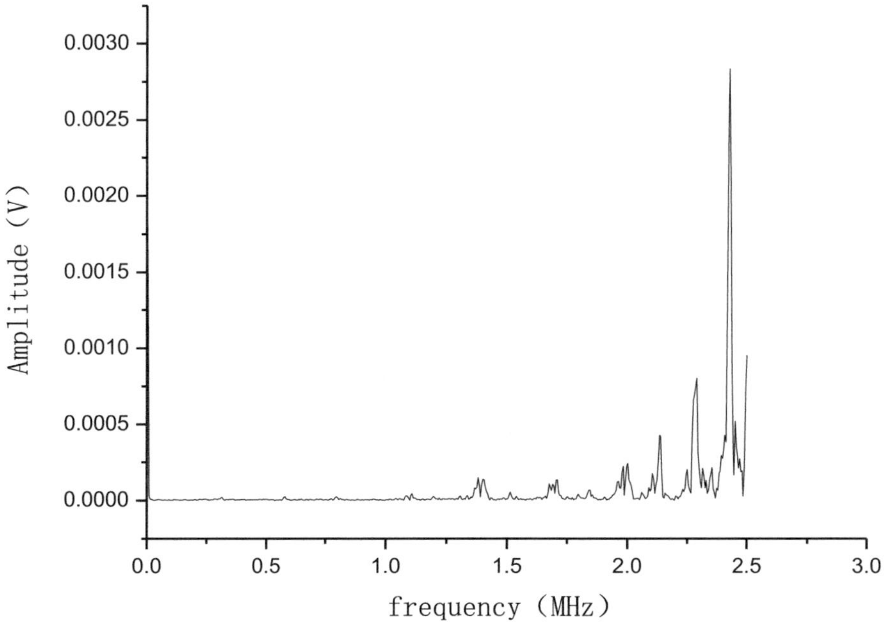

Fig. 2.44 Spectrum of acoustic memory signal

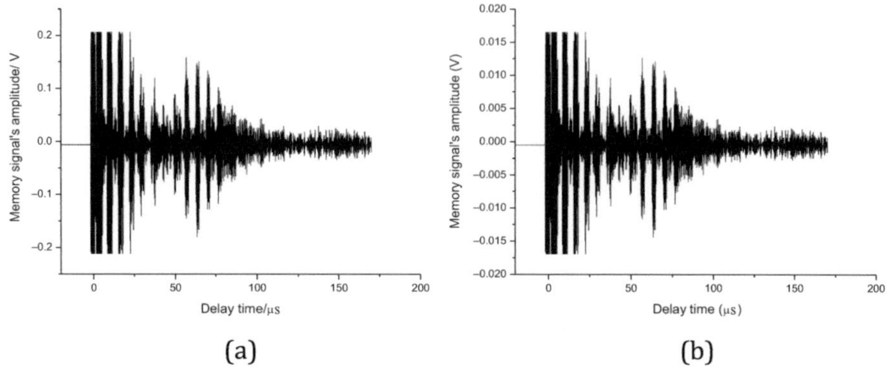

Fig. 2.45 The waveform of the acoustic memory signal when the excitation signal frequency is 5 MHz, and the receiving transducers are respectively 5 MHz (**a**) and 10 MHz (**b**)

2.6.2.3 Hysteresis Relationship of Acoustic Memory Signal Amplitude

The acoustic memory signal changes with the change in transmission amplitude. When the amplitude of the excitation signal increases from 100 mv to 1 v in steps of 100 mv, the change in the amplitude of the acoustic memory signal is shown by the dashed line in Fig. 2.47. Then, as the amplitude of the excitation signal gradually decreases, the change in the amplitude of the acoustic memory signal is shown by the solid line in Fig. 2.47. As can be seen, the two curves do not coincide, indicating a hysteresis phenomenon, which is a significant feature of non-classical nonlinear acoustics. Due to the presence of dislocations and other phenomena in lithium niobate, it exhibits acoustic characteristics different from those of conventional metal materials.

2.6.2.4 Relationship Between the Amplitude of the Acoustic Memory Signal and Temperature

As the temperature increases, the memory signal gradually strengthens, and vice versa. As shown in Fig. 2.48, the dashed line and solid line represent the changes in the acoustic memory signal during the process of temperature gradually increasing and decreasing, respectively. In the process of temperature gradually increasing, when the experimental temperature gradually increases from 5 to 20 °C, the acoustic memory signal strengthens slowly; when the experimental temperature gradually increases from 20 to 50 °C, the acoustic memory signal strengthens dramatically; when the experimental temperature increases from 50 to 55 °C, the acoustic memory signal strengthens dramatically. In the process of temperature decreasing, the memory signal gradually weakens. The rate of change of the acoustic memory signal is different in different temperature ranges. However, at the same temperature, the intensity of the memory signal during the temperature decrease process is much larger than the size

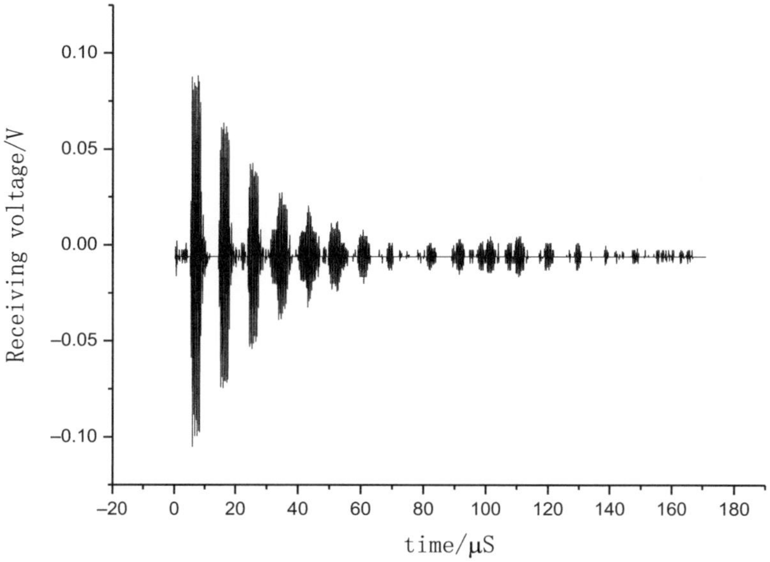

(a)

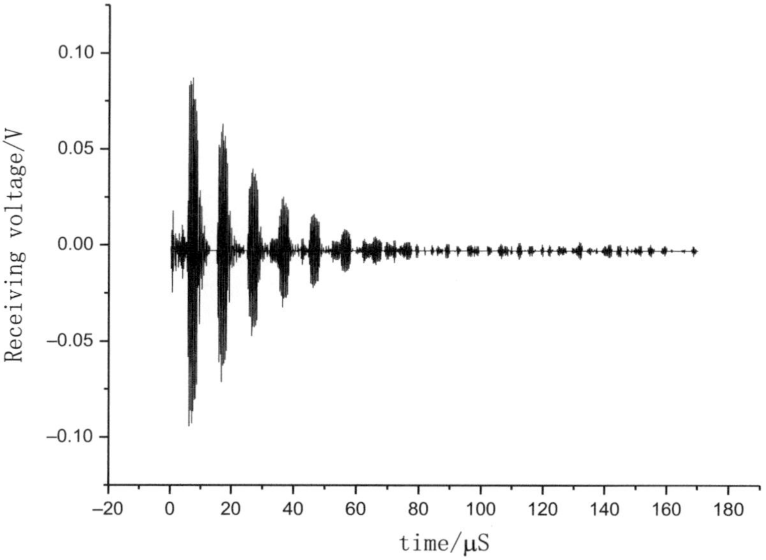

(b)

Fig. 2.46 Acoustic memory phenomenon in a rectangular lithium niobate sample. **a** Acoustic memory phenomenon in the Z-axis direction. **b** Acoustic memory phenomenon in the Y-axis direction

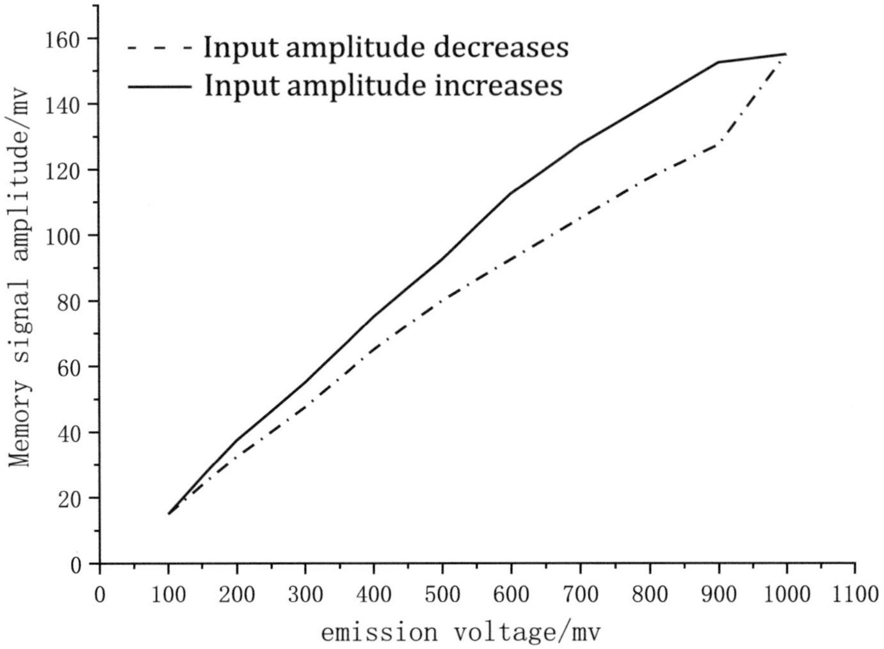

Fig. 2.47 Hysteresis relationship between the amplitude of the acoustic memory signal and the transmission voltage

of the memory signal during the temperature increase process. We found that the temperature relationship in this frequency band is different from that in the literature [64], where the amplitude of the acoustic memory signal decreases as the temperature increases, while our experimental results show that as the temperature increases, the acoustic memory signal increases. The transmission and reception characteristics of the experimental system also change with temperature, mainly due to the performance of the transceiver transducer being related to temperature. Using an aluminum block as the experimental sample, we measured the changes in the experimental system during the heating and cooling process, as shown in Fig. 2.49.

2.7 Multi-segment Acoustic Memory Signal

When the excitation voltage is large enough or each pulse signal contains enough cycles, we found the second and third memory signals. Figure 2.50 describes how the acoustic memory signal changes with the number of pulse cycles at 10 MHz. We used a pair of PZT transducers with a center frequency of 10 MHz as the transceiver transducers. In the experiment, the amplitude of the excitation signal was kept at 200 mv, and the pulse repetition frequency was kept at 100 Hz. Figure 2.50a–f

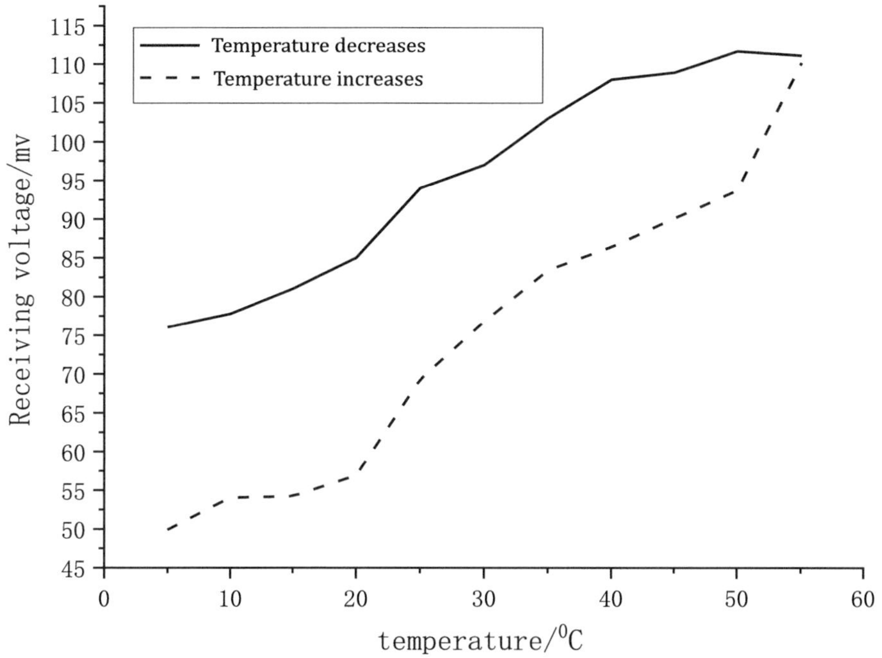

Fig. 2.48 The relationship between the amplitude of the acoustic memory signal and temperature changes

describes the waveform of the memory signal when the number of cycles is 4, 16, and 64. When the number of cycles is 4, a clear memory signal envelope can be observed, as shown in Fig. 2.50a–b. However, the second and third segments are very weak, so we believe that only the first memory signal exists at this time. When the number of cycles is 16, we clearly observed the second acoustic memory signal, as shown in Fig. 2.50c–d. Although the intervals between the first and second memory signals are not very obvious, their responses to the number of cycles are very different (as shown in Fig. 2.11), so we believe these are two different memory signals. Finally, when the number of cycles is 64, the intervals between the first and second memory signals are very obvious, and the third memory signal can be clearly observed, as shown in Fig. 2.10e–f. The amplitudes of the three memory signals are also increasing, and their rates of change are different, that is, each memory signal responds differently to the number of cycles, as shown in Fig. 2.51.

As shown. In a sufficiently thick lithium niobate crystal, the fourth and fifth memory signals will also be observed as the number of periods continues to increase. When the excitation voltage is large enough, we also found similar multi-segment memory signals. Although narrow-band PZT can also produce good multi-segment memory signals, in fact, I first observed multi-segment memory signals using a self-made PVDF transducer. If we delve deeper into the acoustic memory phenomenon of lithium niobate, using a PVDF transducer may lead to more discoveries. When the

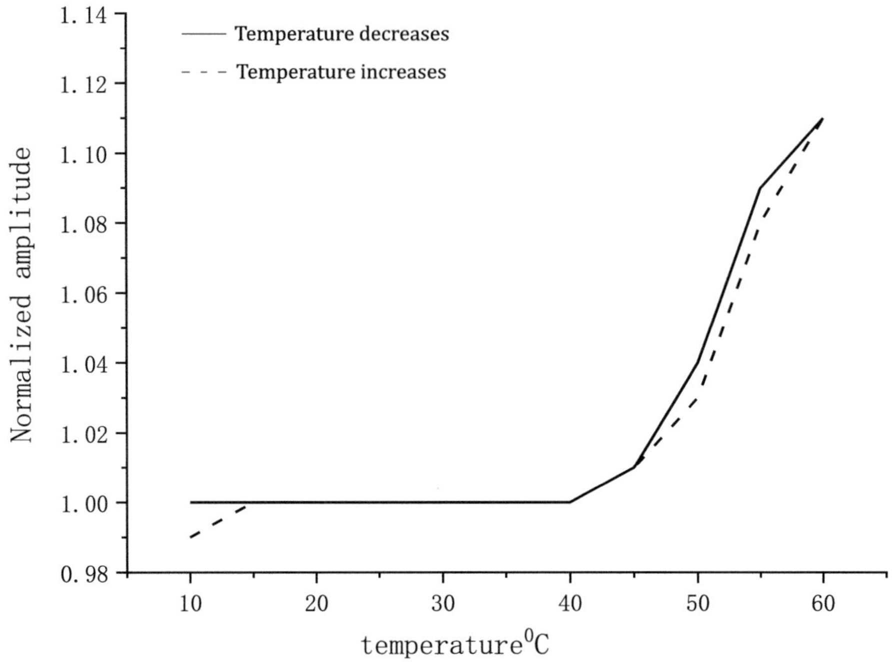

Fig. 2.49 The change in the transmission and reception performance of the experimental system with temperature, the dashed line represents the heating process, and the solid line represents the cooling process

PVDF transducer receives signals, we also observed the waveform of the memory signal when the number of periods is 200 in the cylindrical sample, as shown in Fig. 2.52.

As the number of periods increases, the direct wave weakens, as shown in Fig. 2.53, but the change is small. When the experimental sample is an aluminum block, as the number of periods increases, the amplitude of the direct wave will increase, as shown in Table 2.5. Therefore, in lithium niobate crystals, as the number of periods increases, some energy is transferred to the acoustic memory signal, causing the amplitude of the direct wave to decrease.

2.7.1 Non-exponential Decay Phenomenon of Echo Signals

Not only did we analyze various characteristics of the memory signal, but we also conducted some research on the attenuation characteristics of lithium niobate crystals. Figure 2.54 analyzes the amplitude ratio relationship of the 1st to 8th echoes, the signal after eight echoes is very poor, and the noise is large. Therefore, the echoes after eight times are not considered. It can be seen that the attenuation of lithium

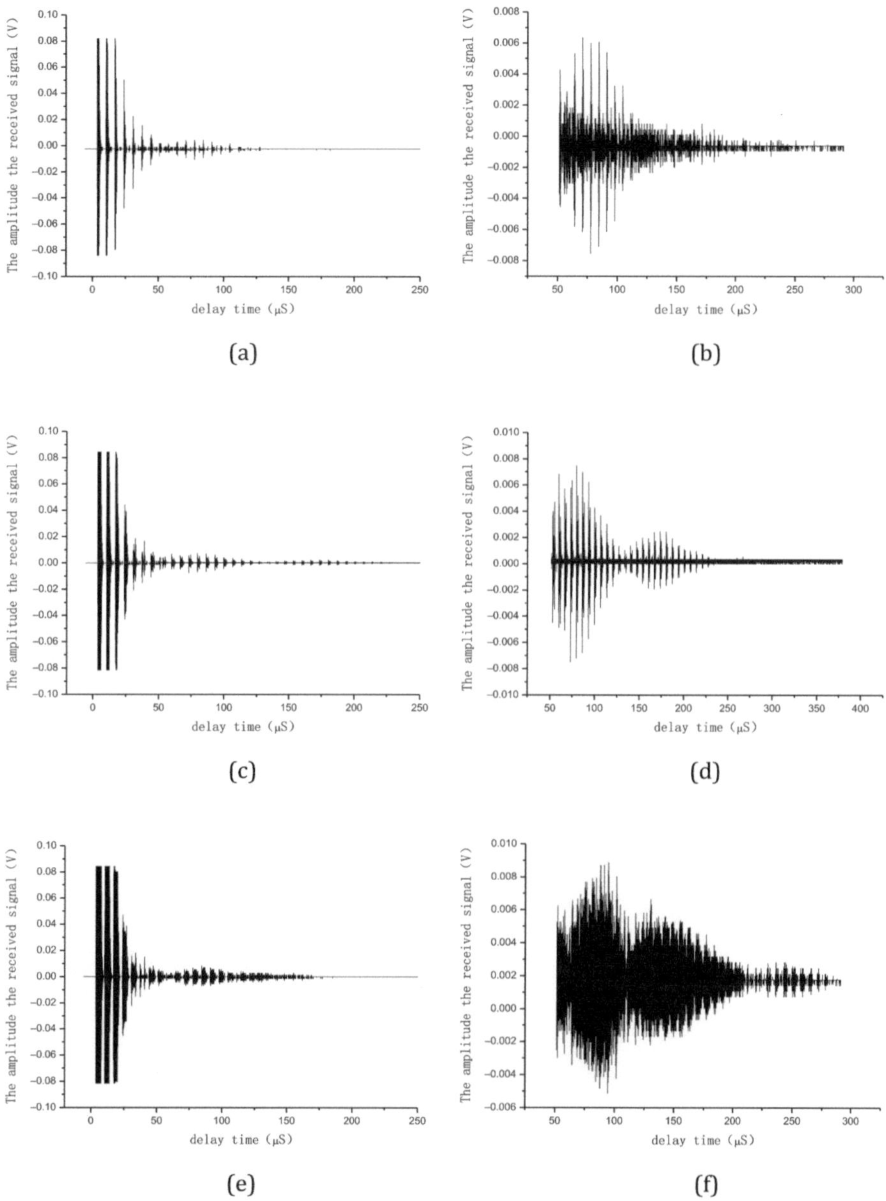

Fig. 2.50 Waveform of the memory signal when each pulse signal contains 4 (**a–b**), 16 (**c–d**), 64 (**e–f**) periods. Figure **b**, **d**, **f** are the magnified waveforms of the memory signal

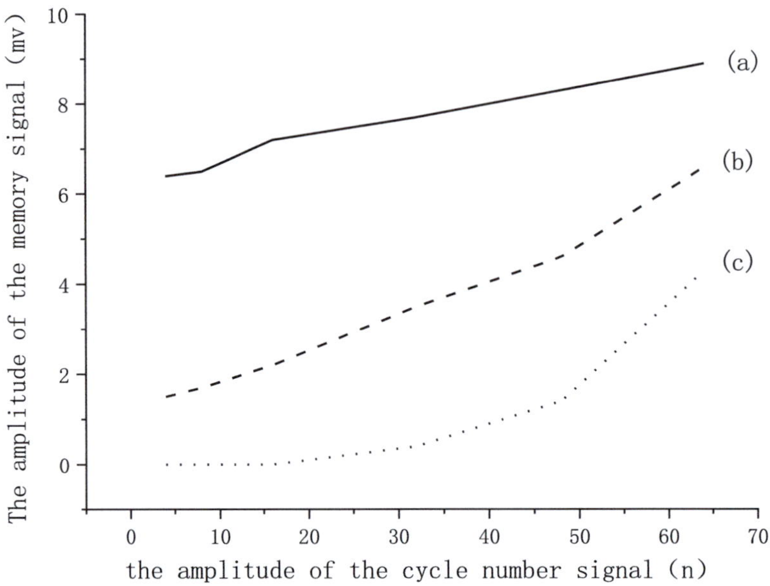

Fig. 2.51 The amplitude of each segment of the memory signal changes with the number of periods (**a** the first segment of the acoustic memory signal; **b** the second segment of the acoustic memory signal; c the third segment of the acoustic memory signal)

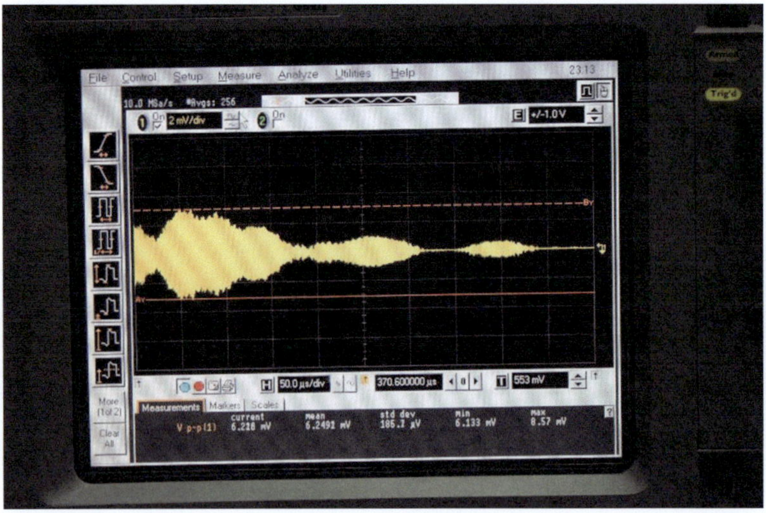

Fig. 2.52 PVDF as a receiving transducer, the waveform of the acoustic memory signal when the number of periods is 200

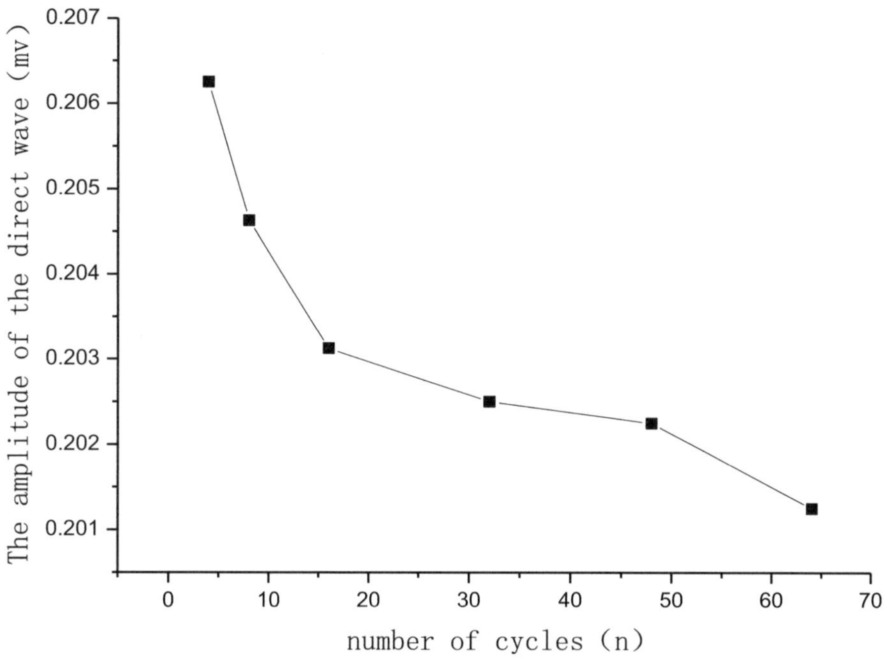

Fig. 2.53 The amplitude of the direct wave in lithium niobate changes with the number of periods

Table 2.5 Relationship between the amplitude of the direct wave and the number of periods in the aluminum block sample

Period number	4	8	16	32	48	64
Amplitude of direct wave (mv)	759.4	759.8	760.1	762.3	763.9	761.1

niobate changes with distance is non-exponential, and the amplitude change of the fourth to fifth echoes is very small. The attenuation coefficient of lithium niobate crystals increases with the increase of the excitation signal, as shown in Fig. 2.55. The dashed line represents the ultrasonic attenuation curve when the excitation signal is 100 mv, and the solid line represents the ultrasonic attenuation curve when the excitation signal is 1 v. For easy comparison, the received signals under two different amplitudes have been normalized with their respective maximum values. It can be seen that the smaller the excitation signal, the faster the ultrasound decays; the larger the excitation signal, the slower the ultrasound decays. This is an important feature of lithium niobate.

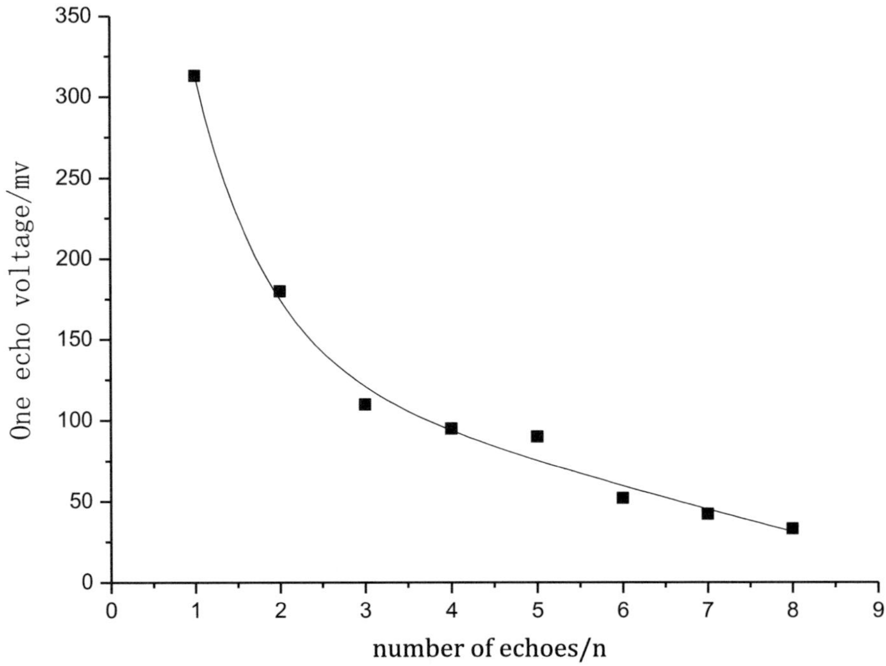

Fig. 2.54 Amplitude characteristics of ultrasonic attenuation

2.7.1.1 The Amplitude Hysteresis Phenomenon of the First Harmonic and Second Harmonic of the Direct Wave

Lithium niobate crystals are a non-classical, highly nonlinear medium. From the spectral analysis of the direct ultrasonic waves from lithium niobate crystals, it can be seen that the third harmonic is stronger than the second harmonic, which is an important feature of non-classical nonlinear acoustics, as shown in Fig. 2.56. Figure 2.57 describes the frequency response curve when there is no lithium niobate crystal emitting transducer. It can be seen that the nonlinearity generated by the emitting transducer itself is very weak, and the second and third harmonics are more than 35 dB weaker than the first harmonic. Therefore, in the experiment, the influence of the emitting transducer on the measurement of the second and third harmonics can be ignored. When the frequency of the excitation signal is 2.5 MHz, digital filters with bandwidths of 2–3 MHz and 4–6 MHz are used to extract the first and second harmonics of the received direct waves. By changing the excitation voltage, the amplitude hysteresis phenomenon of the first and second harmonics was obtained, as shown in Fig. 2.58. The hysteresis phenomenon once again proves that lithium niobate is a highly non-classical nonlinear material, and provides a method for explaining multi-stage memory signals.

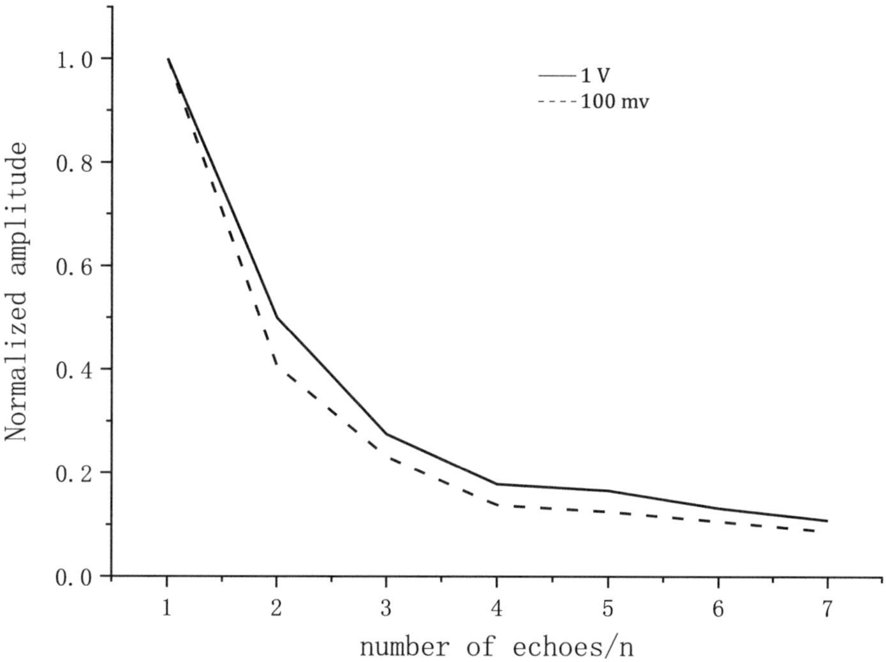

Fig. 2.55 Ultrasonic attenuation characteristics under different emission amplitudes

2.7.2 Analysis and Discussion

Acoustic stress causes the redistribution of ferroelectric domains. The physical mechanism of the interaction between ultrasound and domains originates from the piezoelectric field caused by mechanical stress and sound pressure. The piezoelectric effect causes the vibration of the electric domain and defects. After the ultrasonic wave passes through, the vibration of the electric domain and defects continues. This phenomenon only occurs in crystals, because there is no domain structure in non-crystals. The inhomogeneity of the lattice causes the interaction between the domain wall and the sound wave. It is worth noting that the acoustic memory signal is closely related to the microscopic structure of the crystal. Square crystals, due to different polarization situations in different directions, lead to completely different acoustic memory signals. Therefore, the acoustic memory signal can reflect the microscopic characteristics of the crystal. When defects exist in the substructure inside the crystal, part of the sound energy will be transferred to the irregularities of the crystal and the substructure of the superlattice, becoming a secondary sound source. After the primary sound source and the secondary sound source interact nonlinearly, the sound energy reappears, generating an acoustic memory signal, causing changes in the size of the echo signal. For this reason, the change of the echo with distance is no longer exponential.

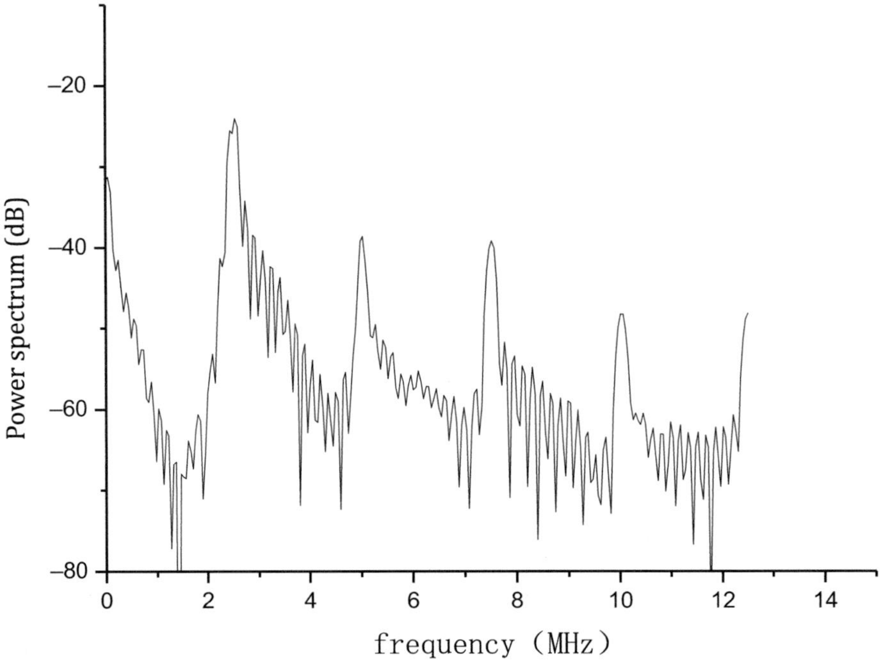

Fig. 2.56 Spectral analysis of direct wave signals from lithium niobate crystals

In the frequency band we studied, it was found that the acoustic memory signal increased with the increase in temperature, while in the 25.9 MHz used by Breazeale, the acoustic memory signal decreased with the increase in temperature. The movement of the ferroelectric domain is related to the acoustic memory signal, and the movement of the ferroelectric domain is mainly determined by external forces and viscous forces [70]. The viscosity coefficient is a function of temperature and frequency. Therefore, the acoustic memory signal will change with changes in temperature and frequency.

When ultrasonic waves propagate in crystals, the domain structure of the ferroelectric body will change due to the influence of ultrasonic waves. When the ultrasonic energy in the domain structure increases, the viscous action of the ferroelectric body will decrease, and the activity of the domain wall will increase. When the number of cycles increases or the excitation voltage increases, more domain walls will be affected by external forces and start to move, thereby generating the second and third stage memory signals.

The hysteresis effect of the acoustic memory signal in lithium niobate crystals reflects the defects in the crystal. Whether it is a natural crystal or a laboratory-prepared crystal, there will be more or less defects. Due to the existence of these defects, the stress–strain relationship in the crystal is no longer a linear relationship, but there is a certain hysteresis relationship, which makes the lithium niobate crystal show non-classical nonlinear acoustic characteristics. The hysteresis phenomenon

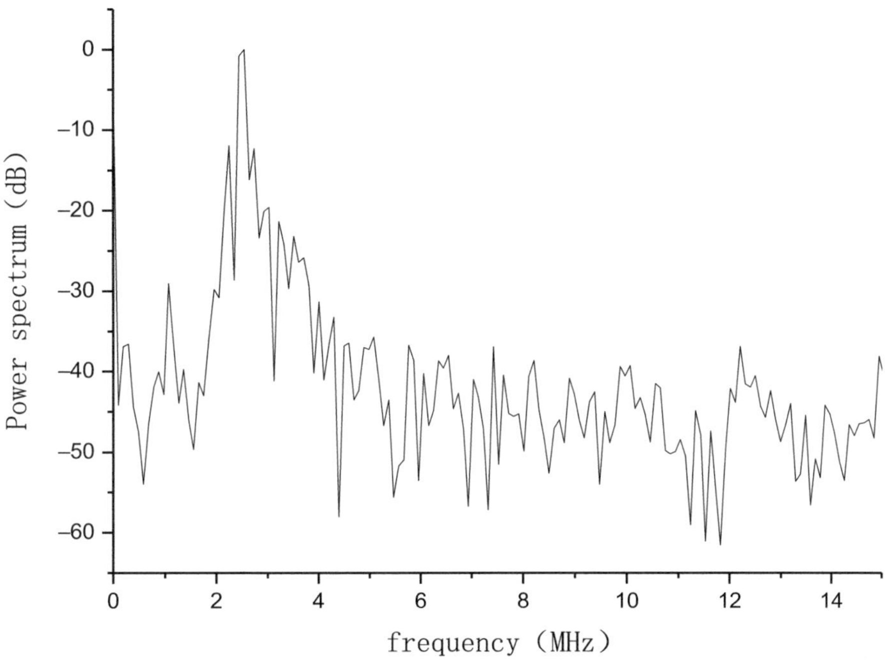

Fig. 2.57 Power spectrum of the emitting transducer

is one of the characteristics of non-classical nonlinear acoustics. In addition, the frequency of the acoustic memory signal obtained by spectral analysis of the acoustic memory signal is 2.44 MHz, indicating that the acoustic memory signal has a frequency offset of 60 kHz, which is also a feature of non-classical nonlinear acoustics.

The acoustic memory signal has a clear hysteresis phenomenon, which is very similar to the hysteresis loop in magnetism. The existence of the hysteresis loop reflects a memory phenomenon in magnetism, and this phenomenon is widely applied to various aspects of our life. Therefore, the existence of the hysteresis phenomenon of the acoustic memory signal also reflects a certain memory function of sound. Its mechanism and application research still need our further in-depth study.

This section investigates the acoustic memory phenomenon in lithium niobate within the frequency range of 2.5–10 MHz, observing the relationship between the acoustic memory signal and the crystal polarization direction. The echo signal was observed to change with distance in a non-exponential form, and the acoustic memory signal lagged behind the transmitted excitation signal. This lagging relationship reflects the characteristics of non-classical nonlinear acoustics, and the lagging phenomenon of the direct wave signal's fundamental and second harmonic confirmed this characteristic. Especially, the relationship between the acoustic memory signal and temperature changes with the different applied ultrasonic frequencies. When the number of periods is very large or the excitation voltage is very large, we found

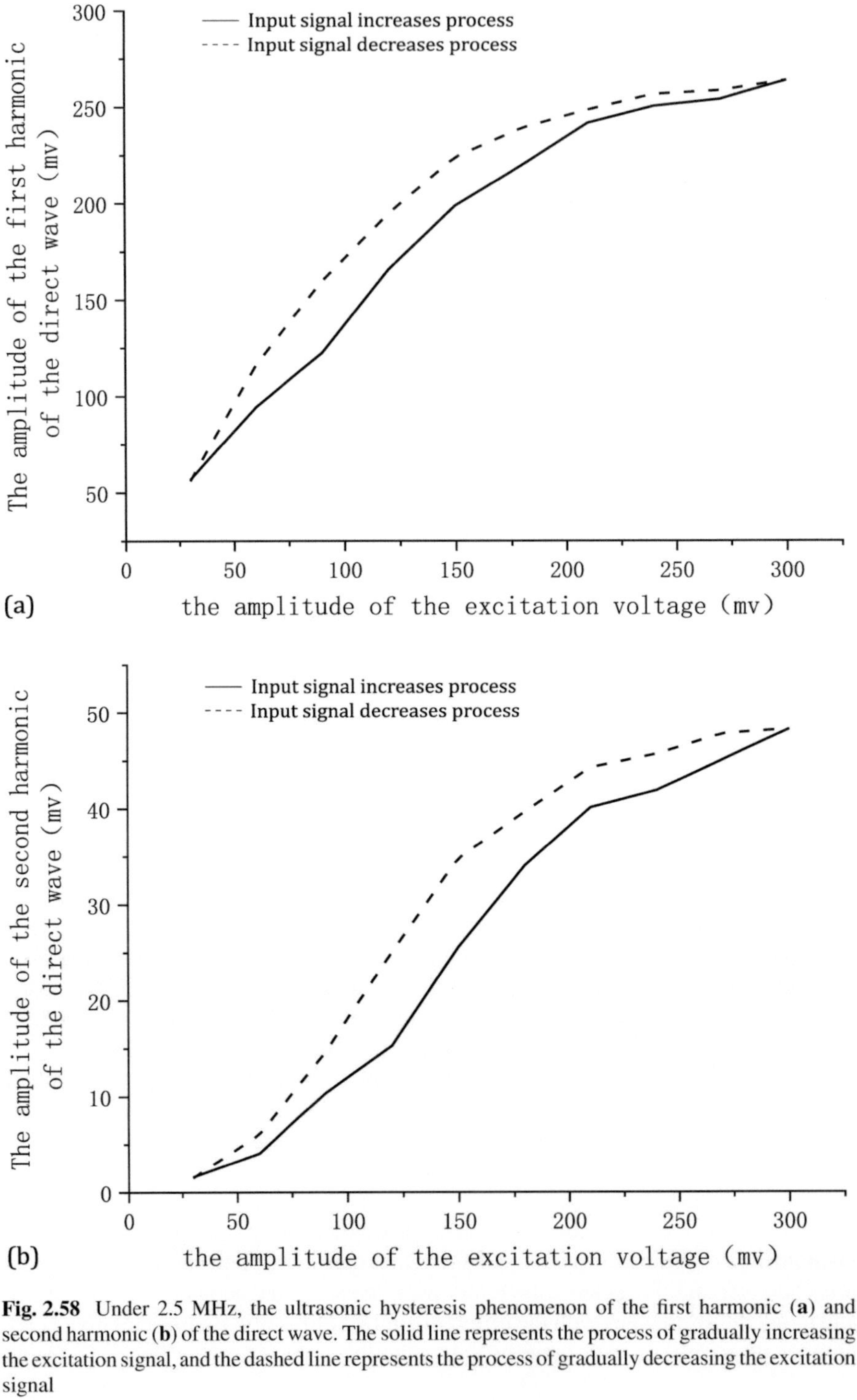

Fig. 2.58 Under 2.5 MHz, the ultrasonic hysteresis phenomenon of the first harmonic (**a**) and second harmonic (**b**) of the direct wave. The solid line represents the process of gradually increasing the excitation signal, and the dashed line represents the process of gradually decreasing the excitation signal

multi-segment memory signals. This section provides a qualitative explanation of the phenomena found in the experiment. Further research on this project will help reveal the physical mechanism of the acoustic memory phenomenon and use the acoustic memory signal to analyze the internal structure of the crystal, promoting the application of the crystal in the field of acoustic memory.

References

1. Jones GL, Kobett DR. Interaction of elastic waves in an isotropic solid. J Acoust Soc Am. 1963;35(1):5–10.
2. Childress JD, Hambrick CG. Interactions between elastic waves in an isotropic solid. Phys Rev. 1964;136:A411–8.
3. Krasilnikov VA, Zarembo LK. Nonlinear interaction of elastic waves in solids. IEEE Trans Sonics Ultrason. 1967;14(1):12–7.
4. Korneer VA, Nihei KT, Myer LR. Nonlinear interaction of plane elastic waves. Earth Science Division; 1998.
5. Domanski W. Propagation and interaction of weakly nonlinear elastic plane waves in a cubic crystal. Wave Motions. 2008;45:337–49.
6. Goldberg ZA. Interaction of plane longitudinal and transverse elastic waves. Soviet Phys Acoust. 1960;6:306–10.
7. Holt AC, Ford J. Theory of ultrasonic pulse measurements of third-order elastic constants for cubic crystals. J Appl Phys. 1967;38(1):42–50.
8. Duquesne JY, Perrin B. Interaction of collinear transverse acoustic waves in cubic crystals. Phys Rev B. 2001;63(6):811–20.
9. Du G. Theory of pure mode nonlinear acoustic wave propagation in crystals. Sci China. 1989;3:276–82.
10. Jiang WH, Du GH. Quasilongitudinal wave along Y-direction of $LiNbO_3$ and ultrasonic nonlinearity parameters. Sci Chin Ser A. 1991;34(3):346–53.
11. Philip J, Breazeale MA. Temperature variation of some combinations of third-order elastic constants of silicon between 300 and 3 °K. J Appl Phys. 1981;52(5):3383–7.
12. Breazeale MA, Philip J, Zarembowitch A, et al. Acoustical measurement of solid state nonlinearity: application to CsCdF3 and $KZnF_3$. J Sound Vib. 1983;88(1):133–40.
13. Duquesne JY, Perrin B. Anisotropic nonlinear elastic properties of an icosahedral quasicrystal. Phys Rev Lett. 2000;85(20):4301.
14. Duquesne YJ, Perrin B. Elastic wave interaction in icosahedral AlPdMn. Physica B. 2002;316–317:317–20.
15. Jiang WH, Cao WW. Second harmonic generation of shear waves in crystals. IEEE Trans Ultrason Ferroelectr Freq Control. 2004;51(2):153–162.
16. Jiang WH, Cao WW, Du GH. Third harmonic generation of transverse acoustic waves in crystals and ceramics. Acta Acust Acust. 2002;88(2):163–7.
17. Domanski W. Propagation and interaction of weakly nonlinear elastic plane waves in a cubic crystal. Wave Motion. 2008;45(3):337–49.
18. Holt AC, Ford J. Theory of ultrasonic three-phonon interactions in single-crystal solids. J Appl Phys. 1969;40:142–8.
19. Duquesne YJ, Perrin B. Interaction of collinear transverse acoustic waves in cubic crystals. Phys Rev B. 2001;63:064303.
20. Zhang SJ, Li F. High performance ferroelectric relaxor-$PbTiO_3$ single crystals: Status and perspective. J Appl Phys. 2012;111(3):2–27.
21. Baumhauer JC, Tiesten HF. Nonlinear electroelastic equations small fields superposed on a bias. J Acoust Soc Am. 1973;54:1017–34.

22. Brugger K. Pure modes for elastic waves in crystals. J Appl Phys. 1965;36(3):759–68.
23. Breazeale MA, Philip J. Physical acoustics, vol. XVII. Mason WP, Thurston RN, editors. Orlando: Academic Press, Inc.; 1984. p. 1–61.
24. Auld BA. Acoustic waves and fields in solids. New York: Wiley; 1973.
25. Liu XZ, Jiang WH, Cao WW, et al. Interactions of collinear acoustic waves propagating along pure mode directions of crystals. J Appl Phys. 2014;115:064909.
26. Cross LE. Overview of relaxor ferroelectrics. Ferroelectrics. 1994;151:305–20.
27. Park SE, Shrout TR. Characteristics of relaxor-based piezoelectric single crystals for ultrasonic transducers. IEEE Trans Ultrason Ferroelectr Freq Control. 1997;44:1140–7.
28. Choi SW, Shrout TR, Jang SJ, et al. Dielectric and pyroelectric properties in the $Pb(Mg_{1/3}Nb_{2/3})O_3$-$PbTiO_3$ system. Ferroelectrics. 1989;100:29–38.
29. Noblanc O, Gaucher P, Galvarin G. Structure and dielectric studies of $Pb(Mg_{1/3}Nb_{2/3})O_3$-$PbTiO_3$ ferroelectric solid solution around the morphotropic boundary. J Appl Phys. 1996;79(8):4291–7.
30. Feng X, Xi Y. Piezoelectric performance of relaxor ferroelectrics at the pseudo-homotype boundary. Funct Mater. 1999;30(6):582–4.
31. Gang C, Liji L. Fundamentals of crystal physics. Beijing: Science Press; 1992. p. 168–84.
32. Auld B A. Acoustic fields and waves in solids. New York: Wiley; 1973. p. 213–41.
33. Tingting Z, Xiaozhou L, Xiufen G. Study on the acoustic characteristics of lithium tantalate crystals. Acoust Technol. 2011;30(4):267–9.
34. Wang H, Jiang W, Cao W. Characterization of lead zirconate titanate piezoceramic using high frequency ultrasonic spectroscopy. J Appl Phys. 1999;85(12):8083–91.
35. Cannata JM, Williams JA, Zhou Q, et al. Development of a 35-MHz piezo-composite ultrasound array for medical imaging. IEEE Trans Ultrason Ferroelectr Freq Control. 2006;53(1):224–36.
36. Lau ST, Li H, Wong KS et al. Multiple matching scheme for broadband 0.72Pb ($Mg1/3Nb_2$/3)O3-0.28$PbTiO_3$ single crystal phased-array transducer. J Appl Phys. 2009;105(9):094908.
37. Zhang R, Jiang B, Cao WW. Elastic, piezoelectric, and dielectric properties of multidomain 0.67Pb $(Mg_{1/3}Nb_{2/3})O_3$–0.33$PbTiO_3$ single crystals. J Appl Phys. 2001; 90(7):3471–75.
38. Feng Z, Zhao X, Luo H. Composition and orientation dependence of dielectric and piezoelectric properties in poled $Pb(Mg_{1/3}Nb_{2/3})O_3$-$PbTiO_3$ crystals. J Appl Phys. 2006;100(2):024104.
39. Hosono Y, Yamashita Y, Sakamoto H, et al. Growth of single crystals of high-curie-temperature $Pb(In_1/2Nb_{1/2})O_3$-$Pb(Mg_{1/3}Nb_{2/3})O_3$-$PbTiO_3$ ternary systems near morphotropic phase boundary. Jpn J Appl Phys. 2003;42(9A):5681–6.
40. Tian J, Han P, Huang X, et al. Improved stability for piezoelectric crystals grown in the lead indium niobate-lead magnesium niobate-lead titanate system. Appl Phys Lett. 2007;91(22):222903.
41. Zhang S, Luo J, Hackenberger W, et al. Characterization of $Pb(In_{1/2}Nb_{1/2})O_3$–$Pb(Mg_{1/3}Nb_{2/3})O_3$–$PbTiO_3$ ferroelectric crystal with enhanced phase transition temperatures. J Appl Phys. 2008;104(6):064106.
42. Liu X, Zhang S, Luo J, et al. Complete set of material constants of $Pb(In_{1/2}Nb_{1/2})O_3$-$Pb(Mg_{1/3}Nb_{2/3})O_3$-$PbTiO_3$ single crystal with morphotropic phase boundary composition. J Appl Phys. 2009;106(7):074112.
43. Sun P, Zhou Q, Zhu B, et al. Design and fabrication of PIN-PMN-PT single-crystal high-frequency ultrasound transducers. IEEE Trans Ultrason Ferroelectr Freq Control. 2009;56(2):2760–3.
44. Du GH. Theory of propagation of nonlinear acoustic waves in pure longitudinal mode in crystals. Sci China. 1989;32(9):1084–92.
45. Jiang W, Du G. The depolarization field in piezoelectric bodies and its impact on the measurement of ultrasonic nonlinear parameters. Acoustics. 1993;18(7):241–8.
46. Albert CH, Joseph F. Theory of ultrasonic pulse measurement of third-order elastic constants for cubic crystals. J Appl Phys. 1967;38(1):42–50.
47. Xu Z, Shen H, Wang Y. Anomalous performance of $LiNbO_3$ crystal near 75 °C. Sci Bull. 1980;13:586–8.
48. Jaffe B, Jaffe H, Cook WR. Piezoelectric ceramics. London: Academic Press; 1971.

49. Whatmore RW. Pyroelectric device and materials. Rep Prog Phys. 1986;49:1335–86.
50. Mukherjee BK, Ren W, Liu SF, et al. Nonlinear properties of piezoelectric ceramics, smart structures and materials. Proc SPIE. 2001;4333:41–54.
51. Li S, Cao W, Cross LE. The extrinsic nature of nonlinear behavior observed in lead zirconate titanate ferroelectric ceramic. J Appl Phys. 1991;69:7219.
52. Morozov M, Damjanovic D, Setter N. The nonlinearity and subswitching hysteresis in hard and soft PZT. J Eur Ceram Soc. 2005;25:2483–6.
53. Fan J, Stoll WA, Lynch CS. Nonlinear constitutive behavior of soft and hard PZT: experiments and modeling. Acta Mater. 1999;47:4415–25.
54. Bharadwaja SSN, Hong E, Zhang SJ et al. Nonlinear dielectric response in (1 −x)Pb(Zn1/3Nb2/3)O3-xPbTiO_3 (x=0.045 and 0.08) single crystals. J Appl Phys. 2007;101:104102.
55. Yamada K, Yamazaki D, Nakamura K. A functionally gradient piezoelectric material created by an internal temperature gradient. In: Proceedings of the 12th IEEE international symposium on application of ferroelectrics; 2001. p. 475–8.
56. Smolenskii GA, Isupov VA, Agranovakaya AI et al. New ferroelectrics of complex composition. IV. Soviet Physics Solid State. 1961;2(11):2651–54.
57. Yu Z, Guo R, Bhalla AS. Dielectric behavior of Ba(Ti1−xZrx)O_3 single crystals. J Appl Phys. 2000;88(1):410–5.
58. Hennings D, Schnell A, Simon G. Diffuse ferroelectric phase transition in Ba(Ti1−yZry)O_3 ceramics. J Am Ceram Soc. 1982;65(11):539–44.
59. Weber U, Greuel G, Boettger U et al. Dielectric properties of Ba(Zr,Ti)O3-based ferroelectrics for capacitor applications. J Am Ceram Soc. 2001;84(4):759–66.
60. Tang XG, Chew KH, Chan HLW. Diffuse phase transition and dielectric tunability of Ba(ZryTi1−y)O_3 relaxor ferroelectric ceramics. Acta Mater. 2004;529(17):5177–83.
61. Weis RS, Gaylord TK. Lithium niobate: summary of physical properties and crystal structure. Appl Phys A. 1985;37:191–203.
62. Podivilov EV, Sturman BI, Calvo GF, et al. Effect of domain structure fluctuations on the photorefractive response of periodically poled lithium niobate. Phys Rev B 2000;62:13182.
63. Schwarz UT, Maier M. Asymmetric Raman lines caused by an anharmonic lattice potential in lithium niobate. Phys Rev B. 1997;55:11041.
64. Scott JF. Ferroelectric memory. Berlin: Springer; 2000.
65. McPherson MS, Ostroskii IV, Breazeale MA. Observation of acoustical memory in $LiNbO_3$. Phys Rev Lett. 2002;89:115506.
66. Ped'ko BB, Kislova IL, Volk TR, et al. New memory effects in lithium niobate nanocrystals. Russian Acad Sci Bull Phys. 2000;64:1145.
67. Breazeale MA, Ostrovskii IV, McPherson M. Thermal hysteresis of nonlinear ultrasonic attenuation in lithium niobate. J Appl Phys. 2004;96:2990.
68. Breazeale MA, Ostrovskii IV. Innovations in nonlinear acoustics. New York: AIP, Melville; 2006.
69. Zhou D, Liu XZ, Gong XF. Experimental study of acoustical memory in lithium niobate. Phys Rev E. 2008;78:016602.
70. Huang YN, Wang YN, Shen HM. Internal friction and dielectric loss related to domain walls. Phys Rev B. 1992;46:3290.

Chapter 3
Nonlinear Propagation of Sound Waves in Porous Materials

In recent years, the propagation of nonlinear waves in porous media has attracted increasing attention. Research results show that the nonlinear coefficient of porous media is two to three orders of magnitude larger than that of solid or fluid without pores [1–3]. Biot proposed the theory of elastic propagation in porous materials saturated with low and high frequency fluids [4, 5]. Donskoy and others studied the nonlinear scattering and longitudinal sound wave propagation in porous media [6]. Ostrovsky derived the nonlinear vibration equation of spherical and cylindrical micro-compressible elastic medium cavities [1, 3]. Geeritis also studied the propagation of sound waves through porous media [7]. Dazel and others studied Biot waves in porous media, considering velocity dispersion, frequency dependence, and nonlinearity, their theory was proposed in the case of granular media with rigid boundaries [8]. Considering that the sound absorption effect in pure viscoelastic materials is very small, media with porous structures are introduced to enhance sound attenuation, thus establishing many theoretical models. Oberst [9] and Meyer et al. [10] established a one-dimensional model of the resonance phenomenon of spherical perforations in the viscoelastic layer. Gaunaurd [11, 12] further studied the situation with multiple perforations based on this. Biot [4, 5] considering the influence of certain frequencies in porous materials on viscous forces, changed the motion viscous force term to oscillating viscous force term. His theory sparked a lot of theoretical research on various types of propagation by later generations. Ostrovsky [3] introduced a new medium called "rubber-like", which has high attenuation and strong nonlinearity for the incident harmonics. The literature [13, 14] used the finite element method to study the harmonic propagation in the viscoelastic lattice. Ostrovsky [3] mentioned that nonlinearity is only considered at low frequencies, and the micro-pores in the material oscillate near the resonance frequency as secondary sound sources. Due to the nonlinearity of the oscillation, higher harmonics, especially the second harmonic components, begin to be significant, and the nonlinearity of the medium is significantly enhanced due to the presence of these micro-pores.

X. Liu, *Nonlinear Sound Waves in Solids*,
Advances in Acoustics, https://doi.org/10.1007/978-981-96-5253-2_3

Fan et al. [15] and Liang et al. [16] used the equivalent medium method to theoretically study the nonlinear sound propagation characteristics of viscoelastic materials containing spherical micro-pores, showing that such media have strong nonlinearity even when the porosity is very small. Liang et al. [17] also studied the sound localization phenomenon of weakly compressible elastic media containing spherical air pores. All of the above research is based on the equivalent medium method. However, the "equivalent medium" actually used by the classical equivalent medium method is a degeneration of an "effective medium" containing multiple scatterers and a uniform medium, and the multiple scattering method is also used to study the nonlinear effects of porous media. The multiple scattering method analyzes the medium from a microscopic perspective, assuming that each air pore scatterer is excited by sound waves to scatter, so that the sound pressure value at any point is the sum of the incident sound pressure and the scattered sound pressure of all other scatterers at that point. Foldy pioneered the theory of multiple scattering half a century ago [18]. Weston et al. [19], Miller et al. [20, 21] developed this theory and named the scattering effect as sound interaction. A large number of experiments have also obtained similar results based on this theory [22, 23]. In addition, improvements to these classical theories are also gradually developing, such as considering the impact of higher-order scattering [22–26]. These improvements are also applicable to the equivalent medium method [23] and the multiple scattering method [22, 23].

3.1 Nonlinear Characteristics of Sound Waves in Viscoelastic Materials Containing Micro-pores

The model in this section is a rubber-like medium containing a large number of cylindrical air pores (see Fig. 3.1), we will study the nonlinear sound propagation of sound waves in this medium, especially the sound propagation performance in the frequency band range near the resonance frequency of the pores.

3.1.1 Theory

3.1.1.1 Longitudinal Acoustic Wave Equation in Porous Rubber Materials

In a homogeneous viscoelastic medium, the mechanical equation for longitudinal wave propagation can be described as follows

$$\rho_0\left(\frac{dv}{dt}\right) = \frac{\partial\sigma(\varepsilon)}{\partial x} \tag{3.1}$$

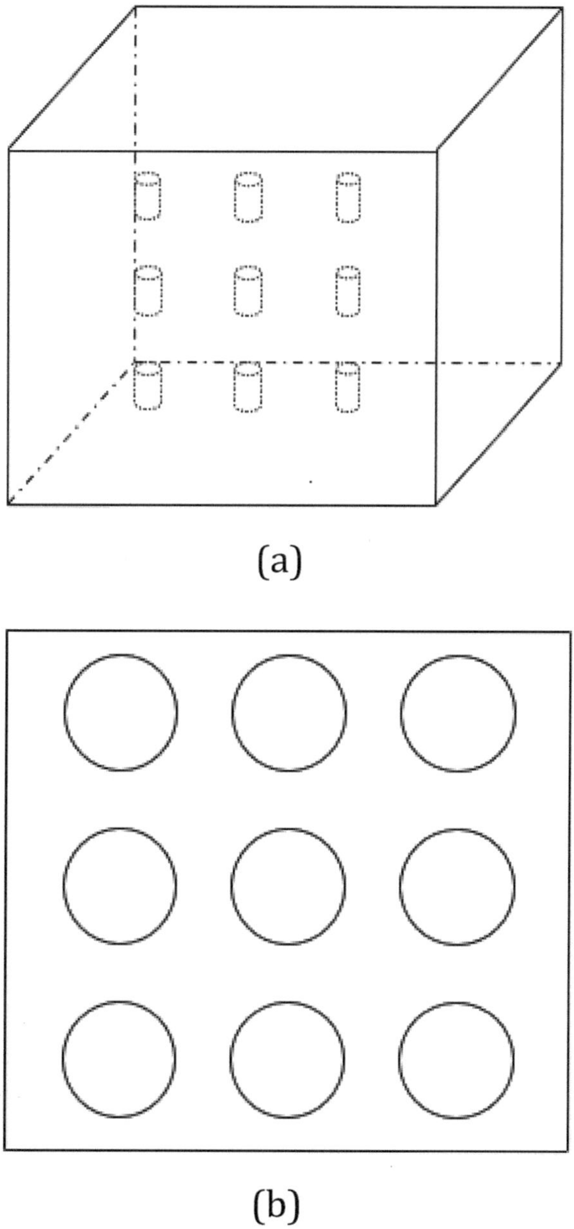

Fig. 3.1 Model of rubber-like medium containing a large number of cylindrical air pores. **a** Three-dimensional legend; **b** Two-dimensional legend

Here $v = v_x = \frac{\partial u_x}{\partial t}$ represents the directional velocity (which can be considered as the velocity component in the x direction in completely symmetrical situations), $\varepsilon = \frac{\partial u_x}{\partial x}$ is the strain corresponding to the stress σ, ρ_0 represents the density of the medium. After some formal transformations, we can obtain the relationship between ε and σ, as shown below

$$\rho \frac{\partial^2 \varepsilon}{\partial t^2} = \frac{\partial^2 \sigma}{\partial x^2}, \tag{3.2}$$

Now, we consider the situation where the medium contains micropores. If the total amount of micropores within each wavelength range is sufficient, such a porous medium can be regarded as an equivalent continuous medium. Therefore, the "equivalent Hooke's law" describing the stress–strain relationship can be written in the form of longitudinal waves as follows:

$$\sigma = (\lambda + 2\mu)(\varepsilon - NU) = \left(K + \frac{4}{3}M\right)(\varepsilon - NU), \tag{3.3}$$

Here λ, μ are Lame coefficients, N represents the total number of micropores, U represents the volume change of a single pore. In this expression, since the pores are very small compared to the entire medium, the strain is approximated directly subtracted from the pore volume change. $Y = m\partial_t$ and $Y = \eta/\kappa + m\partial_t$ represent the bulk modulus and shear modulus of the medium, which can be expressed as

$$\begin{aligned} K &= K' + jK'' = K'(1 + j\eta_k) \\ M &= M' + jM'' = M'(1 + j\eta_M), \end{aligned} \tag{3.4}$$

Here η_k and η_M are dissipation factors. Substituting Eq. (3.3) into Eq. (3.2), we can obtain the longitudinal acoustic wave equation in the viscoelastic medium containing micropores

$$\frac{\partial^2 \sigma}{\partial x^2} - \frac{1}{c_0^2(1 + j\eta)} \frac{\partial^2 \sigma}{\partial t^2} = \rho_0 N \frac{\partial^2 U}{\partial x^2}, \tag{3.5}$$

where $\eta = \frac{\eta_k K' + \frac{4}{3}\eta_M M'}{K' + \frac{4}{3}M'}$ describes the viscous term of the homogeneous medium without pores, its value is generally very small. $c_0 = \sqrt{\frac{K' + \frac{4}{3}M'}{\rho}}$ represents the longitudinal sound speed in the homogeneous medium without pores. To solve this equation, we continue to use the perturbation method, let

$$\sigma = \sigma_1 + \sigma_2 + \cdots \mathrm{U} = \mathrm{U}_1' + U_2' + \cdots, \tag{3.6}$$

Here σ_1, U_1' and σ_2, U_2' represent the first and second order quantities of stress and pore volume change respectively. Substituting them into Eq. (3.5), we get the following expression

$$\begin{aligned}\frac{\partial^2\sigma_1}{\partial x^2}-\frac{1}{c_0^2(1+j\eta)}\frac{\partial^2\sigma_1}{\partial t^2}=\rho N\frac{\partial^2 U_1}{\partial t^2}\\ \frac{\partial^2\sigma_2}{\partial x^2}-\frac{1}{c_0^2(1+j\eta)}\frac{\partial^2\sigma_2}{\partial t^2}=\rho N\frac{\partial^2 U_2}{\partial t^2}\end{aligned} \tag{3.7}$$

3.1.1.2 Nonlinear Acoustic Wave Equation of Cylindrical Porous Rubber Materials

In rubber-like media, the vibration of micropores is oscillatory. The so-called "rubber-like" also known as "hydrophilic" viscoelastic materials, its main feature is weak compressibility, Ostrovsky established an equation to describe the nonlinear oscillation of cylindrical micropores, these "cylindrical microtubes" are essentially consistent with our research object "cylindrical micropores", and we will study the propagation nonlinearity brought about by the nonlinear oscillation of cylindrical micropores, therefore, the theoretical basis of this section adopts Ostrovsky's original equation [1], as follows

$$\frac{1}{4}\rho\left\{\ddot{B}\ln\frac{r_1^2+B}{r_0^2+B}+\frac{1}{2}\dot{B}^2\left(\frac{1}{R_1^2}-\frac{1}{R_0^2}\right)\right\} + 2B\int_{r_0}^{r_1}\frac{2r^2+B}{\left(r^2+B\right)^2}\times\left(\frac{\partial E}{\partial I_1}+\frac{\partial E}{\partial I_2}\right)\frac{dr}{r}=p_0-p_1 \tag{3.8}$$

The parameters we are mainly concerned with are: $B = B_1 + B_2 + \ldots\ldots, B = R^2 - r^2$ ($U = \pi Bl$, l are the sample length), r For Lagrangian initial coordinates. R Represents instantaneous coordinates. $S = \pi B$ represents the change in area with radial direction as reference (or understood as the volume change of a cylindrical oscillator of unit length). r_0, r_1 are the inner and outer diameters of the cylindrical tube, p_0, p_1 are the magnitudes of the internal and external sound pressure of the cylindrical tube, $E(I_1, I_2)$ is the potential energy density, related to the strain tensor I_1, I_2 in the case of radial symmetry, $I_1 = I_2 = \frac{R^2}{r^2} + \frac{r^2}{R^2} + 1 = 3 + \frac{B^2}{(r^2+B)r^2}$. From a series of assumptions, substitutions, and derivations, and we add a term to the equation describing the additional viscous effect brought by the micro-hole oscillation $\gamma\dot{B}$, the following result is finally obtained by the perturbation method

$$\frac{B_1\omega^2}{2}\ln\left(\frac{qr_0\omega}{2c_0}\right)+\frac{\mu B_1}{r_0^2\rho}-\frac{j\pi\omega^2}{4}B_1-\gamma j\omega B_1=\frac{\sigma_1}{\rho} \tag{3.9}$$

$$2\omega^2 B_2 \ln\left(\frac{qr_0\omega}{2c_0}\right) + \frac{\mu B_2}{r_0^2\rho} - j\pi\omega^2 B_2 - j2\omega\gamma B_2 = \frac{3}{2r_0^2}\left(\frac{\mu}{\rho r_0^2} - \frac{\omega^2}{4}\right)B_1^2 + \frac{\sigma_2}{\rho}. \tag{3.10}$$

where r_0 is the micro-hole radius, γ is the parameter we introduced to describe the viscous effect of the medium. $q = 1.78$.

Solving Eq. (3.9), we get the following solution

$$B_1 = \frac{\frac{\sigma_1}{\rho}}{\frac{\omega^2}{2}\ln(\frac{q\omega r_0}{2c_0}) + \frac{\mu}{r_0^2\rho} - \frac{j\pi\omega^2}{4} - \gamma j\omega}. \tag{3.11}$$

Substituting Eq. (3.11) into Eq. (3.10), we can get:

$$B_2 = \frac{\frac{\sigma_2}{\rho}}{2\omega^2\ln\left(\frac{qr_0\omega}{c_0}\right) + \frac{\mu}{r_0^2\rho} - j\pi\omega^2 - j2\omega\gamma} + \frac{\frac{3}{2r_0^2}\left(\frac{\mu}{r_0^2\rho} - \frac{\omega^2}{4}\right)}{2\omega^2\ln\left(\frac{qr_0\omega}{c_0}\right) + \frac{\mu}{r_0^2\rho} - j\pi\omega^2 - j2\omega\gamma}\left(\frac{\frac{\sigma_1}{\rho}}{\frac{\omega^2}{2}\ln\left(\frac{qr_0\omega}{2c_0}\right) + \frac{\mu}{r_0^2\rho} - \frac{j\pi\omega^2}{4} - \gamma j\omega}\right)^2. \tag{3.12}$$

Substituting Eqs. (3.11), (3.12) into Eq. (3.1), we can obtain the sound wave equations for the propagation of the fundamental and second harmonics as follows

$$\frac{\partial^2\sigma_1}{\partial x^2} - \frac{1}{c_0^2(1+j\eta)}\frac{\partial^2\sigma_1}{\partial t^2} = -\left[\frac{Nl\pi\sigma_1}{\frac{1}{2}\ln\left(\frac{q\omega r_0}{2c_0}\right) + \frac{1}{\omega^2}\frac{\mu}{r_0^2\rho} - \frac{j\pi}{4} - j\frac{\gamma}{\omega}}\right]. \tag{3.13}$$

$$\frac{\partial^2\sigma_2}{\partial x^2} - \frac{1}{c_0^2(1+j\eta)}\frac{\partial^2\sigma_2}{\partial t^2} = -\frac{Nl\pi\sigma_2}{2\ln\left(\frac{qr_0\omega}{c_0}\right) + \frac{1}{\omega^2}\frac{\mu}{r_0^2\rho} - j\pi - j\frac{2\gamma}{\omega}} - \rho Nl\pi D\sigma_1^2, \tag{3.14}$$

where:

$$D = \left(\frac{\frac{3}{2r_0^2\rho^2}\left(\frac{\mu}{r_0^2\rho} - \frac{\omega^2}{4}\right)}{2\ln\left(\frac{qr_0\omega}{c_0}\right) + \frac{1}{\omega^2}\frac{\mu}{r_0^2\rho} - j\pi - j\frac{2\gamma}{\omega}}\right)$$

$$\left(\frac{1}{\frac{\omega^2}{2}\ln\left(\frac{q\omega r_0}{2c_0}\right)+\frac{\mu}{r_0^2\rho}-\frac{j\pi\omega^2}{4}-j\gamma\omega}\right)^2.$$

3.1.1.3 Fundamental Wave

To solve Eq. (3.13), we set

$$\begin{aligned}\sigma_1 &= \sigma_{1A}\cdot\exp\left[j\left(\omega t-\overline{k}_1 x\right)\right],\\ \overline{k}_1^2 &= A_1 - jB_1.\end{aligned} \tag{3.15}$$

From this, we solve:

$$A_1=\frac{k_0^2}{1+\eta^2}+Nl\pi\left[\frac{\frac{1}{2}\ln\left(\frac{q\omega r_0}{2c_0}\right)+\frac{1}{\omega^2}\frac{\mu}{r_0^2\rho}}{\left[\frac{1}{2}\ln\left(\frac{q\omega r_0}{2c_0}\right)+\frac{1}{\omega^2}\frac{\mu}{r_0^2\rho}\right]^2+\left(\frac{\pi}{4}+\frac{\gamma}{\omega}\right)^2}\right]. \tag{3.16}$$

$$B_1=\frac{k_0^2\eta}{1+\eta^2}-Nl\pi\left[\frac{\frac{\pi}{4}+\frac{\gamma}{\omega}}{\left[\frac{1}{2}\ln\left(\frac{q\omega r_0}{2c_0}\right)+\frac{1}{\omega^2}\frac{\mu}{r_0^2\rho}\right]^2+\left(\frac{\pi}{4}+\frac{\gamma}{\omega}\right)^2}\right], \tag{3.17}$$

where $k_0=\frac{\omega}{c_0}$, $\overline{k}_1=k_1-j\alpha_1$, $c_1=\omega/k_1$, $\alpha_1=\sqrt{\frac{\sqrt{A_1^2+B_1^2}-A_1}{2}}$.

Here c_1 represents the longitudinal sound speed, α_1 describes the fundamental sound attenuation coefficient.

3.1.1.4 Second Harmonic

Equation (3.14) is a non-homogeneous differential equation. Regarding the solution of this type of equation, the general method is to find a homogeneous solution and a particular solution, and then sum them.

To solve its homogeneous solution, we still let:

$$\begin{aligned}\sigma_2'' &= \sigma_{2A}'\exp\left[j\left(2\omega t-\overline{k}_2 x\right)\right],\\ \overline{k}_2^2 &= A_2 - jB_2.\end{aligned} \tag{3.18}$$

Therefore

$$A_2 = \frac{4k_0^2}{(1+\eta^2)} + \frac{Nl\pi\left[2\ln\left(\frac{qr_0\omega}{c_l}\right) + \frac{1}{\omega^2}\frac{\mu}{r_0^2\rho}\right]}{\left[2\ln\left(\frac{qr_0\omega}{c_0} + \frac{1}{\omega^2}\frac{\mu}{r_0^2\rho}\right)\right]^2 + \left(\pi + \frac{2\gamma}{\omega}\right)^2}, \tag{3.19}$$

$$B_2 = \frac{4k_0^2\eta}{(1+\eta^2)} - \frac{Nl\pi\left(\pi + \frac{2\gamma}{\omega}\right)}{\left[2\ln\left(\frac{qr_0\omega}{c_0}\right) = \frac{1}{\omega^2}\frac{\mu}{\rho r_0^2}\right]^2 + \left(\pi + \frac{2\gamma}{\omega}\right)^2}, \tag{3.20}$$

To find its particular solution, we assume

$$\sigma_2'' = \sigma_{2A}'' \exp\left[j\left(2\omega t - 2\overline{k}_1 x\right)\right],$$

So we get

$$\sigma_{2A}'' = \frac{\rho N\pi l D\sigma_{1A}^2}{4\overline{k}_1^2 - \overline{k}_2^2},$$

The final solution is

$$\sigma_2 = \sigma_2' + \sigma_2'', \tag{3.21}$$

Under rigid boundary conditions: $x = 0$, $\sigma_2 = 0$, Finally, we find the solution to be:

$$\sigma_2 = \frac{\rho N\pi l D\sigma_{1A}^2}{4\overline{k}_1^2 - \overline{k}_2^2} \left[\exp\left(-j\overline{k}_2 x\right) - \exp\left(-2j\overline{k}_1 x\right)\right]\exp(2j\omega t). \tag{3.22}$$

3.1.1.5 Equivalent Nonlinear Parameter

For a non-dissipative elastic medium, the one-dimensional wave equation based on stress containing nonlinear terms can be expressed as follows:

$$\frac{\partial^2\sigma}{\partial x^2} - \frac{1}{c_0^2}\frac{\partial^2\sigma}{\partial t^2} = -\frac{\Gamma}{\lambda + 2\mu}\frac{\partial^2\sigma}{\partial x^2}, \tag{3.23}$$

The solution to the above equation is:

$$\sigma = \sigma_{1A}\exp[j(\omega t - kx)]$$

$$\left[+\frac{\Gamma\omega\sigma_{1A}^2 x}{c_0^3\rho} j\exp[2j(\omega t - kx)]\right], \tag{3.24}$$

Therefore, the relationship between the amplitude of the second harmonic and the fundamental amplitude is as follows:

$$\sigma_{2A} = \frac{\Gamma\omega\sigma_{1A}^2 x}{c_0^3\rho} \tag{3.25}$$

By comparing the above two equations and referring to the solutions of Eqs. (3.15) and (3.22), we derive an "equivalent" nonlinear parameter, which is used to approximately describe the relationship between the second harmonic and the fundamental in the sound propagation in rubber-like porous viscoelastic materials. The expression is as follows:

$$\Gamma = \frac{c_0^3\rho^2\pi NlD\left[\exp\left(-2j\bar{k}_1 x\right) - \exp\left(-j\bar{k}_2 x\right)\right]}{\omega x\left(4\bar{k}_1^2 - \bar{k}_2^2\right)}. \tag{3.26}$$

In the case of short propagation distances, the equation can be simplified using the following approximate inequality: $(\alpha_2 - 2\alpha_1)x \ll 1$, $(k_2 - 2k_1) \ll 1$. From this approximation, we can further derive the amplitude value of the simplified equivalent nonlinear parameter:

$$\Gamma_e = \frac{c_0^3\rho^2\pi NlD}{\omega\sqrt{(2k_1 + k_2)^2 + (2\alpha_1 + \alpha_2)^2}}. \tag{3.27}$$

3.1.1.6 Improved Model Considering Multiple Scattering

Considering the dissipation in porous materials, the longitudinal wave propagation equation can be written as follows:

$$\rho\frac{\partial^2\varepsilon}{\partial t^2} = \frac{\partial^2\sigma}{\partial x^2} + \eta\frac{\partial}{\partial t}\left(\frac{\partial^2\varepsilon}{\partial x^2}\right), \tag{3.28}$$

From Eq. (3.5), we can get:

$$U = \frac{\sigma}{\rho}\pi l / \left[\frac{\omega^2}{2}\ln\left(\frac{qr_0\omega}{2c_l}\right) + \frac{\mu}{r_0^2\rho} - \frac{j\pi\omega^2}{4} - \gamma j\omega\right], \tag{3.29}$$

Substituting Eq. (3.29) into Eq. (3.3), we get:

$$\sigma = (\lambda + 2\mu)\left\{\varepsilon - N\frac{\sigma}{\rho}\pi l / \left[\frac{\omega^2}{2}\ln\left(\frac{qr_0\omega}{2c_l}\right) + \frac{\mu}{r_0^2\rho} - \frac{j\pi\omega^2}{4} - \gamma j\omega\right]\right\}, \tag{3.30}$$

Let

$$A = \frac{\omega^2}{2}\ln\left(\frac{qr_0\omega}{2c_0}\right) + \frac{\mu}{r_0^2\rho} - \frac{j\pi\omega^2}{4} - \gamma j\omega. \tag{3.31}$$

Equation (3.28) can be written as:

$$\begin{aligned} &\left[\frac{(\lambda + 2\mu)\rho A}{\rho A + N\pi l(\lambda + 2\mu)} + \eta(j\omega)\right]\nabla^2\varepsilon + \rho\omega^2\varepsilon = 0 \\ &\nabla^2\varepsilon + \frac{\rho\omega^2}{\frac{(\lambda+2\mu)\rho A}{\rho A+N\pi l(\lambda+2\mu)} + \eta(j\omega)}\varepsilon = 0 \end{aligned} \tag{3.32}$$

Therefore, from Eq. (3.32), the equivalent wave number can be calculated as follows:

$$k^2 = \frac{\rho\omega^2}{\frac{(\lambda+2\mu)\rho A}{\rho A+N\pi l(\lambda+2\mu)} + \eta(j\omega)} \tag{3.33}$$

As discussed earlier, η itself is so small that it can be is so small that it can be ignored, so:

$$\begin{aligned} k^2 &= \frac{\rho\omega^2}{\frac{(\lambda+2\mu)\rho A}{\rho A+N\pi l(\lambda+2\mu)}} \\ &= \frac{\rho\omega^2[\rho A + N\pi l(\lambda + 2\mu)]}{(\lambda + 2\mu)\rho A} \\ &= \frac{\omega^2[\rho A + N\pi l(\lambda + 2\mu)]}{(\lambda + 2\mu)A} \\ &= \frac{\omega^2\rho}{(\lambda + 2\mu)} + \frac{\omega^2 N\pi l(\lambda + 2\mu)}{(\lambda + 2\mu)A} \\ &= k_0^2 + \frac{\omega^2 N\pi l}{A} \\ &= k_0^2 + \frac{N\pi l\omega^2}{\frac{\omega^2}{2}\ln\left(\frac{qr_0\omega}{2c_0}\right) + \frac{\mu}{r_0^2\rho} - \frac{j\pi\omega^2}{4} - \gamma j\omega} \end{aligned}$$

To further derive this equation, we first return to Eq. (3.10), as follows:

$$\frac{B_1\omega^2}{2}\ln\left(\frac{qr_0\omega}{2c_0}\right)+\frac{\mu B_1}{r_0^2\rho}-\frac{j\pi\omega^2}{4}B_1-\gamma j\omega B_1=\frac{\sigma_1}{\rho}.$$

For this equation, we connect the classic form of the physics vibration equation, for example:

$$\ddot{\varepsilon}+\omega_0^2\varepsilon+\cdots=0$$

Connecting such a structural form, we can extract the expression of the resonance frequency from Eq. (3.10):

$$\text{Because: }\frac{B\omega^2}{2}=-\frac{\tilde{B}}{2}$$

$$\text{It follows that: }\omega_0^2=\frac{\frac{\mu}{r_0^2\rho}}{\frac{-\ln(qr_0\omega/2c_0)}{2}}$$

$$\text{Thus, we get: }\frac{\mu}{r_0^2\rho}=-\frac{\omega_0^2}{2}\ln\left(\frac{qr_0\omega}{2c_0}\right)$$

To make this equation hold, the right side ω must be equal to ω_0, that is:

$$\frac{\mu}{r_0^2\rho}=-\frac{\omega_0^2}{2}\ln\left(\frac{qr_0\omega_0}{2c_0}\right)\tag{3.34}$$

Substituting the above equation into the expression of the wave number k^2 above, we get:

$$\begin{aligned}k^2&=k_0^2+\frac{N\pi l\omega^2}{\frac{\omega^2}{2}\ln\left(\frac{qr_0\omega}{2c_0}\right)-\frac{\omega_0^2}{2}\ln\left(\frac{qr_0\omega_0}{2c_0}\right)-\frac{j\pi\omega^2}{4}-\gamma j\omega}\\&=k_0^2+\frac{N\pi l\omega^2}{\frac{\omega^2}{2}\ln\left(\frac{qr_0k_0}{2}\right)-\frac{\omega_0^2}{2}\ln\left(\frac{qr_0\omega_0}{2c_0}\right)-\frac{j\pi\omega^2}{4}-\gamma j\omega}\\&=k_0^2+\frac{2N\pi l\omega^2/\ln(\frac{qr_0k_0}{2})}{\omega^2-\omega_0^2\ln\left(\frac{qr_0\omega_0}{2c_0}\right)/\ln\left(\frac{qr_0k_0}{2}\right)-2j\left(\frac{\pi\omega^2}{4}+\gamma\omega\right)/\ln\left(\frac{qr_0k_0}{2}\right)}\end{aligned}$$

That is:

$$k_m^2=k_0^2+\frac{2N\pi l\omega^2/\ln\left(\frac{qr_0k_0}{2}\right)}{\omega^2-\omega_0^2\ln\left(\frac{qr_0\omega_0}{2c_0}\right)/\ln\left(\frac{qr_0k_0}{2}\right)-2j\left(\frac{\pi\omega^2}{4}+\gamma\omega\right)/\ln\left(\frac{qr_0k_0}{2}\right)}\tag{3.35}$$

According to Kargl's correction idea [24], considering the multiple scattering caused by micropores, we replace all k_0 in the right side of the equation except the first k_0 with k_m. Therefore, the equation becomes an iterative formula

$$k_m^2 = k_0^2 + \frac{2N\pi l\omega^2 / \ln\left(\frac{qr_0k_m}{2}\right)}{\omega^2 - \omega_0^2 \ln\left(\frac{qr_0\omega_0}{2c_0}\right) / \ln\left(\frac{qr_0k_m}{2}\right) - 2j\left(\frac{\pi\omega^2}{4} + \gamma\omega\right) / \ln\left(\frac{qr_0k_m}{2}\right)} \tag{3.36}$$

The final stable k_m obtained from this iterative formula is the corrected wave number we are looking for, which contains the information of the corrected sound speed and the sound attenuation coefficient. Similarly, if we make similar iterative corrections to Eq. (3.27), we will obtain the equivalent sound nonlinearity parameter considering multiple scattering.

3.1.2 Numerical Calculation

We choose the parameters of silicone rubber material as the reference for numerical simulation. Silicone rubber itself has very good rubber-like properties ($K/\mu \gg 1$). Some specific parameters are shown in Table 3.1.

3.1.2.1 Variation of Phase Velocity with Frequency

Figure 3.2 shows the differences in the curves of the normalized phase velocity varying with frequency under different pore radii. In the calculation, the radii are taken as 0.1, 0.2, and 0.3 mm, the volume fraction of the micropores is $NU = 0.1\%$, and the viscous parameter η describing the homogeneous medium without pores is ignored and taken as zero. From the figure, we can see that at very low frequencies, the normalized phase velocities gradually decrease. When the frequency approaches the pore resonance frequency, the phase velocity value increases dramatically until it reaches a peak at the resonance frequency. As the frequency continues to increase, the impact of the pore vibration gradually decreases, and the phase velocity value

Table 3.1 Silicone rubber material parameters

Material	Silicone rubber
Density ρ (Kg/m^3)	1000
Bulk modulus K (Pa)	2.89×10^9
Shear modulus M (Pa)	4×10^5
K/M	7225
Young's modulus E (Pa)	1.20×10^6
Poisson's ratio σ	0.49976

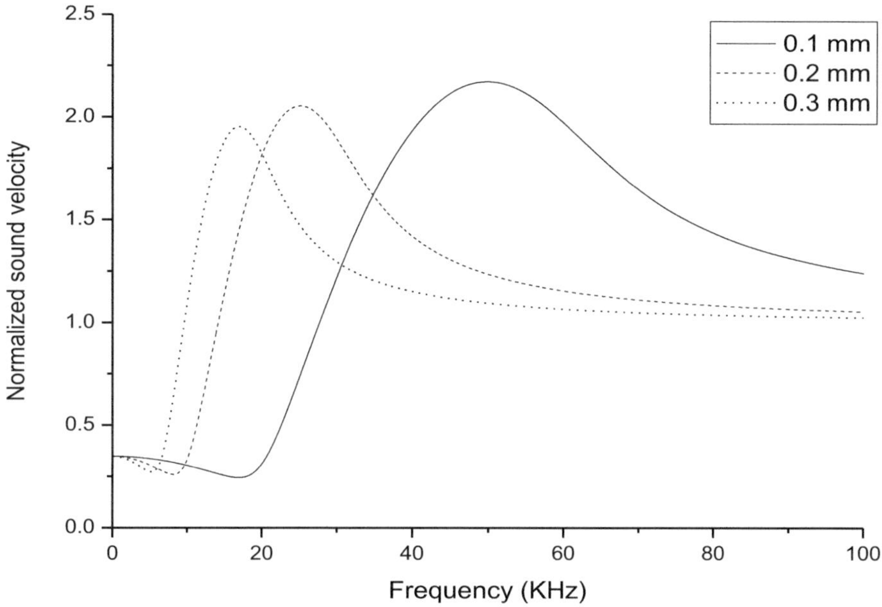

Fig. 3.2 The different normalized phase velocities when different radii are taken under the condition of constant volume fraction. The volume fraction is 0.1%, and the radii are 0.1, 0.2, and 0.3 mm, respectively

drops again, eventually tending towards the propagation phase velocity when the medium does not contain pores. At the same time, as the pore radius increases, the resonance frequency shifts towards lower frequencies, and the absolute value of the phase velocity at the resonance frequency also becomes smaller. In contrast, Fig. 3.3 shows that when the pore size remains unchanged (radius is 0.2 mm), as the volume fraction of the pores gradually increases, the resonance frequency gradually shifts to the right, and the absolute value of the phase velocity at the resonance frequency also gradually increases.

3.1.2.2 Variation of Sound Attenuation Coefficient with Frequency

Similar to Fig. 3.2, Fig. 3.4 shows the sound attenuation coefficient curves under different radii when the volume fraction is fixed, with the parameters taken the same as those in Fig. 3.2. From the figure, we can see that because the nonlinear vibration of the pores reaches its maximum near the resonance frequency, the sound attenuation coefficient reaches its maximum value near the resonance frequency. The farther away from the resonance frequency, the smaller the vibration amplitude of the pores, so the impact of their vibration on propagation can be ignored, and therefore the sound attenuation coefficient is smaller, eventually coinciding with the sound attenuation coefficient curve of the homogeneous medium without pores.

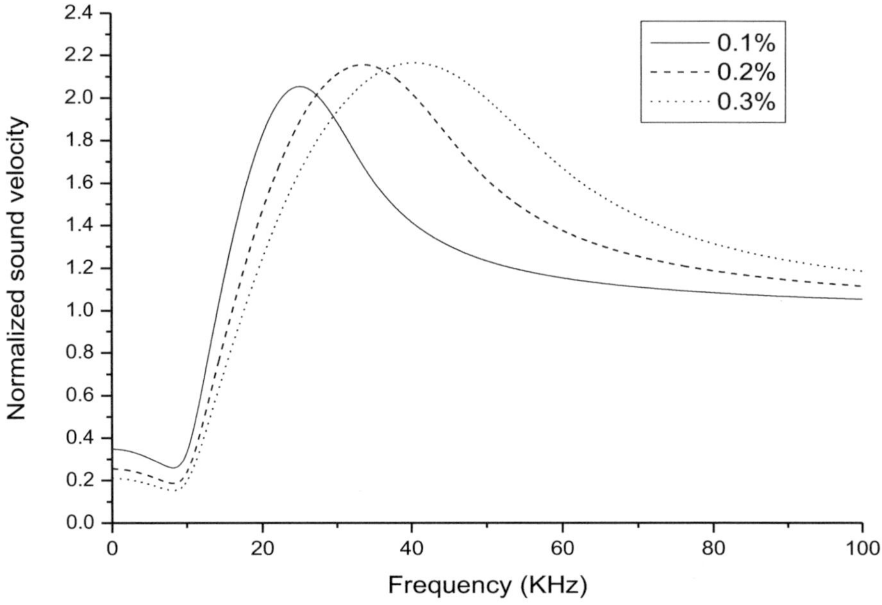

Fig. 3.3 The different normalized phase velocities when different volume fractions are taken under the condition of constant radius. The radius is 0.2, and the volume fractions are 0.1, 0.2, and 0.3%, respectively

As the radius gradually increases, the resonance frequency decreases, so the peak frequency of the sound attenuation coefficient also decreases, and from the figure, its peak absolute value also decreases. Figure 3.5, the description shows the change in the sound attenuation coefficient as the volume fraction occupied by the hole changes, with the radius remaining constant. Clearly, the larger the volume fraction occupied by the hole, the stronger the sound attenuation.

3.1.2.3 Equivalent Nonlinear Parameters

The numerical simulation of the equivalent nonlinear parameter defined by Eq. (3.27) is shown in Fig. 3.6. From the figure, we can see that the frequency at which this parameter reaches its peak is not at the resonance frequency of the hole, but at half of its resonance frequency. For example, when the radius of the micro-hole is 0.2 mm, the corresponding resonance frequency is about 10 kHz, while the frequency at which the equivalent nonlinear parameter reaches its peak is about 5 kHz. In fact, from the figure, we can see that the area where this parameter value is significantly enhanced does indeed include the range near the resonance frequency of the micro-hole, but the enhancement is far less than the half-frequency area. Similarly, as the hole radius increases, the peak frequency of the equivalent nonlinear parameter gradually shifts

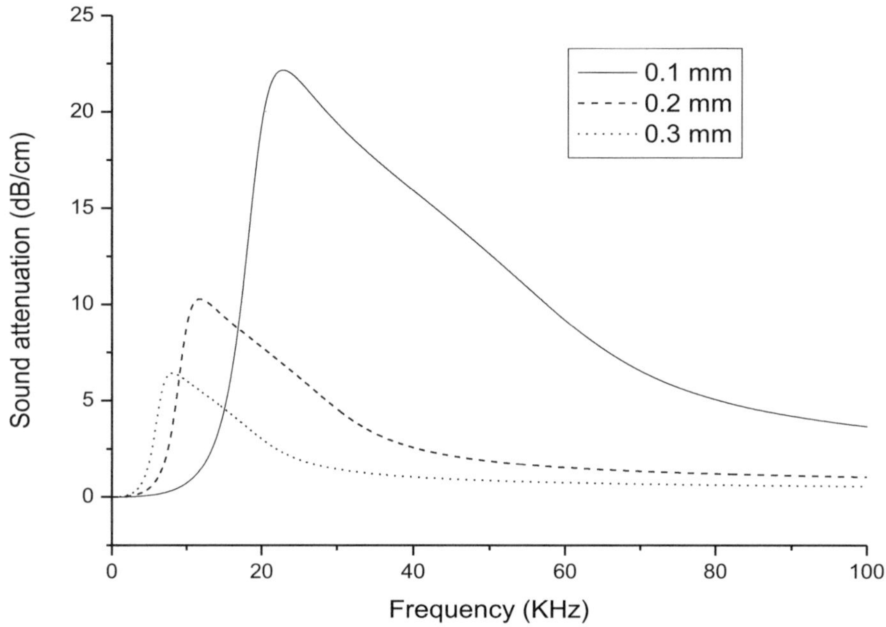

Fig. 3.4 The change in the sound attenuation coefficient with different radii, with the volume fraction remaining constant. The volume fraction is 0.1%, and the radii are 0.1, 0.2, and 0.3 mm respectively

to the low frequency, and the absolute value of the peak also decreases; another obvious trend is that the amplitude of this parameter increases with the increase of the hole volume fraction.

3.1.2.4 The Impact of Multiple Scattering

For the impact of multiple scattering within the medium (also known as hole-to-hole interaction), we first consider the case where the volume fraction is very small, such as $NU = 0.0001\%$, at this time the two acoustic parameters: sound speed and sound attenuation are different from the expressions of Eqs. (3.36) and (3.35) (expressed here by the wave number). Figures 3.7 and 3.8 compare these two parameters with two different theoretical simulations. The conclusion we get from these two figures is: when the volume fraction occupied by the micro-holes is very small, the impact of multiple scattering between the holes is negligible, so the sound speed and sound attenuation values obtained without considering the hole-to-hole interaction are basically consistent with the sound speed and sound attenuation values obtained after considering the hole-to-hole interaction. However, this effect increases with the increase of the volume fraction, and begins to have a substantial impact when the volume fraction reaches a certain value, as seen in Figs. 3.9and3.10, where

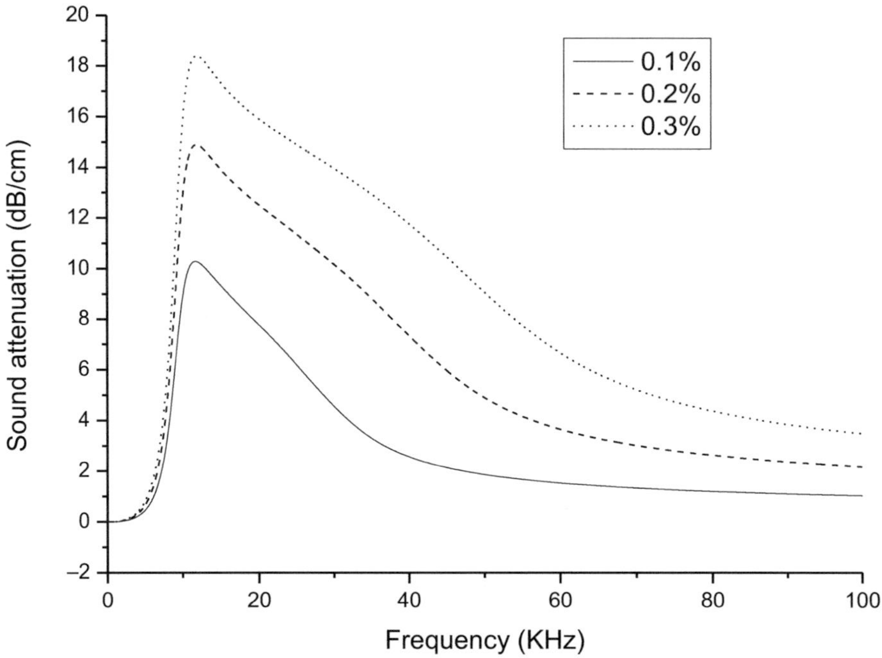

Fig. 3.5 The effect of the volume fraction on the sound attenuation coefficient when the radius remains constant. The radius is 0.2 mm, and the volume fractions are 0.1, 0.2, and 0.3% respectively

the volume fraction is 0.005%. As the volume fraction continues to increase, the interaction between the pores becomes more significant, and the values of the corresponding parameters (especially at the resonance frequency) also change significantly. In Figs. 3.11and3.12, we see that when the volume fraction of the microholes reaches 0.1%, the huge impact of multiple scattering between the holes on these two acoustic parameters cannot be ignored, which reflects the necessity of model correction. Figure 3.13 is a comparison of the corrected equivalent nonlinear parameter with the original one without considering multiple scattering. The obvious difference in the figure shows that this parameter is also affected by the hole-to-hole interaction. In general, the impact of multiple scattering on the propagation of sound waves in this porous material can be ignored when the volume fraction of the microholes is very small (i.e., the arrangement is very sparse); however, when the volume fraction gradually increases to a certain value, the impact of multiple scattering on some acoustic parameters becomes significant, and the consideration of hole-to-hole interaction must be added to the model to better simulate the actual situation. At the same time, we can also see that the impact of multiple scattering is always very small and can be ignored in areas far from the resonance frequency.

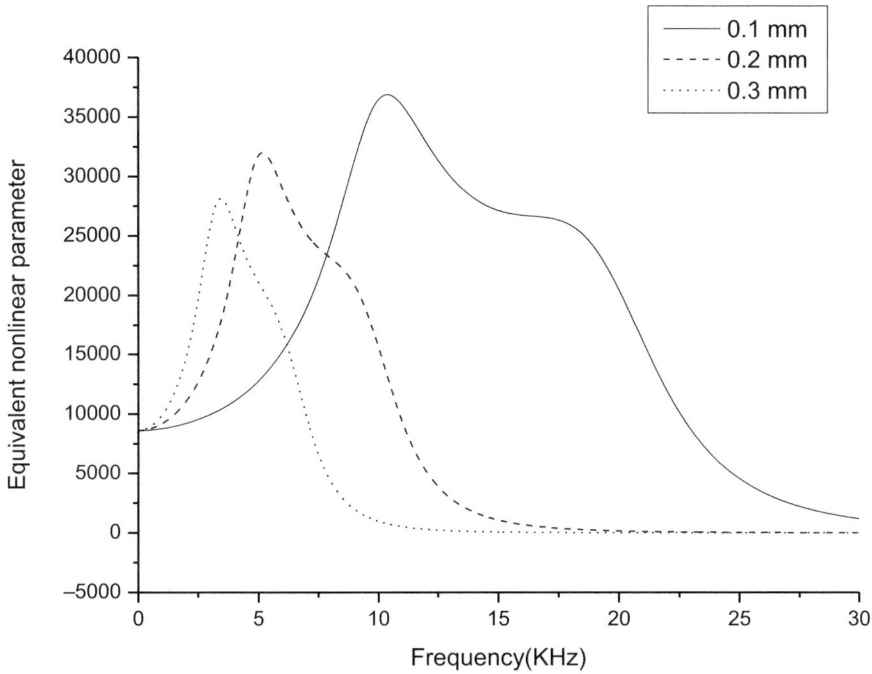

Fig. 3.6 The change in the equivalent nonlinear parameter with frequency for different hole radii, with the volume fraction remaining constant. The volume fraction is 0.1%, and the radii are 0.1, 0.2, and 0.3 mm respectively

3.2 Nonlinear Acoustics of Porous Media Considering Dissipation Effects

Donskoy and others studied the nonlinear acoustics in porous media under the theory of Biot, but they did not consider the dissipation effects of the medium [2]. Since porosity is often related to dissipation, it is more realistic to study wave propagation in porous media. This section is based on Biot's theory, derives the nonlinear equation of wave propagation in dissipative porous media, and conducts calculations and experimental verification. Specifically, we analyzed the effect of dissipation on the second harmonic behavior in such porous media.

3.2.1 *Theory*

Let the vectors $\boldsymbol{u} := (u_1, u_2, u_3)^T$ and $\boldsymbol{w} := (w_1, w_2, w_3)^T$ represent the solid displacement and the fluid displacement relative to the solid, respectively. According to Biot's theory, the general form of the motion equation is [2]

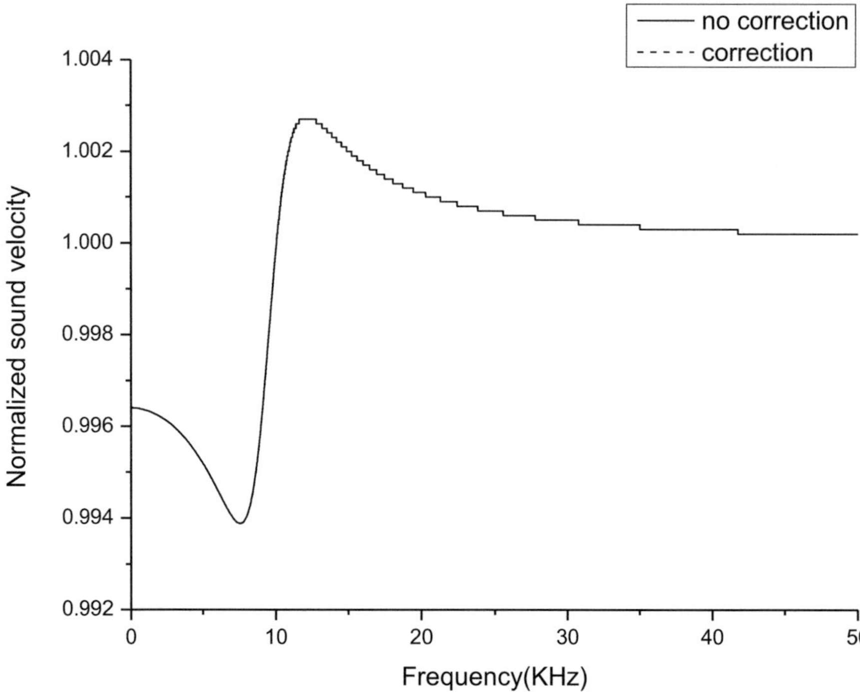

Fig. 3.7 Comparison of normalized sound speed in two models. The volume fraction is 0.0001%, and the radius is 0.2 mm. The solid line represents the uncorrected simulation value, and the dashed line represents the corrected simulation value. As can be seen from the figure, the two basically coincide

$$\begin{aligned} \partial_{x_j}\sigma_{ij} &= \rho\partial_{tt}u_i + \rho_f\partial_{tt}w_i, \\ -\partial_{x_i}p &= \rho_f\partial_{tt}u_i + Y\partial_t w_i. \end{aligned} \tag{3.37}$$

Here, $Y = \eta/\kappa + m\partial_t$ is a visco-dynamic operator. In reference [2], $Y = m\partial_t$ studied the non-dissipative system, while in this section, the study of the dissipative system, $Y = \eta/\kappa + m\partial_t$.

The constitutive relations of porous media are as follows:

$$\begin{aligned} \sigma_{ij} &= 2\mu\varepsilon_{ij} + \delta_{ij}\left[\left(\lambda + \alpha^2 M\right)\varepsilon - \alpha M\xi\right] \\ &\quad + \partial_{\overline{\varepsilon}_{ij}}H|_{p=p^l} - \alpha\theta M\delta_{ij}\delta_{lm}\partial_{\overline{\varepsilon}_{lm}}H|_{p=p^l}, \\ p &= M(\xi - \alpha\varepsilon) + \delta_{ij}\theta M\partial_{\overline{\varepsilon}_{ij}}H|_{p=p^l}. \end{aligned} \tag{3.38}$$

Here, $\overline{\varepsilon}_{ij} = \varepsilon_{ij} + \theta\delta_{ij}p$

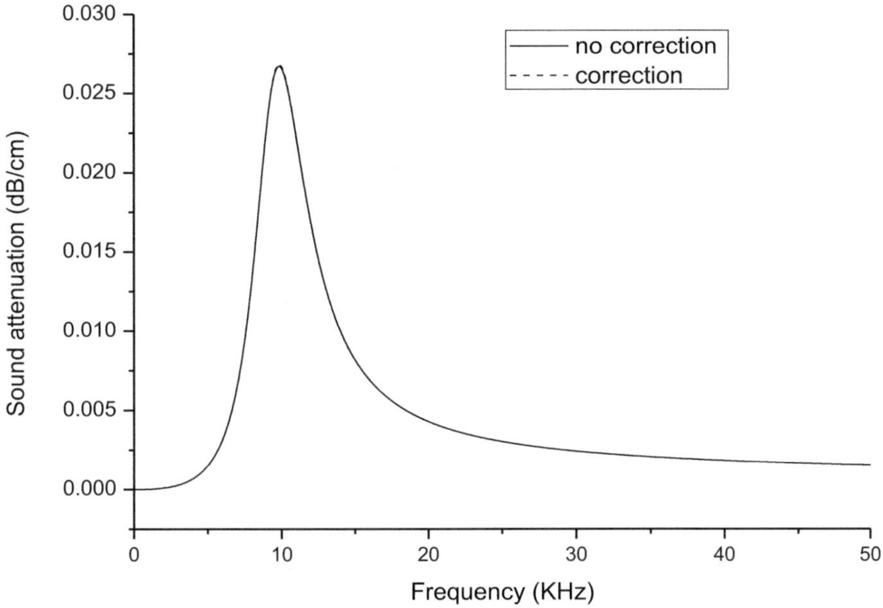

Fig. 3.8 Comparison of sound attenuation coefficients in two models. The volume fraction is 0.0001%, and the radius is 0.2 mm. The solid line represents the uncorrected simulation value, and the dashed line represents the corrected simulation value. As can be seen from the figure, the two basically coincide

Substitute (3.38) into (3.37), the following equation can be obtained:

$$\begin{aligned}&2\mu u_{xx} + \left(\lambda + \alpha^2 M\right)u_{xx} - \alpha M w_{xx}\\&\quad + \partial_{x_j}\left\{\partial_{\overline{\varepsilon}_{ij}} H\left|_{p=p^l} - \alpha\theta M\,\delta_{ij}\delta_{lm}\partial_{\overline{\varepsilon}_{lm}} H\right|_{p=p^l}\right\}\\&= \rho u_{tt} + \rho_f w_{tt},\ M w_{xx} + \alpha M u_{xx}\\&\quad + \partial_{x_j}\left\{\delta_{ij}\theta M\,\partial_{\overline{\varepsilon}_{ij}} H\left|_{p=p^l}\right.\right\} = \rho_f u_{tt} + Y w_t.\end{aligned} \tag{3.39}$$

Therefore, the motion equation is as follows:

$$\begin{aligned}&\left(\lambda_c + \alpha^2 M\right)u_{xx} + \alpha M w_{xx}\\&= \rho M u_{tt} + \rho_f w_{tt} + N_1(u_x, u_{xx}, w_x, w_{xx}),\\&\alpha M u_{xx} + M w_{xx}\\&= \rho_f u_{tt} + Y w_t + N_2(u_x, u_{xx}, w_x, w_{xx}),\end{aligned} \tag{3.40}$$

where, $\lambda_c = \lambda + 2\mu$, N_1 and N_2 are nonlinear functions, defined as follows:

$$\begin{aligned}N_1 &= c_{11}u_x u_{xx} + c_{12}u_x u_{xx} + d_{11}w_x u_{xx} + d_{12}w_x w_{xx},\\N_2 &= c_{21}u_x u_{xx} + c_{22}u_x u_{xx} + d_{21}w_x u_{xx} + d_{22}w_x w_{xx}.\end{aligned} \tag{3.41}$$

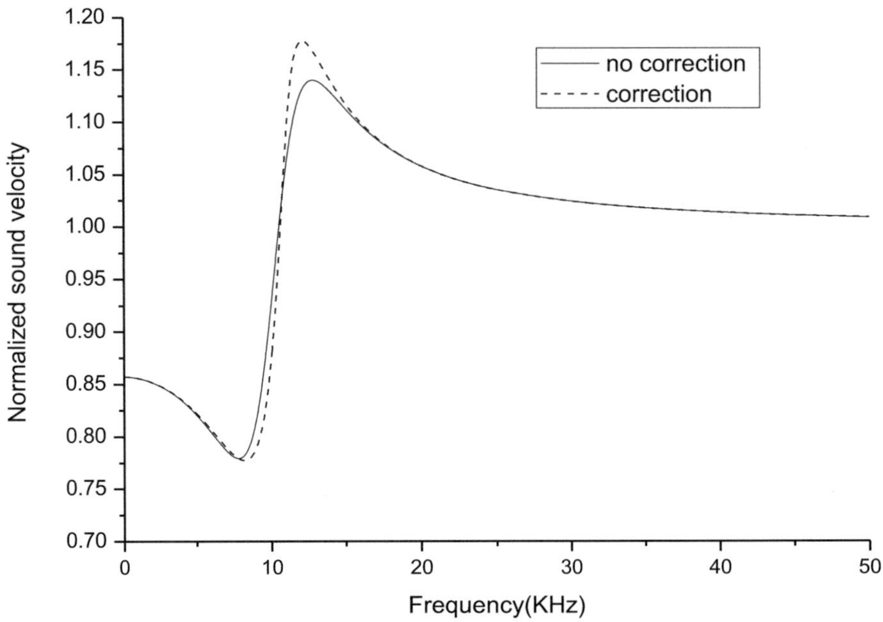

Fig. 3.9 Comparison of normalized sound speed in two models. The volume fraction is 0.005%, and the radius is 0.2 mm. The solid line represents the uncorrected simulation value, and the dashed line represents the corrected simulation value. As can be seen from the figure, the two begin to have visible differences, but far from the resonance frequency, the two still basically coincide

Let $K := D + 6F + G, P := D + 2F$, then

$$\begin{aligned} c_{11} &= 2(\alpha\theta M - 1)\big(D - 2\theta MP\alpha + K\theta^2 M^2\alpha^2\big), \\ c_{12} &= 2(\alpha\theta M - 1)\big(-P\theta M + \alpha K\theta^2 M^2\big), \\ c_{21} &= 2M\theta\big(P - 2\alpha K\theta M + 3\alpha^2 K\theta^2 M^2\big), \\ c_{22} &= 2K\theta^2 M^2(3\theta M\alpha - 1), \\ d_{11} &= c_{12}, d_{12} = 2\theta^2 M^2 K(\theta M\alpha - 1), \\ d_{21} &= c_{22}, d_{22} = 6K\theta^3 M^3. \end{aligned} \tag{3.42}$$

If the dissipation effect is not considered, the characteristic equation of formula (3.40) without considering N_1 and N_2 is

$$\begin{aligned} &\lambda_c M k^4 + \big[\big(2\alpha\rho_f - \rho\big)M - m\big(\lambda_c + \alpha^2 M\big)\big]\omega^2 k^2 \\ &\quad + \left(\rho m - \rho_f^2\right)\omega^4 = 0. \end{aligned} \tag{3.43}$$

Therefore, the linear wave solution is

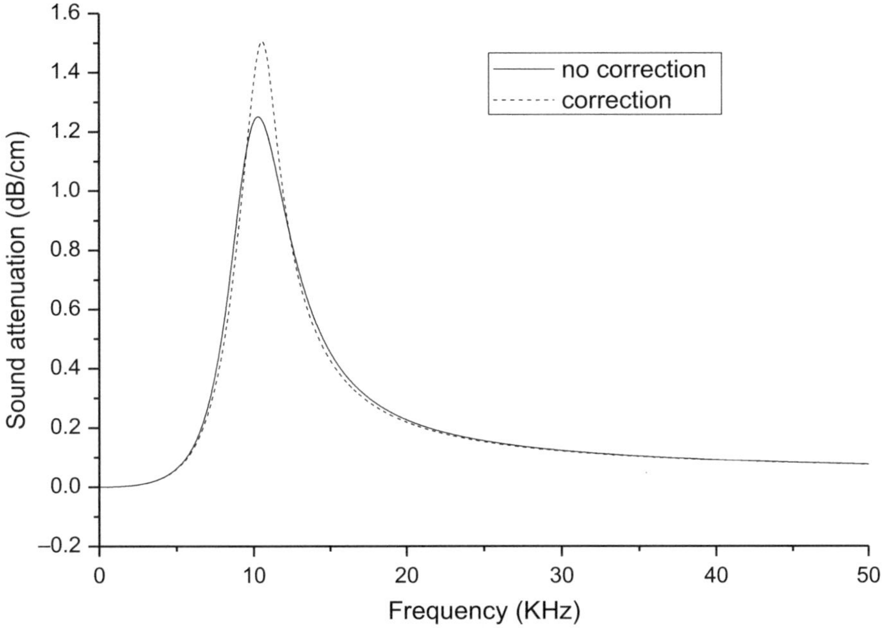

Fig. 3.10 Comparison of sound attenuation coefficients in two models. The volume fraction is 0.005%, and the radius is 0.2 mm. The solid line represents the uncorrected simulation value, and the dashed line represents the corrected simulation value. As can be seen from the figure, the two begin to have visible differences, but far from the resonance frequency, the two still basically coincide

$$\boldsymbol{v}_L(x, t) = \boldsymbol{a}_1 v_1 + \boldsymbol{a}_2 v_2, \tag{3.44}$$

Here: $v_1 = e^{i(\omega t - k_1 x)}$, $v_2 = e^{i(\omega t - k_2 x)}$, $\boldsymbol{a}_1 = (a_1, a_3)^T$, $\boldsymbol{a}_2 = (a_2, a_4)^T$.

If the dissipation effect is considered, the corresponding linear equation of formula (3.45) is as follows:

$$\begin{aligned} &\left(\lambda_c + \alpha^2 M\right) u_{xx} + \alpha M w_{xx} = \rho u_{tt} + \rho_f w_{tt}, \\ &\alpha M u_{xx} + M w_{xx} = \rho_f u_{tt} + m w_{tt} + (\eta/\kappa) w_t. \end{aligned} \tag{3.45}$$

The characteristic equation of formula (3.43) is

$$\begin{aligned} &\lambda_c M k^4 + \{\left(2\alpha\rho_f - \rho\right)M - m\left(\lambda_c + \alpha^2 M\right)]\omega^2 \\ &- \left(\lambda_c + \alpha^2 M\right)(\eta/\kappa) i\omega\} k^2 + \left(\rho m - \rho_f^2\right)\omega^4 \\ &+ \rho\omega^2(\eta/\kappa) i\omega = 0. \end{aligned} \tag{3.46}$$

Then,

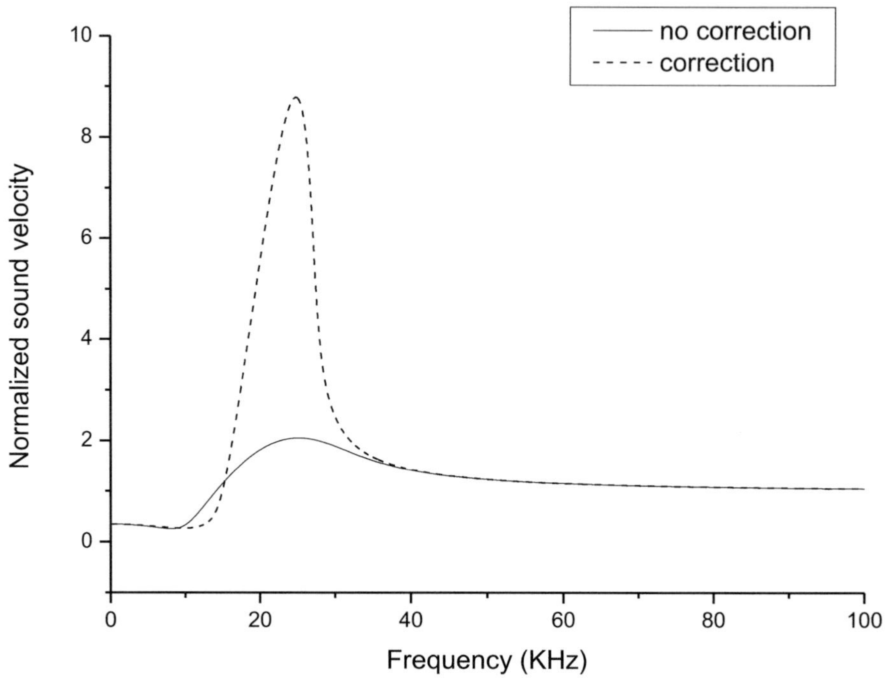

Fig. 3.11 Comparison of normalized sound speed in two models. The volume fraction is 0.1%, and the radius is 0.2 mm. The solid line represents the uncorrected simulation values, and the dashed line represents the corrected simulation values. As can be seen from the figure, there is a very obvious difference between the two, especially near the resonance frequency, there is a multiple difference, but far from the resonance frequency, the two still basically coincide

$$k^* = k_j + \alpha_j i, j = 1, 2. \tag{3.47}$$

The nonlinear wave solution is

$$\begin{aligned} \boldsymbol{v}_N(x,t) &= \boldsymbol{b}_1(x)v_1^2 + \boldsymbol{b}_2(x)v_2^2 + \boldsymbol{b}_3(x)v_1v_2 \\ &= \boldsymbol{b}_1(x)e^{i2(\omega t - k_1 x)}e^{-2\alpha_1 x} + \boldsymbol{b}_2(x)e^{i2(\omega t - k_2 x)}e^{-2\alpha_2 x} \\ &\quad + \boldsymbol{b}_3(x)e^{i[2\omega t - (k_1+k_2)x]}e^{-(\alpha_1+\alpha_2)x}. \end{aligned} \tag{3.48}$$

Here: $\boldsymbol{b}_1 = (b_1(x), b_4(x))^T, \boldsymbol{b}_2 = (b_2(x), b_5(x))^T, \boldsymbol{b}_3 = (b_3(x), b_6(x))^T$.

The linear wave solution is now operated by the nonlinear operator N, and the nonlinear wave is acted upon by the linear operator L, $N[\boldsymbol{v}] = [\boldsymbol{C}u_x + \boldsymbol{D}w_x]v_{xx}, L[\boldsymbol{v}] = [\boldsymbol{A}\partial_{xx} - \boldsymbol{R}\partial_{tt}]v$, where A, R, C and D are the same as in reference [2].

$$L[\boldsymbol{v}_N] = N[\boldsymbol{v}_L]. \tag{3.49}$$

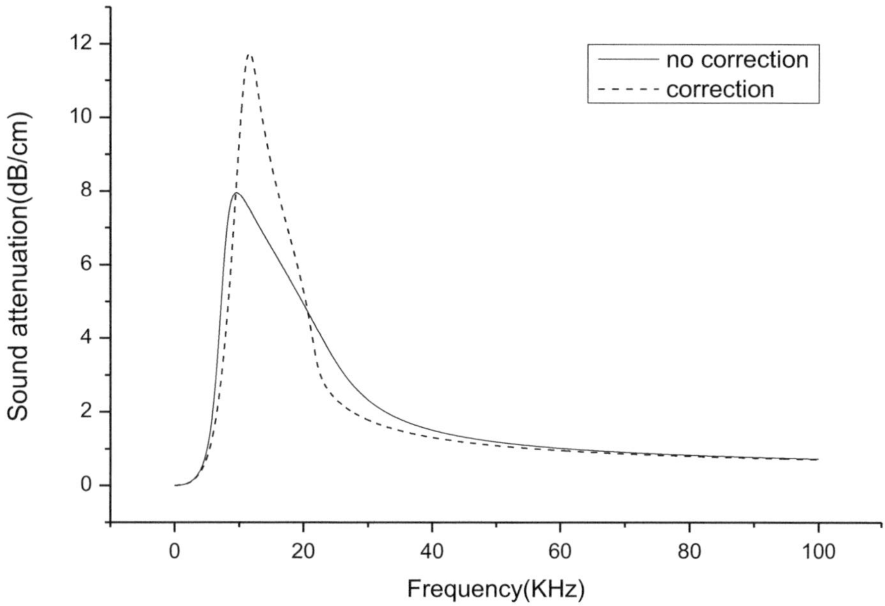

Fig. 3.12 Comparison of sound attenuation coefficients in two models. The volume fraction is 0.1%, and the radius is 0.25 mm. The solid line represents the uncorrected simulation values, and the dashed line represents the corrected simulation values. As can be seen from the figure, there is a very obvious difference between the two, especially near the resonance frequency, there is a multiple difference, but far from the resonance frequency, the two still basically coincide

From the coefficient v_1^2

$$\begin{aligned} &A\Big[\boldsymbol{b}_1''(x) - 4(ik_1 + \alpha_1)\boldsymbol{b}_1'(x)\Big] \\ &+ 4\big[\boldsymbol{R}\omega^2 + \big(ik_1 + \alpha_1)^2\boldsymbol{A}\big)\big]\boldsymbol{b}_1(x) \\ &= -(ik_1 + \alpha_1)^3(a_1\boldsymbol{C} + a_3\boldsymbol{D})a_1. \end{aligned} \tag{3.50}$$

We consider the slow variation of the second harmonic amplitude, b_j (j=1, 2, 3) the second derivative can be ignored.

$$\begin{aligned} &A\Big[\boldsymbol{b}_1''(x) - 4(ik_1 + \alpha_1)\boldsymbol{b}_1'(x)\Big] + 4\big[\omega^2\boldsymbol{R}^2 + (ik_1 + \alpha_1)^2\big]\boldsymbol{A} \\ &= -(ik_1 + \alpha_1)^3(a_1\boldsymbol{C} + a_3\boldsymbol{D})a_1, \\ \boldsymbol{b}_1'(x) &= \boldsymbol{A}^{-1}\frac{\omega^2\boldsymbol{R}^2 + (ik_1 + \alpha_1)^2 b_1(x)\boldsymbol{A}}{ik_1 + \alpha_1} \\ &- \boldsymbol{A}^{-1}\frac{(ik_1 + \alpha_1)^3}{4(ik_1 + \alpha_1)}(a_1\boldsymbol{C} + a_3\boldsymbol{D})a_1 \end{aligned}$$

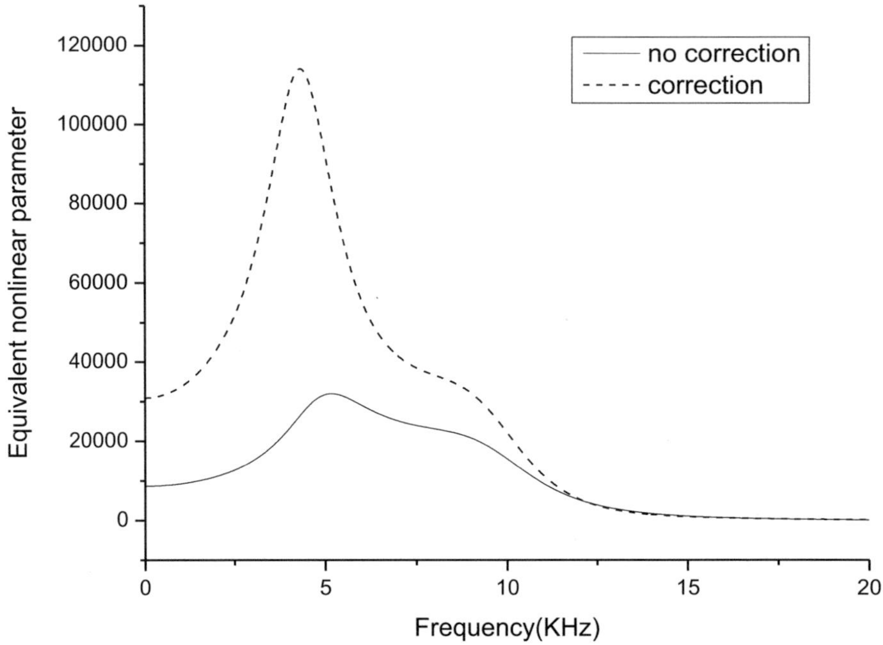

Fig. 3.13 Comparison of equivalent nonlinear parameters in two models. The volume fraction is 0.1%, and the radius is 0.2 mm. The solid line represents the uncorrected simulation values, and the dashed line represents the corrected simulation values. As can be seen from the figure, there is a very obvious difference between the two, especially near the resonance frequency, there is a multiple difference, but far from the resonance frequency, the two still basically coincide

$$\begin{aligned} &= \boldsymbol{A}^{-1}\frac{\left[\omega^2\boldsymbol{R}^2 + (ik_1+\alpha_1)^{\boldsymbol{2}}\boldsymbol{A}\right]b_1(x)}{ik_1+\alpha_1} \\ &- \boldsymbol{A}^{-1}\frac{(ik_1+\alpha_1)^2(a_1\boldsymbol{C}+a_3\boldsymbol{D})a_1}{4}. \end{aligned} \tag{3.51}$$

From the coefficient v_2^2

$$\begin{aligned} &\boldsymbol{A}\left[\boldsymbol{b}_2''(x) - 4(ik_2+\alpha_2)b_2'(x)\right] + 4\left[\omega^2\boldsymbol{R}^2 + (ik_2+\alpha_2\boldsymbol{A})^2\right] \\ &= -(ik_2+\alpha_2)^3(a_2\boldsymbol{C}+a_4\boldsymbol{D})a_2, \\ \boldsymbol{b}_2'(x) &= \boldsymbol{A}^{-1}\frac{\omega^2\boldsymbol{R}^2 + (ik_2+\alpha_2)^2b_2(x)\boldsymbol{A}}{ik_2+\alpha_2} \\ &- \boldsymbol{A}^{-1}\frac{(ik_2+\alpha_2)^3}{4(ik_2+\alpha_2)}(a_2\boldsymbol{C}+a_4\boldsymbol{D})a_2 \\ &= \boldsymbol{A}^{-1}\frac{\left[\omega^2\boldsymbol{R}^2 + (ik_2+\alpha_2)^2A\right]\boldsymbol{b}_2(x)}{ik_2+\alpha_2} \end{aligned}$$

$$- \boldsymbol{A}^{-1} \frac{(ik_2 + \alpha_2)^2 (a_2 \boldsymbol{C} + a_4 \boldsymbol{D}) a_2}{4}. \tag{3.52}$$

From the coefficient $v_1 v_2$:

$$\begin{aligned} &\boldsymbol{A}\big[\boldsymbol{b}_3''(x) + 2(ik_1 + \alpha_1 + ik_2 + \alpha_2)\boldsymbol{b}_3'(x)\big] \\ &\quad + \big[4\omega^2 \boldsymbol{R}^2 + (ik_1 + ik_2 + \alpha_1 + \alpha_2)^2 \boldsymbol{A}\big]\boldsymbol{b}_3(x) \\ &= -(ik_2 + \alpha_2)(ik_1 + \alpha_1)^2 (a_2 \boldsymbol{C} + a_4 \boldsymbol{D}) a_1 \\ &\quad + (ik_1 + \alpha_1)(ik_2 + \alpha_2)^3 (a_2 \boldsymbol{C} + a_4 \boldsymbol{D}) a_2, \\ \boldsymbol{b}_3'(x) &= A^{-1} \frac{\{\big[4\omega^2 \boldsymbol{R}^2 - (ik_1 + ik_2)^2 \boldsymbol{A}\big]\boldsymbol{b}_3(x)}{2(ik_1 + \alpha_1 + ik_2 + \alpha_2)} \\ &\quad + \big[(ik_2 + \alpha_2)(ik_1 + \alpha_1)^2 (a_2 \boldsymbol{C} + a_4 \boldsymbol{D}) a_1 \\ &\quad + (ik_1 + \alpha_1)(ik_2 + \alpha_2)^2 (a_1 \boldsymbol{C} + a_3 \boldsymbol{D}) a_2\big]. \end{aligned} \tag{3.53}$$

The general form of the system is as follows:

$$\boldsymbol{y}'(x) = \boldsymbol{B}y(x) + \boldsymbol{f}, \tag{3.54}$$

For v_1^2:

$$\boldsymbol{B} = \frac{\boldsymbol{A}^{-1}(\omega^2 \boldsymbol{R}) + (ik_1 + \alpha_1)^2}{ik_1 + \alpha}, \tag{3.55}$$

$$\boldsymbol{f} = -\boldsymbol{A}^{-1} \frac{(ik_1 + \alpha_1)^2 (a_1 \boldsymbol{C} + a_3 \boldsymbol{D}) a_1}{4}. \tag{3.56}$$

Here,

$$\boldsymbol{A}^{-1} = \begin{pmatrix} \frac{1}{\lambda_c} & -\frac{\alpha}{\lambda_c} \\ -\frac{\alpha}{\lambda_c} & \frac{\lambda_c + \alpha^2 M}{\lambda_c M} \end{pmatrix}. \tag{3.57}$$

So the matrix B has the following form:

$$\begin{aligned} \boldsymbol{B} &= \frac{\boldsymbol{A}^{-1}(\omega^2 \boldsymbol{R}) + (ik_1 + \alpha_1)^2}{ik_1 + \alpha_1} = \frac{\boldsymbol{A}^{-1}\omega^2 \boldsymbol{R}}{ik_1 + \alpha_1} + (ik_1 + \alpha_1) \\ &= \frac{1}{ik_1 + \alpha_1} \begin{pmatrix} \frac{\omega^2}{\lambda_c}(\rho - \alpha \rho_f) + (ik_1 + \alpha_1)^2 & \frac{\omega^2}{\lambda_c}(\rho_f - \alpha m) \\ \omega^2 \rho_f \frac{\lambda_c + \alpha^2 M}{\lambda_c M} - \omega^2 \frac{\alpha}{\lambda_c} \rho & \begin{matrix} -\omega^2 \dfrac{\alpha}{\lambda_c} \rho_f + \omega^2 \dfrac{\lambda_c + \alpha^2 M}{\lambda_c M} m \\ + (ik_1 + \alpha_1)^2 \end{matrix} \end{pmatrix}. \end{aligned} \tag{3.58}$$

The matrix F has the following form:

$$\begin{pmatrix} f_1 \\ f_4 \end{pmatrix} = \frac{(ik_1+\alpha_1)^2}{4\lambda_c}$$
$$\begin{pmatrix} a_1^2(c_{11}-\alpha c_{12}) + 2a_1a_3(c_{12}-\alpha c_{22}) + a_3^2(d_{12}-\alpha d_{22}) \\ a_1^2\left(c_{21}\frac{\lambda_c+\alpha^2 M}{M} - \alpha c_{11}\right) + 2a_1a_3\left(\frac{\lambda_c+\alpha^2 M}{M}c_{22} - \alpha c_{12}\right) \\ + a_3^2\left(\frac{\lambda_c+\alpha^2 M}{M}d_{22} - \alpha d_{12}\right) \end{pmatrix}. \quad (3.59)$$

Therefore:

$$b_1(x) = -\frac{b_{12}f_2 - b_{22}f_1}{\kappa}x + \left[\frac{1}{\kappa^2}(b_{12}f_2 + b_{11}f_1)\right](e^{\kappa x} - 1) \quad (3.60)$$

$$b_4(x) = -\frac{b_{11}}{b_{12}}\frac{b_{22}f_1 - b_{12}f_2}{\kappa}x + \frac{b_{22}}{b_{12}}\left(\frac{b_{12}f_2 + b_{11}f_1}{\kappa^2}\right)(e^{\kappa x} - 1) \quad (3.61)$$

Here:

$$\begin{pmatrix} b_{11} & b_{12} \\ b_{21} & b_{22} \end{pmatrix}$$
$$= \frac{1}{ik_1+\alpha_1}\begin{pmatrix} \frac{\omega^2}{\lambda_c}(\rho - \alpha\rho_f) + (ik_1+\alpha_1)^2 & \frac{\omega^2}{\lambda_c}(\rho_f - \alpha m) \\ \omega^2\rho_f\frac{\lambda_c+\alpha^2 M}{\lambda_c M} - \omega^2\frac{\alpha}{\lambda_c}\rho & -\omega^2\frac{\alpha}{\lambda_c}\rho_f + \omega^2\frac{\lambda_c+\alpha^2 M}{\lambda_c M}m + (ik_1+\alpha_1)^2 \end{pmatrix}. \quad (3.62)$$

The solutions of the coefficients v_2^2 and are derived similarly.

3.2.2 Numerical Results

The parameters used in the calculation are the same as those in the literature [2]. The other parameters used for calculation are the center frequency of the emission transducer is 25 kHz, and the fluid viscosity of the sample is 0.01. The function of the change of the second harmonic amplitude with distance can be seen in Fig. 3.13. In addition, when the porosity of the sample is 0.125, we find that the amplitude increases with distance until it reaches a maximum at about 0.064 m, and then the amplitude decays with the increase of distance. This is explained as the interaction between nonlinearity and attenuation. At first, nonlinearity dominates compared to attenuation until the amplitude reaches its maximum, then attenuation becomes the dominant factor, and the amplitude gradually approaches zero. The physical mechanism is quite clear, the initial nonlinearity dominates compared to dissipation, causing the amplitude to grow with distance until dissipation is stronger than nonlinearity. Since the attenuation effect is exponential, it strengthens with the increase of distance, and eventually dominates, causing the amplitude to asymptotically decay to zero with distance. In addition, since attenuation is proportional to frequency, and since the second harmonic frequency is higher than the fundamental harmonic frequency, the attenuation of the nonlinear medium is higher. When nonlinearity and attenuation balance, the second harmonic reaches its peak. The higher the frequency, the greater the nonlinearity and attenuation, the maximum second harmonic occurs at a distance of 0.064 m, Fig. 3.14 shows the amplitude of the second harmonic changes with porosity, we find that nonlinearity does not monotonically change with porosity, at $x = 0.1$ m, we find the maximum occurs at a porosity of 0.3, it provides a method to obtain the porosity value of porous media. Figure 3.15 shows the change of the second harmonic amplitude with the viscosity of the medium fluid. The amplitude decreases with the increase of fluid viscosity, which is easy to understand. Due to the increase of fluid viscosity, the sound attenuation also increases, which leads to the decrease of the second harmonic. The change of the second harmonic amplitude with the shear coefficient of the medium is shown in Fig. 3.16. The change of the amplitude with the Poisson's ratio of the medium is shown in Fig. 3.17, we find that, under the shear coefficient 8×10^7, the second harmonic amplitude is the largest, the second harmonic amplitude gradually increases with the increase of the sample's Poisson's ratio, and decreases after reaching 0.28.

3.2.3 Experimental System and Experimental Results

The experimental system is shown in Fig. 3.18, an arbitrary waveform generator (Agilent 33,250, Loveland, CO, USA), emits a pulse modulated signal (center frequency of 25kHz, 40 cycles, pulse repetition frequency of 100 Hz), the signal is added to the power amplifier (ENI A150, Rochester, NY, USA), the amplifier's signal is added to the ultrasonic transducer, the sound wave passes through the porous sample and is received by the accelerometer (PCB PIEZOTRONICS, ICP, USA), and is sampled by the oscilloscope and stored on the computer for processing. The sample is cubic, 6 cm wide, 6 cm high, and of different lengths (5, 10, 15, and 20 cm respectively). The volume ratio of the sample's cement sand is 1 : 3. In order to simulate a porous medium with viscosity, some steel balls with a radius of 6 mm are embedded in the sample. The porosity of the sample is 0.125. For comparison, we also made samples without steel balls. We put all the samples in water for two weeks until the weight of the samples did not increase.

The power spectrum of the received signal of the sample with a porosity of 0.125 is shown in Fig. 3.19, the amplitude of the second harmonic acceleration of these porous samples changes with distance as shown in Fig. 3.20a. We also provide the experimental results of samples without pores (Fig. 3.20b). The measured sound speed is 2500 m/s, the sample density is 2452 kg/m^3, the experimental results are consistent with the theoretical analysis of the change trend of the second harmonic with

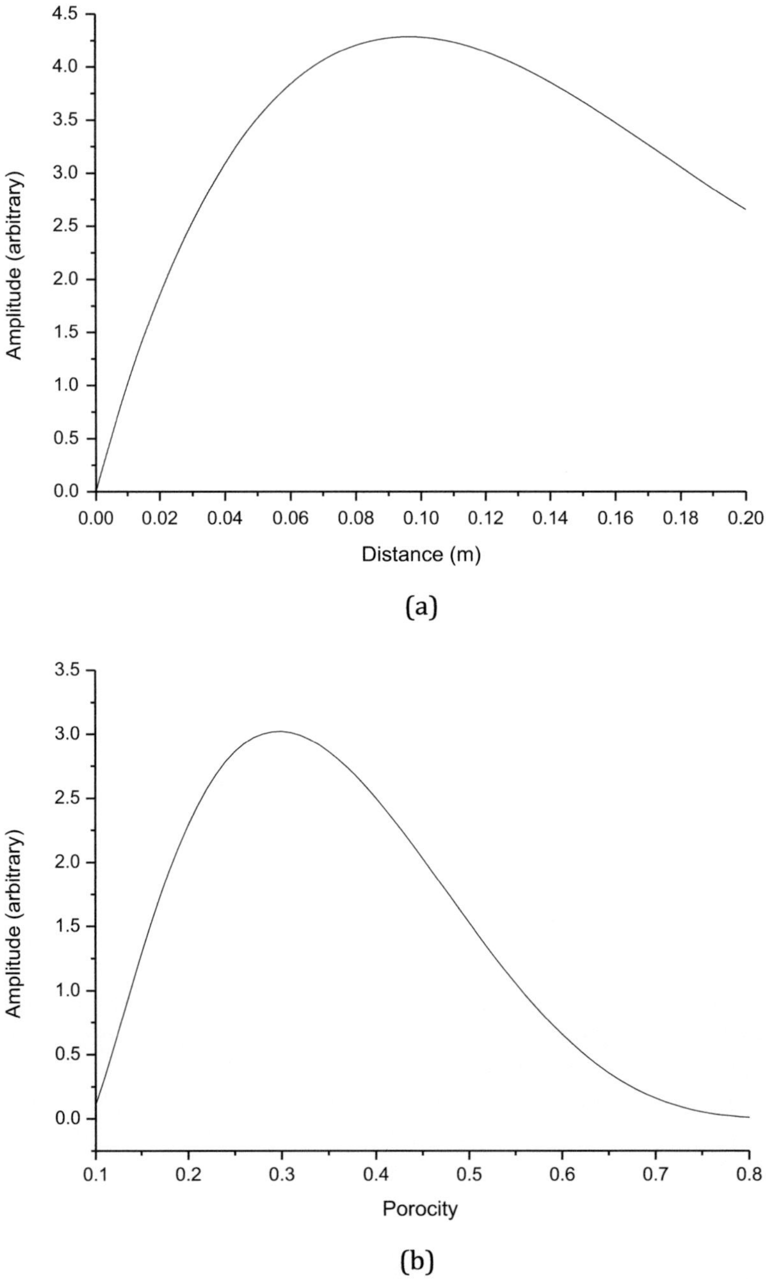

Fig. 3.14 Changes in the second harmonic amplitude with distance (**a**) and porosity (**b**)

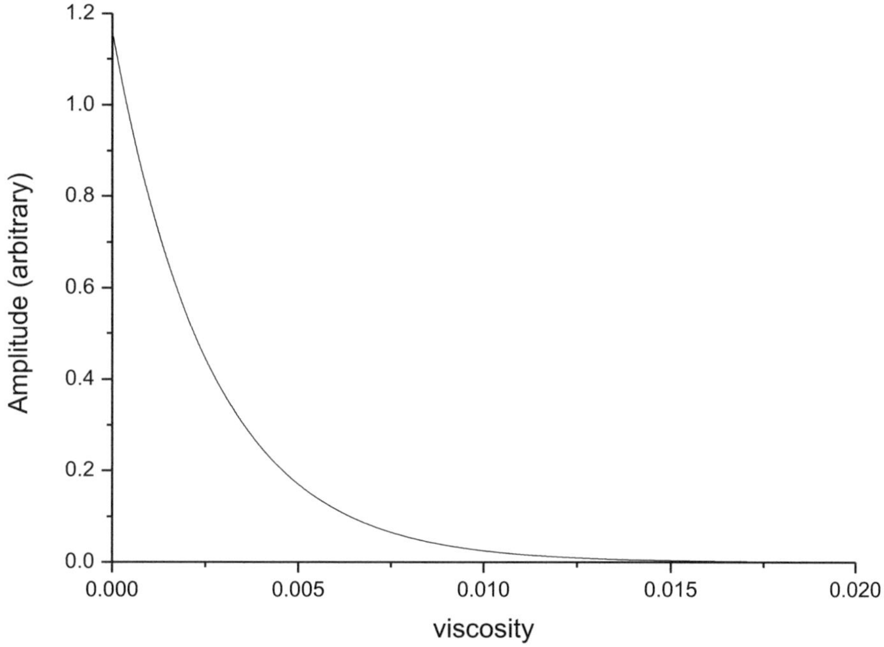

Fig. 3.15 Changes in the second harmonic amplitude with the fluid viscosity of the medium

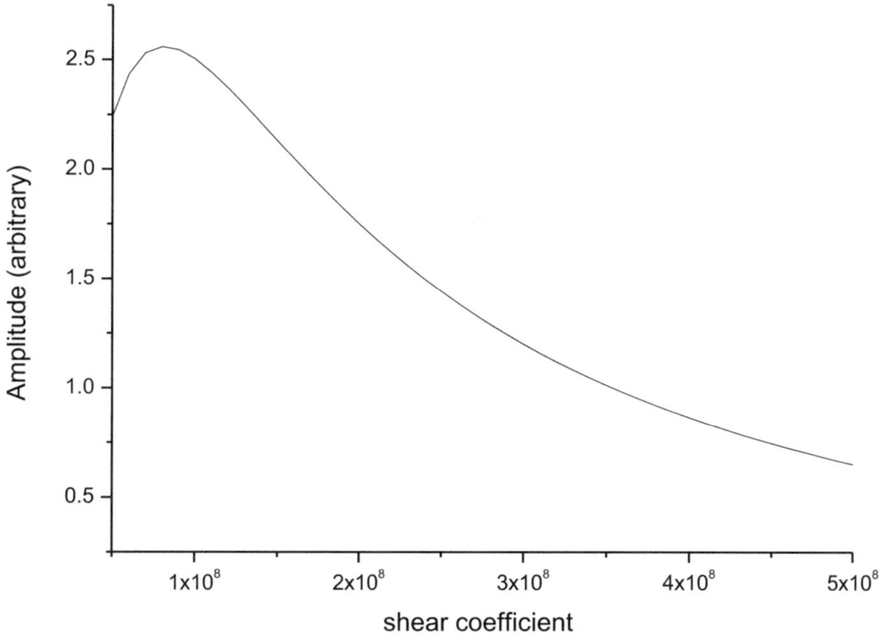

Fig. 3.16 Changes in the second harmonic amplitude with the shear modulus of the medium

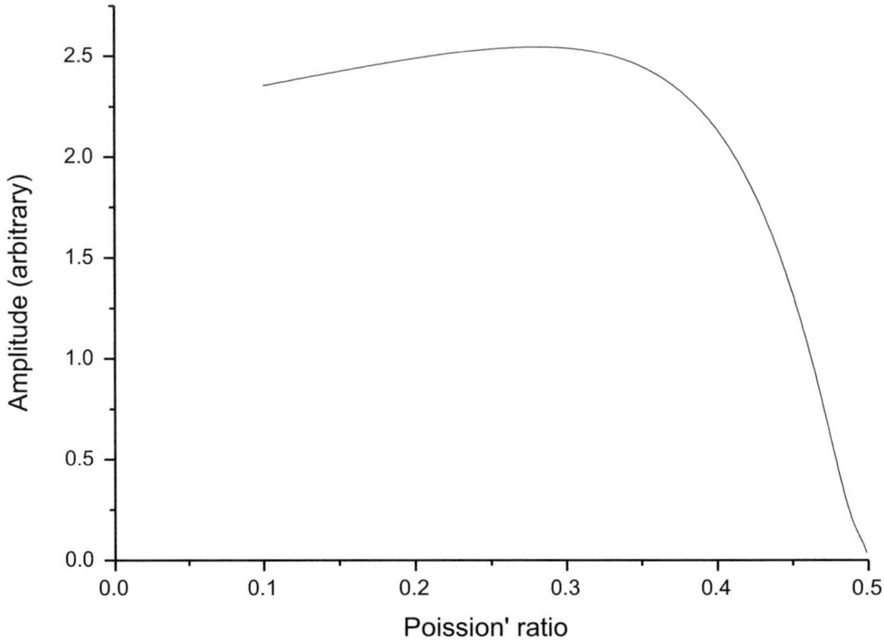

Fig. 3.17 Changes in the second harmonic amplitude with the Poisson's ratio

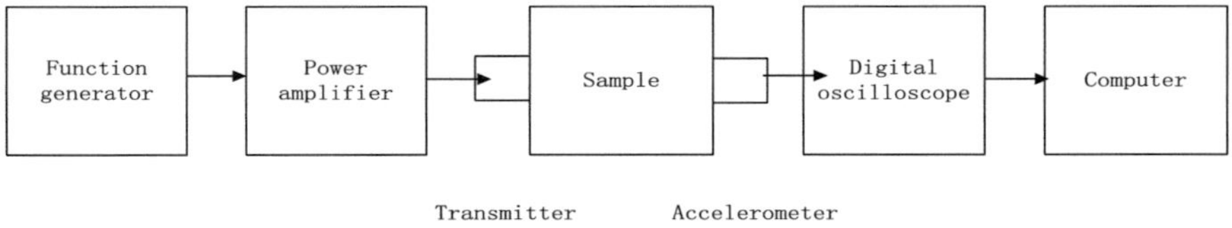

Fig. 3.18 Experimental system

distance (Fig. 3.13). For samples with pores, the amplitude of the second harmonic is larger. Although we cannot determine all the physical parameters of the sample, the experimental results can qualitatively prove the validity of the theoretical model.

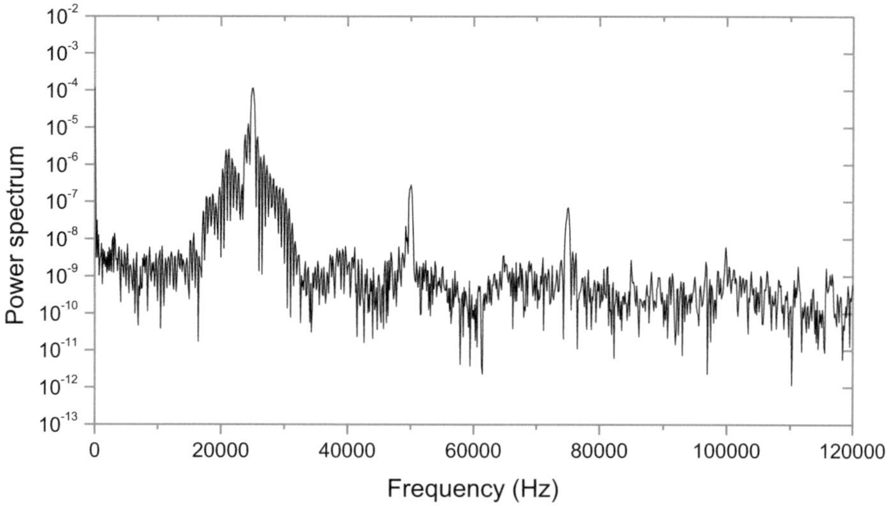

Fig. 3.19 Power spectrum of the received signal of the sample

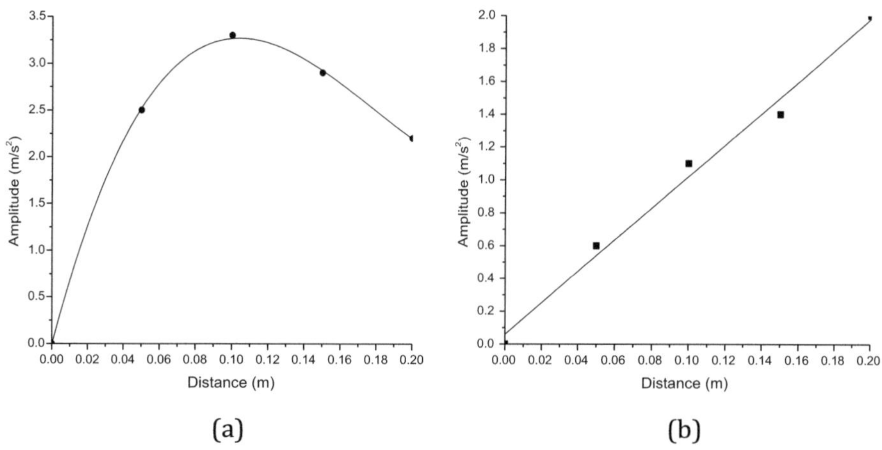

Fig. 3.20 Changes in the second harmonic amplitude with distance, **a** porous sample (0.125), **b** non-porous sample

References

1. Ostrovsky LA. Nonlinear properties of an elastic medium with cylindrical pores. Soviet Phys Acoust. 1989;35(3):286–9.
2. Donskoy DM, Khashanah K, McKee TG, et al. Nonlinear acoustic wave in porous media in the context of Biot's theory. J Acoust Soc Am. 1997;102:2521–28.
3. Ostrovsky LA. Wave progresses in media with strong acoustic nonlinearity. J Acoust Soc Am. 1991;90:3332–7.

4. Biot MA. Theory of propagation of elastic waves in a fluid-saturated porous solid, part I: low frequency range. J Acoust Soc Am. 1956;28(2):168–78.
5. Biot MA. Theory of propagation of elastic waves in a fluid-saturated porous solid, part I: higher frequency range. J Acoust Soc Am. 1956;28(2):179–91.
6. Donskoy DM, Sutin AM. Nonlinear scattering and propagation of longitudinal acoustic waves in porous media. Soviet Phys Acoust. 1984;30:358–61.
7. Geerits TW. Acoustic propagation through porous media revisited. J Acoust Soc Am. 1996;100:2949–59.
8. Dazel V. Tournat DV. Nonlinear Biot waves in porous media with application to unconsolidated granular media. J Acoust Soc Am. 2010;127:692–702.
9. Obsert H. Resonant sound absorbers, technical aspects of sound. Richardon EG, editor; 1957.
10. Meyer E, Brendel K, Tamm KJ. Pulsation oscillation of cavities in rubber. J Acoust Soc Am. 1958;30:1116–24.
11. Gaunaurd GC. One-dimensional model for acoustic absorption in a viscoelastic medium considering cylindrical cavities. J Acoust Soc Am. 1977;62:298–307.
12. Gaunaurd GC, Uberall H. Theory of resonant scattering from spherical cavities in elastic and viscoelastic media. J Acoust Soc Am. 1978;63(6):1699–972.
13. Hladky-Hennion AC, Decarpigny JN. Analysis of the scattering of a plane wave by a doubly periodic structure using the finite element method: application to Alberich anechoic coatings. J Acoust Soc Am. 1991;90:3356–67.
14. Easwaran V, Munjal ML. Analysis of reflection characteristics of a normal incidence plane wave on resonant sound absorbers: a finite element approach. J Acoust Soc Am. 1993;93:1308–18.
15. Fan Z, Ma J, Liang B, et al. The nonlinear acoustic wave propagation in porous rubberlike medium. Acta Acust Acust. 2006;92:217–34.
16. Liang B, Zhu ZM, Cheng JC. The propagation of acoustic wave in viscoelastic medium permeated with air-filled bubbles. Chin Phys B. 2006;15:412–21.
17. Liang B, Cheng JC. Acoustic localization in weakly compressible elastic media containing random air bubbles. Phys Rev E. 2007;75:016605.
18. Foldy LL. The multiple scattering of waves. Phys Rev. 1945;67:107–19.
19. Weston DE. Acoustic interaction effects in arrays of small spheres. J Acoust Soc Am. 1966;39:316–22.
20. Miller DL. Acoustical interaction of spherical and cylindrical bubbles on plane sheets and ribbons. Ultrasonics. 1980;18:277–82.
21. Miller DL. Ultrasonic detection resonant cavitation bubbles flow tube second-harmonic emission. Ultrasonics. 1981;19:217–24.
22. Ye Z, Ding L. Acoustic dispersion and attenuation relations in bubbly mixture. J Acoust Soc Am. 1995;98:1629–36.
23. Henyey FS. Corrections to Foldy's effective medium theory for propagation in bubble clouds and other collections of very small scatterers. J Acoust Soc Am. 1999;105:2149–54.
24. Kargl SG. A modification of Church's model was made by combining it with Kargl's model. J Acoust Soc Am. 2002;111:168–73.
25. Ma J, Yu JF, Fan Z, et al. Acoustic nonlinearity of liquid containing encapsulated microbubbles. J Acoust Soc Am. 2004;116:186–93.
26. Feng YL, Liu XZ, Liu JH, et al. The nonlinear propagation of acoustic waves in a viscoelastic medium containing cylinder micropores. Chin Phys B. 2009;18(9):3909–17.

Chapter 4
Non-classical Nonlinear Propagation of Acoustic Waves in Solids with Cracks

Ultrasonic nondestructive evaluation has two methods, linear and nonlinear. Research in the linear aspect is relatively abundant both domestically and internationally, and has achieved certain results and has been applied in production inspection. Nonlinear detection, however, is a new method that has developed in recent years and has attracted the attention of many scientists [1].

The linear method is a relatively traditional detection technology, which mainly includes ultrasonic guided wave technology, new acoustic emission technology, new non-contact ultrasonic transducer methods, and ultrasonic information processing and pattern recognition methods. Guided wave technology conducts nondestructive testing through a brief analysis of the SH mode guided wave, Lamb wave, and rod wave in materials. Tubular structures are objects where ultrasonic guided waves can show their strengths, and this technology can be used for long-distance one-time inspection of various pipelines [2, 3]. Acoustic emission technology is a passive detection technology, which is still used for hydrostatic testing of missile shells and submarines, and dynamically monitors and evaluates the safety performance and failure behavior of components. There is also a new type of non-contact ultrasonic transducer method, which mainly includes electromagnetic acoustic methods, electrostatic coupling methods, air coupling and laser ultrasonic methods. Because traditional methods require the use of coupling agents or water immersion methods to reduce the loss of ultrasonic waves in the air, many items cannot be tested by traditional methods. This is when non-contact ultrasonic testing is needed, which has the characteristics of non-contact, non-invasive, and completely non-destructive [4]. This technology has a good application prospect. Digital signal processing and digital recognition in testing can separate some complex signals and reduce many errors.

For a long time, when studying various problems in acoustics, it has always been carried out and developed within the theoretical framework of linear acoustics. However, in some cases, the results based on linear acoustic theory will bring large errors and nonlinear phenomena that cannot be explained by linear acoustics

X. Liu, *Nonlinear Sound Waves in Solids*,
Advances in Acoustics, https://doi.org/10.1007/978-981-96-5253-2_4

will appear. Therefore, in recent years, nonlinear acoustics has received widespread attention from scientists and has developed rapidly.

Nonlinear acoustics is an ancient and young discipline, but with the continuous discovery of the information value of nonlinear sound waves, material defect detection technology based on nonlinear acoustics has gained more and more applications [5].

Fatigue can cause changes in the microstructure inside the material, that is, heterogeneity, and there are many interfaces due to the emergence of cracks. Recent theoretical and experimental studies have shown that all of these will contribute to the nonlinear excitation of sound second harmonics. Therefore, it is reasonable to believe that the nonlinear acoustic characteristics of materials will change significantly with different degrees of material fatigue. Nonlinear second harmonic excitation technology is a promising non-destructive fatigue detection method [6].

Under the condition of second-order approximation, the nonlinearity of solids is described by several third-order elastic constants. For isotropic elastic bodies, there are three independent third-order elastic constants. The third-order or higher-order elastic constants are macro parameters characterizing the properties of solids, which are not only closely related to the structure of solids, but will also become new parameters for non-destructive evaluation of defects and fatigue. In addition, the nonlinear parameter B/A of the medium is defined as the ratio of the second-order term coefficient to the linear term coefficient in the Taylor expansion of the medium state equation under adiabatic conditions. It is a parameter describing the characteristics of the second-order nonlinear acoustic medium, indicating the size of the nonlinear effect of the medium. Since the nonlinear parameter B/A of the medium is related to the organization structure of the medium, it is possible to detect medium defects and evaluate the related quality of the medium [7, 8].

4.1 Classical Nonlinear Effects of Cracks

The classical nonlinear effect of cracks is obtained by expanding the stress–strain relationship of cracks according to a power series. Its constitutive relationship is described as follows [9]

$$\sigma = Y_0\left(\frac{\partial u}{\partial x}\right) + Y_1\left(\frac{\partial u}{\partial x}\right)^2 + Y_2\left(\frac{\partial u}{\partial x}\right)^3 \tag{4.1}$$

where, σ is the stress, u is the vibration displacement of the material point in the material, x represents the established coordinates, Y_0, Y_1, Y_2 are the second-order, third-order, and fourth-order elastic constants in the material, respectively. According to the law of conservation of momentum, we obtained its wave equation in the Lagrangian coordinate system

$$\rho_0 \frac{\partial^2 u}{\partial t^2} = Y_0 \frac{\partial^2 u}{\partial x^2} + Y_1 \frac{\partial}{\partial x}\left[\left(\frac{\partial u}{\partial x}\right)^2\right] + Y_2 \frac{\partial}{\partial x}\left[\left(\frac{\partial u}{\partial x}\right)^3\right] \tag{4.2}$$

where ρ_0 is the initial static density, t is the time. Generally, the step-by-step approximation method is used to solve Eqs. (4.1) and (4.2), assuming that the solution has three terms: $u = u_1 + u_2 + u_3$, where u_1 is a first-order approximate solution, u_2 and u_3 are the second-order and third-order approximate solutions respectively. The boundary condition at (x = L) is linear excitation, and the boundary condition at (x = 0) is free. Its linear solution u_1 is

$$u_1(x, t) = -u_0 \cos \omega t \cos k_0 x \tag{4.3}$$

where u_0 is the particle displacement at the excitation point, ω is the vibration angular frequency, k_0 is the wave number. Through the second and third order perturbation methods, we obtain the second order approximate solution and the third order approximate solution respectively.

$$u_2(x, t) = \frac{Y_1 u_0^2}{Y_0}(A(x) + B(x)\cos 2\omega t) \tag{4.4}$$

$$u_3(x, t) = u_0^3\left[F\left(x, \left(\frac{Y_1}{Y_2}\right)^2, \frac{Y_2}{Y_0}\right)\cos \omega t + G\left(x, \left(\frac{Y_1}{Y_0}\right)^2, \frac{Y_2}{Y_0}\right)\cos 3\omega t\right] \tag{4.5}$$

where $A(x)$ and $B(x)$ are coordinate functions, $F\left(x, \left(\frac{Y_1}{Y_2}\right)^2, \frac{Y_2}{Y_0}\right)$ and $G\left(x, \left(\frac{Y_1}{Y_2}\right)^2, \frac{Y_2}{Y_0}\right)$ are spatial functions and the third and fourth order elastic constants. The resonance frequency offset is

$$k = k_0 + \Delta k \tag{4.6}$$

$$\Delta k = -\frac{k^3 u_0^2}{64}\left(21\left(\frac{Y_1}{Y_0}\right)^2 - 18\left(\frac{Y_2}{Y_0}\right)\right) \tag{4.7}$$

where k is the nonlinear wave number, k_0 is the wave number under small amplitude conditions. It is worth noting that: in this nonlinear stress–strain situation, the size of the third harmonic is directly proportional to the cube of the fundamental amplitude. The offset of the resonance frequency is directly proportional to the square of the fundamental amplitude.

4.2 Non-classical Nonlinear Effects of Cracks

4.2.1 *Elastic Hysteresis Effect of Cracks*

First, we analyze the elastic hysteresis effect and its constitutive equation. Qualitatively speaking, the elastic hysteresis is similar to the Granato-Lucke hysteresis effect of dislocations [10]

$$\sigma = (\varepsilon, \mathrm{sg}\dot{\mathrm{n}}\,\varepsilon, \dot{\varepsilon}) = E\left[\varepsilon - f(\varepsilon, \mathrm{sg}\dot{\mathrm{n}}\,\varepsilon)\right] + \alpha\rho\dot{\varepsilon} \tag{4.8}$$

$$f(\varepsilon, \mathrm{sg}\dot{\mathrm{n}}\,\varepsilon) = \frac{1}{n}\begin{cases} \gamma_1\varepsilon^n, \varepsilon > 0, \dot{\varepsilon} > 0 \\ (\gamma_1 + \gamma_2)\xi_m^{n-1}\varepsilon - \gamma_2\varepsilon^n, \varepsilon > 0, \dot{\varepsilon} < 0 \\ -\gamma_3\varepsilon^n, \varepsilon < 0, \dot{\varepsilon} < 0 \\ (-1)^n(\gamma_3 + \gamma_4)\xi_m^{n-1}\varepsilon + \gamma_4\varepsilon^n, \varepsilon < 0, \dot{\varepsilon} > 0 \end{cases} \tag{4.9}$$

where: σ, ε and $\dot{\varepsilon}$ are the stress, strain, and strain relative to the rod ratio of the longitudinal wave. E is the Young's modulus, $f = f(\varepsilon, \mathrm{sg}\dot{\mathrm{n}}\,\varepsilon)$ is the hysteresis function, and: $|f_\varepsilon(\varepsilon, \mathrm{sg}\dot{\mathrm{n}}\,\varepsilon)| << 1$, γ_{1-4} is the nonlinear hysteresis parameter, $\xi_m = \xi_m(x)$ is the strain amplitude at x, and $\xi_m < |\xi_{th}|$, ξ_{th} is the limit of yield strength (beyond this limit, irreversible plastic deformation will occur). For most materials, $|\xi_{th}| > \left(10^{-4} - 10^{-3}\right)$, $\alpha = C_0^2/\Omega_p Q_p$ is the viscosity coefficient, ρ is the density. Under quasi-static conditions, when $\alpha\rho|\dot{\varepsilon}| << E|f(\varepsilon, \mathrm{sg}\dot{\mathrm{n}}\,\varepsilon)|$, we have when $\varepsilon = 0$, $\sigma(\varepsilon, \mathrm{sg}\dot{\mathrm{n}}\,\varepsilon, \dot{\varepsilon}) = 0$, therefore, this type of hysteresis effect is called the elastic hysteresis effect. In the I area ($\varepsilon < \varepsilon^*$) n = 3, in the II area ($\varepsilon > \varepsilon^*$) n = 2, as shown in Fig. 4.1.

4.2.2 *Inelastic Hysteresis Effect of Cracks*

The nonlinear ADIF effect can not only be described as an elastic hysteresis effect, but also as an inelastic hysteresis effect. Its nonlinear hysteresis function can be represented as [10]

$$f(\varepsilon, \mathrm{sg}\dot{\mathrm{n}}\,\varepsilon) = \beta\varepsilon\left(3\xi_m^2 + \varepsilon^2\right) + 3\beta\xi_m\begin{cases} \varepsilon^2 - \xi_m^2 & \dot{\varepsilon} > 0 \\ -\varepsilon^2 + \xi_m^2 & \dot{\varepsilon} < 0 \end{cases} \tag{4.10}$$

The qualitative quasi-static inelastic image is shown in Fig. 4.2. It can be clearly seen from the figure that when $\varepsilon = 0$, $\sigma(\varepsilon, sgn\dot{\varepsilon}, \dot{\varepsilon}) \neq 0$.

This type of nonlinear effect has a significant difference from the classical nonlinear effect in that the size of each high-order harmonic caused by the hysteresis effect is squared with the amplitude of the fundamental wave, while in the classical nonlinear effect, the nth harmonic is related to the fundamental wave in the nth power.

Therefore, it is easy to judge whether the nonlinear relationship shown by the crack is classical or non-classical.

4.3 Other Nonlinear Models of Cracks

Other nonlinear stress–strain relationships of cracks also include bilinear stress–strain relationships [11]. As shown in Fig. 4.3, longitudinal waves propagate in a bar-shaped material with cracks, one end is fixed, and the other end is excited by a sine signal,

The bilinear stress–strain relationship at the crack is shown in Fig. 4.4. Usually, when the material is stretched, its strain ε is positive, and when it is compressed, its strain ε is negative. Therefore, mathematically, when $\varepsilon|_{x_c} > 0$ the crack opens, $\varepsilon|_{x_c} < 0$ the crack closes. If the material has a uniform density and is all ρ, then the wave equation in the material can be expressed as

$$\rho\frac{\partial^2 u}{\partial t^2} = \frac{\partial}{\partial x}\left(E\frac{\partial u}{\partial x}\right) \tag{4.11}$$

where the Young's modulus at the crack satisfies a bilinear relationship, and other places satisfy a normal linear relationship

$$E = \begin{cases} E_2 \; for \, x = x_c \; and \; \varepsilon|_{x=x_c} = \left.\frac{\partial u}{\partial x}\right|_{x=x_c} > 0 \\ E_1 \qquad\qquad Others \end{cases} \tag{4.12}$$

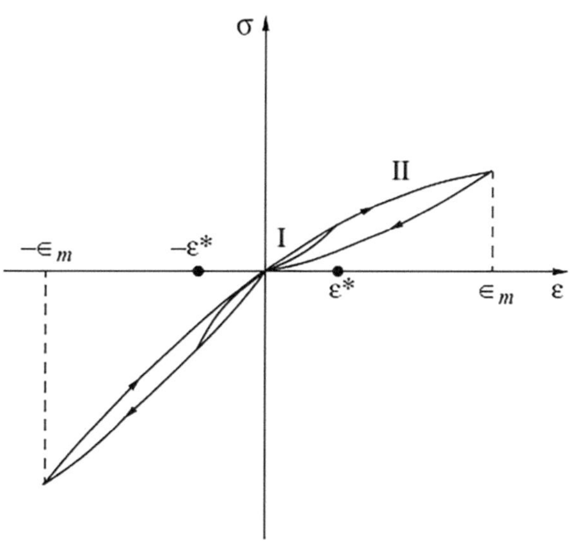

Fig. 4.1 Quasi-static elastic hysteresis stress–strain relationship diagram [10]

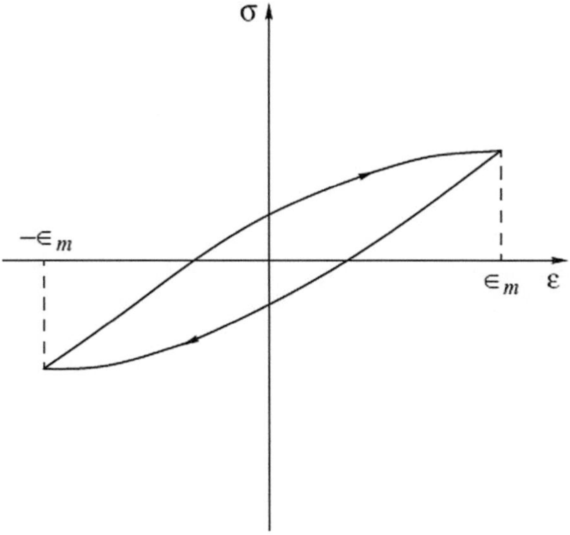

Fig. 4.2 Quasi-static non-elastic hysteresis stress–strain relationship diagram [10]

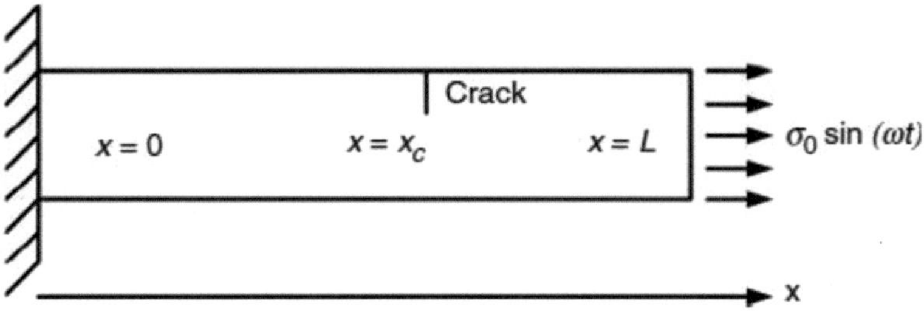

Fig. 4.3 Sine signal excitation diagram with cracks [11]

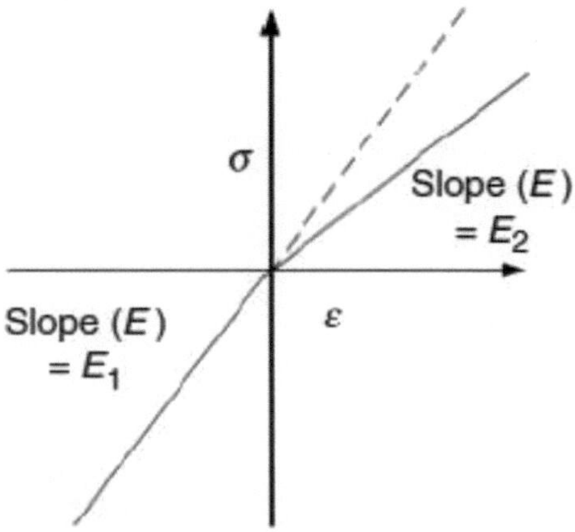

Fig. 4.4 Bilinear stress–strain relationship at the crack [11]

This is because when the crack opens, it is assumed that its interaction is relatively small, that is, the Young's modulus is small, and when the crack closes, it is the same as the situation without cracks. Under this nonlinear stress–strain relationship, its harmonic feature is that only even harmonics exist and odd harmonics do not exist, as shown in Fig. 4.5. And its second harmonic amplitude is linearly related to the excitation amplitude, as shown in Fig. 4.6. At present, there is a wide range of research on this type of nonlinear stress–strain relationship, but its disadvantage is that its theoretical prediction will not produce odd harmonics, but the existence of third harmonics is found in experiments.

In summary, the nonlinearity of solids is divided into classical nonlinearity and non-classical nonlinearity. The nonlinearity in solids is classical nonlinearity, which

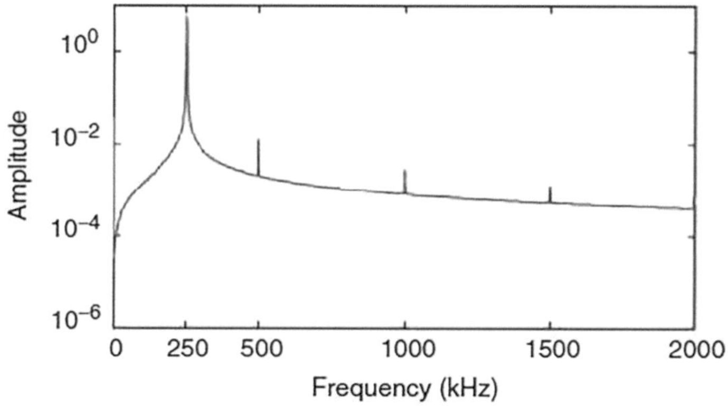

Fig. 4.5 Vibration spectrum under crack conditions [11]

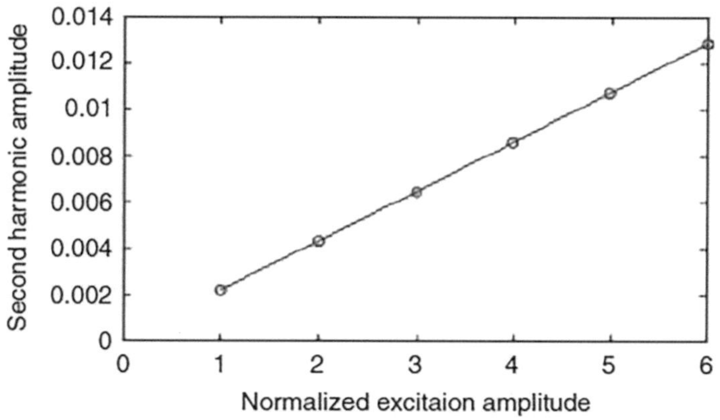

Fig. 4.6 Linear relationship of the second harmonic amplitude to the fundamental amplitude [11]

is caused by the high-order terms of the relationship between stress and strain. Classical nonlinearity refers to the non-classical nonlinear characteristics of solid materials, whether in theory or experiment, it has just begun to develop. The non-classical nonlinearity in solids is mainly manifested in [11, 12]: (1) There is a hysteresis curve in the relationship between stress and strain during the loading and unloading process of nonlinear solid materials, (2) In the resonance response experiment, the offset of the resonance frequency is proportional to the applied strain. (3) When the strain is very small (10^{-6}) When there is a nonlinear phenomenon, its nonlinear parameters are much larger than those of traditional gases, liquids, or solids, which means that the nonlinear effect is very obvious. (4) Sound attenuation and sound speed change with strain, and the elastic modulus depends on the amplitude, temperature, humidity, and pore fluid of the excitation. (5) After the sound wave excitation is stopped, it takes a long time for the elastic modulus to recover, which is called the slow dynamics phenomenon. (6) The odd harmonics of the transmitted wave are in a quadratic relationship with the fundamental wave, etc.

It is now generally believed that the reason for this non-classical nonlinear acoustic phenomenon is due to the microscopic characteristics of solid materials, such as cracks, fissures, particles, etc., which affect the mechanical properties and manifest as hysteresis phenomena on a macroscopic scale. Preisach and others studied this phenomenon early on and proposed a model to explain it [13, 14]. Later, McCall and Guyer conducted a large amount of theoretical and experimental research on this basis, further developed and perfected this model, now known as the Preisach-Mayergoyz (PM) model [15, 16]. Van Den Abeele [17] used the PM model to theoretically explain the detailed analytical solution of non-classical nonlinearity in one dimension, and also provided one-dimensional and two-dimensional numerical simulations [18, 19], The two-dimensional result is based on explosive wave excitation, studying its transient solution, which does not include the effect of attenuation in the solution.

4.4 Preisach-Mayergoyz (PM) Model

By establishing a hysteresis model, the non-classical nonlinearity caused by microcracks in materials can be simulated. Mathematically speaking, hysteresis nonlinearity is a scalar nonlinear function, which has a non-local memory characteristic between its input and output, and is a many-to-many mapping relationship. The sub-hysteresis loop at a certain point in the main hysteresis loop is not only one, but countless, and all are within the main hysteresis loop: which one is chosen as the next moment's motion trajectory of this point not only depends on the current input, but also depends on the historical input, especially related to the extreme value of the historical input.

According to the PM hysteresis model theory, non-classical nonlinearity is manifested at the microscopic level as the step response of particle stress and strain, and the collection of microscopic units is manifested at the mesoscopic level as the

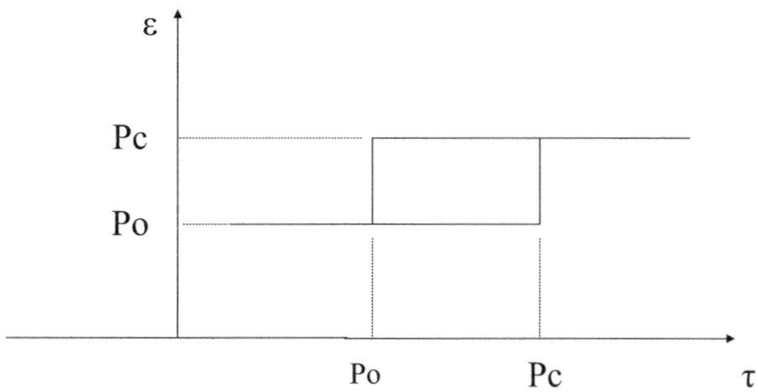

Fig. 4.7 Step response of stress and strain in the PM model

hysteresis curve of solid materials. Mesoscopic units make up macroscopic non-classical nonlinear solid materials, that is, what is commonly referred to as damaged materials. Through statistical analysis of material microscopic units, the impact of non-classical nonlinearity can be attributed to the effect of stress on the modulus at the mesoscopic unit scale. The basic hysteresis unit stress–strain relationship in the original PM model is shown in Fig. 4.7. When the stress increases from Po to Pc, the hysteresis unit strain is 0, and when the stress is less than Po or greater than Pc, the hysteresis unit strain is also 0.

According to the PM model theory, non-classical nonlinearity is manifested at the microscopic level as the step response of particle stress and strain, and the collection of microscopic units is manifested at the mesoscopic level as the hysteresis curve of solid materials. Mesoscopic units make up macroscopic non-classical nonlinear solid materials, that is, what is commonly referred to as damaged materials. Through statistical analysis of material microscopic units, the impact of non-classical nonlinearity can be attributed to the effect of stress on the modulus at the mesoscopic unit scale. Equation (4.13) gives the specific modulus-stress relationship of the mesoscopic unit [19]

$$K^{-1} = \lim_{\Delta\sigma \to 0}\left(\frac{\Delta\varepsilon}{\Delta\sigma}\right) = \frac{1}{K_c(\sigma)} + \hat{r}\frac{df_c}{d\sigma}(-\sigma) \tag{4.13}$$

Here K_c represents the modulus of the material under classical conditions, $\hat{r}$ characterizes the strength of the material's non-classical nonlinearity, σ is the stress received by the mesoscopic unit, f_c refers to the proportion of closed units in the microscopic units of the PM model under the action of stress σ

Since the original PM model can only generate odd harmonics and cannot generate even harmonics, an improved PM model is used, as shown in Fig. 4.8, the stress–strain nonlinear hysteresis relationship in the defect area. It is assumed that the hysteresis units are initially all in an open state. When the stress starting from P_o is still less than P_c, the hysteresis unit maintains linear properties, and its modulus is K_{M1}. When the stress exceeds P_c, the strain jumps, and its jump quantity is r_2. At this time, the hysteresis unit enters a closed state. If the stress is further increased, the hysteresis unit remains linear, with its elastic modulus K_{M2}. Even if the initial stress greater

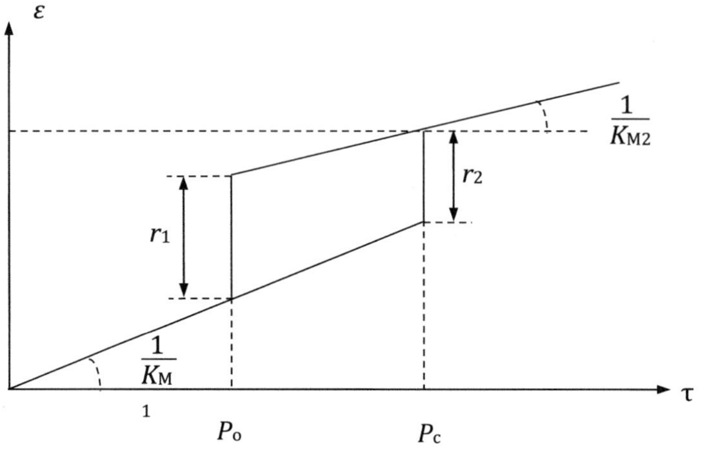

Fig. 4.8 Basic hysteresis unit of the improved PM model

than P_c begins to decrease, the inherently closed hysteresis unit will maintain its linear properties, with its elastic modulus being K_{M2}, until the stress is less than P_o. Correspondingly, when the stress jumps over P_o, the strain will undergo a jump of value r_1 and the hysteresis unit will become open. When the stress continues to decrease, the hysteresis unit maintains linear elastic properties, with its elastic modulus being K_{M1}.

According to the opening and closing modes of the basic hysteresis unit of the PM model described above, when the pore pressure increases ($\frac{\partial P}{\partial t} > 0$), the strain of each hysteresis unit ε_M can be represented as

$$\begin{aligned} \frac{\partial \varepsilon_M}{\partial P} &= -\frac{1}{K_{M1}}, \ \textit{When the delay unit is in the open state} \\ \frac{\partial \varepsilon_M}{\partial P} &= -\frac{1}{K_{M2}}, \ \textit{When the delay unit is in the off state} \end{aligned} \tag{4.14}$$

The nonlinear stress–strain hysteresis relationship in the defect area is obtained through the weighted superposition of these basic hysteresis units. Define $f(P_o, P_c)$ as the density distribution function of the PM space, the strain of all hysteresis units ε_H can be represented as

$$\frac{\partial \varepsilon_H}{\partial P} = -\int_{O_2}^{P} dP_o r_2 f(P_o, P)$$
$$-\frac{1}{K_{M2}} \int_{-\infty}^{P_c} dP_o \int_{P_0}^{+\infty} dP_c f(P_o, P_c)$$
$$-\left(\frac{1}{K_{M1}} - \frac{1}{K_{M2}}\right) \int_{O_2}^{P_c} dP_o \int_{P}^{+\infty} dP_c f(P_o, P_c) \tag{4.15}$$

where O_2 is the P_o value of the hysteresis unit in the PM space. When all hysteresis units are in the open state, the first double integral term in Eq. (4.15) is corrected by the inverse of the bulk modulus; when all hysteresis units are in the closed state, the second double integral term in Eq. (4.15) is corrected by the inverse of the bulk modulus; the first integral term on the right side of the equation is the correction term for the inverse of the bulk modulus of all hysteresis units in the open state under pore pressure P.

Similarly, when the pore pressure decreases ($\frac{\partial P}{\partial t} < 0$), the strain of each hysteresis unit can also be obtained, and the equation for the strain of all hysteresis units ε_H can be derived as follows

$$\frac{\partial \varepsilon_H}{\partial P} = -\int_{P}^{C_2} dP_c r_1 f(P, P_c)$$
$$-\frac{1}{K_{M1}} \int_{-\infty}^{P_c} dP_o \int_{P_0}^{+\infty} dP_c f(P_o, P_c)$$
$$-\left(\frac{1}{K_{M2}} - \frac{1}{K_{M1}}\right) \int_{-\infty}^{P_c} dP_o \int_{P_0}^{C_2} dP_c f(P_o, P_c) \tag{4.16}$$

where C_2 is the P_c value of the hysteresis unit in the PM space.

4.5 Nonlinear Resonance Spectral Method

4.5.1 Free at Both Ends

Assume that when there are defects in a one-dimensional uniform metal rod with both ends free (density ρ, , length interval [0, L]) (Fig. 4.9), the wave equation is represented by the nonlinear unique stress–strain formula, the relationship between stress σ and strain ε is

$$\sigma = K\left(1 + \beta\varepsilon + \delta\varepsilon^2 + \cdots\right)\varepsilon + K\frac{\alpha}{2}\left[sign(\partial_t\varepsilon)\left((\Delta\varepsilon)^2 - \varepsilon^2\right) - 2(\Delta\varepsilon)\varepsilon\right] \quad (4.17)$$

where: $\varepsilon = \partial_x u$, $\Delta\varepsilon$ is the amplitude of the strain, K is the linear strength coefficient, α represents the strength of the hysteresis nonlinearity, β and δ are combinations of the third and fourth order elastic constants representing classical nonlinearity.

Use $u(x, t)$ Representing the relationship between the wave and the horizontal axis coordinate, time, substitute into Eq. (4.17), decompose u into the product of the functions of variable x and time t. $\psi_i(x)$ Representing the phase, $z_i(t)$ Representing the amplitude (at the same time $\psi_i(x)$ satisfies the orthogonal conditions and derivative orthogonal conditions of [0, L], $i \neq j$)

$$u(x, t) = \sum_i z_i(t)\psi_i(x) \quad (4.18)$$

$$\psi_i(x) = \cos\left(\frac{i\pi x}{L}\right) \quad (4.19)$$

Substitute into (4.17) to get the equation satisfied by $z_n(t)$

$$\partial_{tt}^2 z_n + \frac{\Omega}{Q}\partial_t z_n + \omega_n^2 z_n = F\cos(\Omega t) - \sum_{j,k} B_{njk} z_j z_k - \sum_{j,k,l} D_{njkl} z_j z_k z_l$$
$$H_{mm}\left[sign(\partial_t z_m)\left(A_m^2 - z_m^2\right) - 2A_m z_m\right] \quad (4.20)$$

where

$$F = 2\tilde{F} * /\rho L \quad (4.21a)$$

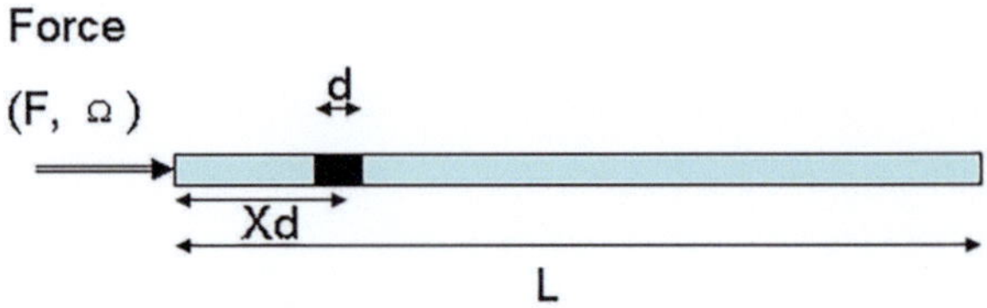

Fig. 4.9 Geometric model of a one-dimensional metal rod with a single crack

$$H_{mm} = \frac{K}{\rho L} \int_0^L dx\alpha(x)\partial_x\Psi_m|\partial_x\Psi_m|\partial_x\Psi_m \tag{4.21b}$$

$$B_{njk} = \frac{2K}{\rho L} \int_0^L dx\beta(x)\partial_x\Psi_n\partial_x\Psi_j\partial_x\Psi_k \tag{4.21c}$$

$$D_{njkl} = \frac{2K}{\rho L} \int_0^L dx\delta(x)\partial_x\Psi_n\partial_x\Psi_j\partial_x\Psi_k\partial_x\Psi_l \tag{4.21d}$$

ω_m represents the resonance frequency of the object under linear defects and small amplitudes; H, B, D coefficients depend on higher-order tensors. Therefore, once it is considered that there are nonlinear characteristics in the medium, then all existing linear modes and their disturbances will interact to a nonlinear degree, so there will be non-zero strain where nonlinear characteristics exist.

To solve the nonlinear problem, let's now recall the solution when there are linear defects.

If the defects contained in a one-dimensional object are linear, then B, D, H are all 0, Eq. (4.20) can be transformed into

$$\partial_{tt}^2 z_n + \frac{\Omega}{Q}\partial_t z_n + \omega_n^2 z_n = F\cos(\Omega t) \tag{4.22}$$

4.5.1.1 When One-Dimensional Defects Are Linear (Only Coefficient K Exists)

Solving the partial differential Eq. (4.22) can know the relationship between the resonance frequency and ω_m, Q. It can be found that when A_m reaches its maximum value (when resonance occurs), the frequency of the external force F is

$$\Omega_{res} = \omega_m/\sqrt{1 + 1/Q^2}\left(\text{for example, } \omega_m = m\frac{\pi c}{L}\right) \tag{4.23}$$

This is just the solution under linear conditions. It can be imagined that if the object has nonlinear defects, then the formula for the resonance frequency will change to a certain extent. The following analyzes two nonlinear situations: (1) Classic cubic nonlinearity (2) Nonlinear hysteresis nonlinearity.

4.5.1.2 Solution of Classic Cubic Nonlinearity

The frequency offset at mode m under sinusoidal excitation (only contains δ):

Equation (4.20) becomes

$$\partial_{tt}^2 z_m + \frac{\Omega}{Q}\partial_t z_m + \omega_m^2 z_m = F\cos(\Omega t) - D_{mmmm} z_m^3 \tag{4.24}$$

$$D_{mmmm} = \frac{2}{\rho L}\int_0^L dx K\delta(x)(\partial_x \psi_m)^4$$

The resonance frequency is solved as

$$\Omega_{res}(\varepsilon_m) \approx \Omega(0)\left[1 + \frac{3}{4l}\varepsilon_m^2 \int_0^l dx\delta(x)\sin^4\left(\frac{m\pi x}{l}\right)\right]$$

$$\Omega_{res,local}(\varepsilon_m) \approx \Omega_{res}(0)\left[1 + C_{1,m}\varepsilon_m^2\right] \tag{4.25}$$

$\delta(x)$ We give the following definition: $\delta(x) = \overline{\delta} G_{local} x : x_d, d$

Where $\overline{\delta} = \hat{\delta}\frac{d}{L}$ and $G_{local} = 0$, $if\, x \notin \left[x_d - \frac{d}{2}, x_d + \frac{d}{2}\right]$,

Otherwise, $G_{local} = L/d$, So we can get:

$$C_{1,m} = \frac{3}{4}\overline{\delta}\sin^4\left(m\frac{\pi}{L}x_d\right) \tag{4.26}$$

So the drift of the center frequency is $\Delta\omega/\omega_m = C_{1,m}\varepsilon_m^2$, which is a square relationship with the strain.

The variation of $C_{1,m}$, $C_{3,m}$ and $C_{5,m}$ with xd/L (Fig. 4.10).

As can be seen, for different positions, the offset corresponds to different reactions, but overall, the coefficient C is symmetrical about $X_d = L/2$.

Now suppose Z_m is the fundamental wave ($m = 1$), as can be known from Eq. (4.24), the 2nd ($m = 2$), 3rd ($m = 3$) harmonics can be solved by the following differential equation

$$\partial_{tt}^2 z_{2m} + \frac{\Omega}{Q}\partial_t z_{2m} + \omega_{2m}^2 z_{2m} = -D_{2mmmm} z_m^3 \text{ (2nd harmonic)} \tag{4.27}$$

$$\partial_{tt}^2 z_{3m} + \frac{\Omega}{Q}\partial_t z_{3m} + \omega_{3m}^2 z_{3m} = -D_{3mmmm} z_m^3 \text{ (3rd harmonic)} \tag{4.28}$$

(a)

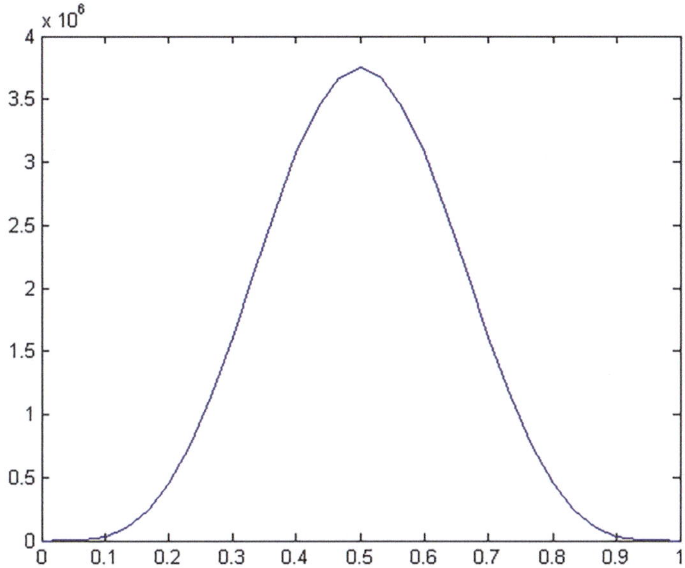

(b)

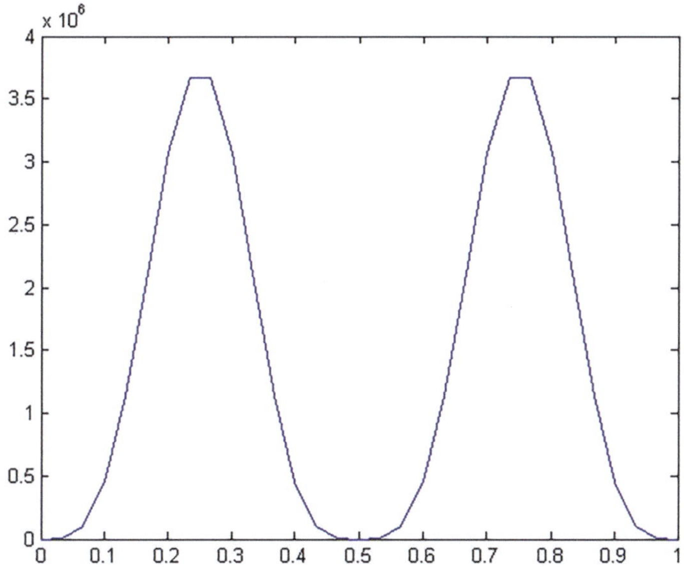

Fig. 4.10 **a** C(1,m) when m = 1, the horizontal coordinate is X_d/L, **b** C(1,m) when m = 2, the horizontal coordinate is X_d/L, **c** C(1,m) when m = 3, the horizontal axis is X_d /L, **d** C(3,m) when m = 1, the horizontal axis is Xd/L, **e** C(3,m) when m = 2, the horizontal axis is X_d/L, **f** C(5,m) when m = 1, the horizontal axis is X_d/L, **g** C(5,m) when m = 2, the horizontal axis is X_d/L

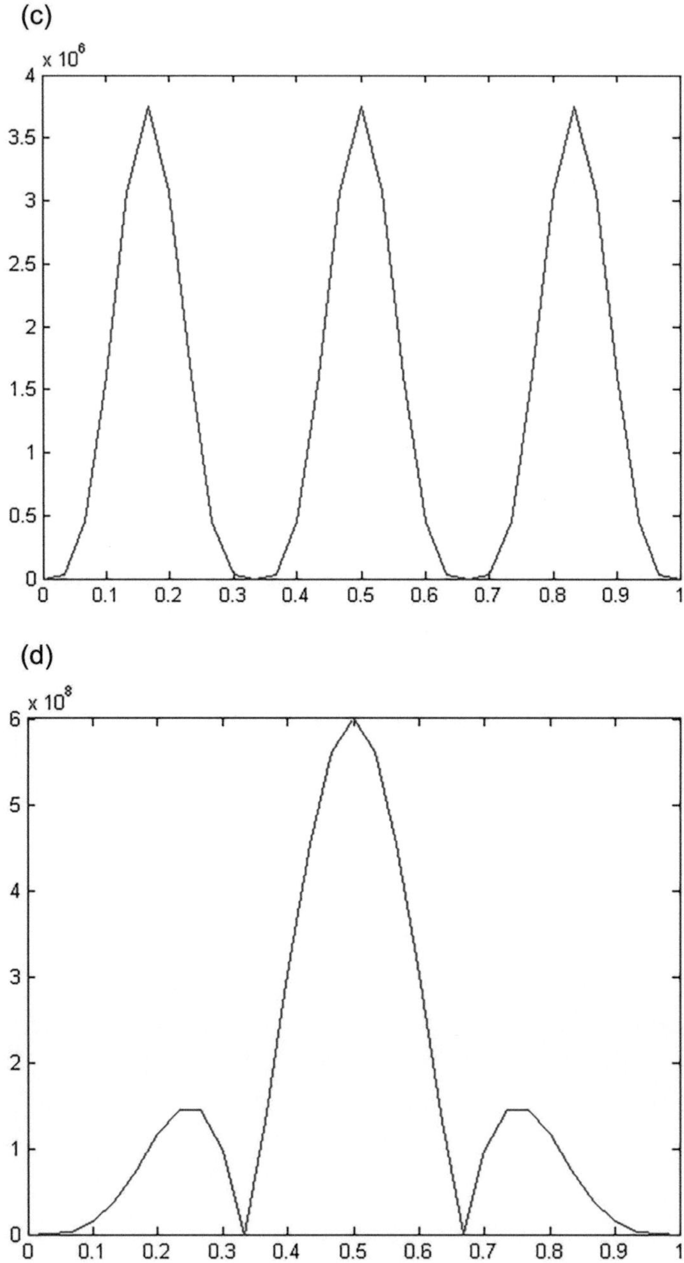

Fig. 4.10 (continued)

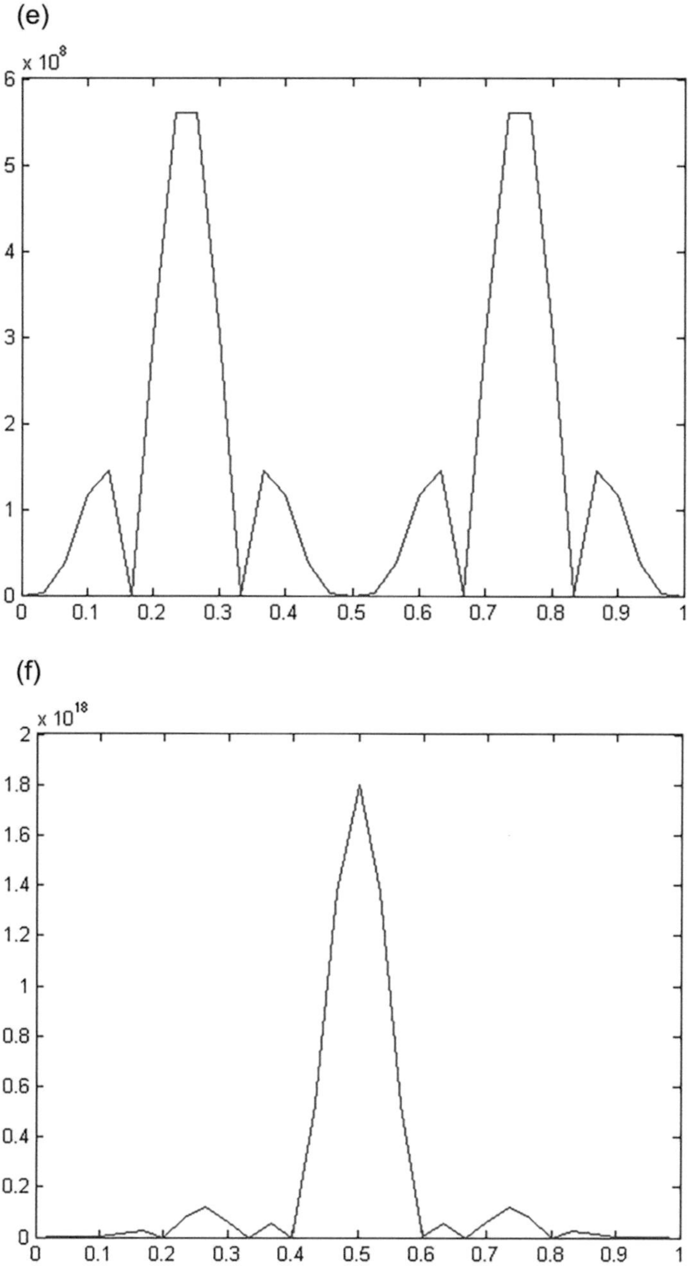

Fig. 4.10 (continued)

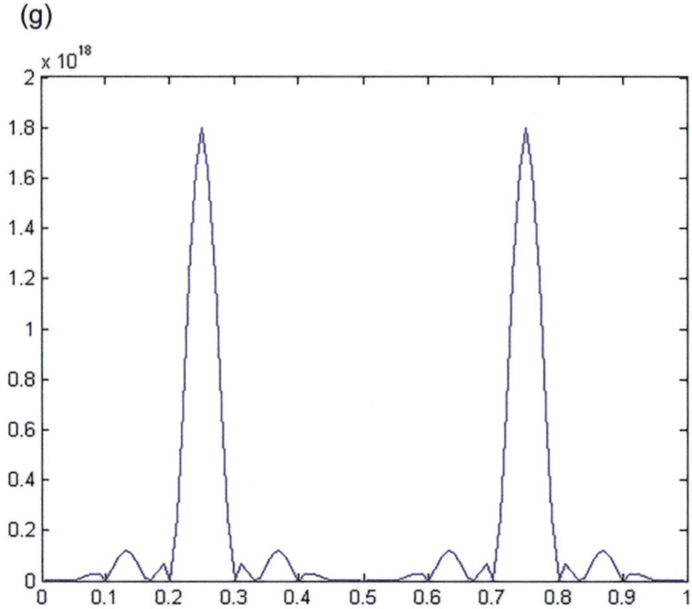

Fig. 4.10 (continued)

Assume the frequency of Z_m is ω, the right side of the differential equation listed for the 2nd harmonic is in the form of Z_m cubed, which means it can only generate the 3rd harmonic, and the 2nd harmonic is 0. Solve the differential equation of the 3rd harmonic, according to the form of (4.25), find out $C_{5,m}$

$$C_{3,m} = \frac{3}{2} Q \left| \bar{\delta} \sin\left(\frac{3m\pi}{L} x_d\right) \sin^3\left(\frac{m\pi}{L} x_d\right) \right| \tag{4.29}$$

$$\varepsilon_{3,m} = \frac{3}{2} Q \left| \bar{\delta} \sin\left(\frac{3m\pi}{L} x_d\right) \sin^3\left(\frac{m\pi}{L} x_d\right) \right| \varepsilon_m^3 \tag{4.30}$$

The frequency drift of the third harmonic and the stress present a cubic relationship, and the coefficient C(3,m) is still symmetrical about Xd = L/2. The graph of $C_{3,m}$ is as follows.

Since there is no 2nd harmonic, there is no 4th harmonic, and the 5th harmonic relationship is given by the following formula:

$$\partial_{tt}^2 z_{5m} + \frac{\Omega}{Q} \partial_t z_{5m} + \omega_{5m}^2 z_{5m} = -D_{5m3mmm} z_m^3 - D_{5mm3mm} z_m^3 - D_{5mmm3m} z_m^3 \tag{4.31}$$

Solution:

$$C_{5,m} = \frac{45}{4} Q^2 \left| \overline{\delta}^2 \sin\left(\frac{5m\pi}{L} x_d\right) \sin^2\left(\frac{3m\pi}{L} x_d\right) \sin^5\left(\frac{m\pi}{L} x_d\right) \right| \tag{4.32}$$

$$\varepsilon_{5,m} = \frac{45}{4} Q^2 \left| \overline{\delta}^2 \sin\left(\frac{5m\pi}{L} x_d\right) \sin^2\left(\frac{3m\pi}{L} x_d\right) \sin^5\left(\frac{m\pi}{L} x_d\right) \right| \varepsilon_m^5 \tag{4.33}$$

In summary, for classical nonlinear defects, the resonance frequency (high harmonics) all drift with the center frequency, and the frequency drift is proportional to the square, cube, and fifth power of the strain ε_m (respectively for the fundamental, 3rd, 5th harmonics), and the coefficient C is symmetrical about L/2 in the interval [0, L], which brings trouble to the judgment of the defect position below, because the frequency shift form and value of the two points symmetrical about L/2 are exactly the same.

4.5.1.3 Solution of Non-classical Nonlinearity

Frequency offset at mode m under sinusoidal excitation:

Under this condition, let all classical nonlinear variables be 0 ($\beta = 0$, $\delta = 0$), only consider the non-classical nonlinear coefficient α, when the sinusoidal excitation is in the n = m mode, formula (4.20) can be transformed into Fig. 4.11

$$\begin{aligned} \partial_{tt}^2 z_m + \frac{\Omega}{Q} \partial_t z_m + \omega_m^2 z_m &= F \cos(\Omega t) \\ - H_{mm} - &\left[\frac{8}{3\pi} A_m^2 \sin(\Omega t + \phi_m) - 2A_m^2 \cos(\Omega t + \phi_m) + \cdots \right] \end{aligned} \tag{4.34}$$

Solution

$$\begin{aligned} \Omega_{res,Local}(\varepsilon_m) &\approx \Omega_{res}(0) \left[1 - \overline{\alpha} \left(1 + \frac{4}{3\pi Q} \right) \left| \sin^3 (m \frac{\pi}{L} x_d) \right| \varepsilon_m \right] \\ &= \Omega_{res}(0) \left[1 - X_{1,m} \varepsilon_m \right] \end{aligned} \tag{4.35}$$

As can be seen, the frequency drift of the fundamental wave of non-classical nonlinearity is directly proportional to the strain, and the coefficient X is still symmetrical about L/2. Similarly, X(3,m) and X(5,m) can be plotted, and the high harmonics are related to the square of the strain.

4.5.1.4 Reverse Modeling

Global multi-mode NRUS method:

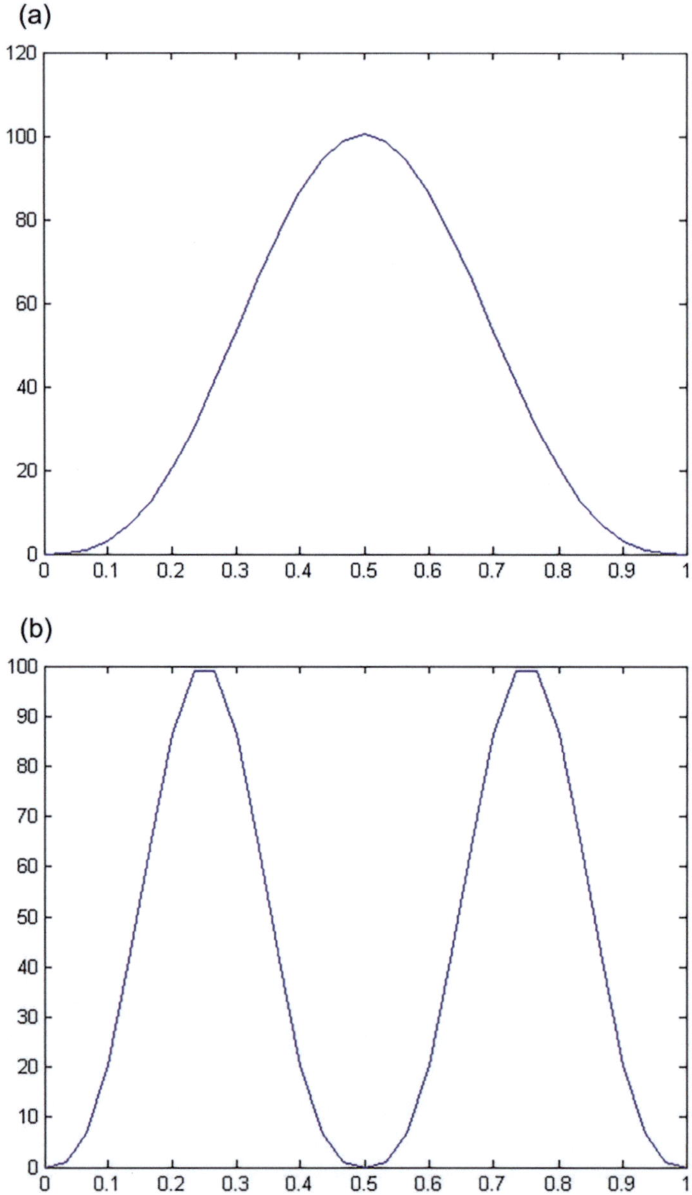

Fig. 4.11 **a** X(1,m) when m = 1, the horizontal axis is X_d/L, **b** X(1,m) when m = 2, the horizontal axis is X_d/L, **c** X(3,m) when m = 1, the horizontal axis is X_d/L, **d** X(3,m) when m = 2, the horizontal axis is X_d/L, **e** X(5,m) when m = 1, the horizontal axis is X_d/L, **f** X(5,m) when m = 2, the horizontal axis is X_d/L

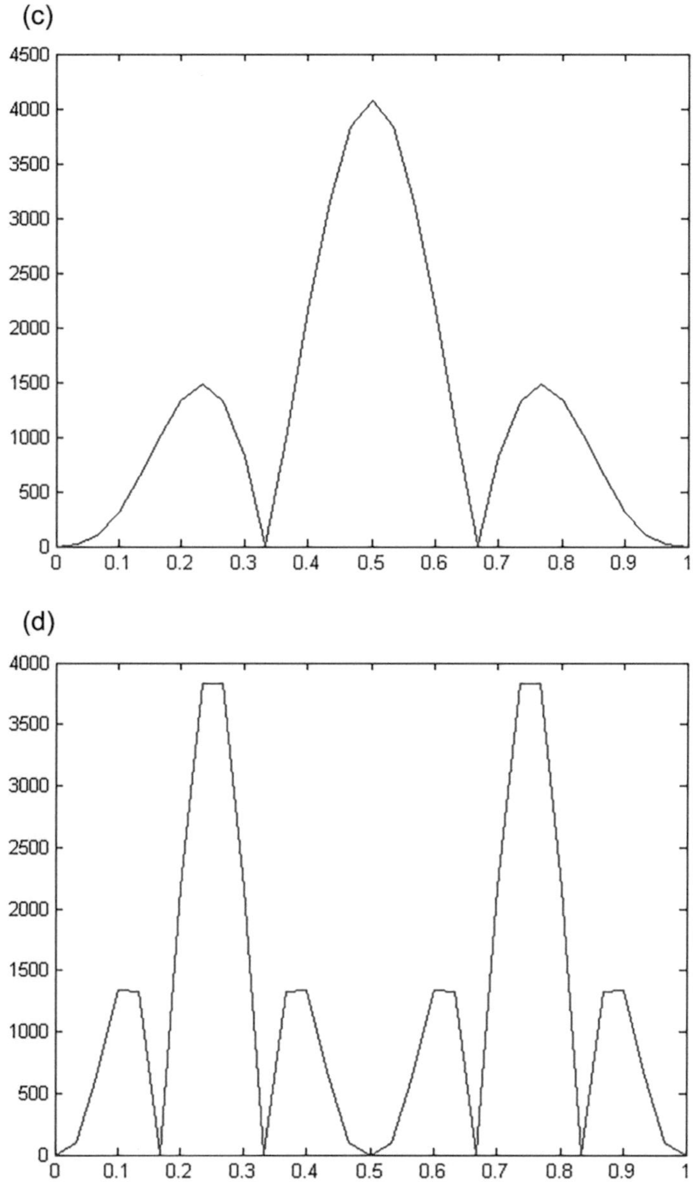

Fig. 4.11 (continued)

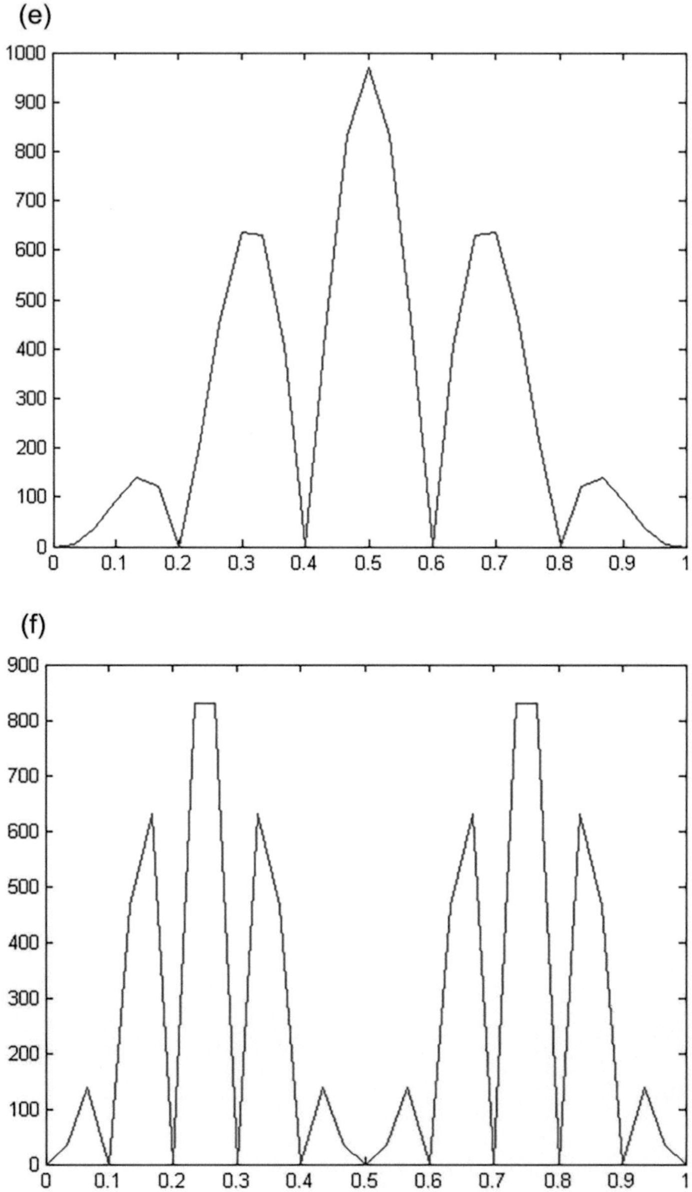

Fig. 4.11 (continued)

If the object contains multiple types of defects, then the resonance frequency can be written in the form of classical nonlinearity and non-classical nonlinearity as follows

$$\Omega_{res,Local}(\varepsilon_m) \approx \Omega_{res}(0)\left[1 - \Upsilon_m \varepsilon_m - \Theta_m \varepsilon_m^2\right] \tag{4.36}$$

The intensity function can be represented as

$$S_M(x) = \frac{2}{M}\sum_{m=1}^{M}\left[|\Upsilon_m| + \sqrt{|\Theta_m|}\right]\sin^2\left(\frac{m\pi}{L}x\right) \tag{4.37}$$

M represents the number of modes, according to the derivation above

$$\Upsilon_m = \overline{\alpha}\left(1 + \frac{4}{3\pi Q}\right)\left|\sin^3\left(m\frac{\pi}{L}x_d\right)\right| \tag{4.38}$$

$$\Theta_m = \frac{3}{4}\overline{\delta}\sin^4\left(m\frac{\pi}{L}x_d\right) \tag{4.39}$$

Therefore, from Fig. 4.12, the larger the number of modes, the clearer the defects can be seen, and it can be seen that the position of the defects is symmetrical about the midpoint. For the defects, we artificially set two positions: the non-classical nonlinearity defect is at $x_d/L = 72/250$, and the classical nonlinearity is at 50/250. It is clear that no matter what kind of defect, even if there is only a unique defect position, because of the free symmetry of the boundary conditions, two positions will still be given. These two positions are the positions symmetrical about the center. It is difficult to distinguish from the naked eye whether the defect is on the left or the right, which brings trouble to the positioning. Next, we discuss the defect positioning method under non-symmetric boundary conditions, which does not bring us this trouble.

4.5.2 One End Free, One End Fixed

Here we modify the boundary conditions: one end free, one end fixed, which can be represented mathematically as

$$u_x(0,t) = u(l,t) = 0 \tag{4.40}$$

Then the eigenvalue:

$$\lambda = (2n+1)^2\pi^2/4l^2 (\mathrm{n} = 0, 1, 2, 3...) \tag{4.41}$$

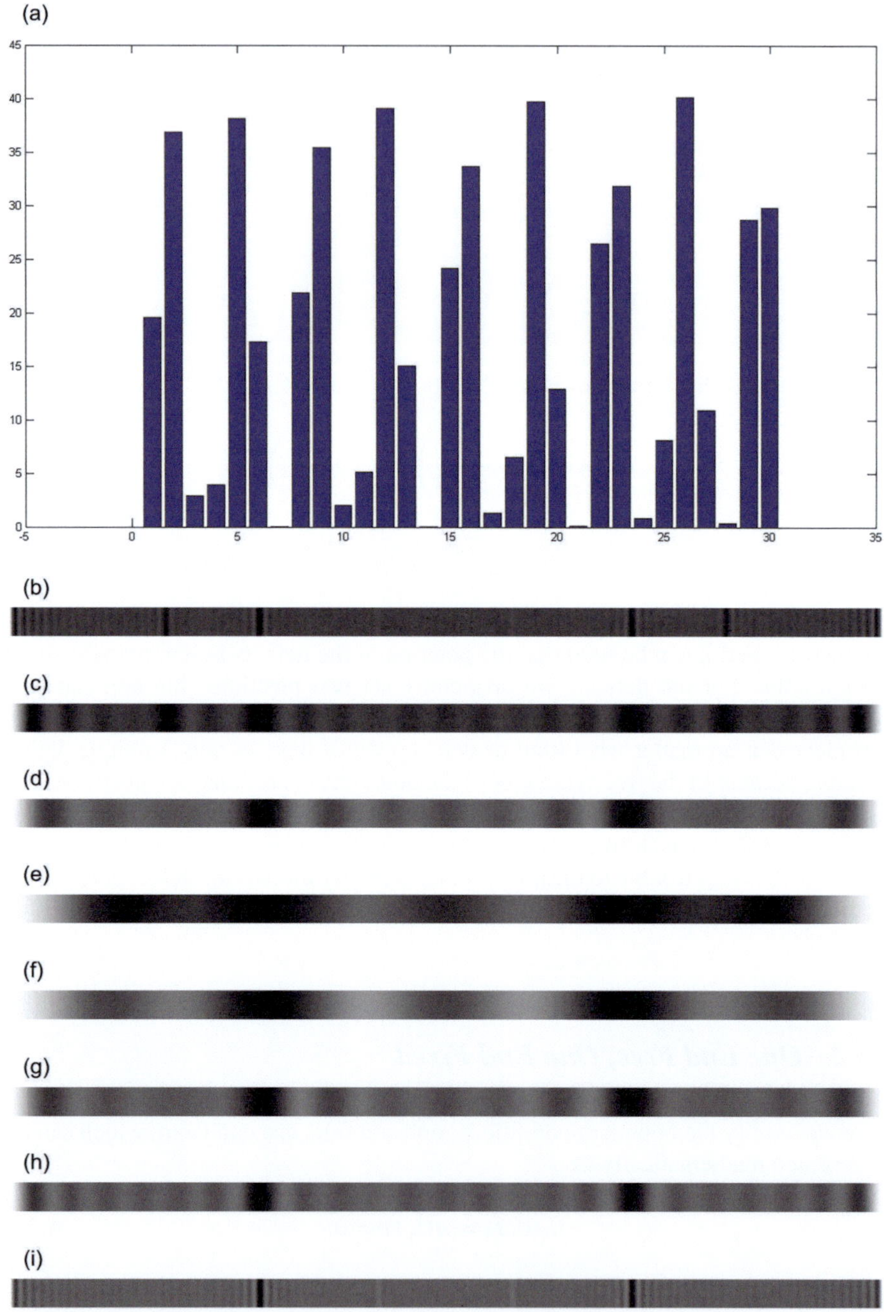

Fig. 4.12 (**a**) The variation of Ym with modulus M, (**b**)–(**i**) The spatial distribution of Sm with respect to the position in the rod, (**b**) M = 100, xd/L = 72/250,50/250 (**c**) M = 25, xd/L = 72/250,50/250 (**d**) M = 15, xd/L=72/250,50/250 (**e**) M = 5, xd/L = 72/250,50/250 (**f**) M = 5, xd/L = 72/250 (**g**) M = 15, xd/L = 72/250 (**h**) M = 25, xd/L = 72/250 (**i**) M = 100. xd/L = 72/250

$$\Psi_m = C_2 \cos\frac{(2m+1)\pi x}{2l}(\mathrm{m} = 0, 1, 2\ldots) \tag{4.42}$$

Get $z_n(t)$ satisfying the equation:

$$\partial_{tt}^2 z_n + \frac{\Omega}{Q}\partial_t z_n + \omega_n^2 z_n = F\cos(\Omega t) - \sum_{j,k} B_{njk} z_j z_k - \sum_{j,k,l} D_{njkl} z_j z_k z_l$$
$$H_{mm}\left[sign(\partial_t z_m)\left(A_m^2 - z_m^2\right) - 2A_m z_m\right] \tag{4.43}$$

4.5.2.1 Solution When Containing Classical Cubic Nonlinearity (Only Containing δ):

Equation (4.43) becomes:

$$\partial_{tt}^2 z_m + \frac{\Omega}{Q}\partial_t z_m + \omega_m^2 z_m = F\cos(\Omega t) - D_{mmmm} z_m^3 \tag{4.44}$$

At this time

$$D_{mmmm} = \frac{2k}{\rho L}\left(\frac{(2m+1)\pi}{2l}\right)^4 \int_0^L dx\delta(x)\sin^4\left(\frac{(2m+1)\pi}{2l}x\right) \tag{4.45}$$

And

$$\Omega_{res}(\varepsilon_m) \approx \Omega(0)\left[1 + \frac{3}{4l}\varepsilon_m^2 \int_0^l dx\delta(x)\sin^4\left(\frac{(2m+1)\pi x}{2l}\right)\right] \tag{4.46}$$

It can be concluded that:

$$C_{1,2m+1} = \frac{3}{4}\overline{\delta}(x)\sin^4\left(\frac{(2m+1)\pi x_d}{2l}\right) \tag{4.47}$$

When m = 0, m = 1, m = 2, m = 3, $\left|C_{1,2m+1}\right|$ changes with x_d/L shown in Fig. 4.13.

It can be calculated that when the defect extends to the entire one-dimensional coordinate [0, L], at this time:

$$\Omega_{res,Globle}(\varepsilon_m) \approx \Omega_{res}(0)\left[1 + 0.2814\overline{\delta}\varepsilon_m^2\right] \tag{4.48}$$

The equation satisfied by the third harmonic is

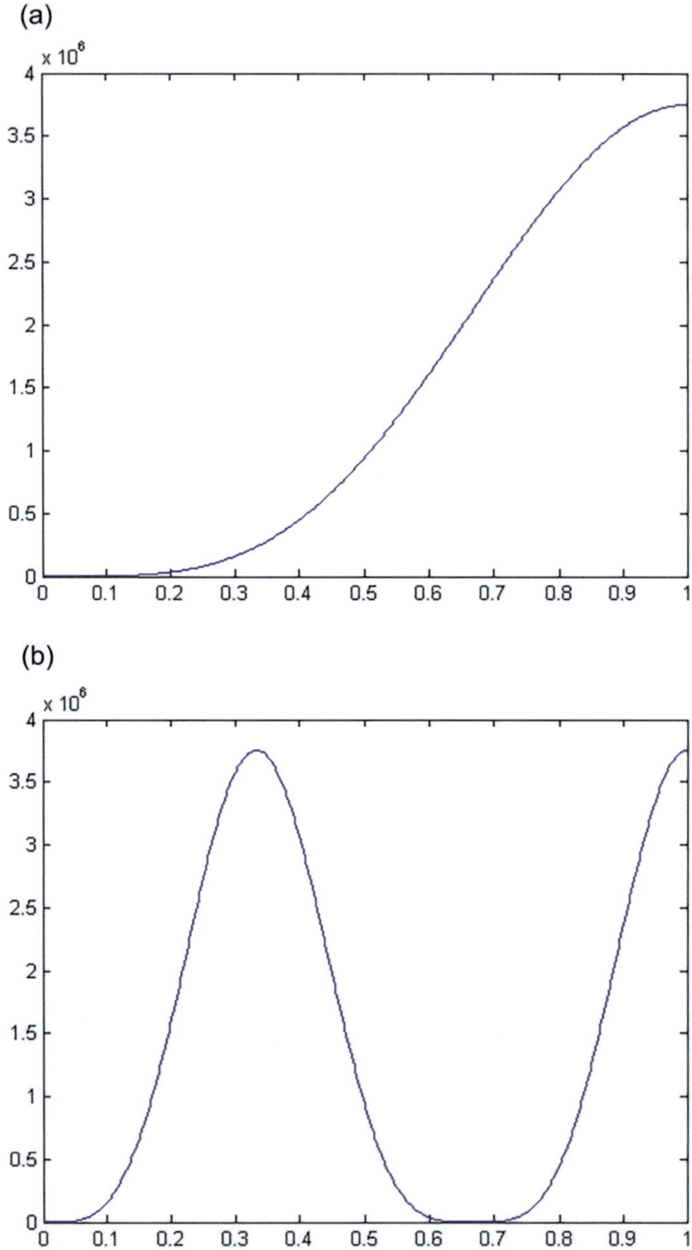

Fig. 4.13 **a** C(1,2m + 1) when m = 0, the horizontal axis is X_d/L, **b** C(1,2m + 1) when m = 1, the horizontal axis is X_d/L, **c** C(1,2m + 1) when m = 2, the horizontal axis is X_d/L, **d** C(1,2m + 1) when m = 3, the horizontal axis is X_d/L

(c)

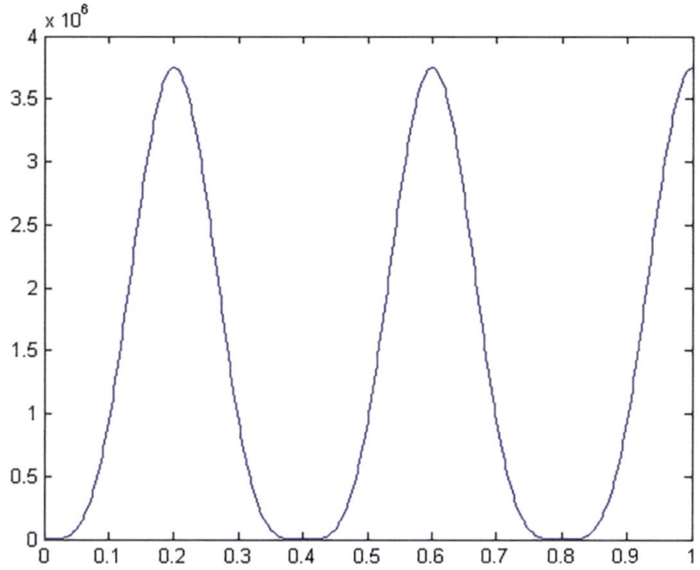

(d)

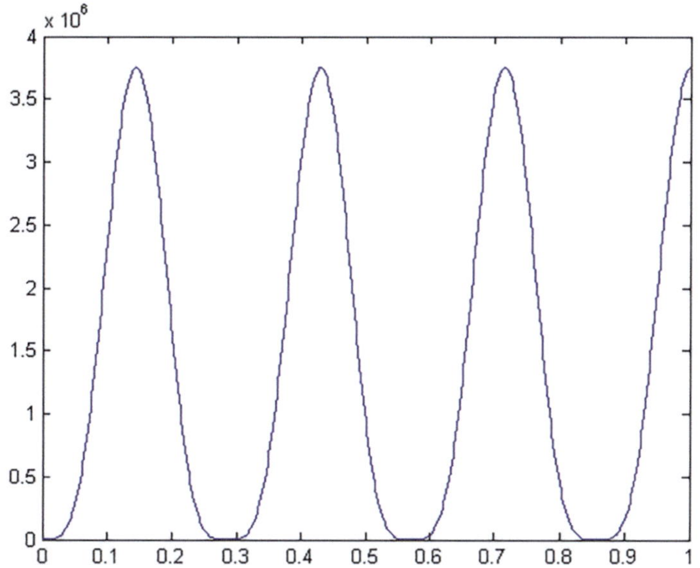

Fig. 4.13 (continued)

$$\partial_{tt}^2 z_{3,m} + \frac{\Omega}{Q}\partial_t z_{3,m} + \omega_{3,m}^2 z_{3,m} = F\cos(\Omega t) - D_{3mmmm} z_m^3 \tag{4.49}$$

The solution to (4.49) is

$$A_{3,m} = \frac{|D_{3mmmm}|}{4} * \frac{A_m^3}{\sqrt{((3\Omega)^2 - \omega_{3,m}^2)^2 + \Omega^2(3\Omega)^2/Q^2}} \tag{4.50}$$

where

$$\varepsilon_m = A_{m,max} * \frac{(2m+1)\pi}{2l} \tag{4.51}$$

$$\omega_{3,m} = 3 * \frac{(2m+1)\pi c}{2l} \tag{4.52}$$

Similarly, we can get:

$$C_{3,2m+1} = \frac{3}{2}Q\left|\overline{\delta}\sin\left(\frac{3(2m+1)\pi}{2L}x_d\right)\sin^3\left(\frac{(2m+1)\pi}{2L}x_d\right)\right| \tag{4.53}$$

$$\varepsilon_{3,2m+1} = \frac{3}{2}Q\left|\overline{\delta}\sin\left(\frac{3(2m+1)\pi}{2L}x_d\right)\sin^3\left(\frac{(2m+1)\pi}{2L}x_d\right)\right|\varepsilon_m^3 \tag{4.54}$$

$$C_{5,2m+1} = \frac{45}{4}Q^2\left|\overline{\delta}^2\sin\left(\frac{5(2m+1)\pi}{2L}x_d\right)\sin^2\left(\frac{3(2m+1)\pi}{2L}x_d\right)\sin^5\left(\frac{(2m+1)\pi}{2L}x_d\right)\right| \tag{4.55}$$

$$\varepsilon_{5,2m+1,local} = C_{5,2m+1}\varepsilon_m^5 \tag{4.56}$$

The changes of $C_{3,2m+1}$, $C_{5,2m+1}$ with xd/L are shown in Fig. 4.14.

4.5.2.2 Only Contains Non-classical Nonlinearity

The resonance frequency offset when the sine excitation is in the n = m mode:
Equation (4.20) can be transformed into

$$\partial_{tt}^2 z_m + \frac{\Omega}{Q}\partial_t z_m + \omega_m^2 z_m = F\cos(\Omega t) - H_{mm}\left[-\frac{8}{3\pi}A_m^2\sin(\Omega t + \phi_m - 2A_m^2\cos(\Omega t + \phi_m) + \cdots)\right] \tag{4.57}$$

For the boundary conditions at this time:

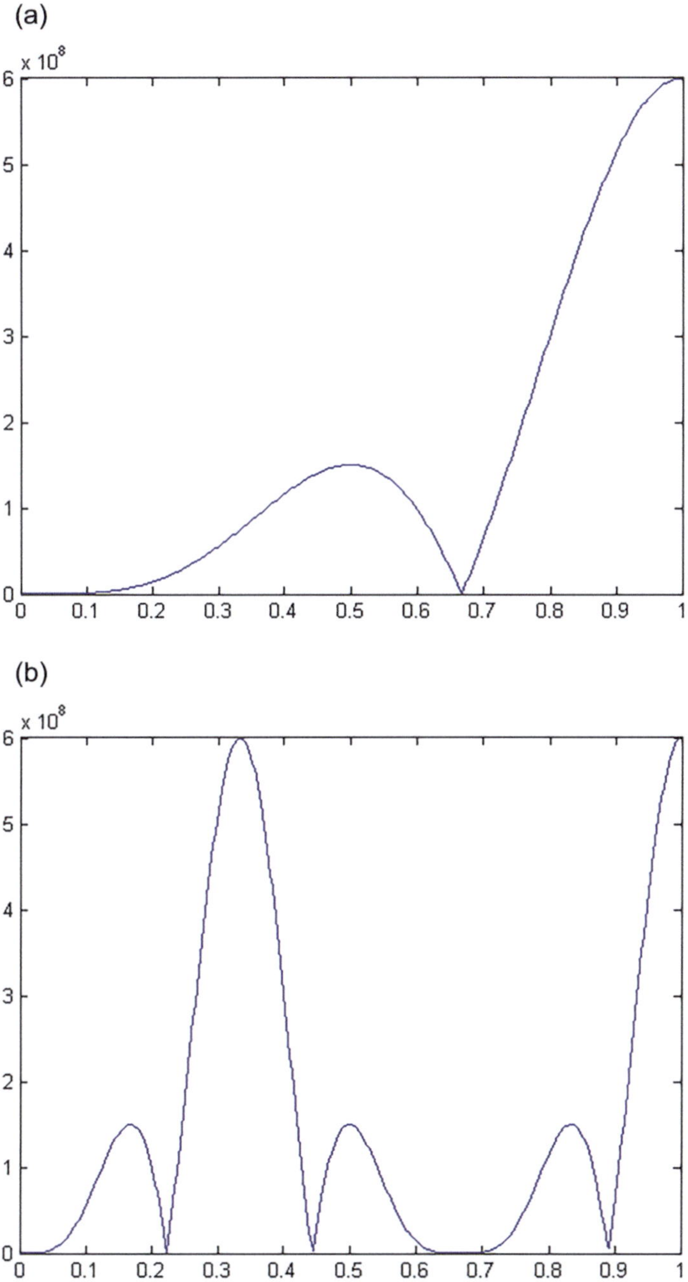

Fig. 4.14 **a** C(3,2m + 1) when m = 0, the horizontal axis is X_d/L, **b** C(3,2m + 1) when m = 1, the horizontal axis is X_d/L, **c** C(3,2m + 1) when m = 2, the horizontal axis is X_d/L, **d** C(3,2m + 1) when m = 3, the horizontal axis is X_d/L, **e** C(5,2m + 1) when m = 0, the horizontal axis is X_d/L, **f** C(5,2m + 1) when m = 1, the horizontal axis is X_d/L, **g** C(5,2m + 1) when m = 2, the horizontal axis is X_d/L, **h** C(5,2m + 1) when m = 3, the horizontal axis is X_d/L

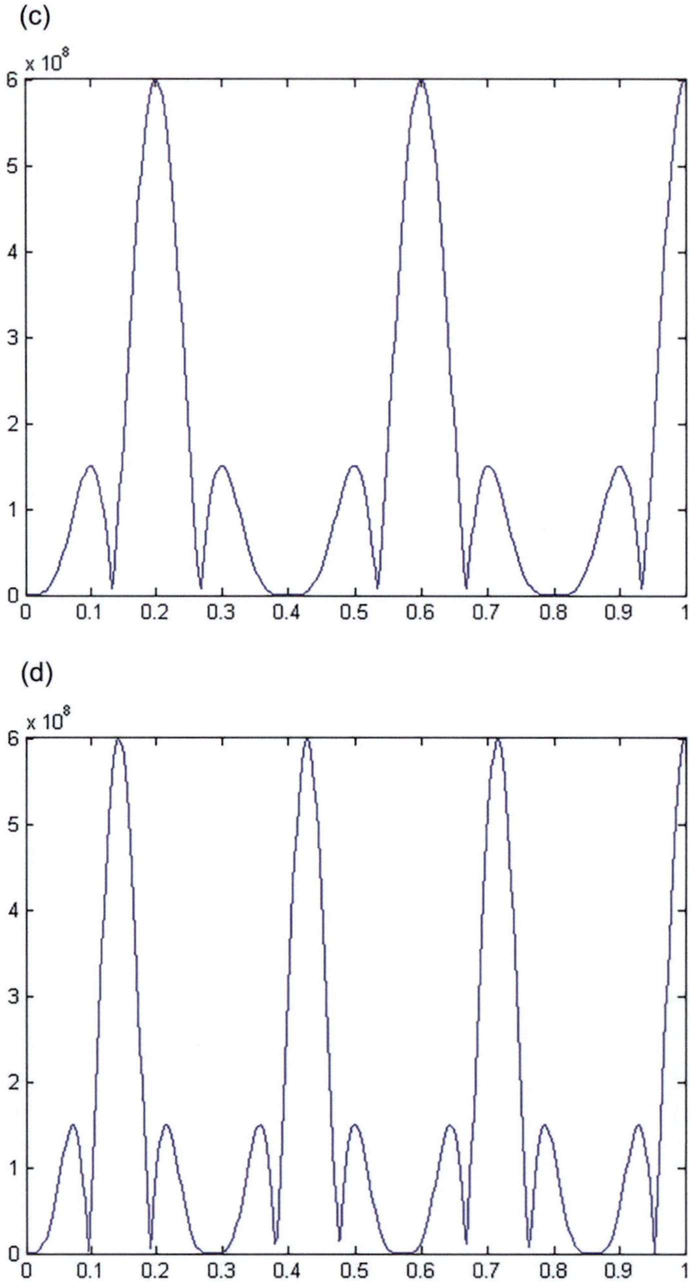

Fig. 4.14 (continued)

(e)

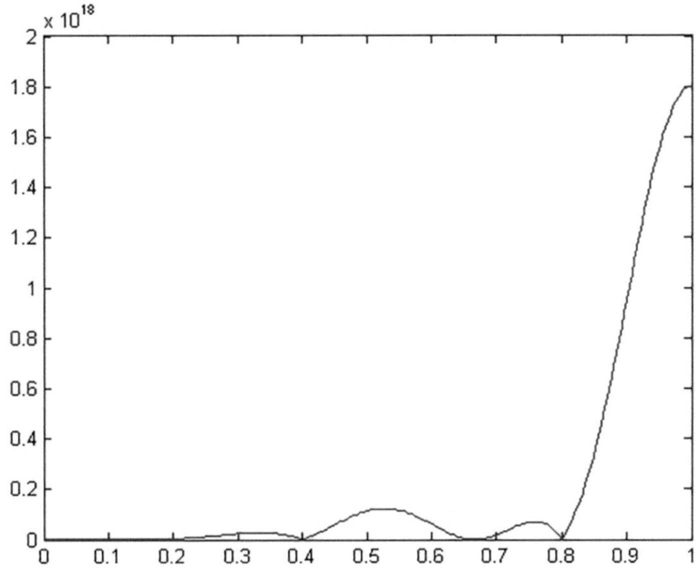

(f)

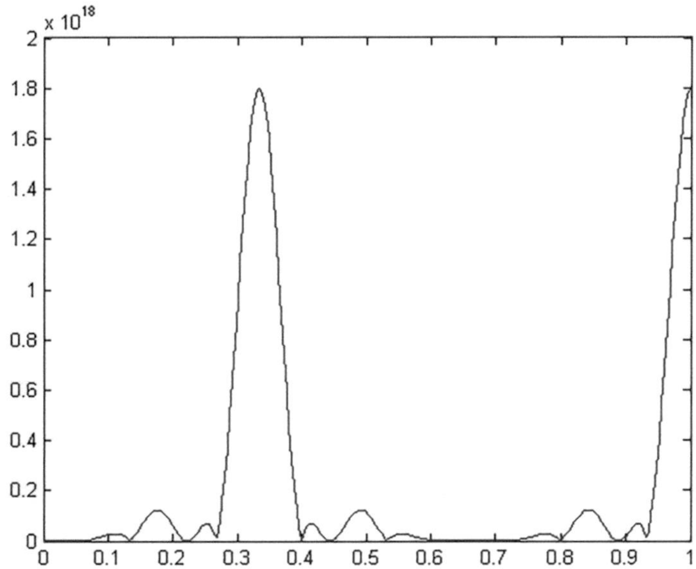

Fig. 4.14 (continued)

(g)

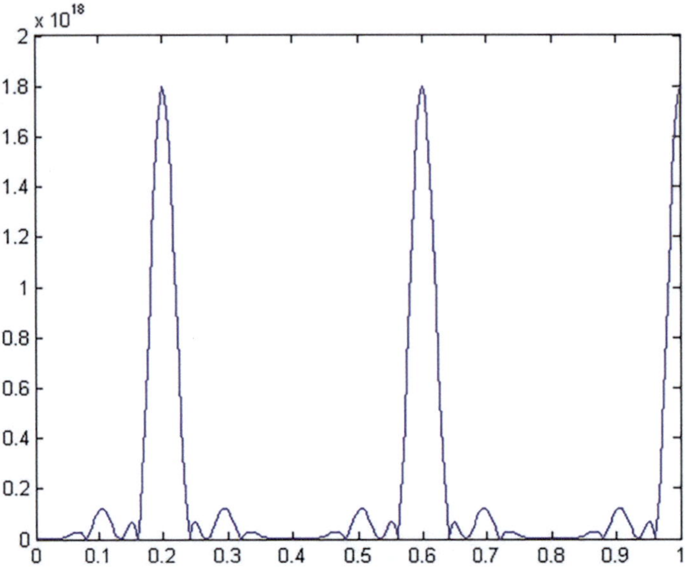

(h)

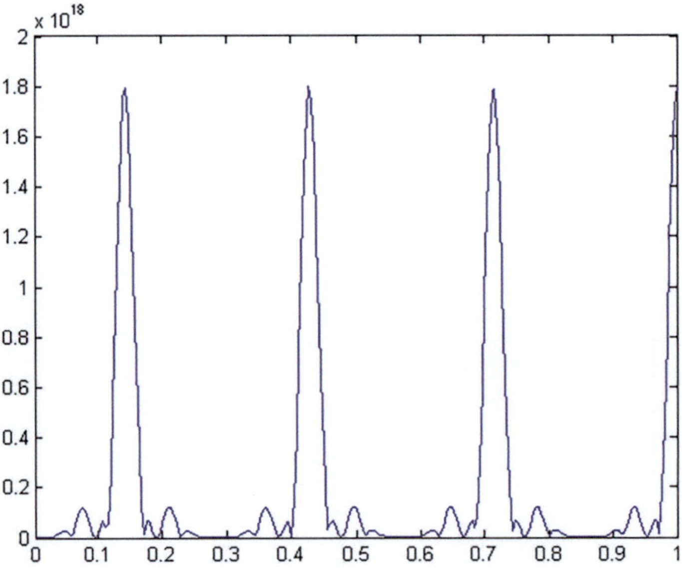

Fig. 4.14 (continued)

$$u_x(0,t) = u(l,t) = 0 \tag{4.58}$$

$$H_{mm} = \frac{K}{\rho L}\int_0^L dx\alpha(x)\partial_x\Psi_m|\partial_x\Psi_m|\partial_x\Psi_m \tag{4.59}$$

The frequency drift is obtained.

$$\begin{aligned}\Omega_{res,local}(\varepsilon_m) &\approx \Omega_{res}(0)\left[1-\left(1+\frac{4}{3\pi Q}\right)\frac{cH_{mm}\varepsilon_m}{\omega_m^3}\right]\\ &= \Omega_{res}(0)\left[1-\overline{\alpha}\left(1+\frac{4}{3\pi Q}\right)\left|\sin^3\left(\frac{(2m+1)\pi x_d}{2L}\right)\right|\varepsilon_m\right]\\ &= \Omega_{res}(0)\left[1-X_{1,2m+1}\varepsilon_m\right]\end{aligned} \tag{4.60}$$

Similarly, we can calculate:

$$X_{3,2m+1} = \frac{8Q}{5\pi}\frac{\hat{\alpha}d}{L}\left|\sin\left(\frac{3(2m+1)\pi}{2L}x_d\right)\right|\sin^2\left(\frac{(2m+1)\pi}{2L}x_d\right) \tag{4.61}$$

$$X_{5,2m+1} = \frac{8Q}{21\pi}\frac{\hat{\alpha}d}{L}\left|\sin\left(\frac{5(2m+1)\pi}{2L}x_d\right)\right|\sin^2\left(\frac{(2m+1)\pi}{2L}x_d\right) \tag{4.62}$$

The frequency drift is related to the square of the strain. The variation of $X_{1,2m+1}$, $X_{3,2m+1}$ and $X_{5,2m+1}$ with xd/L is shown in Fig. 4.15.

Under these boundary conditions, the frequency drift in different modes, whether it is classical nonlinearity or non-classical nonlinearity, the corresponding change coefficients are asymmetric, that is, any two frequency shift relations can determine the position of the defect, and the position of the defect is unique and will not present a symmetrical relationship. This can be reflected in the following inverse modeling.

4.5.2.3 Inverse Modeling

(1) x_d Determination of position

Derivation of classical cubic nonlinearity

$$\text{let } \alpha = \frac{\pi x_d}{2l}$$

$$\frac{\sin(3\alpha)}{\sin(\alpha)} = \sqrt[4]{C_{1,3}/C_{1,1}}$$

$$\cos\alpha = \frac{\sqrt{\sqrt[4]{\frac{C_{1,3}}{C_{1,1}}}+1}}{2} = \frac{\sqrt{\sqrt[4]{\frac{\varepsilon_1^2\Delta\omega_3/\omega_3}{\varepsilon_3^2\Delta\omega_1/\omega_1}}+1}}{2}$$

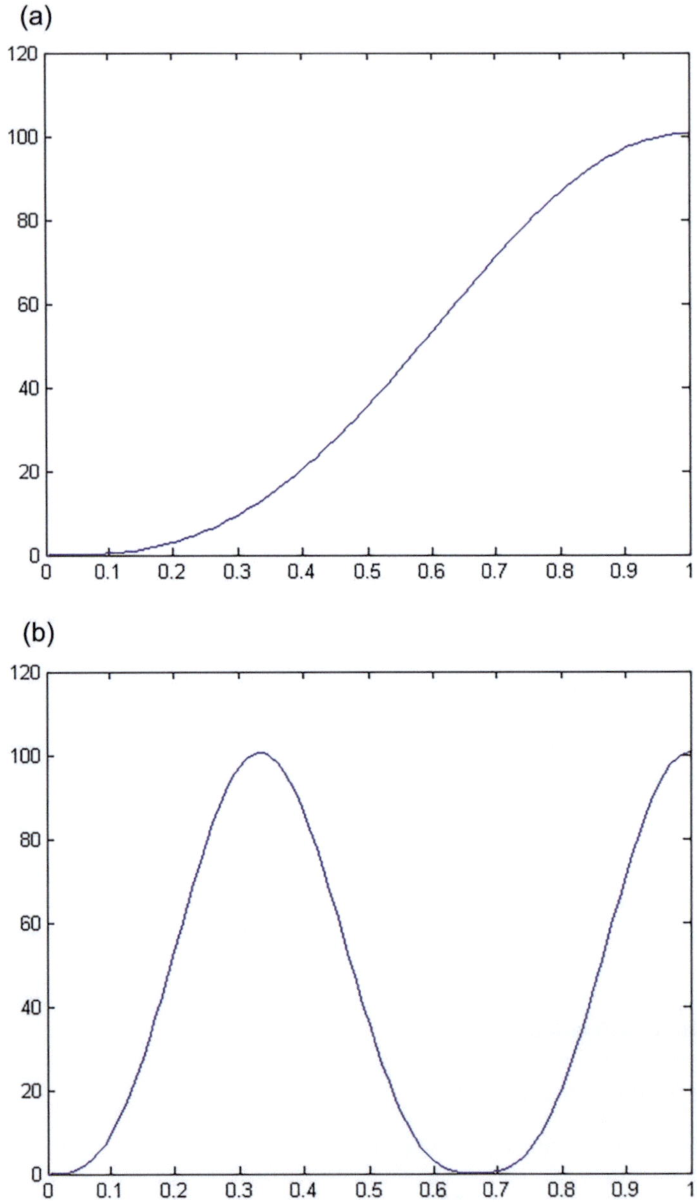

Fig. 4.15 **a** X(1,2m + 1) when m = 0, the horizontal axis is X_d/L, **b** X(1,2m + 1) when m = 1, the horizontal axis is X_d/L, **c** X(1,2m + 1) when m = 2, the horizontal axis is X_d/L, **d** X(3,2m + 1) when m = 0, the horizontal axis is X_d/L, **e** X(3,2m + 1) when m = 1, the horizontal coordinate is X_d/L, **f** X(3,2m + 1) when m = 2, the horizontal coordinate is X_d/L, **g** X(5,2m + 1) when m = 0, the horizontal coordinate is X_d/L, **h** X(5,2m + 1) when m = 1, the horizontal coordinate is X_d/L, **i** X(5,2m + 1) when m = 2, the horizontal coordinate is X_d/L

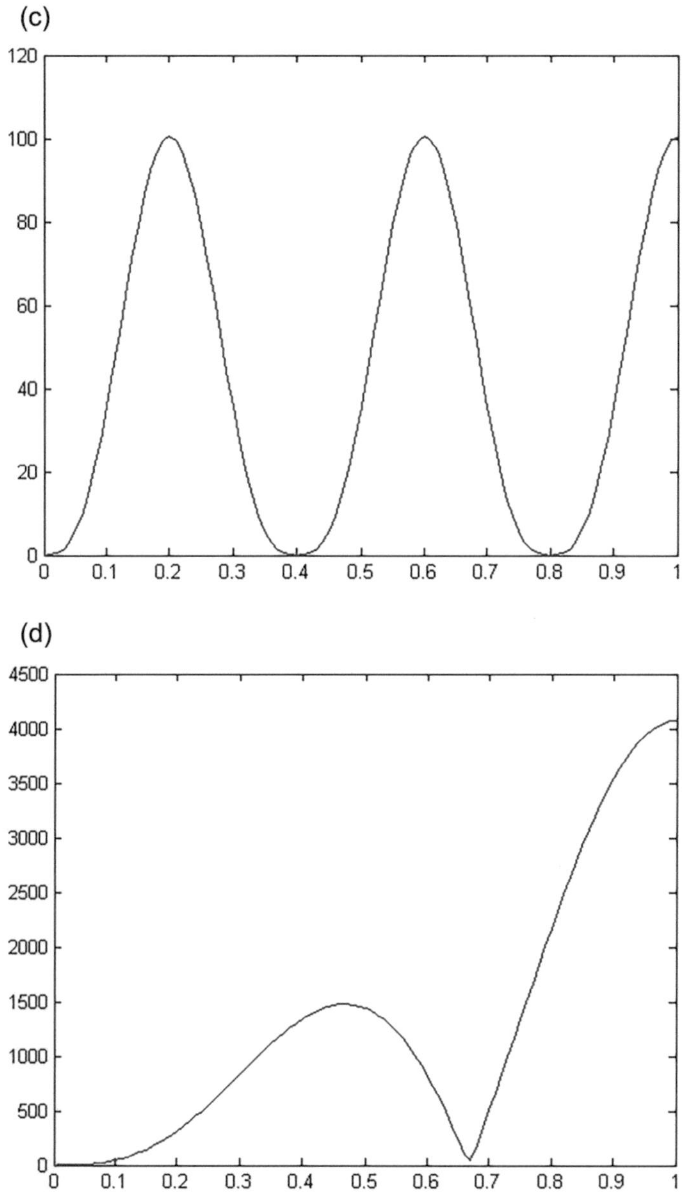

Fig. 4.15 (continued)

(e)

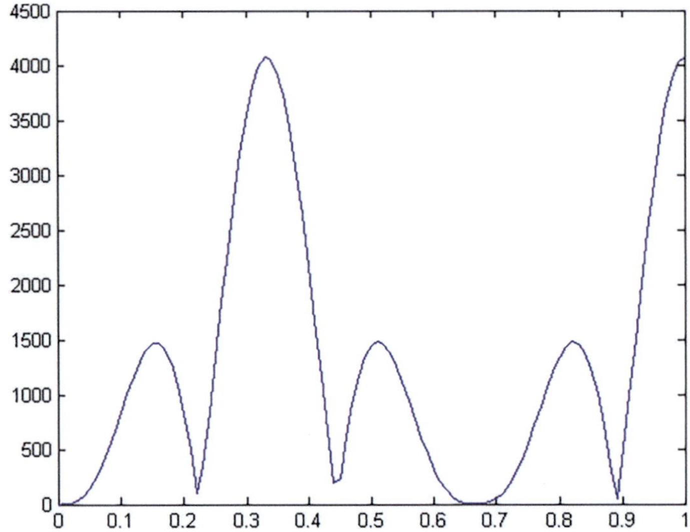

(f)

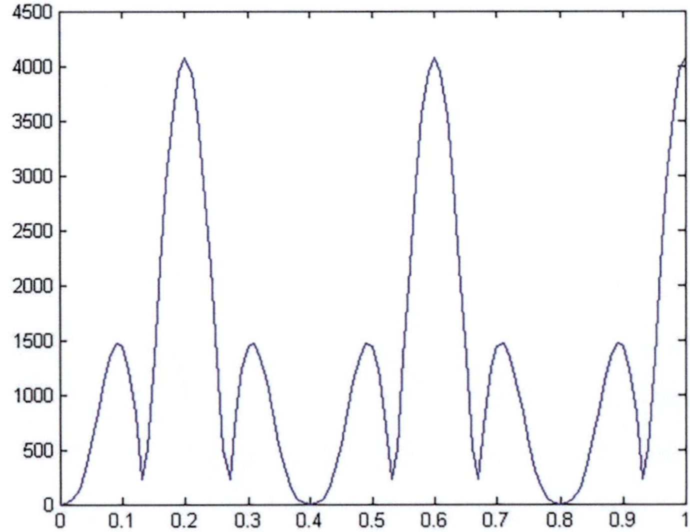

Fig. 4.15 (continued)

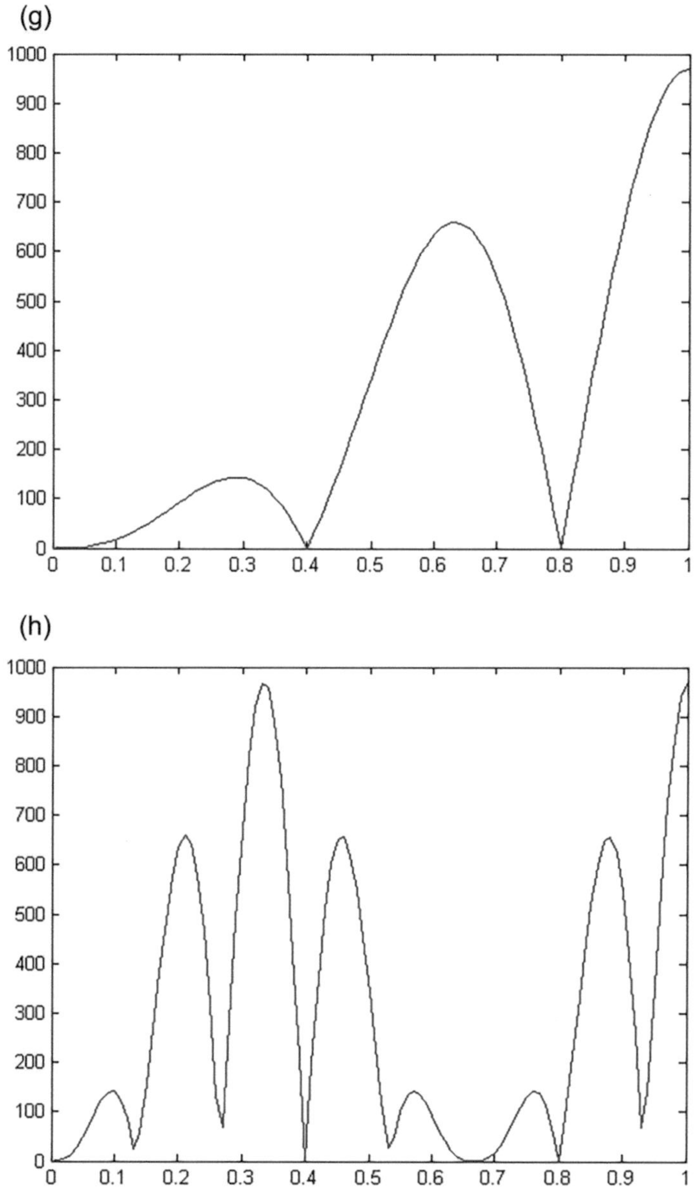

Fig. 4.15 (continued)

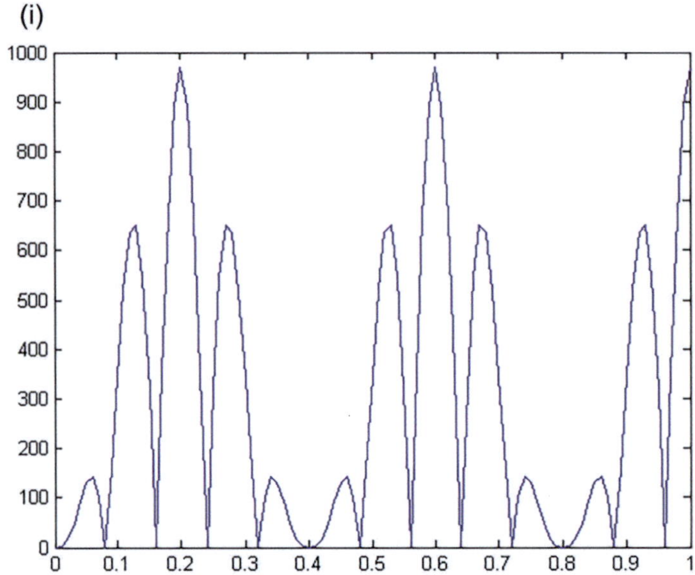

Fig. 4.15 (continued)

$$\frac{\pi x_d}{2l} = \arccos \frac{\sqrt{\sqrt[4]{\frac{\varepsilon_1^2 \Delta\omega_3/\omega_3}{\varepsilon_3^2 \Delta\omega_1/\omega_1}} + 1}}{2}$$

$$\frac{x_d}{l} = \frac{2}{\pi} \arccos \frac{\sqrt{\sqrt[4]{\frac{\varepsilon_1^2 \Delta\omega_3/\omega_3}{\varepsilon_3^2 \Delta\omega_1/\omega_1}} + 1}}{2} \tag{4.63}$$

Similarly

$$\frac{\sin(5\alpha)}{\sin(\alpha)} = \sqrt[4]{C_{1,5}/C_{1,1}}$$

$$16\cos^4\alpha - 12\cos^2\alpha + 1 = \sqrt[4]{C_{1,5}/C_{1,1}}$$

$$\text{let } K = 1 - \sqrt[4]{C_{1,5}/C_{1,1}}$$

$$\cos^2\alpha = \frac{3 + \sqrt{5 + 4K}}{8}$$

$$\cos\alpha = \sqrt{\frac{3 + \sqrt{5 + 4K}}{8}}$$

$$\alpha = \frac{\pi x_d}{2l} = \arccos\sqrt{\frac{3 + \sqrt{5 + 4K}}{8}}$$

$$\frac{x_d}{l} = \frac{2}{\pi}\arccos\sqrt{\frac{3+\sqrt{5+4K}}{8}}$$

$$\frac{x_d}{l} = \frac{2}{\pi}\arccos\sqrt{\frac{3+\sqrt{5+4\left(1-\sqrt[4]{C_{1,5}/C_{1,1}}\right)}}{8}}$$

$$\frac{x_d}{l} = \frac{2}{\pi}\arccos\sqrt{\frac{3+\sqrt{5+4\left(1-\sqrt[4]{C_{1,5}/C_{1,1}}\right)}}{8}} \tag{4.64}$$

The derivation for higher orders is as follows

$$\cos\alpha = \sqrt{C_{3,1}/(2QC_{1,1})+1}/2$$
$$\alpha = \frac{\pi x_d}{2l} = \arccos\left(\sqrt{C_{3,1}/(2QC_{1,1})+1}/2\right)$$
$$\frac{x_d}{l} = \frac{2}{\pi}\arccos\left(\sqrt{C_{3,1}/(2QC_{1,1})+1}/2\right) \tag{4.65}$$

Derivation of non-classical nonlinearity

$$\textit{Because}\ X_{1,2m+1} = \overline{\alpha}\left(1+\frac{4}{3\pi Q}\right)\left|\sin^3\left(\frac{(2m+1)\pi x_d}{2L}\right)\right|$$
$$\sqrt[3]{\frac{X_{1,3}}{X_{1,1}}} = 4\cos^2\left(\frac{\pi x_d}{2l}\right) - 1$$
$$\cos\left(\frac{\pi x_d}{2l}\right) = \frac{\sqrt{\sqrt[3]{\frac{X_{1,3}}{X_{1,1}}}+1}}{2}$$
$$\frac{\pi x_d}{2l} = \arccos\left(\frac{\sqrt{\sqrt[3]{\frac{X_{1,3}}{X_{1,1}}}+1}}{2}\right)$$
$$\frac{x_d}{l} = \frac{2}{\pi}\arccos\left(\frac{\sqrt{\sqrt[3]{\frac{X_{1,3}}{X_{1,1}}}+1}}{2}\right) \tag{4.66}$$

The derivation for higher orders

$$\begin{aligned}
&\frac{X_{3,1}\times 5\pi\left[1+4/(3\pi Q)\right]}{X_{1,1}\times 8Q}=\frac{\sin 3\alpha}{\sin\alpha}\\
&\frac{X_{3,1}\times 5\pi\left[1+4/(3\pi Q)\right]}{X_{1,1}\times 8Q}=4\cos^2\alpha-1\\
&\cos\alpha=\frac{\sqrt{\frac{X_{3,1}\times 5\pi\left[1+4/(3\pi Q)\right]}{X_{1,1}\times 8Q}+1}}{2}\\
&\alpha=\frac{\pi x_d}{2l}=\arccos\left(\frac{\sqrt{\frac{X_{3,1}\times 5\pi\left[1+4/(3\pi Q)\right]}{X_{1,1}\times 8Q}+1}}{2}\right)\\
&\frac{x_d}{l}=\frac{2}{\pi}\arccos\left(\frac{\sqrt{\frac{X_{3,1}\times 5\pi\left[1+4/(3\pi Q)\right]}{X_{1,1}\times 8Q}+1}}{2}\right)
\end{aligned}\tag{4.67}$$

4.5.2.4 Overall Multi-mode NRUS Method for Detecting Defect Location

$$\Omega_{res,Local}(\varepsilon_m)\approx\Omega_{res}(0)\left[1-\Upsilon_m\varepsilon_m-\Theta_m\varepsilon_m^2\right]\tag{4.68}$$

The intensity function can be expressed as:

$$S_M(x)=\frac{2}{M}\sum_{m=1}^{M}\left[|\Upsilon_m|+\sqrt{|\Theta_m|}\right]\sin^2\left(\frac{(2m+1)\pi}{2L}x\right)\tag{4.69}$$

$$\Upsilon_m=\overline{\alpha}\left(1+\frac{4}{3\pi Q}\right)\left|\sin^3\left[(2m+1)\frac{\pi}{2L}x_d\right]\right|\tag{4.70}$$

$$\Theta_m=\frac{3}{4}\overline{\delta}\sin^4\left[(2m+1)\frac{\pi}{2L}x_d\right]\tag{4.71}$$

Therefore, it can be seen that this mode can clearly determine the position of the defect, will not cause confusion, and the larger the number of modes considered, the clearer the position of the defect can be reflected (Fig. 4.16).

This section uses the overall multi-mode method to study nonlinear defects, and analyzes the relationship between the resonance frequency drift when this defect exists and the original linear resonance frequency, and derives the drift coefficient of the fundamental and higher harmonics under the existence of classical nonlinearity and non-classical nonlinearity defects and the relationship with strain. By measuring these relationships through experiments, the position of the defect can be derived. Inverse modeling uses the method of intensity function to realize the observation and

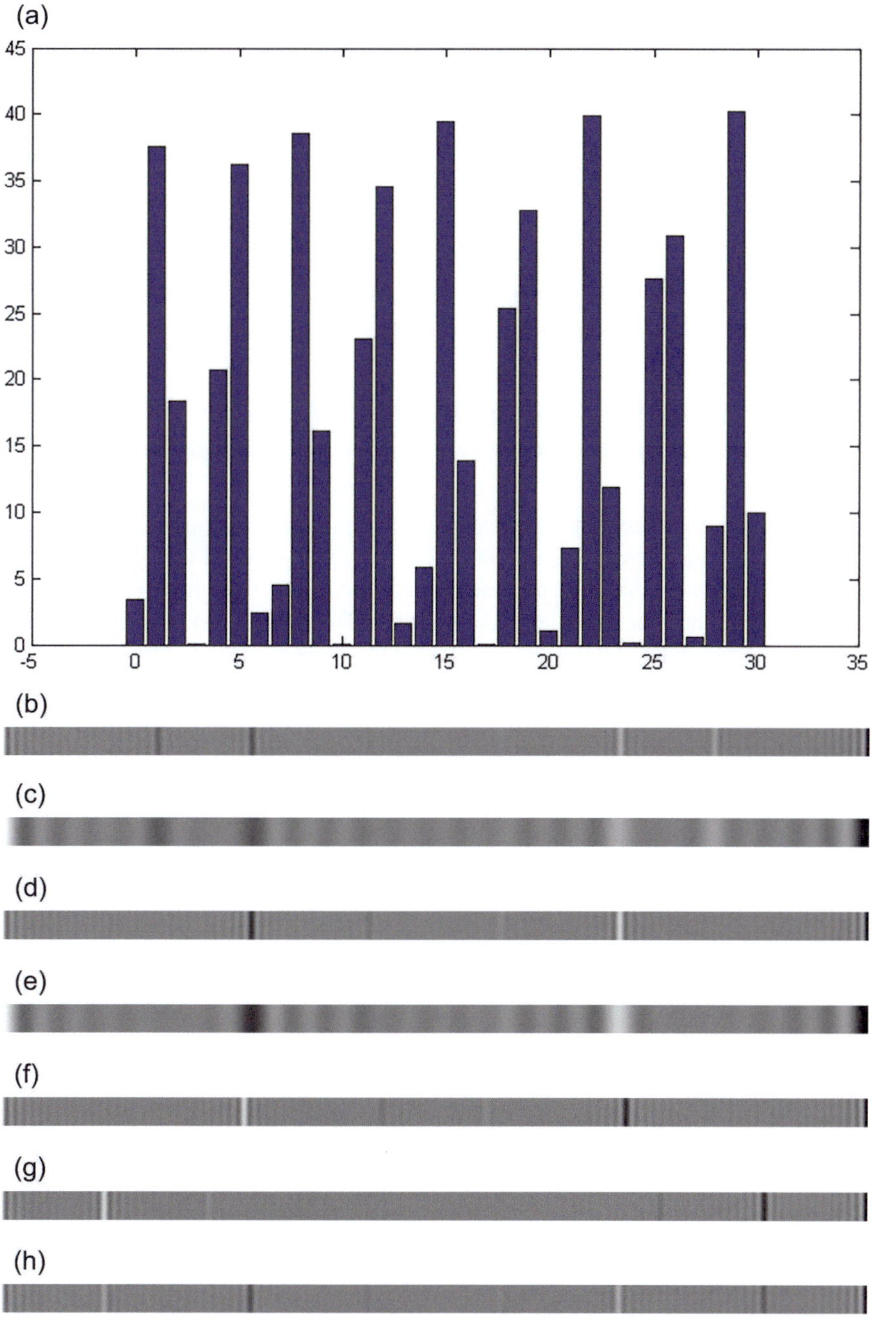

Fig. 4.16 (**a**) The variation of Ym with modulus M, (**b**)–(**h**) The spatial distribution of Sm with respect to the position in the rod, (**b**) M = 100, xd/L = 72/250,50/250 (**c**) M = 25, xd/L = 72/250,50/250 (**d**) M = 100, xd/L = 72/250 (**e**) M = 25, xd/L = 72/250 (**f**) M = 100, xd/L = 72/250 (**g**) M = 100, xd/L = 220/250 (**h**) M = 100, xd/L = 72/250,220/250

positioning of defects. Because the initial modeling uses the boundary condition of symmetry at both ends, it causes the trouble of the symmetry problem of the defect position. By modifying the boundary conditions to be free at one end and fixed at the other, a more accurate positioning method is derived. This positioning method will not produce a symmetry problem, because the initial boundary adjustment is asymmetric. This solution has strong experimental value and practical significance.

4.6 Analysis of High-Order Harmonics of One-Dimensional Non-classical Nonlinearity

4.6.1 Nonlinear Stress–Strain Relationship and Its Wave Equation of One-Dimensional Cracked Metal Rod

In this section, we will consider the nonlinear stress–strain relationship and its differential equation caused by multiple cracks. There are two types of nonlinear effects in the case of cracks, the classical nonlinear effect is caused by the high-order effect of the power series, and the non-classical nonlinear effect corresponds to the hysteresis effect. In micro-cracks, the hysteresis effect is the main nonlinear effect, so here we will consider the non-classical nonlinear effect caused by micro-cracks and propose the differential equation in the one-dimensional model.

As shown in Fig. 4.17, the metal rod with a length of L, a density of ρ, a total mass of M, a cross-sectional area of S, a Young's modulus of E, one end free, one end with a mass load of m, excited by a sound source with a circular frequency of $\Omega\omega$, there are n micro-cracks in the rod, the midpoints of the cracks are located at $x = x_1, x_2, \cdots x_n$, the effective lengths of the cracks are $d_1, d_2, \cdots d_n$, and they satisfy $d_i << L(i = 1, 2, \cdots n)$.

According to PM theory, the non-classical stress–strain relationship in the metal is [17]

$$\sigma = E\varepsilon + E\frac{\alpha(x)}{2}\left[sign(\partial_t\varepsilon)\left((\Delta\varepsilon)^2 - \varepsilon^2\right) - 2(\Delta\varepsilon)\varepsilon\right] \tag{4.72}$$

where $\varepsilon = \partial_x\xi$ is the strain in the metal rod, $\Delta\varepsilon$ is the amplitude of the strain, the parameter $\alpha(x)$ indicates the size of the hysteresis effect and the size of the hysteresis loop opening in the stress–strain relationship, and it satisfies

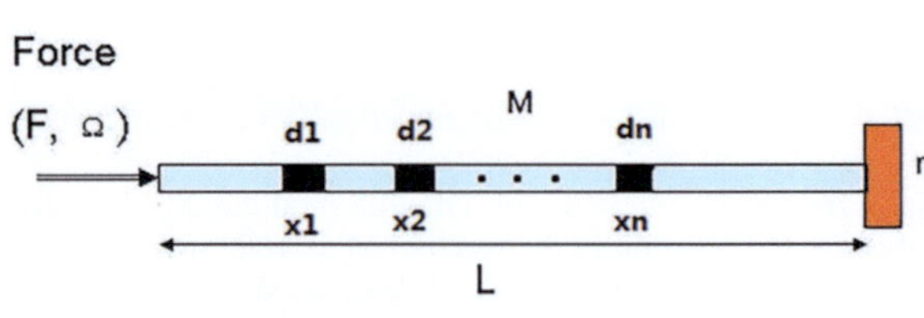

Fig. 4.17 Geometric model of a one-dimensional metal rod with multiple cracks

$$\alpha(x) = \begin{cases} \alpha \, for\left(x_i - \frac{d_i}{2} < x < x_i + \frac{d_i}{2}\right) & i = 1, 2, \ldots n \\ 0 & others \end{cases} \tag{4.73}$$

The above formula assumes that the nonlinear stress–strain relationship only occurs where there are cracks, and for the position without cracks,

It still satisfies the linear stress–strain relationship. We have obtained the differential equation with multiple cracks in the one-dimensional rod model considering the damping effect

$$\rho \partial_{tt}^2 \xi = \partial_x \left\{ E\varepsilon + E\frac{\alpha(x)}{2}\left[sign(\partial_t \varepsilon)\left((\Delta\varepsilon)^2 - \varepsilon^2\right) - 2(\Delta\varepsilon)\varepsilon)\right]\right\} - \rho\frac{\omega}{Q}\partial_t \xi \tag{4.74}$$

where Q is the quality factor, representing the strength of the material damping.

4.6.2 Wave Equation and Its Linear Approximate Solution

The wave Eq. (4.78) is a nonlinear equation, we let $\xi = \xi_1 + \xi_2$, where ξ_1 is a first-order approximate solution, ξ_2 is a second-order approximate solution. In the case of linear approximation, the first-order approximate vibration equation can be obtained

$$\rho \partial_{tt}^2 \xi_1 = E\partial_{xx}^2 \xi_1 - \rho\frac{\omega}{Q}\partial_t \xi_1 \tag{4.75}$$

The model is excited by a sinusoidal sound source at one end, with an amplitude of F, an angular frequency of ω, and a load with a mass of m at the other end, its boundary conditions can be written as

$$\begin{cases} SE\left.\frac{\partial \xi_1}{\partial x}\right|_{x=0} = Fe^{j\omega t} \\ SE\left.\frac{\partial \xi_1}{\partial x}\right|_{x=L} = -m\left.\frac{\partial^2 \xi_1}{\partial t^2}\right|_{x=L} \end{cases} \tag{4.76}$$

Assume its solution is

$$\xi_1 = \left(A\cos k'x + B\sin k'x\right)e^{j(\omega t + \varphi)} \tag{4.77}$$

where, k' is the complex wave number in the rod, A and B represent the amplitude of its sine component and the amplitude of its cosine component, respectively. φ represents its phase relative to the excitation sound source. Substituting (4.77) into the boundary condition (4.76) gives

$$B = \frac{Fe^{-j\varphi}}{SEk'} \tag{4.78}$$

Substituting (4.77), (4.78) into (4.75) gives

$$\frac{1}{kQ}e^{j\omega t}\left[Ek'^2Q-(Q-j)\rho\omega^2\right]\left(Ak'e^{j\varphi}\cos k'x+F\sin k'x\right)=0 \tag{4.79}$$

Since the above equation holds for any x, we need to find

$$Ek'^2Q-(Q-j)\rho\omega^2=0 \tag{4.80}$$

We let $c=\sqrt{\frac{E}{\rho}}$, $k=\frac{\omega}{c}$, $k'=a+bj$, where a and b represent the real and imaginary parts of the complex frequency, respectively. Substituting them into (4.80) and requiring both the real and imaginary parts to be zero, we get

$$\begin{cases} a^2=b^2+k^2 \\ ab=-\frac{k^2}{2Q} \end{cases} \tag{4.81}$$

In general solids, the attenuation is small and the quality factor Q is large, so the real and imaginary parts of the complex frequency satisfy a $\gg$ b. Hence, we can obtain the approximate solution of (4.81) and the expression of k'

$$\begin{cases} a=k \\ b=-\frac{k}{2Q} \end{cases} \tag{4.82}$$

$$k'=k\left(1-\frac{j}{2Q}\right) \tag{4.83}$$

To obtain the expression of the amplitude A, we substitute (4.77), (4.78) into (4.76) to get

$$A=\frac{F\left(SEk'-m\omega^2\tan k'L\right)}{SEk\left(m\omega^2+SEk'\tan k'L\right)}e^{-j\varphi} \tag{4.84}$$

Thus, we obtain the expression for the displacement of the particle under the first-order approximation ξ

$$\xi_1=\frac{F\left[SEk'\cos k'(L-x)-m\omega^2\sin k'(L-x)\right]}{SEk'\left(m\omega^2\cos k'L+SEk'\sin k'L\right)}e^{j\omega t} \tag{4.85}$$

We use the resonant frequency to excite, and assume that at resonance, the driving force F is very small, that is, $B=0$ in Eq. (4.78), and $k'\approx k=\frac{\omega}{c}$. Substituting Eq. (4.77) into Eq. (4.76), we get

$$\tan kl=-\frac{\omega mc}{SE} \tag{4.86}$$

The total mass of the rod is $M = SL\rho$, simplifying (4.86) gives its natural frequency equation

$$\frac{\tan kL}{kL} = -\frac{m}{M} \tag{4.87}$$

The expression for the actual displacement of the particle is $\xi_1 = \mathrm{Re}(\xi_1)$. Using Eq. (4.87) and the approximation that the quality factor Q is large, we can obtain the approximate solution

$$\xi_1 = \frac{2QF(M\cos kL - mkL\sin kL)}{SL\rho\omega^2[m(kL\sin kL - \cos kL) - M(2 + \cos kL)]}\cos kx \sin \omega t \tag{4.88}$$

The numerical simulations in the following sections all use the following data: $\frac{m}{M} = 0.5$, which gives the first natural frequency $k_1L = 2.289$ and the second natural frequency $k_2L = 5.087$. Under the first natural frequency, take $L = 0.2\,\text{m}$, then $k_1 = 11.44\,\text{m}^{-1}$. Under the second natural frequency, take $L = 0.2\,\text{m}$, then $k_2 = 25.43\,\text{m}^{-1}$, the quality factor $Q = 80$, the cross-sectional area $S = 2.5 \times 10^{-3}\,\text{m}^2$, the speed of sound in the rod $c = 6.0 \times 10^3$ m/s, the density of the rod $\rho = 7.8 \times 10^3\,\text{kg/m}^3$, the amplitude of the force $F = 10^5 N$, and the nonlinear parameter $\alpha = 2000$. The fundamental wave amplitude distribution in the rod is shown in Fig. 4.18. It can be seen that due to the influence of the mass load boundary conditions, at x = L, its amplitude is neither the maximum nor the minimum.

4.6.3 Harmonic Components Caused by Hysteresis Effect

4.6.3.1 Analytical Solution of High-Order Harmonics Under Perturbation Approximation

Under the first or second type of boundary conditions, the eigenfrequencies are proportional to each other, and the high-order harmonics produced by the nonlinear effect are exactly equal to their higher eigenfrequencies, so they can satisfy the boundary conditions. However, for the mass load boundary conditions, the eigenfrequencies are not proportional to each other, and the high-order harmonics produced do not equal their eigenfrequencies. What will be the behavior of the high-order harmonics at this time? In this section, we will discuss the vibration distribution under this situation in detail.

Let $H = \frac{2QF(M\cos kL - mkL\sin kL)}{SL\rho\omega^2[m(kL\sin kL - \cos kL) - M(2+\cos kL)]}$, so Eq. (4.88) becomes

$$\xi_1 = H\cos kx \sin \omega t \tag{4.89}$$

Here, H is the amplitude of the fundamental particle vibration displacement, and after taking the partial derivative of Eq. (4.89) with respect to x, we get:

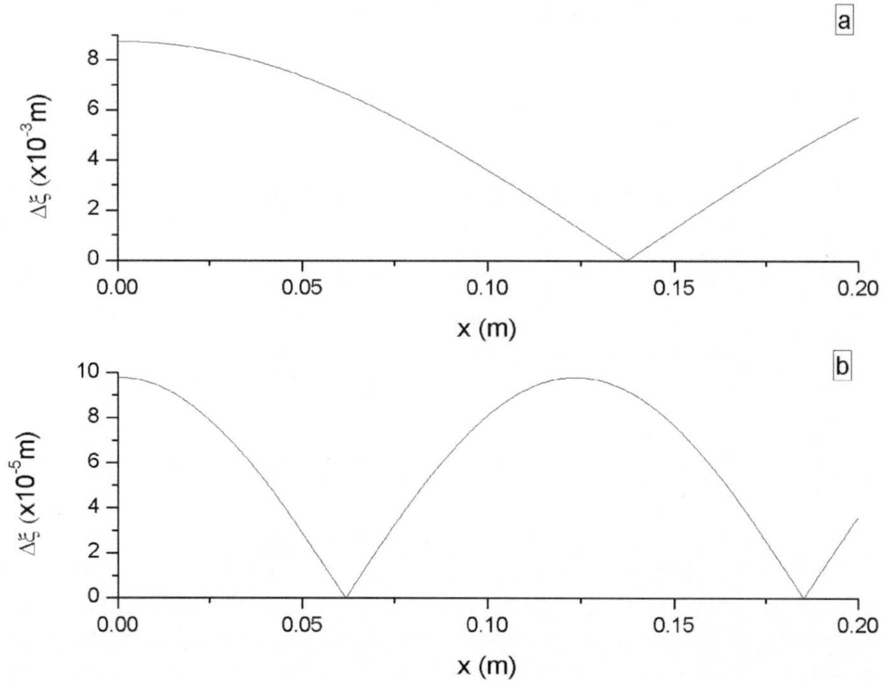

Fig. 4.18 **a** The distribution of the amplitude of the fundamental particle vibration displacement under the first natural frequency. **b** The distribution of the amplitude of the fundamental particle vibration displacement under the second natural frequency

$$\varepsilon_1 = \partial_x \xi_1 = -kH \sin kx \sin \omega t \tag{4.90}$$

ε_1 is the strain, and its amplitude $\Delta\varepsilon_1$ satisfies:

$$\Delta\varepsilon_1 = \Delta(\partial_x \xi_1) = |kH \sin kx| \tag{4.91}$$

In the case of second-order approximation, Eq. (4.74) can be transformed into:

$$\begin{aligned} &\rho \partial_{tt}^2 \xi_2 - E \partial_{xx}^2 \xi_2 + \rho \frac{\omega}{Q} \partial_t \xi_2 \\ &= \partial_x \left\{ E \frac{\alpha(x)}{2} \left[sign(\partial_t \varepsilon_1) \left((\Delta\varepsilon_1)^2 - \varepsilon_1^2 \right) - 2(\Delta\varepsilon_1)\varepsilon_1) \right] \right\} \end{aligned} \tag{4.92}$$

The right side of the above equation is the force source term of the high-order harmonics caused by the micro-crack. After performing a Fourier transform on $sign(\partial_t \varepsilon_1)$, we can get

$$sign(\partial_t \varepsilon_1) = sign(H \sin kx) \sum_{n=0}^{\infty} \left(-1 \right)^{n+1} \frac{4}{(2n+1)\pi} \sin[(2n+1)\omega t] \Big] \tag{4.93}$$

Substituting into Eq. (4.92) and simplifying, we get

$$\rho \partial_{tt}^2 \xi_2 - E \partial_{xx}^2 \xi_2 + \rho \frac{\omega}{Q} \partial_t \xi_2 = Ek^3 H^2 \frac{\alpha(x)}{2} sign(H \sin kx) \sin 2kx$$
$$\times \left[\left(2 - \frac{2}{3\pi} \right) \sin \omega t - \frac{8}{15\pi} \sin 3\omega t + \frac{8}{105\pi} \sin 5\omega t - \frac{8}{315\pi} \sin 7\omega t + \cdots \right] \tag{4.94}$$

It can be seen that due to the influence of the hysteresis effect, harmonic terms appear in the wave equation. However, it is clear that the wave equation only contains odd harmonic components and no even harmonic components. Therefore, under the influence of the hysteresis effect, only odd harmonics will be excited in the rod. We first study the relationship between the third harmonic and the crack, and Eq. (4.94) can be simplified to

$$\begin{aligned} &\rho \partial_{tt}^2 \xi_2 - E \partial_{xx}^2 \xi_2 + \rho \frac{\omega}{Q} \partial_t \xi_2 \\ &= -\frac{8}{15\pi} Ek^3 H^2 \frac{\alpha(x)}{2} sign(H \sin kx) \sin 2kx \sin 3\omega t \end{aligned} \tag{4.95}$$

Because the eigenfrequencies of the system are not proportional to each other, but the third harmonic needs to satisfy the boundary conditions, its particle vibration distribution will be different from the general situation. But in those positions without cracks, the vibration still satisfies the linear differential equation, so the vibration distribution between the crack and the boundary should satisfy the trigonometric function distribution. Similarly, we can set the solution of this wave equation as

$$\begin{cases} \xi_{231} = (A_{31} \cos 3kx + B_{31} \sin 3kx) \sin 3\omega t \\ \xi_{232} = (A_{32} \cos 3kx + B_{32} \sin 3kx) \sin 3\omega t \\ \vdots \\ \xi_{23(n+1)} = \left(A_{3(n+1)} \cos 3kx + B_{3(n+1)} \sin 3kx \right) \sin 3\omega t \end{cases} \tag{4.96}$$

where, ξ_{ijk} represents i order approximation j harmonic in the $k-1$ and k cracks in the rod particle vibration displacement. And it is required to satisfy the boundary conditions of the rod

$$\begin{cases} \left. \frac{\partial \xi_{231}}{\partial x} \right|_{x=0} = 0 \\ SE \left. \frac{\partial \xi_{233}}{\partial x} \right|_{x=L} = -m \left. \frac{\partial^2 \xi_{233}}{\partial t^2} \right|_{x=L} \end{cases} \tag{4.97}$$

Since the cracks produced at the beginning of material fatigue are micro-cracks, it is required that the particle vibration displacement of the rod is continuous at the crack [20]. That is

$$\begin{cases} \xi_{231}|_{x=x_1} = \xi_{232}|_{x=x_1} \\ \xi_{232}|_{x=x_2} = \xi_{233}|_{x=x_2} \\ \vdots \\ \xi_{23n}|_{x=x_n} = \xi_{23(n+1)}\big|_{x=x_n} \end{cases} \tag{4.98}$$

The right side of Eq. (4.95) is equivalent to a force source. Since the crack width is very small, we can write Eq. (4.95) at the i crack using the derivative definition:

$$\begin{aligned} &\left(\rho\partial_{tt}^2\xi_2 - E\frac{\left(\partial_x\xi_{23(i+1)} - \partial_x\xi_{23i}\right)}{d_i} + \rho\frac{\omega}{Q}\partial_t\xi_2\right)\Bigg|_{x=x_i} \\ &= \left(-\frac{8}{15\pi}Ek^3H^2\frac{\alpha(x)}{2}sign(H\sin kx)\sin 2kx\sin 3\omega t\right)\Bigg|_{x=x_i} \end{aligned} \tag{4.99}$$

Multiply both sides of Eq. (4.99) by d_i. Since $d_i << L$, we can ignore the influence of $\rho\partial_{tt}^2\xi_2 + \rho\frac{\omega}{Q}\partial_t\xi_2$, and obtain the approximate expression of Eq. (4.99):

$$\partial_x\xi_{23(i+1)} - \partial_x\xi_{23i} = \frac{8}{15\pi}d_ik^3H^2\frac{\alpha(x)}{2}sign(H\sin kx_i)\sin 2kx_i\sin 3\omega t \tag{4.100}$$

The physical meaning of Eq. (4.100) is the strain relationship at the crack. The stress difference at the ends of the crack is caused by the nonlinear source. According to Eq. (4.97), we can write all the stress continuity equations:

$$\begin{cases} \partial_x\xi_{232}|_{x=x_1} - \partial_x\xi_{231}|_{x=x_1} = \frac{8}{15\pi}d_1k^3H^2\frac{\alpha(x)}{2}\text{sign}(H\sin kx_1)\sin 2kx_1\sin 3\omega t \\ \partial_x\xi_{233}|_{x=x_2} - \partial_x\xi_{232}|_{x=x_2} = \frac{8}{15\pi}d_2k^3H^2\frac{\alpha(x)}{2}\text{sign}(H\sin kx_2)\sin 2kx_2\sin 3\omega t \\ \vdots \\ \partial_x\xi_{23(n+1)}\big|_{x=x_n} - \partial_x\xi_{23n}|_{x=x_n} = \frac{8}{15\pi}d_nk^3H^2\frac{\alpha(x)}{2}\text{sign}(H\sin kx_n)\sin 2kx_n\sin 3\omega t \end{cases} \tag{4.101}$$

For a certain crack position x_i and crack size d_i, Eq. (4.96) has $2(n+1)$ unknowns, while Eqs. (4.97) to (4.99) have $2(n+1)$ equations. Therefore, the solution of equations A_i and B_i can be uniquely determined.

4.6.3.2 Crack Location Method Using Third Harmonic

For simplicity, this section discusses and numerically simulates two cracks. We rewrite the wave Eq. (4.96) and its solution (4.97), boundary condition (4.97), micro-crack particle displacement continuity condition (4.98), and stress continuity condition on both sides of the crack (4.101) as follows:

$$\rho\partial_{tt}^2\xi_{23i} - E\partial_{xx}^2\xi_{23i} + \rho\frac{\omega}{Q}\partial_t\xi_{23i} = 0\ \ (i = 1, 2, 3) \tag{4.102}$$

$$\begin{cases} \xi_{231} = (A_{31}\cos 3kx + B_{31}\sin 3kx)\sin 3\omega t \\ \xi_{232} = (A_{32}\cos 3kx + B_{32}\sin 3kx)\sin 3\omega t \\ \xi_{233} = (A_{33}\cos 3kx + B_{33}\sin 3kx)\sin 3\omega t \end{cases} \tag{4.103}$$

$$\begin{cases} \left.\frac{\partial\xi_{231}}{\partial x}\right|_{x=0} = 0 \\ SE\left.\frac{\partial\xi_{233}}{\partial x}\right|_{x=L} = -m\left.\frac{\partial^2\xi_{233}}{\partial t^2}\right|_{x=L} \end{cases} \tag{4.104}$$

$$\begin{cases} \xi_{231}|_{x=x_1} = \xi_{232}|_{x=x_1} \\ \xi_{232}|_{x=x_2} = \xi_{233}|_{x=x_2} \end{cases} \tag{4.105}$$

$$\begin{cases} \partial_x\xi_{232}|_{x=x_1} - \partial_x\xi_{231}|_{x=x_1} \\ \quad = \frac{8}{15\pi}d_1k^3H^2\frac{\alpha(x)}{2}sign(H\sin kx_1)\sin 2kx_1\sin 3\omega t \\ \partial_x\xi_{233}|_{x=x_2} - \partial_x\xi_{232}|_{x=x_2} \\ \quad = \frac{8}{15\pi}d_2k^3H^2\frac{\alpha(x)}{2}sign(H\sin kx_2)\sin 2kx_2\sin 3\omega t \end{cases} \tag{4.106}$$

Let $D_{31} = \frac{8}{30\pi}d_1k^3H^2\alpha sign(H\sin kx_1)\sin 2kx_1$, $D_{32} = \frac{8}{30\pi}d_2k^3H^2\alpha sign(H\sin kx_2)\sin 2kx_2$, then Eq. (4.106) can be simplified as:

$$\begin{cases} \partial_x\xi_{232}|_{x=x_1} - \partial_x\xi_{231}|_{x=x_1} = D_{31}\sin 3\omega t \\ \partial_x\xi_{233}|_{x=x_2} - \partial_x\xi_{232}|_{x=x_2} = D_{32}\sin 3\omega t \end{cases} \tag{4.107}$$

From Eq. (4.97), we can get $B_{31} = 0$, that is:

$$\xi_{231} = A_{31}\cos 3kx\sin 3\omega t \tag{4.108}$$

From the frequency Eq. (4.87), we can get:

$$\tan 3kL = \frac{(kLm)^3 - 3kLM^2}{M^3 - 3(kL)^2M} = \frac{1 - 3\beta^2}{\beta^3 - 3\beta} \tag{4.109}$$

where $\beta = \frac{M}{kLm}$. From the boundary condition (4.108), we can get:

$$\frac{M}{3mkL}(B_{33} - A_{33}\tan 3kL) = A_{33} + B_{33}\tan 3kL \tag{4.110}$$

Substituting Eq. (4.109) into Eq. (4.110), we can get:

$$B_{33} = \frac{8\beta}{3 - 6\beta^2 - \beta^4} A_{33} = \chi_3 A_{33} \tag{4.111}$$

where $\chi_3 = \frac{8\beta}{3-6\beta^2-\beta^4}$. From Eqs. (4.103), (4.105), (4.109) to (4.111), we can get that the amplitude of the cosine component should satisfy the following relationship:

$$\begin{cases} A_{31}\cos 3kx_1 = A_{32}\cos 3kx_1 + B_{32}\sin 3kx_1 \\ A_{32}\cos 3kx_2 + B_{32}\sin 3kx_2 = A_{33}\cos 3kx_2 + \chi_3 A_{33}\sin 3kx_2 \\ B_{32}\cos 3kx_1 + (A_{31} - A_{32})\sin 3kx_1 = \frac{D_{31}}{3k} \\ (\chi_3 A_{33} - B_{32})\cos 3kx_2 + (A_{32} - A_{33})\sin 3kx_2 = \frac{D_{32}}{3k} \end{cases} \tag{4.112}$$

By solving the above equations, we can get the relationship between the wave coefficient and the crack size and position:

$$\begin{cases} A_{31} = \frac{D_{31}\cos 3kx_1 + D_{32}\cos 3kx_2 + D_{31}\chi_3\sin 3kx_1 + D_{32}\chi_3\sin 3kx_2}{3k\chi_3} \\ A_{32} = \frac{D_{31}\cos 3kx_1 + D_{32}(\cos 3kx_2 + \chi_3\sin 3kx_2)}{3k\chi_3} \\ B_{32} = \frac{D_{31}\cos 3kx_1}{3k} \\ A_{33} = \frac{D_{31}\cos 3kx_1 + D_{32}\cos 3kx_2}{3k\chi_3} \\ B_{33} = \frac{D_{31}\cos 3kx_1 + D_{32}\cos 3kx_2}{3k} \end{cases} \tag{4.113}$$

4.6.3.3 Crack Location Method Using Fifth Harmonic

Similarly, we can study the relationship between the size of the fifth harmonic in the rod and the crack position and size. Its wave equation is:

$$\rho\partial_{tt}^2\xi_2 - E\partial_{xx}^2\xi_2 + \rho\frac{\omega}{Q}\partial_t\xi_2 = \frac{8}{105\pi}Ek^3H^2\frac{\alpha(x)}{2} \\ sign(H\sin kx)\sin 2kx \sin 5\omega t \tag{4.114}$$

Assume its solution is:

$$\begin{cases} \xi_{251} = (A_{51}\cos 5kx + B_{51}\sin 5kx)\sin 5\omega t \\ \xi_{252} = (A_{52}\cos 5kx + B_{52}\sin 5kx)\sin 5\omega t \\ \xi_{253} = (A_{53}\cos 5kx + B_{53}\sin 5kx)\sin 5\omega t \end{cases} \tag{4.115}$$

Satisfying the boundary condition:

$$\begin{cases} \left.\frac{\partial\xi_{251}}{\partial x}\right|_{x=0} = 0 \\ SE\left.\frac{\partial\xi_{253}}{\partial x}\right|_{x=L} = -m\left.\frac{\partial^2\xi_{253}}{\partial t^2}\right|_{x=L} \end{cases} \tag{4.116}$$

Displacement Continuity Condition

$$\begin{cases} \xi_{251}|_{x=x_1} = \xi_{252}|_{x=x_1} \\ \xi_{252}|_{x=x_2} = \xi_{253}|_{x=x_2} \end{cases} \tag{4.117}$$

Strain Relation Condition

$$\begin{cases} \partial_x\xi_{252}|_{x=x_1} - \partial_x\xi_{251}|_{x=x_1} = D_{51}\sin 5\omega t \\ \partial_x\xi_{253}|_{x=x_2} - \partial_x\xi_{252}|_{x=x_2} = D_{52}\sin 5\omega t \end{cases} \tag{4.118}$$

where $D_{51} = -\frac{8}{210\pi}d_1k^3H^2\alpha sign(H\sin kx_1)\sin 2kx_1$, $D_{52} = -\frac{8}{210\pi}d_2k^3H^2\alpha sign(H\sin kx_2)\sin 2kx_2$, can be determined by the boundary conditions: $B_{51} = 0$, $B_{53} = \chi_5 A_{53}$, where $\chi_5 = \frac{8\beta(3-5\beta^2)}{5-45\beta^2+15\beta^4+\beta^6}$ Thus, the relationship between the crack position, size, and coefficients can be obtained.

$$\begin{cases} A_{51}\cos 5kx_1 = A_{52}\cos 5kx_1 + B_{52}\sin 5kx_1 \\ A_{52}\cos 5kx_2 + B_{52}\sin 5kx_2 = A_{53}\cos 5kx_2 + \chi_5 A_{53}\sin 5kx_2 \\ B_{52}\cos 5kx_1 + (A_{51} - A_{52})\sin 5kx_1 = \frac{D_{51}}{5k} \\ (\chi_5 A_{53} - B_{52})\cos 5kx_2 + (A_{52} - A_{53})\sin 5kx_2 = \frac{D_{52}}{5k} \end{cases} \tag{4.119}$$

Easily, we can obtain the solution of the above equation, that is, the expression of the coefficient about the crack position and size.

$$\begin{cases} A_{51} = \frac{D_{51}\cos 5kx_1 + D_{52}\cos 5kx_2 + D_{51}\chi_5\sin 5kx_1 + D_{52}\chi_5\sin 5kx_2}{3k\chi_3} \\ A_{52} = \frac{D_{51}\cos 5kx_1 + D_{52}(\cos 5kx_2 + \chi_5\sin 5kx_2)}{5k\chi_5} \\ B_{52} = \frac{D_{51}\cos 5kx_1}{5k} \\ A_{53} = \frac{D_{51}\cos 5kx_1 + D_{52}\cos 5kx_2}{5kx_5} \\ B_{53} = \frac{D_{51}\cos 5kx_1 + D_{52}\cos 5kx_2}{5k} \end{cases} \tag{4.120}$$

4.6.4 Inversion and Location of Multiple Cracks

4.6.4.1 Crack Inversion Problem

Usually during detection, we can use an accelerometer to scan along the direction of the rod to determine the relationship between the vibration amplitude and phase in the rod with the position x, from which we can obtain the fluctuation coefficients A_{ij}, B_{ij}.

When using the 3rd harmonic for detection, from Eq. (4.112) we can obtain the relationship between the crack position and the 3rd harmonic fluctuation coefficient.

$$\begin{cases} x_1 = \frac{1}{3k}\arctan\frac{A_{31}-A_{32}}{B_{32}} \\ x_2 = \frac{1}{3k}\arctan\frac{A_{32}-A_{33}}{\chi_3 A_{33}-B_{32}} \end{cases} \tag{4.121}$$

After obtaining the crack position, using the last two equations of (4.112), we can obtain the size of the crack.

$$\begin{cases} d_1 = \left| \frac{90\pi [B_{32} \cos 3kx_1 + (A_{31} - A_{32}) \sin 3kx_1]}{8k^2 H^2 \alpha \sin 2kx_1} \right| \\ d_2 = \left| \frac{90\pi [(\chi_3 A_{33} - B_{32}) \cos 3kx_2 + (A_{32} - A_{33}) \sin 3kx_2]}{8k^2 H^2 \alpha \sin 2kx_2} \right| \end{cases} \tag{4.122}$$

When using the 5th harmonic for detection, from Eq. (4.119) we can obtain the relationship between the crack position and the 5th harmonic fluctuation coefficient:

$$\begin{cases} x_1 = \frac{1}{5k} \arctan \frac{A_{51} - A_{52}}{B_{52}} \\ x_2 = \frac{1}{5k} \arctan \frac{A_{52} - A_{53}}{\chi_5 A_{53} - B_{52}} \end{cases} \tag{4.123}$$

And the size of the crack:

$$\begin{cases} d_1 = \left| \frac{1050\pi [B_{52} \cos 5kx_1 + (A_{51} - A_{52}) \sin 5kx_1]}{8k^2 H^2 \alpha \sin 2kx_1} \right| \\ d_2 = \left| \frac{1050\pi [(\chi_5 A_{53} - B_{52}) \cos 5kx_2 + (A_{52} - A_{53}) \sin 5kx_2]}{8k^2 H^2 \alpha \sin 2kx_2} \right| \end{cases} \tag{4.124}$$

4.6.4.2 Numerical Calculation of Crack Inversion

In the numerical simulation of the next section, two main problems are solved and verified: (a) whether the position and size of multiple cracks can be uniquely determined by the method discussed above. (b) In the case of only one crack, whether the position and size of this crack can be uniquely determined by the method of detecting multiple cracks.

Numerical Simulation of Multiple Crack Location

We use the first eigenfrequency excitation, $k_1 = 11.44\,\text{m}^{-1}$, $\omega_1 = 6.867 \times 10^4$ rad/s, $f_1 = 10.93$ kHz, and assume that two cracks are at $x_1 = 6$ cm, $x_2 = 11$ cm, the sizes of the cracks are $d_1 = 0.5$ mm, $d_2 = 1.8$ mm, we substitute the above parameters into Eqs. (4.117), (4.124) to calculate the coefficients in the fluctuation solution: the fundamental amplitude $H = 8.75 \times 10^{-3}$ m, the coefficients on both sides of each crack of the 3rd harmonic $A_{31} = 8.07 \times 10^{-5}$ m, $A_{32} = -1.65 \times 10^{-4}$ m, $B_{32} = -1.31 \times 10^{-4}$ m, $A_{33} = 1.89 \times 10^{-4}$ m, $B_{33} = -6.10 \times 10^{-4}$ m, the coefficients on both sides of the 5th harmonic crack $A_{51} = -9.35 \times 10^{-5}$ m, $A_{52} = -1.00 \times 10^{-4}$ m, $B_{52} = 2.28 \times 10^{-5}$ m, $A_{53} = -9.98 \times 10^{-5}$ m, $B_{53} = -2.83 \times 10^{-5}$ m. The fundamental amplitude can be seen. H is almost two orders of magnitude larger than A_{ij} and B_{ij}, so the perturbation method can meet the accuracy requirements. The positive and negative signs of the coefficients in the above numerical simulation can be determined by the phase during the scanning measurement.

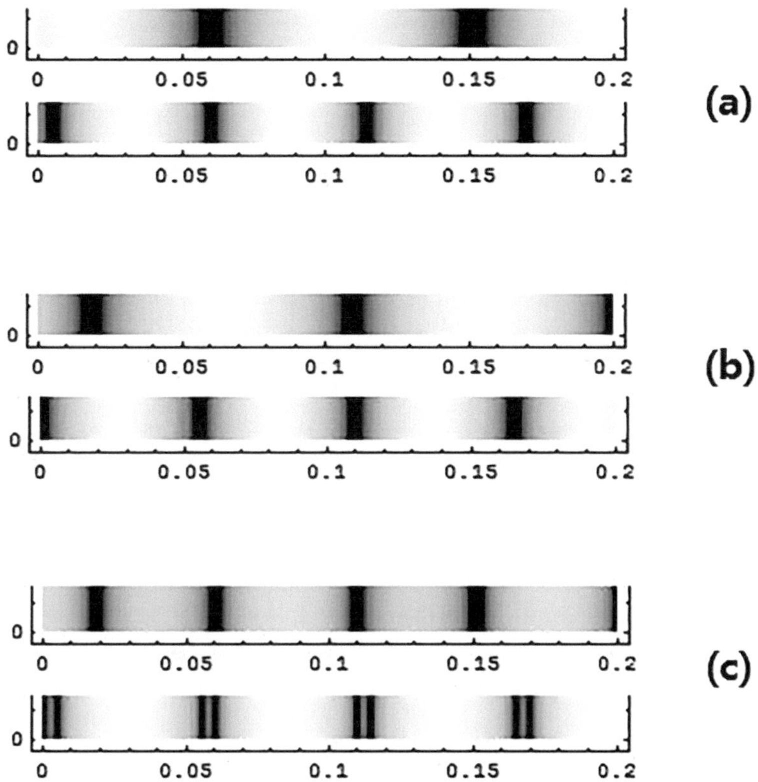

Fig. 4.19 Comparison diagram of the possible location distribution of the first crack (**a**) and the second crack (**b**), as well as a composite diagram of crack comparison (**c**), through the third harmonic (as shown in the figure above) and the fifth harmonic (as shown in the figure below)

Next, we verify whether we can infer the position and size of the crack by solving the coefficients through fluctuations. By substituting the above coefficients into Eqs. (4.121) and (4.123), we can solve for the 3rd harmonic location of the first crack: $x_{131} = 6.00$ cm, $x_{132} = 15.15$ cm, and the 5th harmonic location: $x_{151} = 0.51$ cm, $x_{152} = 6.00$ cm, $x_{153} = 11.49$ cm, $x_{154} = 16.98$ cm. Here, x_{ijk} represents the i possible solution calculated by the j harmonic of the k crack. Since the position of the first crack in the actual situation is unique, it is obvious from the above data that the actual position of the first microcrack is $x_1 = 6.00$ cm. Figure 4.19a is a possible position distribution diagram of the first crack. The upper figure is obtained through the 3rd harmonic, and the lower figure is obtained through the 5th harmonic. It is easy to find that only at $x_1 = 6.00$ cm the possible positions of the cracks are completely coincident. Similarly, we can calculate the 3rd harmonic location of the second crack: $x_{231} = 1.85$ cm, $x_{232} = 11.00$ cm, and the 5th harmonic location: $x_{251} = 0.02$ cm, $x_{252} = 5.51$ cm, $x_{253} = 11.00$ cm, $x_{254} = 16.49$ cm. Since the actual position of the second crack is also unique and should be after the first crack, after comparing the

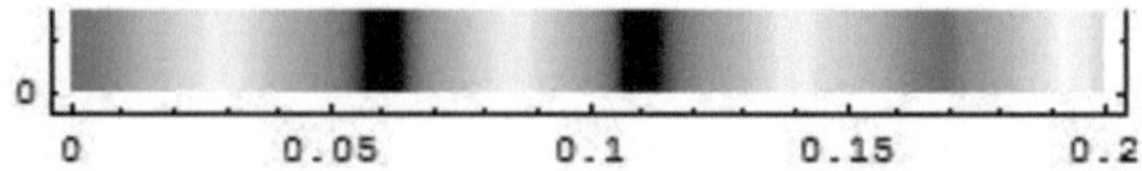

Fig. 4.20 The actual crack distribution diagram after comparison and processing

above data. And after the first crack, it is easy to find $x_2 = 11.00$ cm by comparing the above data, as shown in Fig. 4.19(b). Overlapping Fig. 4.19(a) with Fig. 4.19(b) yields Fig. 4.19(c). It can be seen from this that the positions of the two cracks only overlap at $x_1 = 6.00$ cm and $x_2 = 11.00$ cm. By comparing and processing, we can finally obtain the distribution map of actual cracks, as shown in Fig. 4.20.

After determining the position of the crack, we can calculate the size of the crack by substituting it into Eqs. (4.113) and (4.115): $d_1 = 0.5\,\text{mm}$, $d_2 = 1.8\,\text{mm}$. The calculated value is consistent with the expected value, indicating that this method can effectively determine the position and size of the crack. Figure 4.21 is a distribution diagram of the displacement amplitude and stress amplitude of the particles in the metal rod. It can be seen from Fig. 4.21a, c that the displacement at $x_1 = 6.00\,\text{cm}$ and $x_2 = 11.00\,\text{cm}$ is still continuous, and it can be seen from Figs. 4.21b, d that the stress at $x_1 = 6.00\,\text{cm}$ and $x_2 = 11.00\,\text{cm}$ is discontinuous. This is because a nonlinear source appears in the fluctuation equation. It is obvious that the jump at $x_2 = 11.00\,\text{cm}$ is significantly larger than that at $x_1 = 6.00\,\text{cm}$, which is because the crack length at $x_2 = 11.00\,\text{cm}$ is larger than that at $x_1 = 6.00\,\text{cm}$.

Numerical Simulation of Single Crack Location

The effectiveness of the detection method for multiple cracks has been discussed above. Next, we will discuss the effectiveness of using the multi-crack detection method to reverse engineer and apply it to single crack detection. We also use the first eigenfrequency excitation, but assume that there is only one crack in the metal rod, the crack is located at $x_1 = 6.00\,\text{cm}$, the effective length of the horizontal coordinate is $d_1 = 0.5$mm. Since there is no second crack, it is equivalent to $d_2 = 0$, that is, $D_{32} = 0$, $D_{52} = 0$. Substituting it into Eqs. (4.103) and (4.110) can yield the coefficients of the 3rd harmonic: $A_{31} = 2.86 \times 10^{-4}$ m, $A_{32} = A_{33} = 4.04 \times 10^{-5}$ m, $B_{32} = B_{33} = -1.31 \times 10^{-4}$ m, and the coefficients of the 5th harmonic: $A_{51} = 8.74 \times 10^{-5}$ m, $A_{52} = A_{53} = 8.05 \times 10^{-5}$ m, $B_{52} = B_{53} = 2.28 \times 10^{-5}$ m. It can be seen that the absence of the second crack makes $\xi_2 = \xi_3$.

Conversely, we verify whether the above values measured in the experiment can be used to determine the location and size of the crack. Substituting the above coefficients into Eqs. (4.121) and (4.123) can solve the 3rd harmonic location of the first crack: $x_{131} = 6.00\,\text{cm}$, $x_{132} = 15.15\,\text{cm}$, 5th harmonic location: $x_{151} = 0.51\,\text{cm}$, $x_{152} = 6.00\,\text{cm}$, $x_{153} = 11.49\,\text{cm}$, $x_{154} = 16.98\,\text{cm}$. By comparison, it can be concluded that $x_1 = 6.00\,\text{cm}$, the comparison chart is shown in Fig. 4.22. However, when we substitute the values into Eqs. (4.125) and (4.127) to calculate the 3rd harmonic location of the second crack, we find that $x_2 = \frac{1}{3k} \arctan \frac{A_{32}-A_{33}}{\chi_3 A_{33} - B_{32}}$ in $\frac{A_{32}-A_{33}}{\chi_3 A_{33} - B_{32}} = \frac{0}{0}$, is indeterminate, that is, we cannot get the solution of x_2, the same

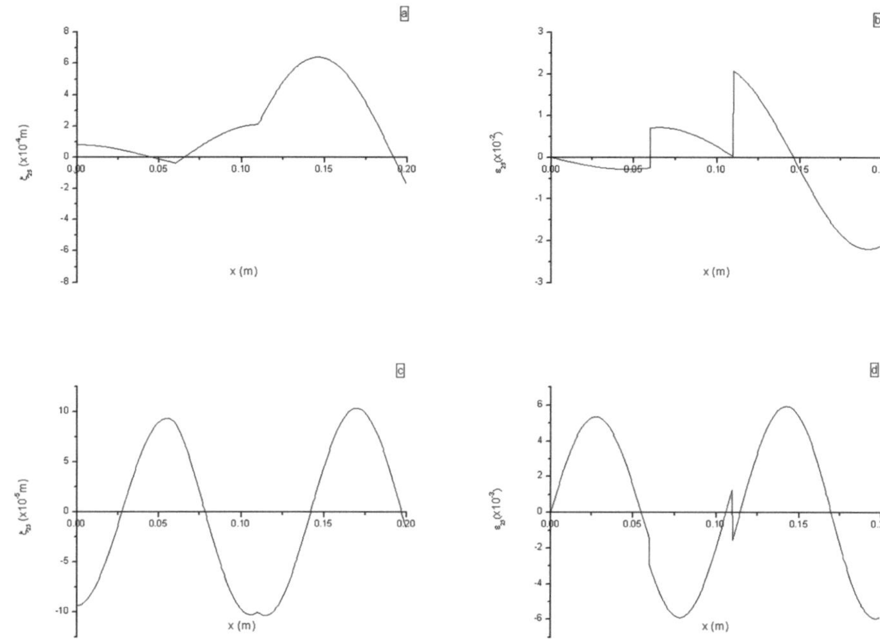

Fig. 4.21 **a** The distribution of the 3rd harmonic particle vibration displacement in the metal rod. **b** shows the distribution of the 3rd harmonic strain in the metal rod. **c** The distribution of the 5th harmonic particle vibration displacement in the metal rod. **d** The distribution of the 5th harmonic strain in the metal rod

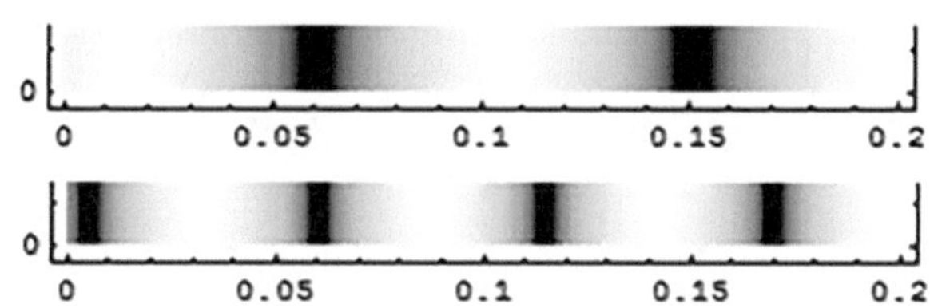

Fig. 4.22 Comparison diagram of the possible location distribution of the first crack obtained through the third harmonic (as shown in the figure above) and the fifth harmonic (as shown in the figure below)

problem also occurs when locating the 5th harmonic. Therefore, we can conclude that x_2 does not exist, that is, there is no second crack. However, in actual situations, the ratio obtained by measurement is a definite number, but there is no overlapping solution obtained by comparing the locations obtained by the 3rd and 5th harmonics, which can also explain the non-existence of the crack. After comparison and processing, we get the actual crack distribution as shown in Fig. 4.23, it is easy to see that there is only one crack in the rod.

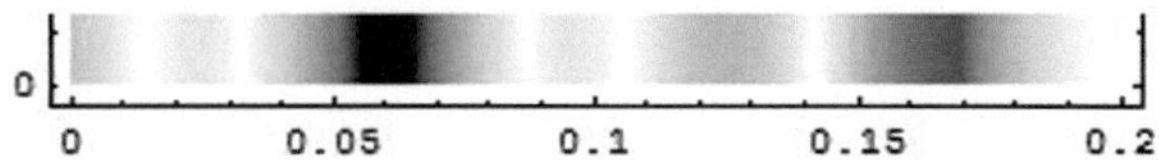

Fig. 4.23 The actual crack distribution after comparison and processing

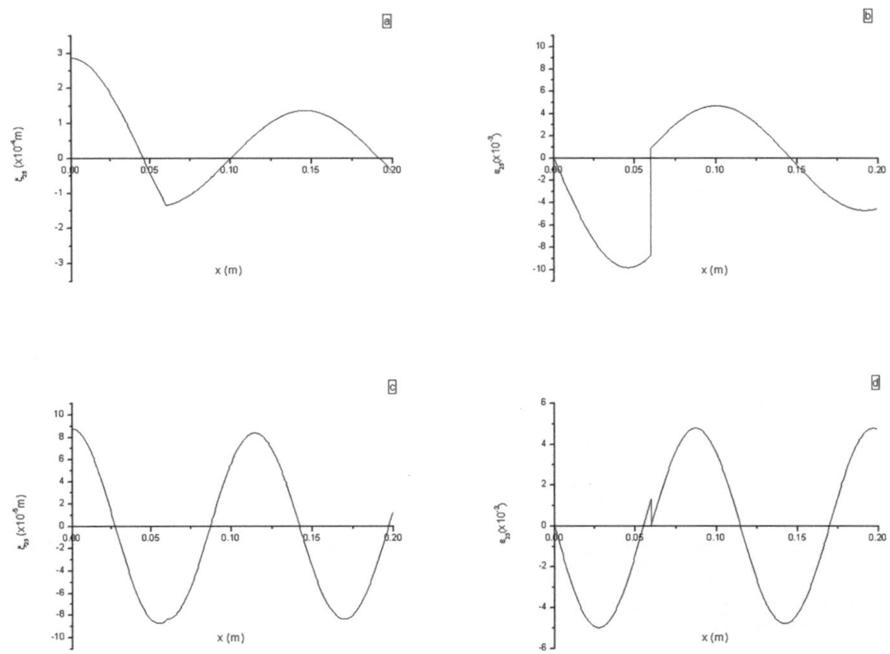

Fig. 4.24 **a** The distribution of the 3rd harmonic particle displacement vibration in the metal rod. **b** The distribution of the 3rd harmonic strain in the metal rod. **c** shows the distribution of the 5th harmonic particle displacement vibration in the metal rod. **d** The distribution of the 5th harmonic strain in the metal rod

After determining the location of the crack, we can calculate the size of the crack by substituting it into Eqs. (4.122) and (4.124): $d_1 = 0.5$ mm, $d_2 = 0.00$ mm. The non-existence of the second crack can also be explained from $d_2 = 0.00$ mm. The amplitude distribution of each harmonic and stress amplitude in the rod is shown in Fig. 4.24. In the figure, only at $x_1 = 6.00$ mm is there a discontinuity in stress, while it is continuous at other locations.

4.6.5 Discussion on the Validity of Crack Location

4.6.5.1 Discussion on the Accuracy and Method of Crack Location

In grayscale images, the darker the color, the greater the likelihood that a crack is located at that position. The wider the dark area, the greater the uncertainty of the crack location, and thus the lower the accuracy. To improve the accuracy of crack location, after determining the crack location, one can excite the material with a higher intrinsic frequency and use the resulting higher harmonics to improve accuracy. This is significant for practical applications because a higher intrinsic frequency causes the amplitude of the sound wave in the rod to change faster with the position of the rod. Although increasing the excitation intrinsic frequency improves the lateral uncertainty, it increases the possible locations of the crack. Figures 4.25 and 4.26 are comparison diagrams of crack inference using different intrinsic frequencies, using the simulation data used in the last two sections. It can be clearly seen that the higher the intrinsic excitation frequency, the more possible crack locations are produced. However, this problem can be solved as follows: first, excite the material with the first intrinsic frequency, determine the approximate location of the crack, and then use a higher intrinsic frequency to excite and improve lateral accuracy. For the "false" positions caused by the higher intrinsic frequency, compare with the first intrinsic frequency to filter out "false" crack information. For example, in Fig. 4.25a, it can be inferred that the crack is located at $x_1 = 6.00\,\text{cm}$, $x_2 = 11.00\,\text{cm}$ nearby. For the crack location inference produced in Fig. 4.25b–e, only consider the value closest to $x_1 = 6.00\,\text{cm}$, $x_2 = 11.00\,\text{cm}$ to improve accuracy without introducing new "false" cracks.

4.6.5.2 Discussion on the Quality Factor and Its Accuracy in the Analytical Solution

From Eqs. (4.81) to (4.82), the approximation of large quality factor Q is used here, and the imaginary part of the complex wave number is considered to be a first-order small quantity relative to the real part. Here, we will discuss the effectiveness of the large Q approximation method and the applicability of the crack location method proposed in this paper. The exact solution of Eq. (4.82) is

$$\begin{cases} a = \dfrac{k^2}{\sqrt{2}Q\sqrt{-k^2+\frac{\sqrt{k^4\left(Q^2+Q^4\right)}}{Q^2}}} \\ b = -\dfrac{\sqrt{-k^2+\frac{\sqrt{k^4\left(Q^2+Q^4\right)}}{Q^2}}}{\sqrt{2}} \end{cases} \tag{4.125}$$

Figure 4.27a is a comparison of the exact value (4.125) and the approximate value (4.82) of the real part of the complex wave number for different Q values when k

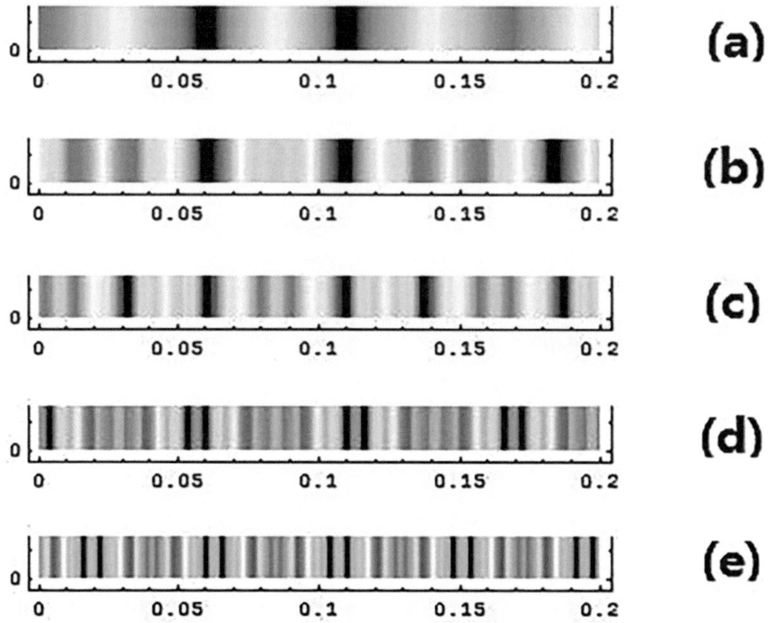

Fig. 4.25 Comparison of crack location and excitation frequency for two cracks. **a** First eigenfrequency. **b** Second eigenfrequency. **c** Third eigenfrequency. **d** Fourth eigenfrequency. **e** Fifth eigenfrequency

$= 11.44\ \mathrm{m}^{-1}$ and $k = 25.43\ \mathrm{m}^{-1}$. Figure 4.27b is a comparison of the exact value (4.125) and the approximate value (4.82) of the imaginary part of the complex wave number for different Q values. From the above two figures, we can see that as the quality factor gradually increases, the approximate value gets closer and closer to the exact value. It is calculated that when the quality factor $Q > 5$, the relative errors of the real and imaginary parts of the complex wave number do not exceed 0.5%, and when $Q > 80$ (the value taken in the numerical calculation in this paper), the relative errors of the real and imaginary parts do not exceed 0.002%. Therefore, the large Q approximation used in this section is effective, and this crack location method can be applied to all low-loss materials.

4.7 Two-Dimensional Non-classical Nonlinear Signal Simulation

For two-dimensional acoustic wave propagation, considering the attenuation of the solid and the propagation of the acoustic wave under sinusoidal wave excitation, the acoustic wave propagation equation is [18–20]:

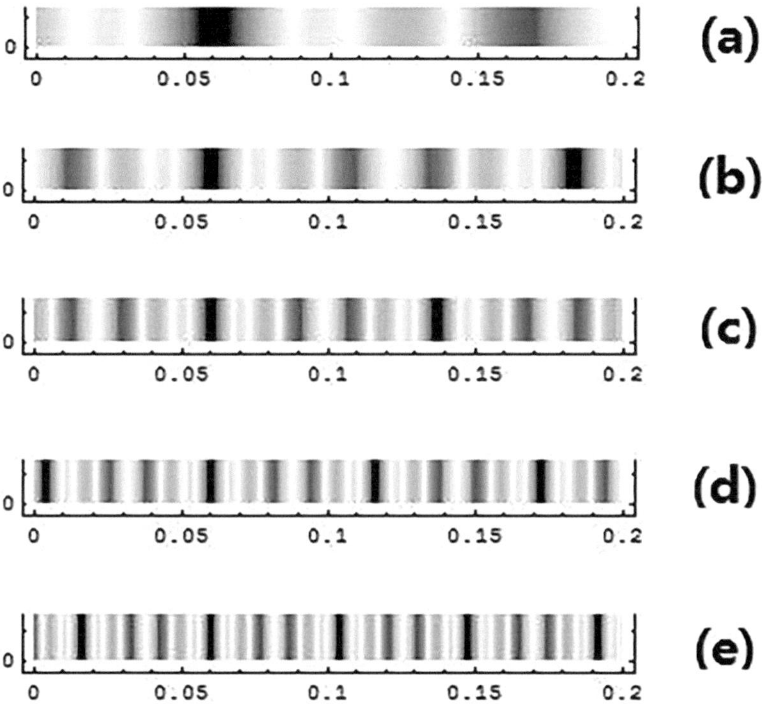

Fig. 4.26 Comparison of crack location and excitation frequency for a single crack. **a** First eigenfrequency. **b** Second eigenfrequency. **c** Third eigenfrequency. **d** Fourth eigenfrequency. **e** Fifth eigenfrequency

$$\dot{v}_x = \frac{1}{\rho}\left(\frac{\partial \sigma_{xx}}{\partial x} + \frac{\partial \sigma_{xy}}{\partial y}\right) + F_x \tag{4.126a}$$

$$\dot{v}_y = \frac{1}{\rho}\left(\frac{\partial \sigma_{xy}}{\partial x} + \frac{\partial \sigma_{yy}}{\partial y}\right) + F_y \tag{4.126b}$$

$$\dot{\sigma}_{xx} = K_1\dot{\varepsilon}_{xx} + K_2\dot{\varepsilon}_{yy} = K_1\frac{\partial v_x}{\partial x} + K_2\frac{\partial v_y}{\partial y} \tag{4.126c}$$

$$\dot{\sigma}_{yy} = K_1\dot{\varepsilon}_{yy} + K_2\dot{\varepsilon}_{xx} = K_1\frac{\partial v_y}{\partial y} + K_2\frac{\partial v_x}{\partial x} \tag{4.126d}$$

$$\dot{\sigma}_{xy} = \mu\left(\dot{\varepsilon}_{xy} + \dot{\varepsilon}_{yx}\right) = \mu\left(\frac{\partial v_y}{\partial x} + \frac{\partial v_x}{\partial y}\right) \tag{4.126e}$$

In the above formula, σ_{xx}, σ_{yy}, σ_{zz} is the stress component received by the particle, ε_{xx}, ε_{yy}, ε_{xy} is the corresponding strain component. v_x, v_y is the velocity component of the particle, F_x, F_y are the stresses received in the x and y directions respectively. K_1, K_2 is the linear elastic modulus of the solid, and their relationship with the Lame

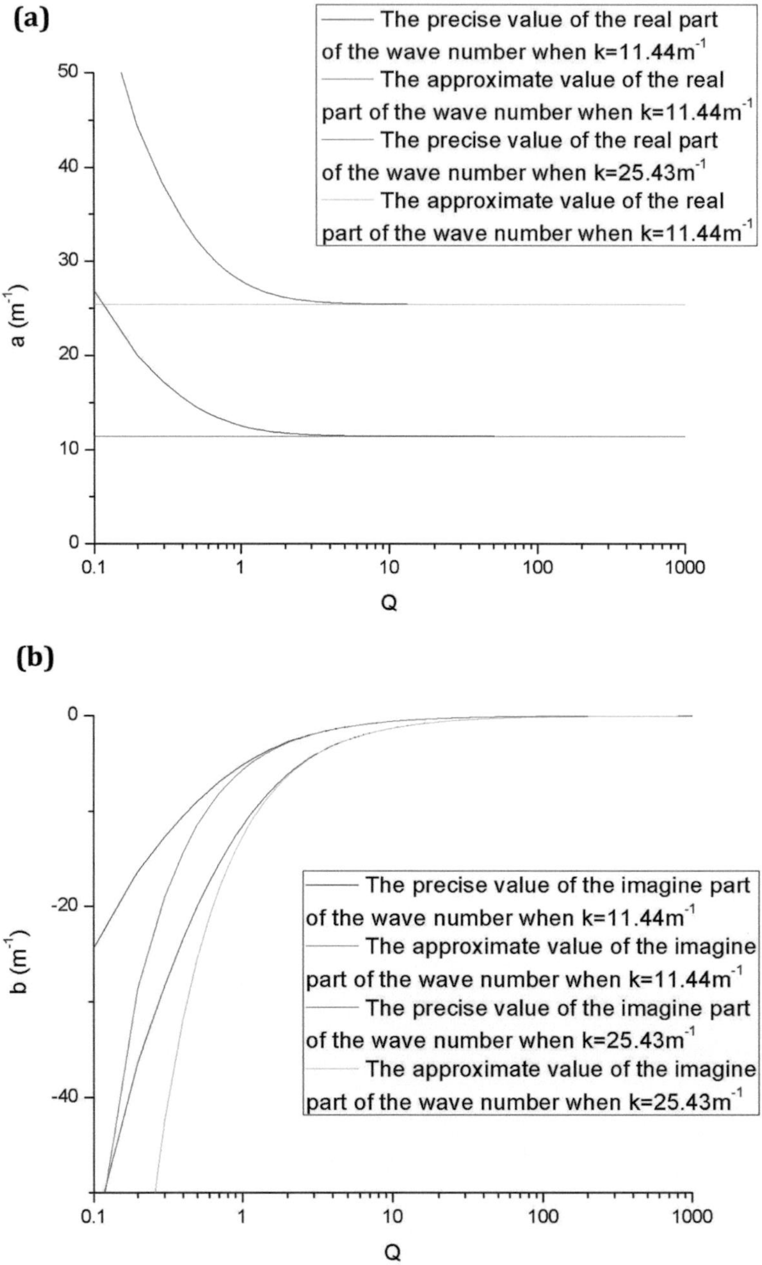

Fig. 4.27 **a** Comparison of the exact value (4.125) and the approximate value (4.82) of the real part of the complex wave number for different Q values when $k = 11.44\ m^{-1}$ and $k = 25.43\ m^{-1}$. **b** Comparison of the exact value (4.125) and the approximate value (4.82) of the imaginary part of the complex wave number for different Q values

constants (λ, μ) is as follows:

$$K_1 = \lambda + 2\mu, K_2 = \lambda \tag{4.127}$$

For the impact of attenuation, a simplified model [21] is used. After each calculation and update of stress and strain, it is multiplied by an attenuation factor α:

$$\alpha = \exp(-\pi f \Delta t/(2Q)) \tag{4.128}$$

In the formula, f is the frequency of the sound source, Δt is the discrete time interval.

Using the discrete formula (4.128) of dynamic elastic finite integral technology, the solid medium is divided into mesoscopic units as suggested by Fellinger [22], as shown in Fig. 4.28. The stress components σ_{xx}, σ_{yy} are calculated at the center of each unit, the velocity components v_x, v_y are sought on the four sides, and σ_{xy} are calculated at the four corners. Ignoring the external source, the recursive equations of stress and velocity of the discrete unit in Fig. 4.28 are as follows:

$$\begin{aligned}
\dot{v}_x^{(n,m)}(t) &= \frac{1}{\Delta}\frac{2}{\left(\rho^{(n,m)} + \rho^{(n+1,m)}\right)}\left(\sigma_{xx}^{(n+1,m)}(t) - \sigma_{xx}^{(n,m)}(t) + \sigma_{xy}^{(n,m)}(t) - \sigma_{xy}^{(n,m-1)}(t)\right)\\
\dot{v}_y^{(n,m)}(t) &= \frac{1}{\Delta}\frac{2}{\left(\rho^{(n,m)} + \rho^{(n,m+1)}\right)}\left(\sigma_{yy}^{(n,m+1)}(t) - \sigma_{yy}^{(n,m)}(t) + \sigma_{xy}^{(n,m)}(t) - \sigma_{xy}^{(n-1,m)}(t)\right)\\
\dot{\sigma}_{xx}^{(n,m)}(t) &= \frac{1}{\Delta}K_1^{(n,m)}\left[v_x^{(n,m)}(t) - v_x^{(n-1,m)}(t)\right] + \frac{1}{\Delta}K_2^{(n,m)}\left[v_y^{(n,m)}(t) - v_y^{(n,m-1)}(t)\right]\\
\dot{\sigma}_{yy}^{(n,m)}(t) &= \frac{1}{\Delta}K_1^{(n,m)}\left[v_y^{(n,m)}(t) - v_y^{(n,m-1)}(t)\right] + \frac{1}{\Delta}K_2^{(n,m)}\left[v_x^{(n,m)}(t) - v_x^{(n-1,m)}(t)\right]\\
\dot{\sigma}_{xy}^{(n,m)}(t) &= \frac{1}{\Delta}\frac{4}{\left(\frac{1}{\mu^{(n,m)}} + \frac{1}{\mu^{(n+1,m)}}\frac{1}{\mu^{(n,m+1)}}\frac{1}{\mu^{(n+1,m+1)}}\right)}\\
&\quad\left[v_x^{(n,m+1)}(t) - v_x^{(n,m)}(t) + v_y^{(n+1,m)}(t) - v_y^{(n,m)}(t)\right]
\end{aligned} \tag{4.129}$$

According to the PM model theory [19], non-classical nonlinearity is manifested at the microscopic level as the step response of particle stress and strain, and at the mesoscopic level as the hysteresis curve of solid materials after the collection of microscopic units. Mesoscopic units constitute macroscopic non-classical nonlinear solid materials, that is, damaged materials in the usual sense. Through statistical analysis of the microscopic units of the material, the impact of non-classical nonlinearity can be attributed to the effect of stress on the modulus at the mesoscopic unit scale. Formula (4.130) gives the specific modulus-stress relationship of the mesoscopic unit [19]:

$$K^{-1} = \lim_{\Delta\sigma\to 0}\left(\frac{\Delta\varepsilon}{\Delta\sigma}\right) = \frac{1}{K_c(\sigma)} + \hat{r}\frac{df_c}{d\sigma}(-\sigma) \tag{4.130}$$

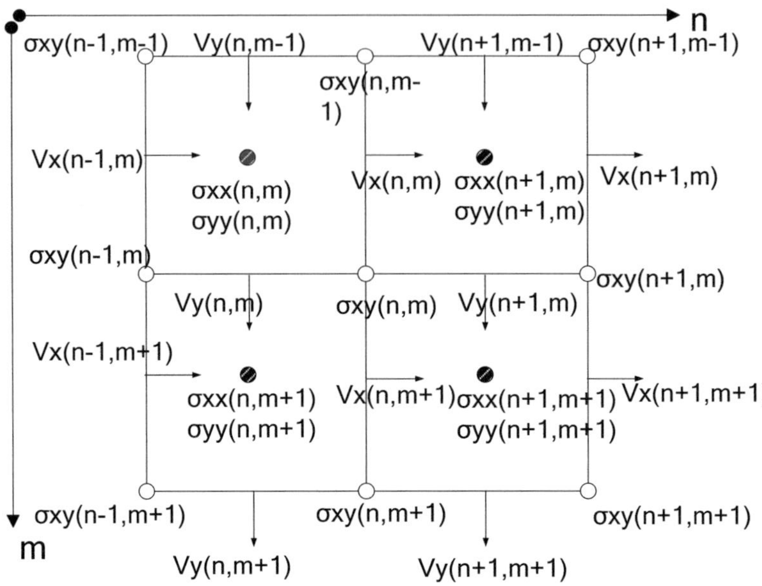

Fig. 4.28 Discrete mode of solid medium

Here K_c represents the modulus of the material under classical conditions, $\hat{r}$ characterizes the size of the material's non-classical nonlinearity, σ is the stress received by the mesoscopic unit, f_c refers to the proportion of closed units in the microscopic units of the PM model under the action of stress σ.

In the two-dimensional case, the force situation of the particle is more complicated, and the characteristic system of the elastic tensor [23] is used to describe the impact of non-classical nonlinearity. The stress is projected onto the three eigenvector spaces of the elastic tensor. Since the eigenvector spaces are orthogonal to each other, the non-classical nonlinearity caused by the pressure added along these three pressure directions can be described by the corresponding three PM spaces.

For simplicity in numerical simulation, the sample is taken as a rectangle, as shown in Fig. 4.29, with geometric dimensions of 250 mm × 150 mm. The center coordinates of the damaged area are (125 mm, 75 mm), and the size is 50 mm × 130 mm. The material parameters are taken as those of iron, $K_1 = 1.412 \times 10^{11}$ Pa, $K_2 = 5.2202 \times 10^{10}$ Pa, $\mu = 4.45 \times 10^{10}$ Pa [7], the Q value is set to 2 [21], and only one of the three PM spaces is selected, with a density of 5×10^{-13} Pa^{-2}. The discrete grid size is 2 mm × 2 mm, and the discrete time interval is 2×10^{-7} s. The sound source is a sine wave, acting on one side of the sample width, with an action range of 50 mm, and the rest of the sides are taken as free boundary conditions [24]. Non-classical nonlinearity only considers one feature vector space, and the PM space density distribution is uniform.

When analyzing the output signal, the data after 4Q cycles of the output signal are taken as stable values. The input signal only has a component in the x direction,

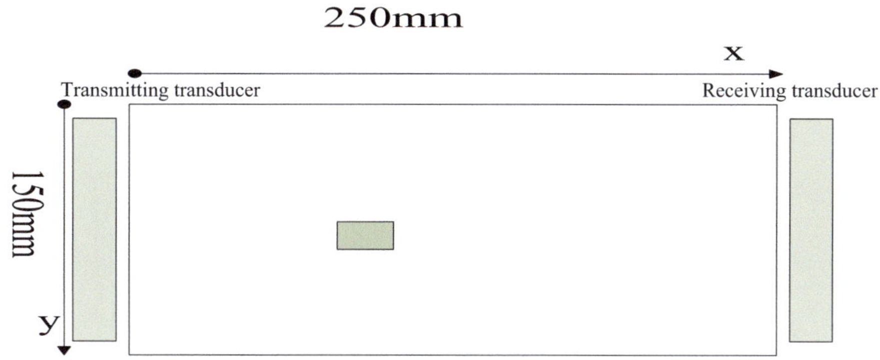

Fig. 4.29 Sample geometry, the gray area in the middle is the damaged area

and the other directions are 0. The stress is taken as $F_x = A\sin(2\pi ft)$, $F_y = 0$, $A = 1 \times 10^8$, $f = 8.5$ KHz, when the nonlinear parameter is $\hat{r} = 10^{-9}$. If there is no damaged area in the sample, that is, only classical nonlinearity exists, to simplify the situation, only consider the change of K_c under classical nonlinearity, and use a simplified formula to represent the relationship between K_c and stress

$$K_c = K_0(1 + \beta\sigma), \beta = 1 \times 10^{-10} \tag{4.131}$$

K_0 is the value of K_c under linear conditions, β is the classical nonlinear parameter. When the transmitted signal is a continuous sine wave, only classical nonlinearity is considered in the sample. Without attenuation, the received signal in the time domain is a modulated signal, and the received signal and its corresponding spectrum are shown in Fig. 4.30a; when there is classical nonlinearity and attenuation in the sample, the received signal in the time domain is a steady-state signal, and the received signal and spectrum are shown in Fig. 4.30b. Without attenuation, the increase in the harmonic of the natural frequency increases the complexity of frequency analysis and interferes with the extraction and study of characteristic signals; after adding attenuation, the harmonics of the natural frequency of the sample are basically attenuated, mainly leaving the harmonics caused by the frequency of the transmitted signal. Comparing the figures, Fig. 4.30b is clearly easier to identify harmonic features than Fig. 4.30a, which is also consistent with the actual situation.

Considering the attenuation of the sample, when the sample produces cracks, only non-classical nonlinearity is considered, and the power spectrum of the output signal is as follows (Fig. 4.31a).

At the same time, the power spectrum of the received signal when classical nonlinearity and non-classical nonlinearity coexist is given, see Fig. 4.31b

In numerical simulation, if classical nonlinearity is not considered, according to PM theory, non-classical nonlinearity only produces odd harmonics, so theoretically there will be no even harmonics. In Fig. 4.31a, the fundamental frequency is 8.5 kHz, and the harmonic components only appear at 25.5 and 42.5 kHz, that is, the third

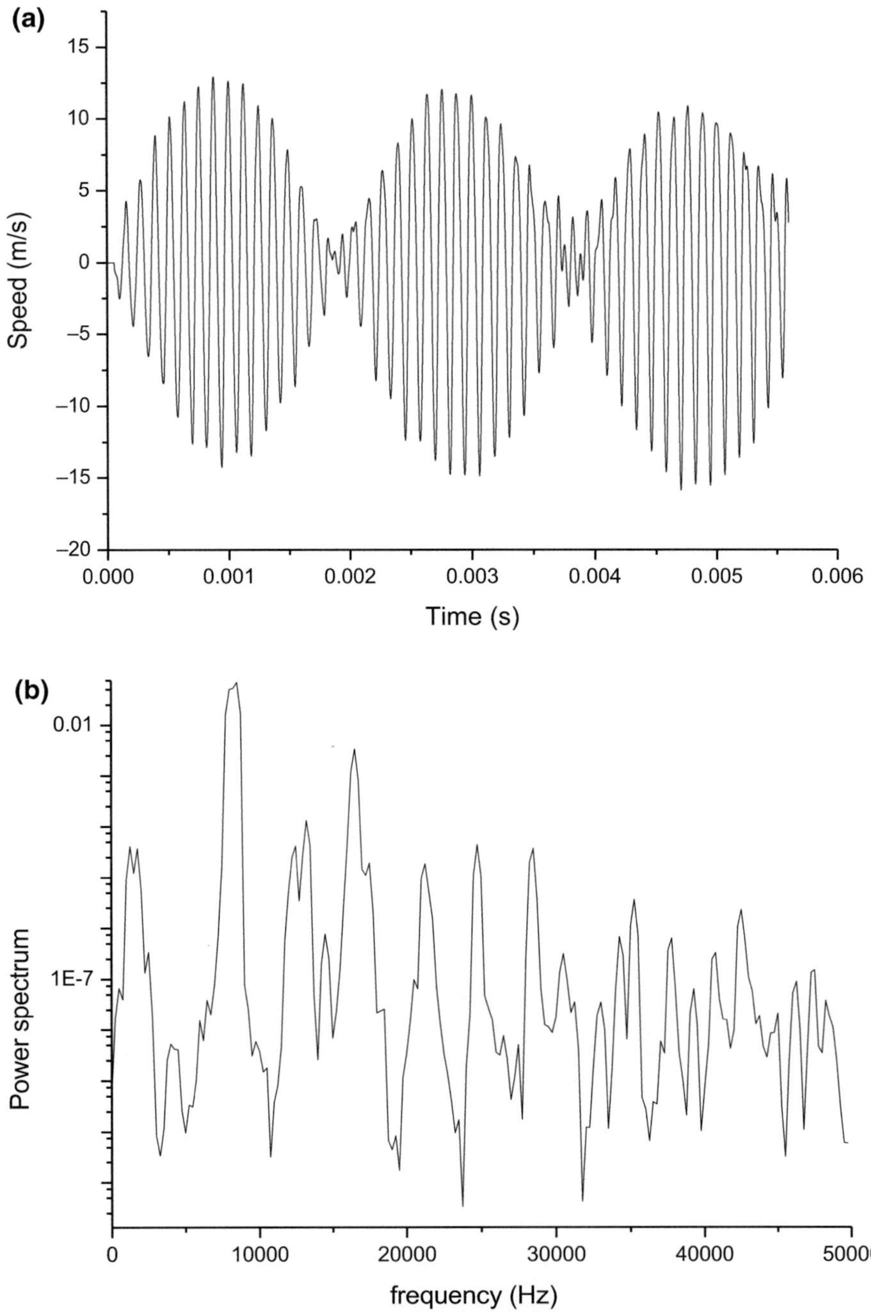

Fig. 4.30 **a**, **b** Received signal and its corresponding power spectrum when only considering classical nonlinearity and not considering attenuation. **c**, **d** Received signal and its corresponding power spectrum when considering classical nonlinearity and attenuation

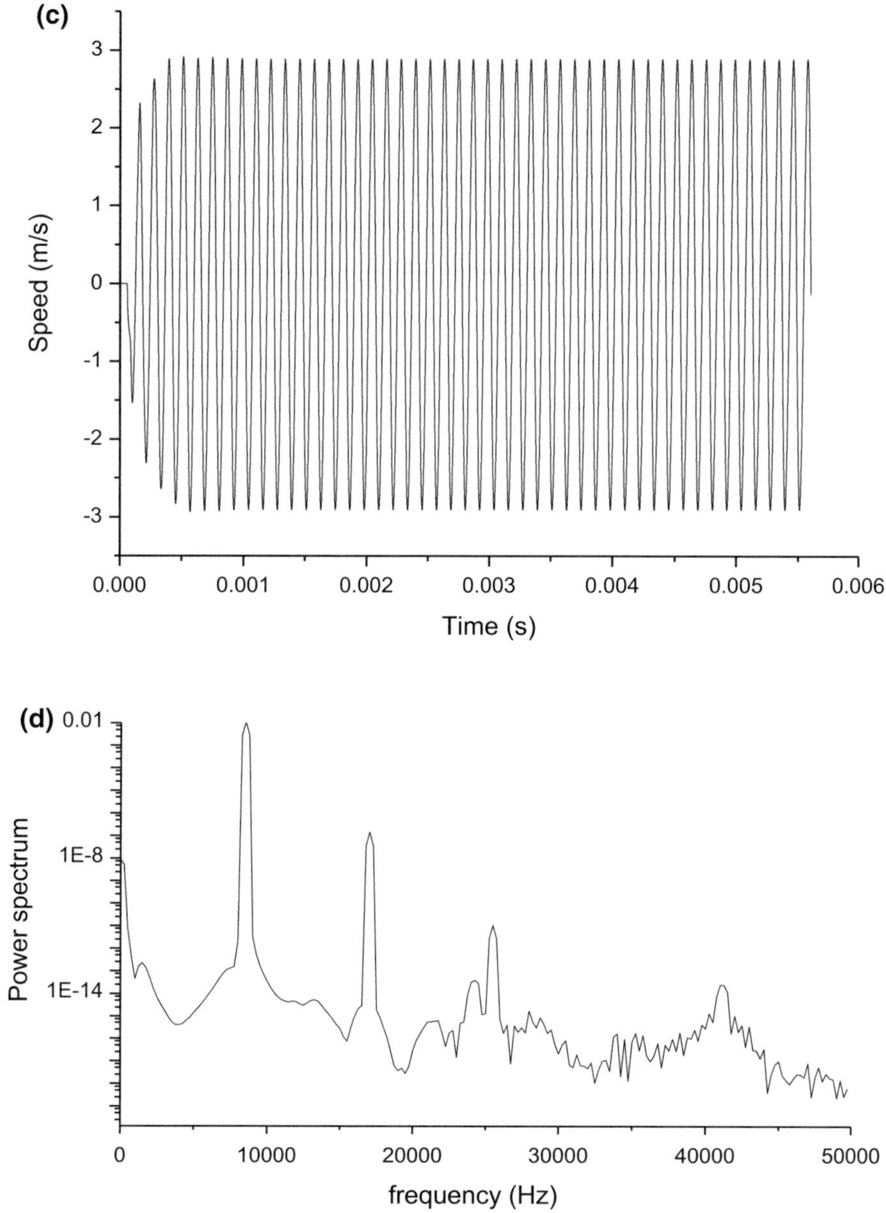

Fig. 4.30 (continued)

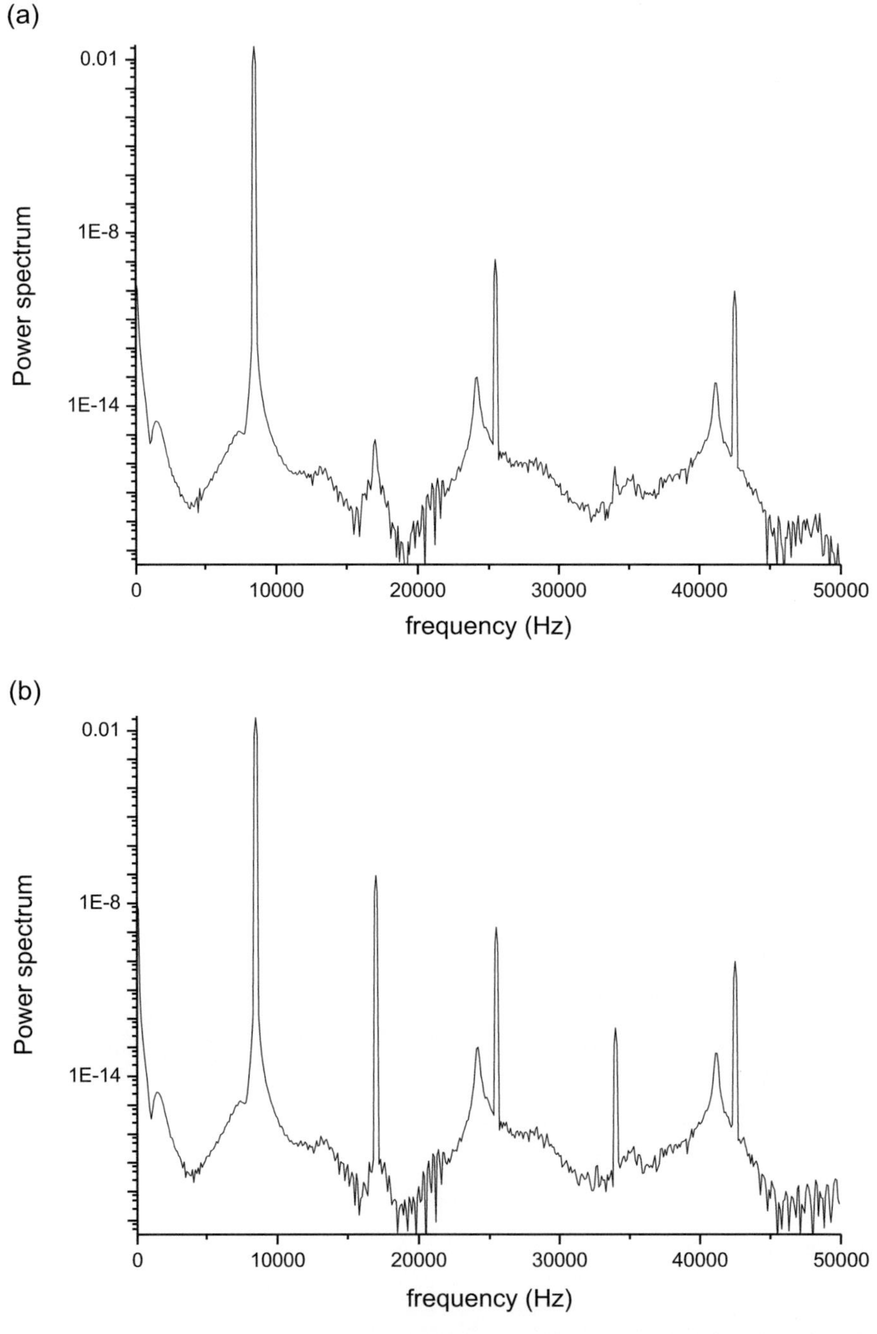

Fig. 4.31 **a** Power spectrum of the received signal when only considering non-classical nonlinearity. **b** Power spectrum of the signal when classical nonlinearity and non-classical nonlinearity coexist

and fifth harmonics, which also conforms to PM theory. In Fig. 4.30b, when only classical nonlinearity exists, both odd and even harmonics exist, but higher harmonics decay quite quickly. Figure 4.31b shows that when non-classical nonlinearity exists, odd harmonics are greatly enhanced. In actual non-destructive testing, it can be considered to use high-order odd harmonics to increase the accuracy of measurement. In Fig. 4.31b, not only the third harmonic and the fifth harmonic are generated, but also the fourth harmonic is generated. This is due to the interaction between the odd harmonics generated by non-classical nonlinearity and the harmonics generated by classical nonlinearity, thus generating other harmonics. The enhancement of the fourth harmonic is the result of this effect. Figure 4.32 is when only considering non-classical nonlinearity, the stress amplitude of the transmitted sound pressure gradually increases from 1×10^8 Pa, and the stress increase at each step is 1×10^6 Pa, the curve fitting the relationship between the amplitude of the odd harmonic of the received signal and the amplitude of the fundamental wave. In Fig. 4.32a, the slope of the fitted line for the third harmonic is 2.0015, and in Fig. 4.32b, the slope of the fitted line for the fifth harmonic is 2.0010. The amplitudes of the third and fifth harmonics are basically squared with the fundamental wave, which on the one hand shows the difference from the classical nonlinearity, and on the other hand is another manifestation consistent with PM theory.

Let the nonlinear acoustic parameter $\alpha_f = A_3/A_1^2, A_3$ be the amplitude of the third harmonic, A_1 be the amplitude of the fundamental wave. By changing the position and size of the damaged area in the sample, the relationship between α_f and the position and range can be calculated. Take the same input signal as Fig. 4.31, the damaged area is a rectangular area as shown in Fig. 4.29, with a length of 130 mm and a width of 2 mm, and the PM spatial density remains unchanged. By changing the position of the damaged area along the length direction of the sample, the relationship between α_f and the position is obtained in Fig. 4.33a. The function relationship of α_f in the figure is relatively complex, but it can still be seen that α_f is more sensitive to the damage of the sample in the two symmetrical areas near the center of the rod and the points closer to the sound source, while the sensitivity is lower in the middle of the rod and the farthest end from the sound source. Figure 4.33b is the relationship between α_f and the width of the damaged area. The center of the damage is in the center of the rod sample, the length is 130 mm, the initial width is 2 mm, and the increased width each time is 4 mm. As the width expands, α_f also begins to increase, but before the damaged area spreads to the edge of the rod, α_f has a downward trend. Figure 4.33a, b are both calculated results at a frequency of 8.5 kHz. To better study the relationship between α_f and position and width, the frequency can be further increased and the grid can be refined.

The two-dimensional numerical simulation results show the influence of classical nonlinearity and non-classical nonlinearity on harmonics, and obtain the rule of the nonlinear acoustic parameter α_f changing with the position and size of the damaged area under non-classical nonlinearity. The relationship between the size of the harmonic and the position and size of the damaged area can be further studied, and the position and size of the damage can be inverted based on this relationship.

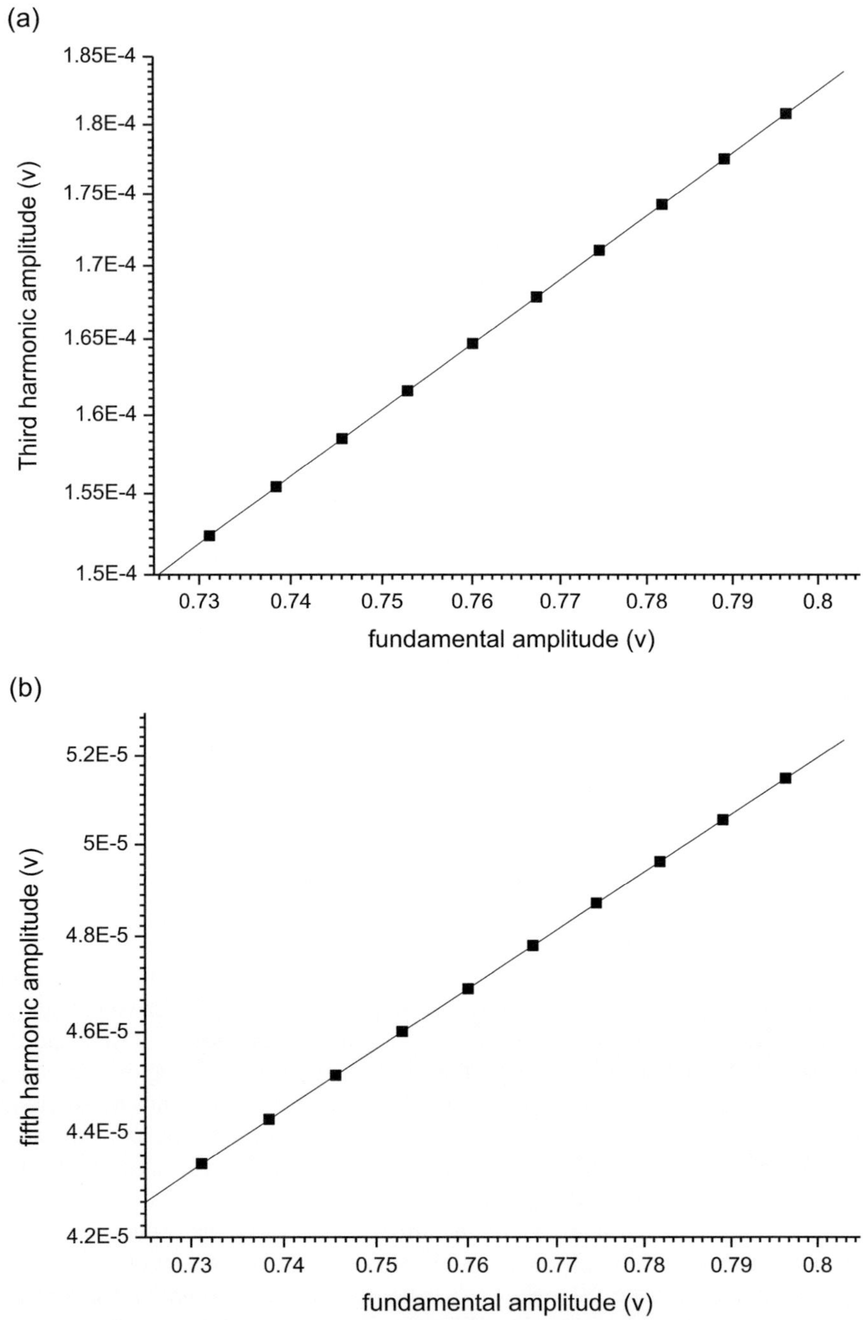

Fig. 4.32 **a** Relationship between the third harmonic and the fundamental wave. **b** Relationship between the fifth harmonic and the fundamental wave

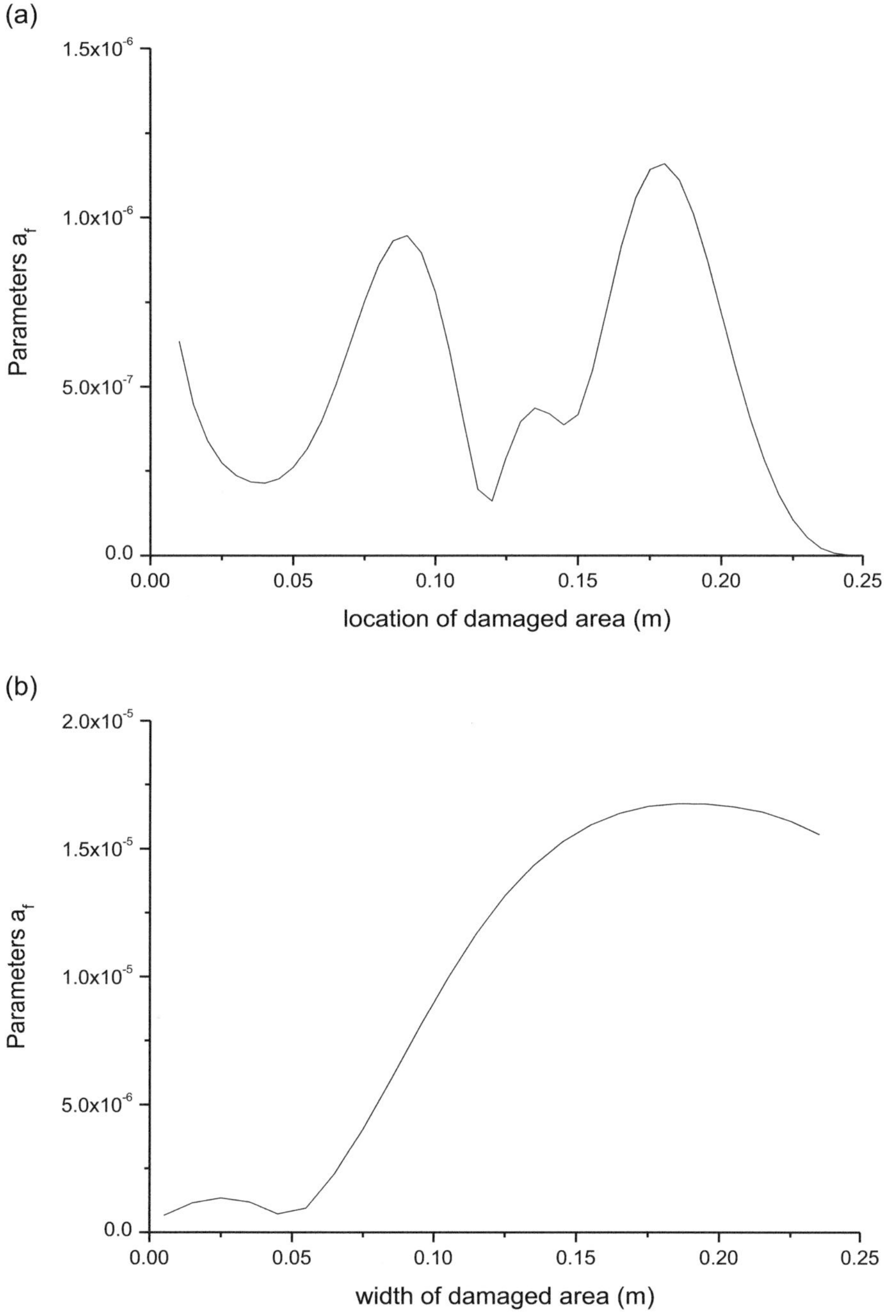

Fig. 4.33 **a** Relationship between parameter α_f and the position of the damaged area. **b** Relationship between parameter α_f and the width of the damaged area

4.8 Two-Dimensional Non-classical Nonlinear Signal Inversion Imaging

4.8.1 Time Reversal Theory

In industrial flaw detection, an effective way is needed to determine the position of defects in the metal in order to carry out effective treatment. Delay technology was initially used for positioning. Although it can display the characteristic information of the target in some cases, it implies an assumption: it requires that the cumulative result of the effect produced by the inhomogeneous target on the wavefront is just a delay in time. In fact, this situation only holds when the inhomogeneous target is very thin and very close to the receiving and transmitting array. In most cases, this assumption does not hold. Therefore, Mathias Fink [25] proposed the time reversal imaging theory, which can overcome the shortcomings of delay technology and achieve better imaging. Consider a lossless fluid medium, its compressibility κ(r) and density ρ(r) change with space, define the local sound speed $c(r) = (\kappa(r)\ \rho(r))^{-1/2}$, the transient sound pressure propagation equation can be obtained:

$$\vec{\nabla} \cdot \left(\frac{\vec{\nabla} p}{\rho} \right) - \frac{1}{\rho c^2} \frac{\partial^2 p}{\partial t^2} = 0 \tag{4.132}$$

Observing this equation, it can be found that it only performs a second-order derivative on time. This means that if p(r, t) is a solution to the equation, then p(r,-t) is another solution, a property that leads to invariance in time reversal operations. It should be noted that all the above assumptions are based on lossless media. If the propagation medium has a frequency-dependent attenuation, the acoustic propagation equation will contain odd derivatives with respect to time, and the invariance in time reversal operations will be lost. However, in practice, if the attenuation is small enough, time reversal is still effective.

Based on the above theory, for a medium with a volume of v, as shown in Fig. 4.33, the black dot in the medium is the sound source, and the initial conditions (sound source and boundary conditions) determine the unique solution p(r, t) to the wave equation. The current goal is to modify the initial conditions in the experiment to obtain another solution p(r, −t). Due to the limitations of causality, p(r, −t) cannot be obtained in practice. Therefore, another method can be used, that is, to solve for p(r, T − t), where T must be long enough so that when t > T, the sound wave has already weakened to 0.

When solving for p(r, T − t), a new problem arises. We must measure p(r, t) in the entire volume of the medium, and then back-propagate p(r, T − t) in the entire volume, which is impossible to achieve in practical applications. A more realistic method is to use Huygens' theory: the sound field at any point within a certain volume can be predicted by the sound field and its normal derivative on a closed surface surrounding this volume. Starting from this point, time reversal focusing on

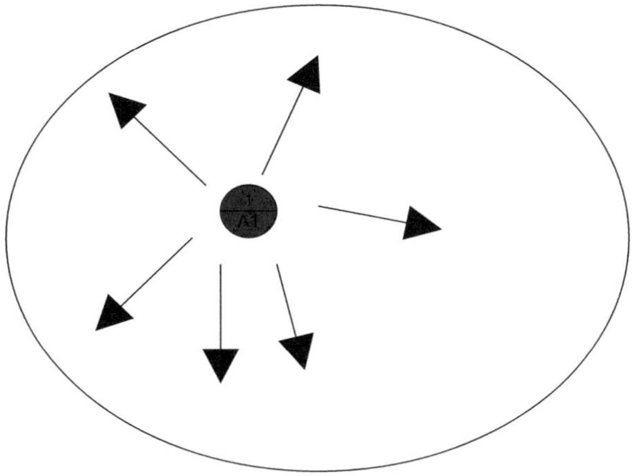

Fig. 4.34 Schematic diagram of time reversal closed curve

a target in a heterogeneous medium can be described as follows: A point source in a heterogeneous medium generates a spherical wavefront, and the wavefront is distorted due to the heterogeneity of the medium during propagation. Considering a closed surface that encompasses the entire medium, suppose we can measure the sound field and its normal derivative at any point on the closed surface, and then we can create a sound source on the closed surface that can generate the time-reversed signal of the received signal. These signals will be transmitted back into the medium and interact with the medium to produce distortion. It can be proven that the reversed signal sound field will eventually focus on the initial sound source (Fig. 4.34).

The closed surface used in time reversal is a perfect implementation conceptually, but in actual operation, the transceiver array cannot achieve a true closed surface, and in most cases, it may only be located on one side of the medium to be detected. Therefore, the reversal surface is generally replaced by a reversal mirror. In this case, time reversal will still form effective focusing.

In general, the time reversal method can be divided into three steps: first, emit a sound wave signal into the heterogeneous medium, the sound wave signal encounters the target to be measured in the medium, and the target scatters the sound wave, and the scattered signal propagates in the medium; second, use the receiving array to receive the scattered signal; finally, time-reverse the received signal and then transmit it back into the medium, and the signal will eventually refocus on the target. Various experimental results have also proven the effectiveness of the time reversal method [26–28].

At the same time, compared with traditional delay techniques, time reversal does not require the heterogeneous medium to be located near the transceiver array, that is, it does not need to assume that the heterogeneous medium only produces a simple delay effect on the transmitted signal, which has unparalleled advantages over traditional delay techniques.

According to the principle of reciprocity, in linear elastic materials, for two transducers A and B, the signal received by transducer B from transmitting transducer A is the same as the signal received by transducer A when the signal source is at transducer B. The time reversal method is based on this time reversal invariance of the acoustic wave equation. The main process is as follows: first, emit a sound signal into the heterogeneous medium (that is, a solid material sample with defects); the sound signal encounters the target to be measured (that is, the defect) in the medium, and the target scatters the sound wave to produce a scattered signal that propagates in the medium; use the receiving array to receive this scattered signal; after time-reversing the received signal, transmit it back into the medium, and the signal will eventually refocus on the target, thereby achieving the positioning of the defect.

The above time reversal method is based on the linear propagation of sound waves. When applied to the analysis of nonlinear harmonic focusing characteristics, when the medium as a whole maintains linear elasticity and the nonlinear effect is small or highly localized, it does not destroy the invariance of time reversal, and the focusing of the time reversal method is still effective [29].

4.8.2 Numerical Simulation of Time Reversal

4.8.2.1 Time Reversal Simulation of Harmonic Signals

Now we consider the numerical simulation of time reversal imaging under non-classical nonlinear effects. The sample model used is exactly the same as the numerical simulation sample, as shown in Fig. 4.29, and the material parameters are taken as those of iron, $K_1 = 1.412 \times 10^{11}$ Pa, $K_2 = 5.2202 \times 10^{10}$ Pa, $\mu = 4.45 \times 10^{10}$ Pa. The propagation of sound waves still uses the finite difference method, with a discrete grid size of 0.5 mm × 0.5 mm and a discrete time interval of 5×10^{-8} s. The gray area in the sample is the damaged non-classical nonlinear area. At the same time, attenuation is added to the sample, with an attenuation parameter Q = 10. Literature [30], In the past, extensive research has been conducted on the time-reversal imaging of short-duration signals in a single nonlinear region within a lossless medium. Therefore, in this numerical simulation, we use a continuous sinusoidal excitation signal as the sound source to observe whether a continuous steady-state signal in a lossy medium can reflect the presence of one or more defects in time-reversal imaging. The transmitting transducer covers the left side of the sample from 5 to 145 mm, and the receiving transducer covers the right side of the sample from 1 to 149 mm.

The input signal only has a component in the x-direction, with the other directions being 0. The stress is $F_x = A\sin(2\pi ft)$, $F_y = 0, A = 1 \times 10^8, f = 250$ KHz, and the nonlinear parameter is $r_1 = 2 * 10^{-3}$, $r_2 = 1 * 10^{-3}$. When there are no defects in the sample, i.e., the entire sample is linearly elastic, the time-domain waveform of the signal received at the center of the receiving end on the right side of the sample is shown in Fig. 4.35a. The Fourier transform of the time-domain signal yields its frequency-domain waveform as shown in Fig. 4.35b. It can be seen that when there are no defects, the signal passes through the entire sample without generating any other harmonics, and the waveform does not distort.

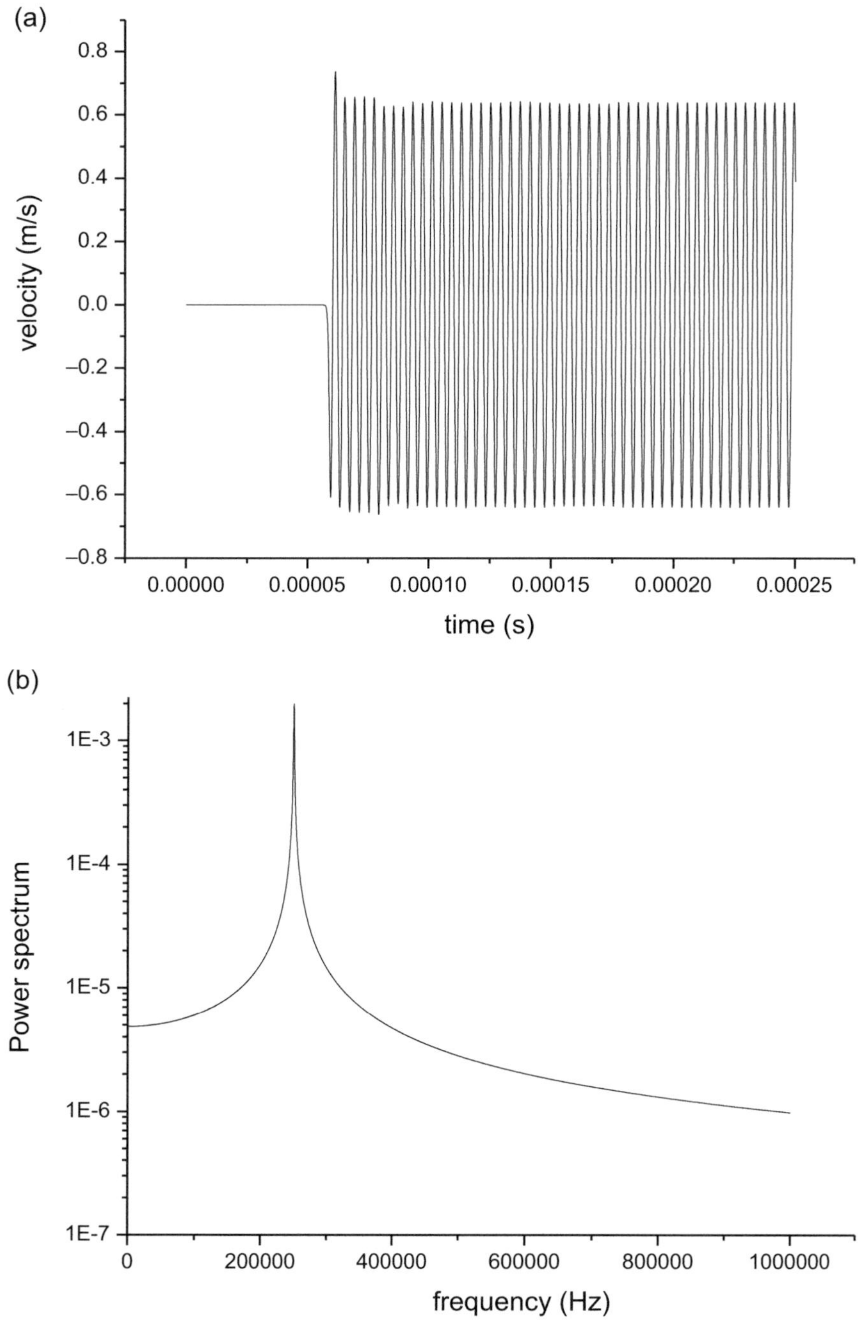

Fig. 4.35 Received signal (**a**) and its spectrum (**b**) when the sample has no defects

Now we consider the changes in the waveform received after adding a single defect. The defect size is 10 mm × 10 mm, centered at (95 mm, 95 mm). Figure 4.35a shows the time-domain waveform of the received signal, and Fig. 4.35b shows the corresponding spectrum. Compared with the original PM model used, after using the improved PM model, the sound wave not only generates odd harmonics after passing through the entire sample, but also generates even harmonics (Fig. 4.36).

After receiving the signal, filter out the nonlinear components to perform time-reversal imaging. Compared with the sound wave signal spectrum in the lossless sample, it can be found that the high-order harmonics are all generated due to nonlinear effects. Therefore, any high-order harmonic component can be filtered out for reversal. First, filter out the 500 kHz harmonic and observe its time-reversal imaging. To make the imaging clearer, we use the method proposed in literature [31], and use the signal energy E for imaging after reversal. The relationship between the signal energy E(x, y) at time t_k and the signal amplitude U(x, y) at the coordinate (x, y) is as follows:

$$E(\mathrm{x,y,t}_k) = \int_{t_k}^{t_k+\Delta t} U^2(x, y, t_k)d\tau \tag{4.133}$$

Figure 4.37a is the time-reversal image of the second harmonic at t = 1.205e−5s. The red box in the figure represents the actual defect size (the red box in all time-reversal images below represents the actual defect size). It can be seen that the dark area of the reversal image extends along the actual defect area, which may be due to the accumulation of nonlinear effects. In the nonlinear defect area, the high-order harmonics gradually increase until they reach the maximum at the edge of the nonlinear area, and then as they leave the defect area, the harmonics generated by the nonlinearity will slowly diffract and decrease, causing the dark area to extend during imaging.

We also filter out other high-frequency components for comparison. Figure 4.37b–d are the reversal images of the 750, 1000, and 1250 kHz harmonic components at the same time, respectively. These images can roughly reflect the existence of defect positions, indicating that time-reversal imaging of continuous steady-state waves in lossy media is feasible if the loss parameters are appropriate. At the same time, we observe that as the harmonic frequency increases, the energy of the harmonic during reversal is more concentrated in the center of the wavefront, and the energy on both sides gradually attenuates. Compared with the actual area, the defect area of the reversal image begins to move back as a whole as the frequency increases. Compared with low-frequency imaging, high-frequency imaging has less interference, but the direct positioning of defects in the horizontal direction is not as accurate as low-frequency. If we consider taking the point with the maximum amplitude during reversal imaging as the boundary point of the defect in the horizontal direction, and then swap the positions of the transmitting and receiving transducers for a second

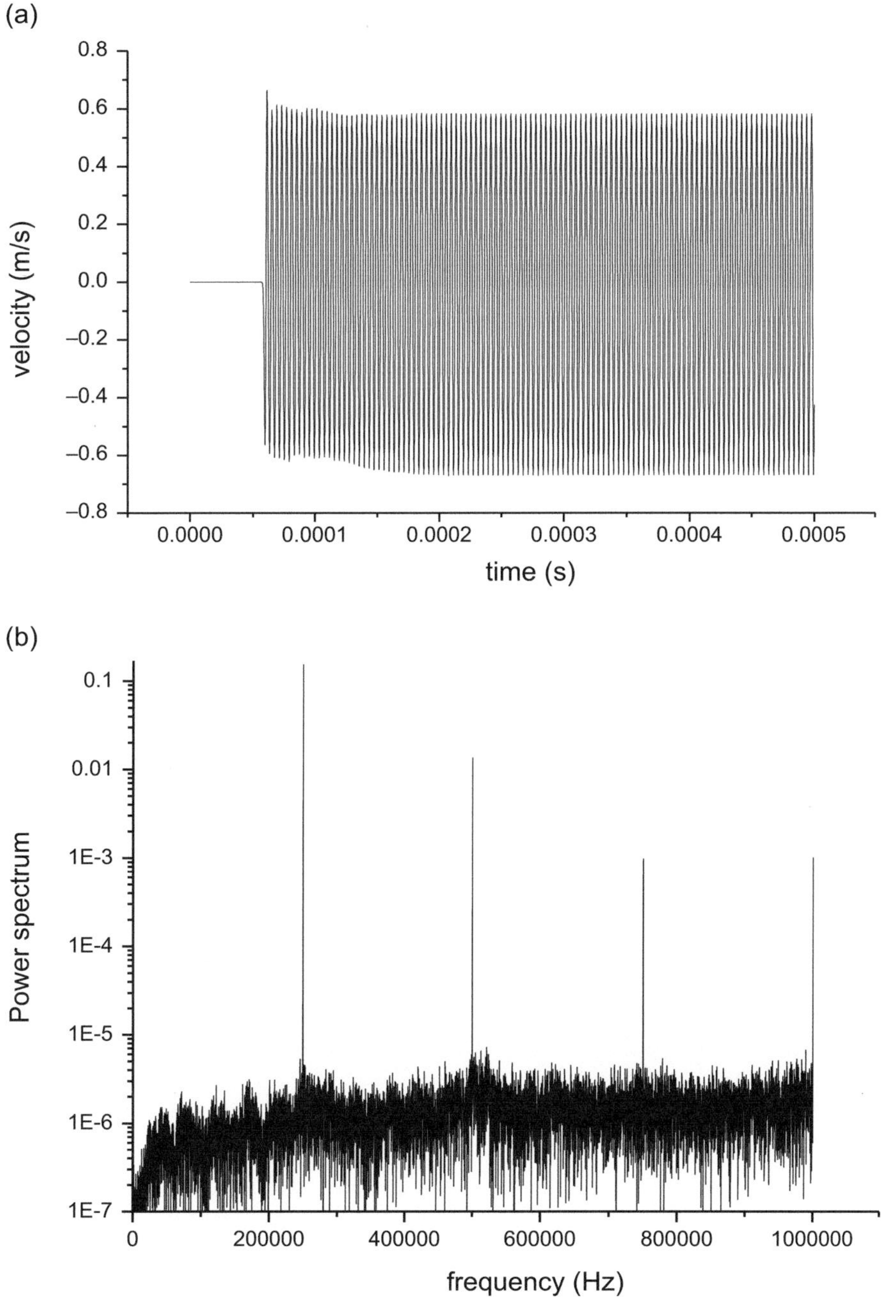

Fig. 4.36 Received signal (**a**) and its spectrum (**b**) received when the sample has a defect, with the defect center at (95 mm, 95 mm)

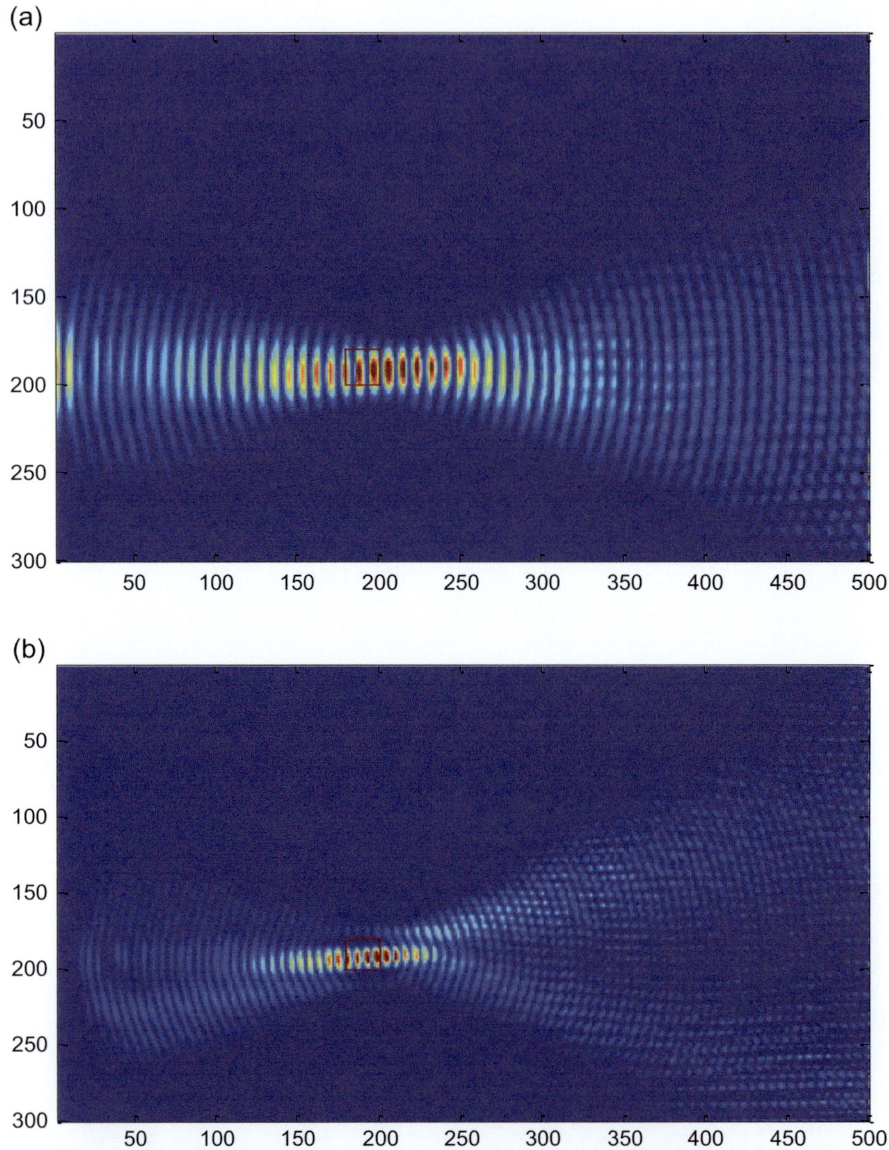

Fig. 4.37 The time reversal imaging of the second harmonic, with the defect center at (95 mm, 95 mm). Time reversal imaging of the second harmonic (**a**), third harmonic (**b**), fourth harmonic (**c**), and fifth harmonic (**d**), with the defect center at (95 mm, 95 mm)

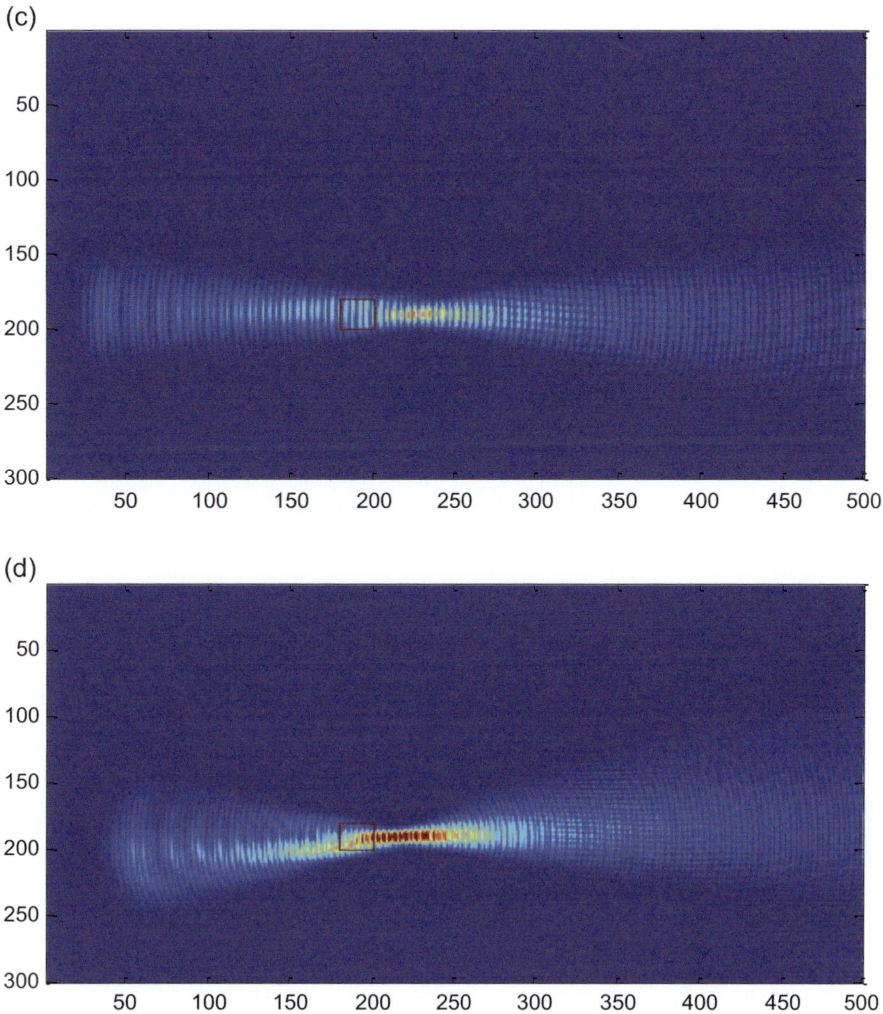

Fig. 4.37 (continued)

imaging to find the other boundary point of the defect, the horizontal positioning of high-frequency imaging may be more accurate.

After observing the imaging of a single defect, we now consider the imaging of two defects, taking the center of the second defect (65 mm, 65 mm), the size remains 10 mm × 10 mm. First, we analyze the signal received in the center of the receiving end on the right side of the sample. Figure 4.38a–c are the received signal and its spectrum analysis, respectively. After the defect is added, the signal frequency component is not affected, but the size will change.

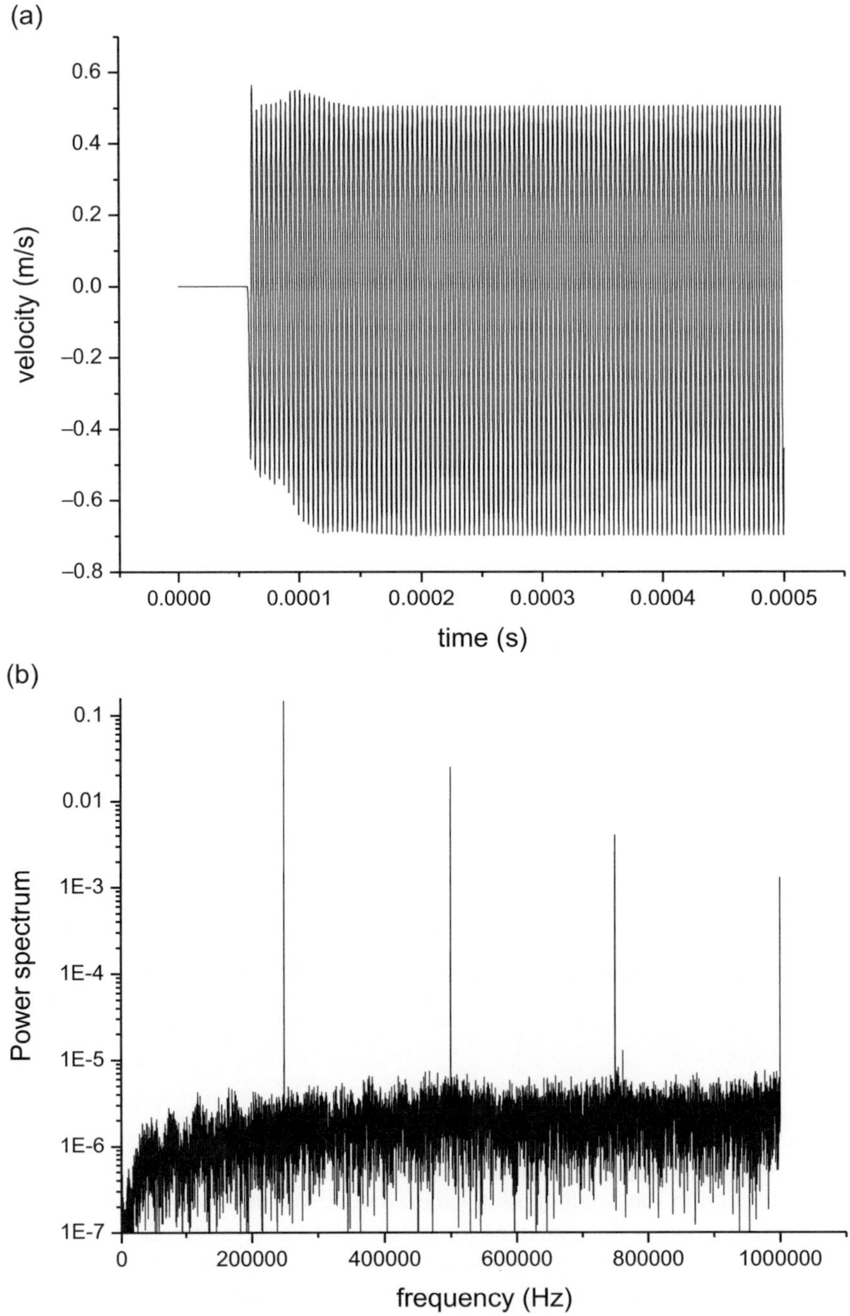

Fig. 4.38 The received signal when the sample has two defects (**a**) and its spectrum (**b**), The centers of the defects are located at (95 mm, 95 mm) and (65 mm, 65 mm)

Then we take the nonlinear harmonic component for reverse imaging. First, we take the 500 kHz harmonic. Figure 4.39a is the second harmonic time reversal image at t = 1.205e−5s. From the figure, although compared with the imaging of a single defect, the sound waves scattered by the two defects have some interference with the imaging, but the defect area can still be distinguished from the imaging. In order to better explain the existence of only two defects, we try to compare with higher frequency harmonic imaging. Figure 4.39b–d are the time reversal images of the 750, 1000, 1250 kHz harmonic signals at t = 1.205e-5s. In these three images, the interference between the defects does indeed decrease with the increase in frequency, but as the frequency increases, the brightness of the two defects begins to differentiate. The time reversal image of the 1000 kHz component can hardly judge the existence of the defect located at (90 mm, 90 mm), and the time reversal image of the 1250 kHz component can only vaguely judge the existence of the defect located at (65 mm, 65 mm). Therefore, we can try to compare multiple high-frequency images to judge the number of defects. Then combine the images of various frequencies to locate the defects.

Figure 4.39a is the time reversal image of the second harmonic, with the centers of the defects located at (95 mm, 95 mm) and (65 mm, 65 mm).

The above numerical simulation shows that in lossy media, using the time reversal imaging of continuous steady-state waves to judge and locate one or more defects is theoretically feasible.

4.8.2.2 Time Reversal Simulation of Frequency Modulated Wave Signal

Literature [26] conducted a time reversal experiment on frequency modulated waves. He input high-frequency wave f_h and low-frequency wave f_l into the damaged sample at the same time, and received the frequency modulated wave $f_h \pm f_l$ at the receiving end. Then filter out the frequency modulated signal, reverse the signal in time, and transmit it back to the sample. A vibration scanning system is used to measure the vibration amplitude of each point of the sample at a certain moment. Finally, calculate the vibration energy of each point and use the energy for time reversal imaging. The results prove that the frequency modulated wave generated by the damaged sample can reflect the existence of defects.

Based on the improved PM model, the generation of frequency modulated waves and their time reversal imaging can be theoretically studied. Take the sample in Fig. 4.28 for numerical simulation, add a set of transmitting arrays on the upper side of the sample to transmit low-frequency waves, as shown in Fig. 4.40. The transmitting array covers the upper side from 50 to 100mm. The input signal of the transmitting array on the left only has a component in the x direction, and the other directions are 0. The stress $F_x = A\sin(2\pi ft)$, $F_y = 0$, $A = 1 \times 10^8$, $f = 250\,\text{KHz}$. The input signal of the transmitting array on the upper side only has a component in the y direction, and the other directions are 0. The stress $F_x = 0$, $F_y = A\sin(2\pi ft)$, $A = 1 \times 10^8$, $f = 4\,\text{KHz}$.

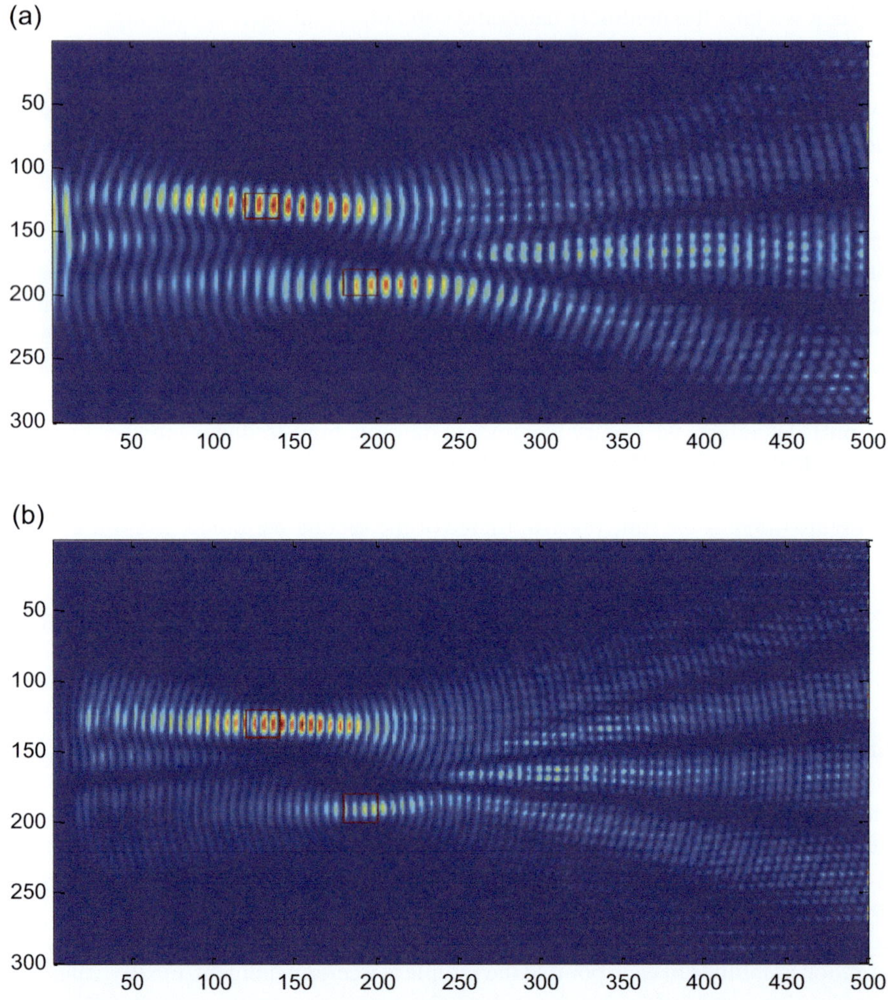

Fig. 4.39 The time reversal images of the second harmonic (**a**), third harmonic (**a**), fourth harmonic (**b**), and fifth harmonic (**c**), The centers of the defects are located at (95 mm, 95 mm) and (65 mm, 65 mm)

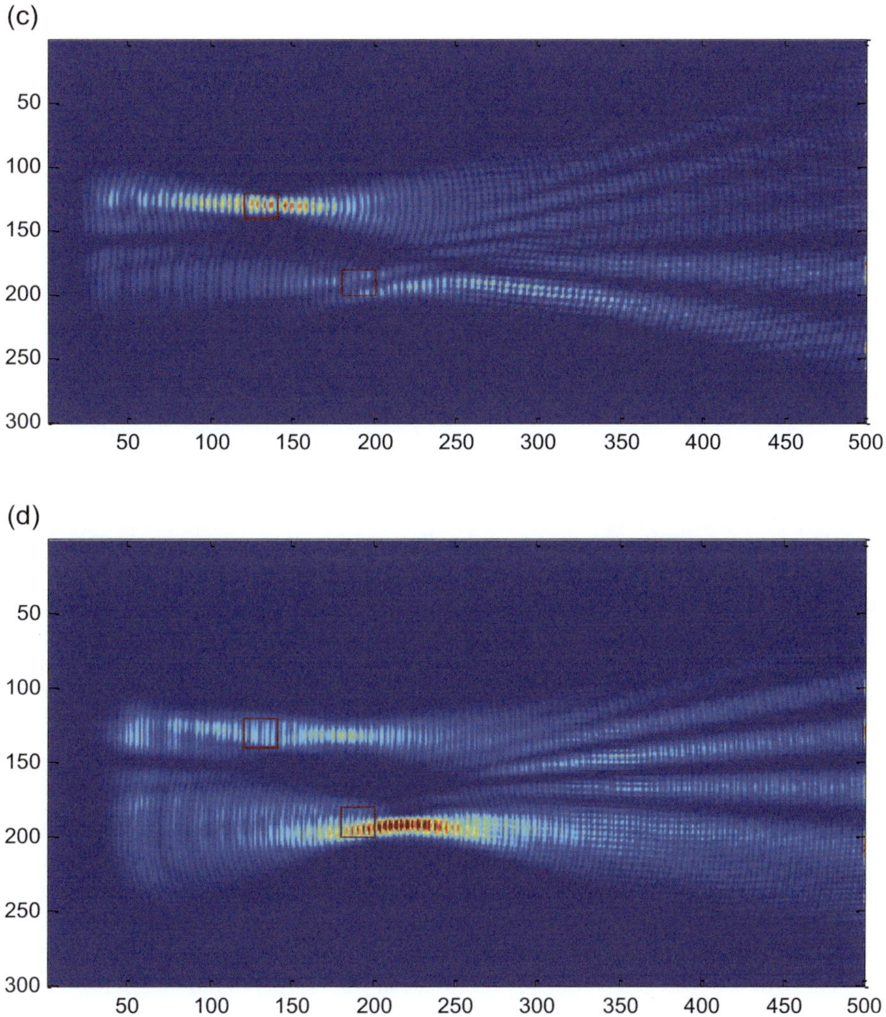

Fig. 4.39 (continued)

First, study the case where there is only one defect. The center of the defect is located at (90 mm, 90 mm), the size is 10 mm × 10 mm, and the time domain waveform of the signal received in the center of the receiving end on the right is shown in Fig. 4.41a, and the corresponding spectrum analysis is shown in Fig. 4.41b. After adding low-frequency waves, the obtained signal will generate frequency modulated wave components near the fundamental and each harmonic frequency, which is consistent with the results obtained in the experiment in literature [31], indicating that the improved PM model can explain the generation of frequency modulated waves. Figure 4.42a–e are the images at t = 1.205e−5s after filtering out signals

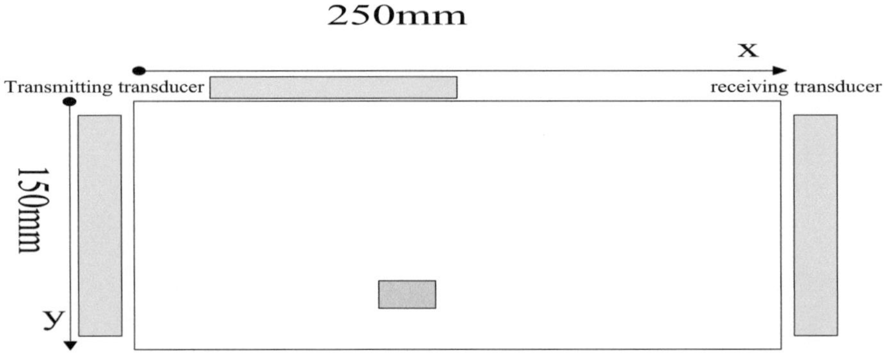

Fig. 4.40 Schematic diagram of modulated wave time reversal imaging

with frequencies of 246, 500, 496, 750, and 746 kHz and reversing them in time. In the experiment, the frequency modulation wave near the fundamental frequency was used, and the results obtained could only roughly judge the existence of defects, but could not determine the size of the defects or accurately locate the defects. In the numerical simulation imaging diagram, when there is only a single defect, the imaging diagram of the frequency modulation wave near the fundamental frequency, as shown in Fig. 4.42a, is quite satisfactory compared to the experimental results.

Figure 4.42b is the reverse imaging of the 500 kHz second harmonic signal, and Fig. 4.42c is the reverse imaging of the 496 kHz frequency modulation wave signal near the second harmonic. Both can reflect the existence of defects. Comparing the two images, it can be seen that the frequency modulation wave has a certain distortion in the propagation direction, which may be caused by the superposition of wave vectors due to the generation of frequency modulation waves in both horizontal and vertical directions. Similarly, the 746 kHz frequency modulation wave imaging diagram near 750 kHz, as shown in Fig. 4.49e, is also the case, and the propagation direction is clearly no longer horizontal as shown in Fig. 4.42d. Observing the imaging of harmonics and frequency modulation waves, it is found that in the case of a single defect, the imaging of the frequency modulation wave near the fundamental frequency can better judge the size and position of the defect.

Adding a second defect, the center of the newly added defect is at (65 mm, 65 mm), the size is 10 mm $\times$ 10 mm, and the time-domain waveform of the signal received in the center of the receiving end on the right is analyzed for spectrum, and it is found that the harmonic component is the same as when there is a single defect, only the size is different. The received signal is filtered, time-reversed imaging, and the imaging effects of different frequency components are compared. Figure 4.43 is the time-reversal imaging diagram of the 246 kHz frequency component. In this numerical simulation, when there are two defects, the time-reversal imaging effect of the frequency modulation wave near the fundamental frequency is not very good, and it can only be roughly seen that the frequency modulation wave converges in two directions, and there may be damage points located in the dark area.

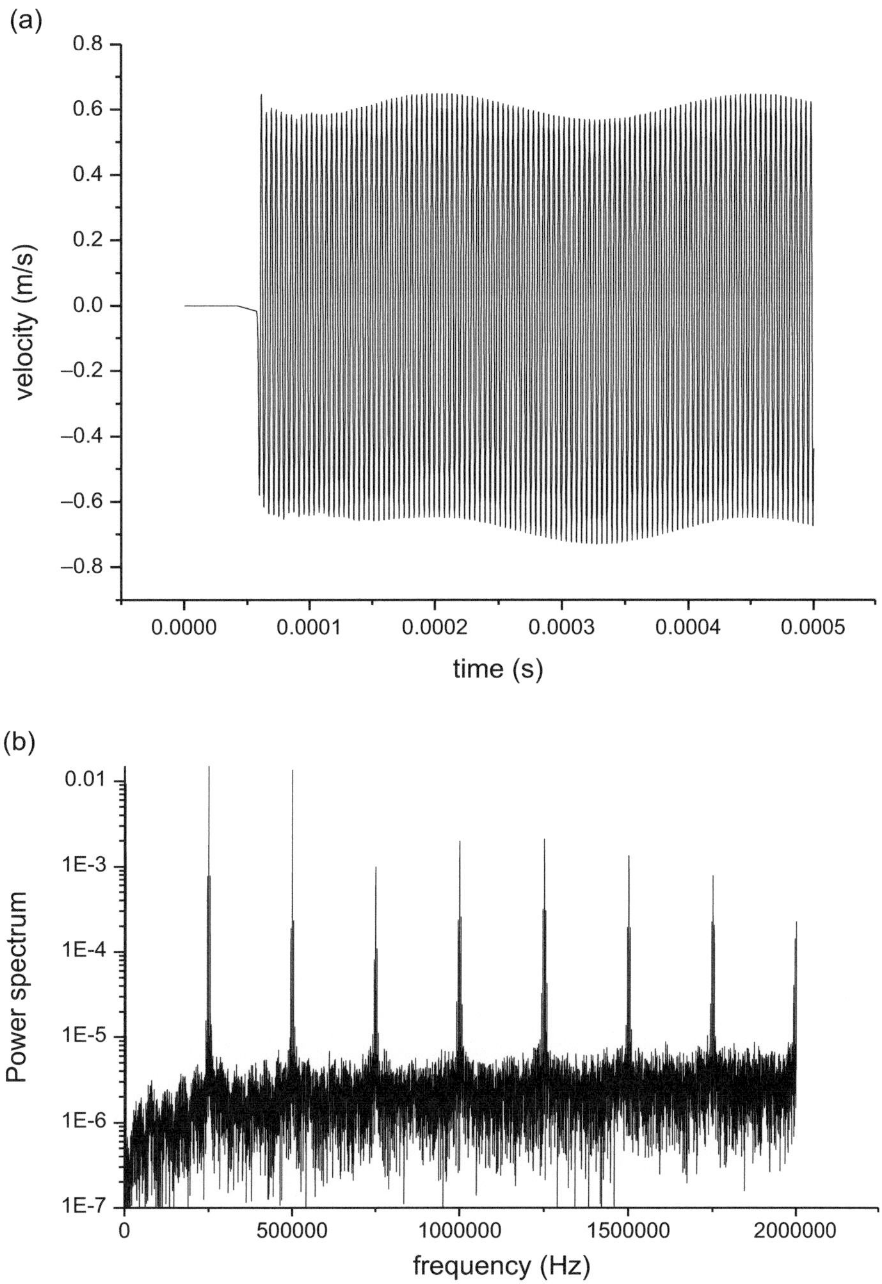

Fig. 4.41 The received signal when the sample has a defect (**a**) and its spectrum (**b**), the center of the defect is at (90 mm, 90 mm)

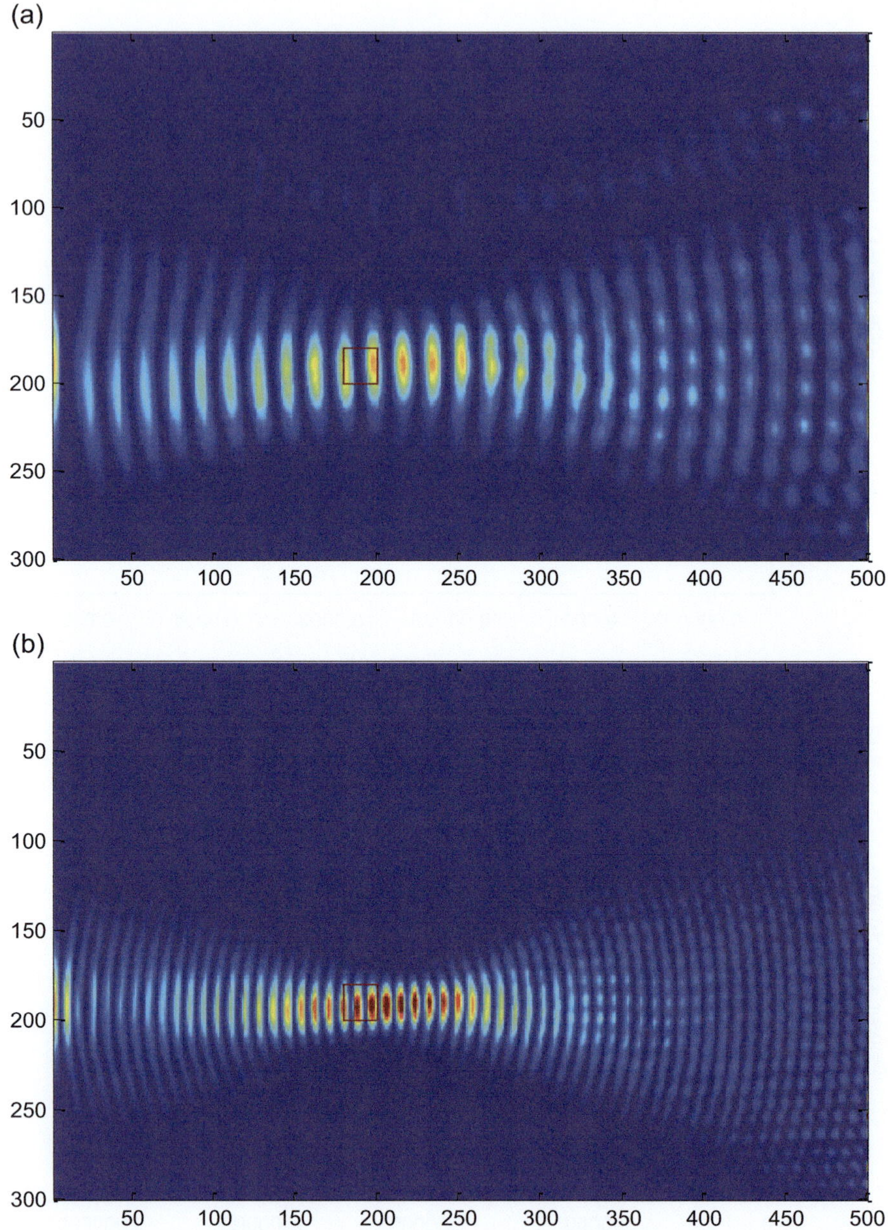

Fig. 4.42 **a**–**f** The time reversal images at frequencies of 246 kHz (**a**), 500 kHz (**b**), 496 kHz (**c**), 750 kHz (**d**), 746 kHz(**e**), respectively, the center of the defect is at (90 mm,90 mm)

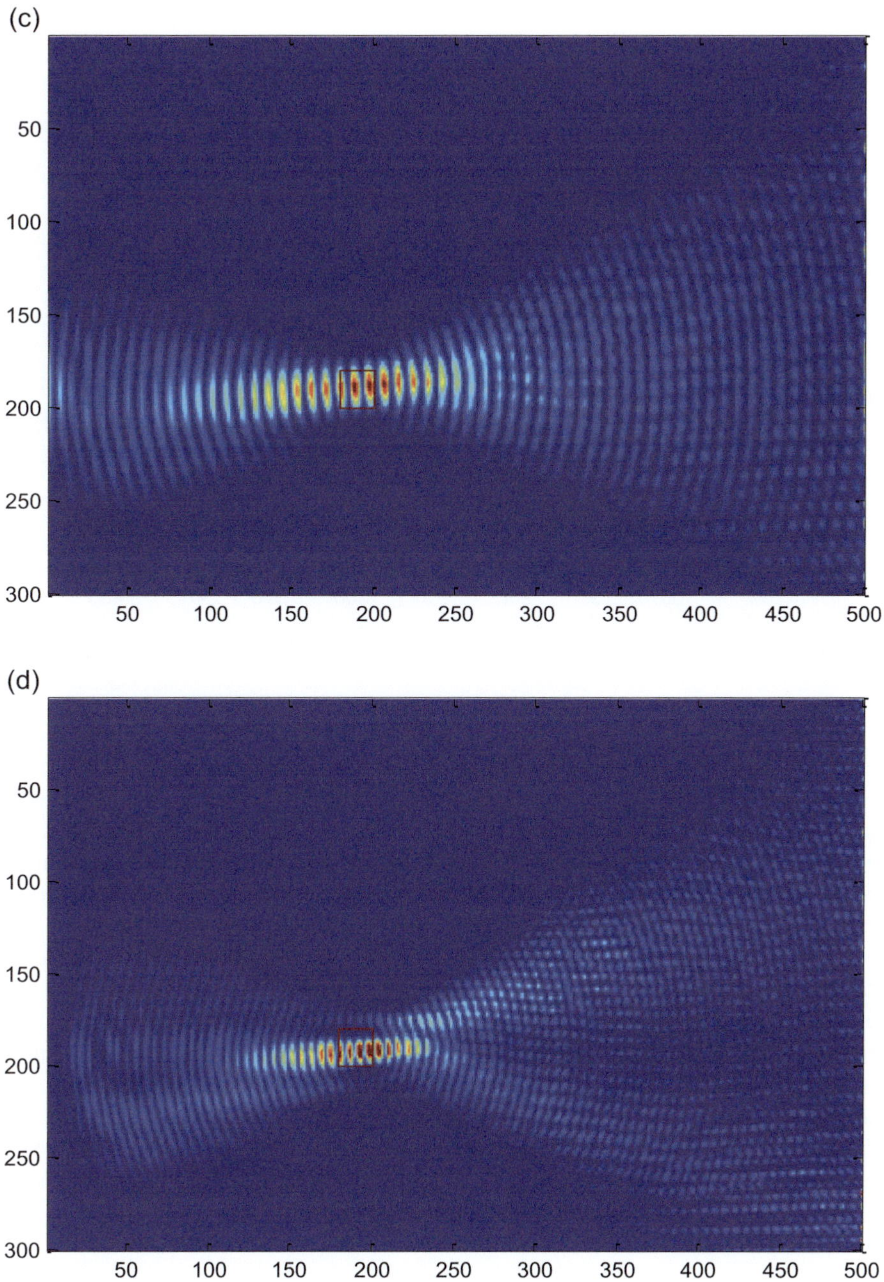

Fig. 4.42 (continued)

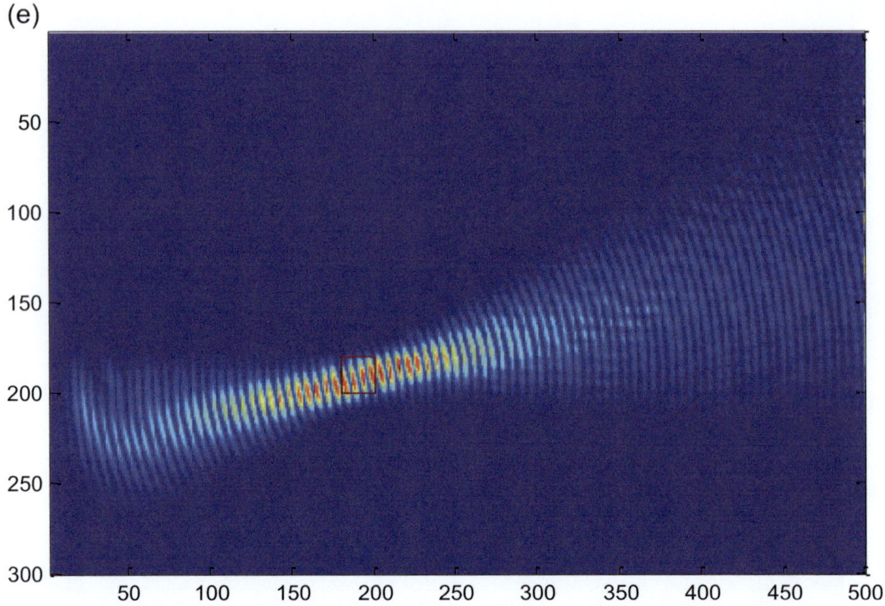

Fig. 4.42 (continued)

Now use different high-frequency harmonic components for imaging and compare the differences in results. Figure 4.43b–g are time-reversal imaging of 500, 496, 750, 746, 1500, 1496 kHz frequency components, respectively. Compared with the imaging of frequency modulation waves near the fundamental frequency, the defect areas displayed by the imaging of frequency modulation waves near high-order harmonics are closer to the actual areas, that is, the imaging accuracy of high-order frequency modulation waves is better than that of frequency modulation waves near the fundamental frequency. As the frequency increases, the interference during imaging becomes smaller, but the signal energy at the two defect sites is not consistent, which will cause one defect to be strong and the other defect to be weak, the most serious of which is Fig. 4.43e, the defect below is almost indistinguishable. But by integrating the time-reversal imaging diagrams of multiple frequency components, the number of defects can basically be accurately judged. As for the determination of the size of the defect area, because the higher the frequency, the energy will concentrate from the sides of the wavefront to the center, so multiple images must be compared to select the appropriate high-frequency component.

4.8.2.3 Maximum Value Imaging Method

In the time reversal process of numerical calculation, this section uses the maximum value imaging method to image the crack [32]. According to the maximum value

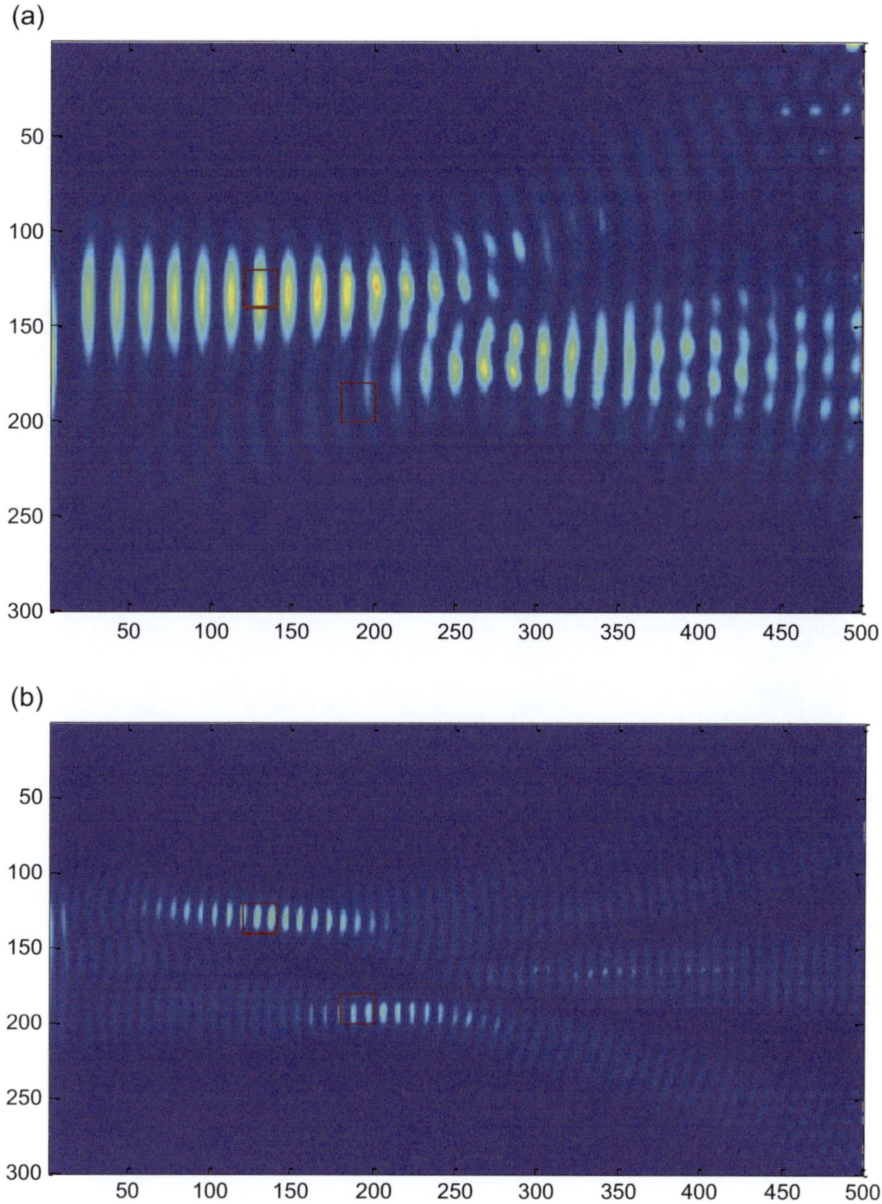

Fig. 4.43 **a–g** The time reversal images at frequencies of 246, 500, 496, 750, 746, 1500, 1496 kHz, respectively, the centers of the two defects are at (90 mm, 90 mm) and (65 mm, 65 mm), respectively

(c)

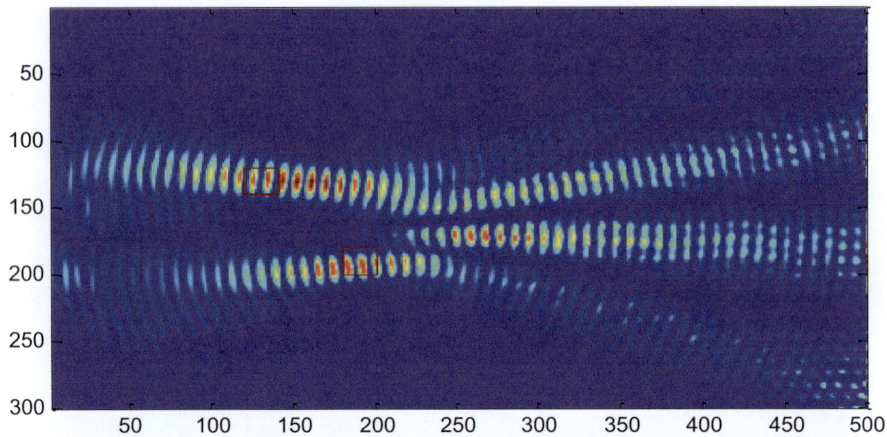

(d)

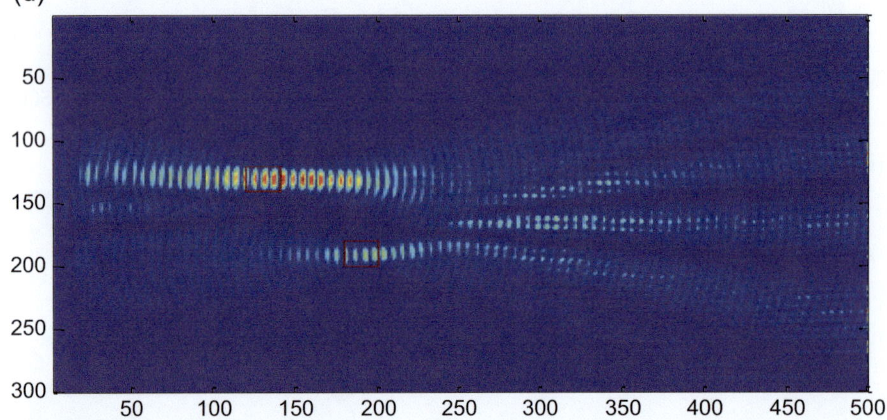

Fig. 4.43 (continued)

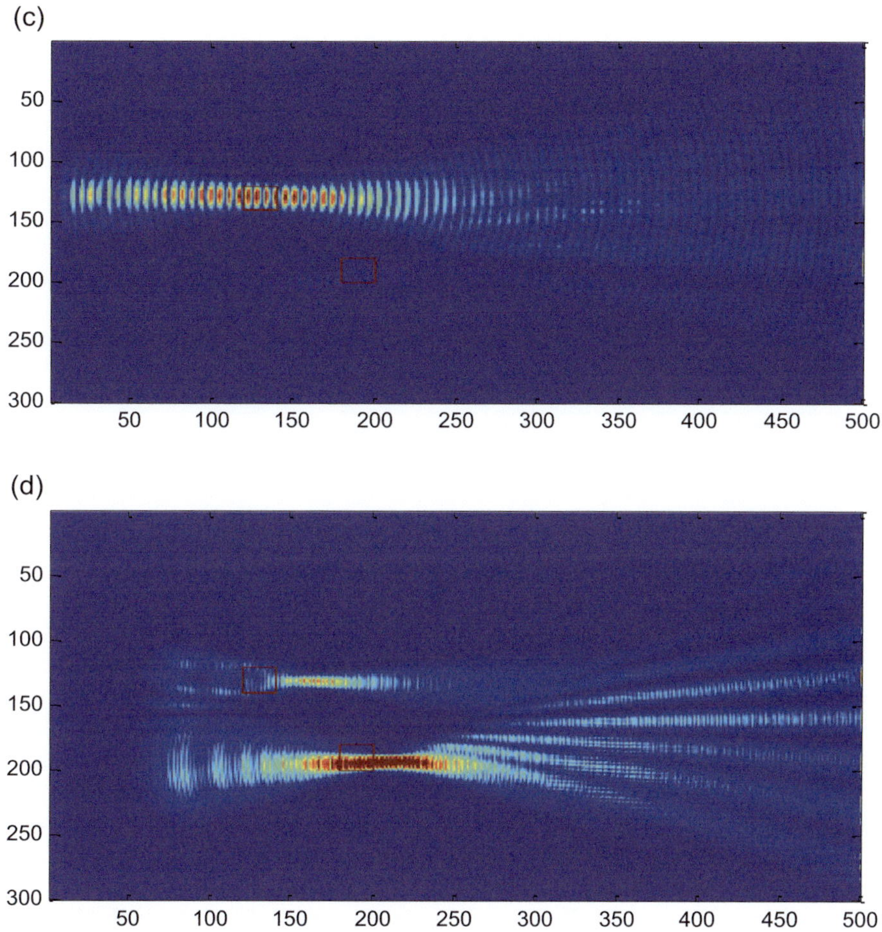

Fig. 4.43 (continued)

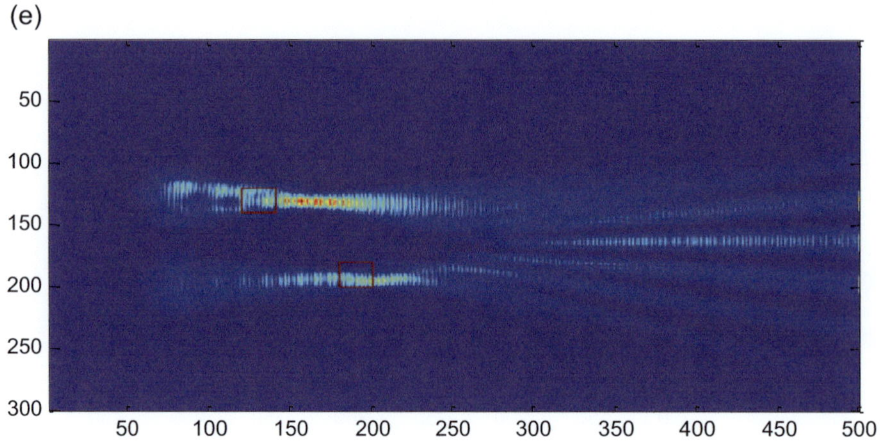

Fig. 4.43 (continued)

imaging method, for the velocity component of the particle at the coordinate (x, y) at time t, $v_x\ (x, y; t)$, the function $M\ (x, y) = \max[v_x\ (x, y; t)]$ is defined to characterize the specified particle at x The maximum value of the directional speed v_x. It can be predicted that it will reach its maximum value in the area where the reversed sound waves converge, M and when the reversed sound waves emitted by N receiving transducers interfere, it will reach the second highest value. From the final image, it can be seen that these second highest value areas form multiple paths, pointing to the convergence area. The matrix M is used for imaging, which is the so-called maximum value imaging method. From the image formed by the maximum value imaging method, we can not only observe the convergence point, but also observe the convergence path of the sound wave.

In addition, this section also uses the maximum value imaging method with a window function to study the sound field of a specific time period, that is, only time reversal processing is performed on the harmonic signal, and it is re-emitted into the medium to be tested. The re-emitted sound wave signal will form interference in the area with cracks, thereby focusing the energy of the sound wave more accurately in the crack area, that is, the position causing the harmonic.

At the same time, through nonlinear spectral analysis and software filtering, the sound signal can be extracted to ensure that only the nonlinear components are focused at the scattering place, and no linear components appear. In the numerical calculations of this section, the pulse reversal filtering method is used to extract the nonlinear components in the received signal [33]. Compared with the traditional harmonic filtering method, the pulse reversal filtering method can obtain defect images more stably. For example, for defects near the boundary, the pulse reversal filtering method can effectively detect them, but the harmonic filtering method cannot. This is because the harmonic filtering method cannot completely filter out the fundamental wave.

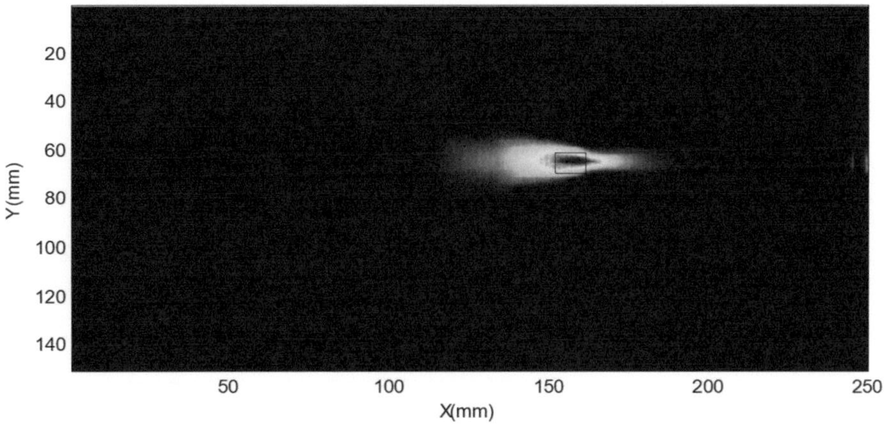

Fig. 4.44 Imaging of a single crack with a center coordinate of (157.5, 64) in the sample

Crack Imaging Based on the Maximum Value Imaging Method

The geometric dimensions of a solid sample with a single crack are shown in Fig. 4.28. Its density is $\rho = 7700$ kg/m ^{3}, elastic modulus $K_1 = 1.412 \times 10^{11}$ Pa, $K_2 = 5.2202 \times 10^{10}$ Pa, $\mu = 4.45 \times 10^{10}$ Pa, quality factor $Q = 10$. This section uses a continuous sound source with a frequency of f to excite the sample with a crack, and applies an external force $F_x = A \sin(2\pi ft)$, $F_y = 0$ in the two-dimensional plane, where $A = 1 \times 10^8$. In the discretization process of the wave equation, the side length of the square grid is 0.5 mm, the time step is 5×10^{-8} s, the sampling frequency is 20 MHz, and the number of samples is 5000.

According to the trial calculation analysis, as the excitation frequency increases, the response amplitude of the signal power spectrum decreases. Since the response amplitude of the signal power spectrum at different excitation frequencies represents the size of the defect's nonlinear response, that is, the larger the signal power spectrum amplitude, the more significant the defect's nonlinear response; on the contrary, as the signal power spectrum amplitude becomes smaller, the defect's nonlinear response also weakens accordingly, and it is easy to miss detection when imaging the defect. Therefore, the excitation frequency should not be chosen too high. However, when the frequency decreases, the wavelength increases, and the resolution of ultrasonic detection decreases. To balance, this section uses an ultrasonic transducer with a center frequency of $f = 250$ kHz to detect cracks in solid materials.

For a solid material sample of the size shown in Fig. 4.28, when it contains a crack, and the center coordinate of the crack is (157.5, 64), the imaging of the crack is shown in Fig. 4.44. According to the maximum value imaging method, the crack can be clearly identified (the area inside the black solid line in the figure is the actual position of the defect).

When there are two cracks in the sample, it is necessary to analyze the distribution of the center coordinates. When the center frequency is $f =$ At 250 kHz, when the

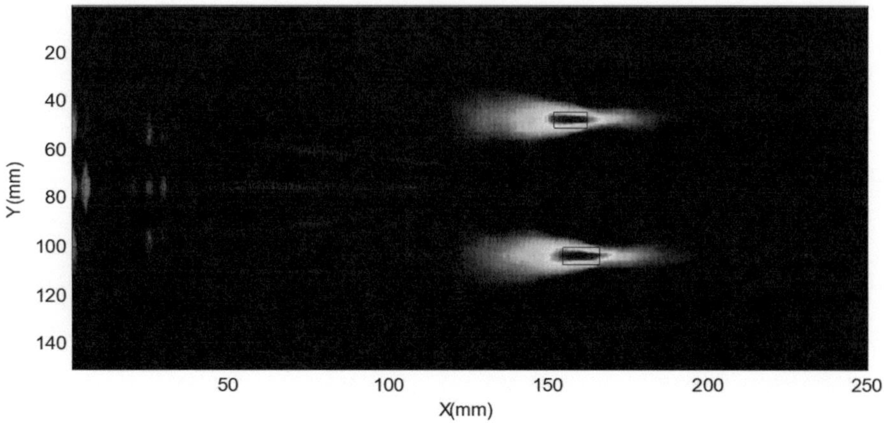

Fig. 4.45 Imaging of two cracks with center coordinates of (157.5, 103.5) and (157.5, 46.5) respectively

x-coordinates of the centers of two cracks are the same and the y-coordinates are more than 25 mm apart, both defects can be clearly identified in the resulting image, as shown in Fig. 4.45. When the x-coordinates of the crack centers are the same and the y-coordinates are closer together, the image will show a pseudo-crack.

The calculation results show that as the frequency of the sound source increases, the pseudo-crack will disappear. For example, when excited with a sine signal with a center frequency of 700 kHz, even if the y-coordinates of the centers of the two cracks are less than 5 mm apart, the cracks can still be clearly identified in the image. However, when the selected excitation source frequency is high and far from the system resonance frequency, the particle vibration becomes smaller and is not easily detected by the receiving transducer. Therefore, the method of improving the imaging results by increasing the excitation source frequency has significant limitations, and it is necessary to propose an effective method to distinguish pseudo-cracks.

When the center positions of the two cracks are relatively close, first consider the situation where the x-coordinates of the crack centers are the same and the y-coordinates are different. Figure 4.46 shows the imaging results of two cracks with center coordinates of (157.5, 64) and (157.5, 86) respectively. Their maximum speeds are located at (157, 64) and (158, 86), and the error between the detection imaging results and the actual crack distribution is small. Therefore, the maximum value imaging method can effectively image two relatively close cracks. However, in Fig. 4.46, a bright spot appears on the left side of the two cracks, which can easily be mistaken for another defect area, that is, the so-called micro-crack. The formation mechanism will be analyzed below using the maximum value imaging method with a time window.

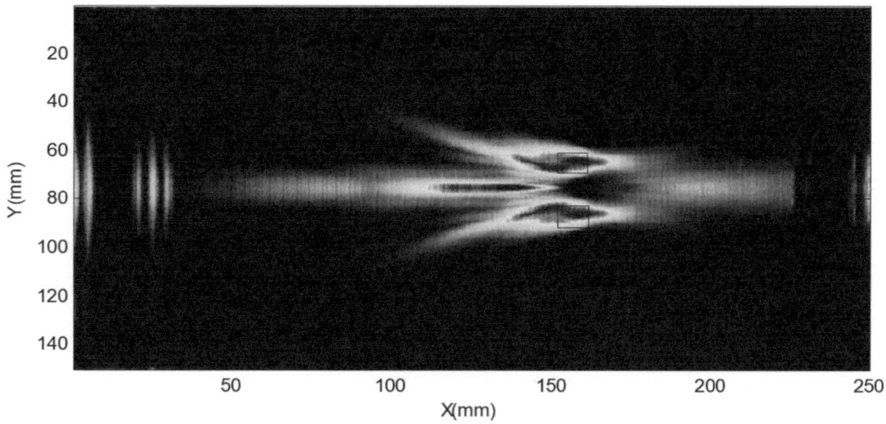

Fig. 4.46 Imaging of two cracks with center coordinates of (157.5, 64) and (157.5, 86) respectively

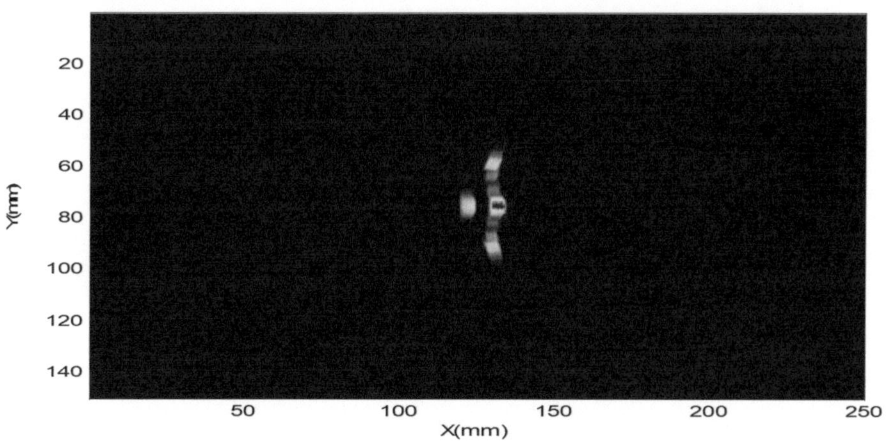

Fig. 4.47 Imaging of two cracks with center coordinates of (157.5, 64) and (157.5, 86) respectively, starting from a time window of 0.2177 ms

Maximum Value Imaging Method with Time Window for Crack Imaging

In this section, a time window of length Δt starting from t_R will be set for the maximum value imaging method. To analyze the formation of pseudo-cracks, a narrow time window span is chosen. When an $t_R = 0.2177$ ms, $\Delta t = 6.25 \times 10^{-4}$ ms time window is added to the imaging of Fig. 4.46, the imaging is as shown in Fig. 4.47.

According to Fig. 4.47, some pseudo-cracks (bright spots on the left side of the actual cracks) are located where the wavefronts of higher harmonics emitted from different receiving transducers intersect. Since the sound wave will continue to propagate after converging at the defect, and its wavefront will spread out, when

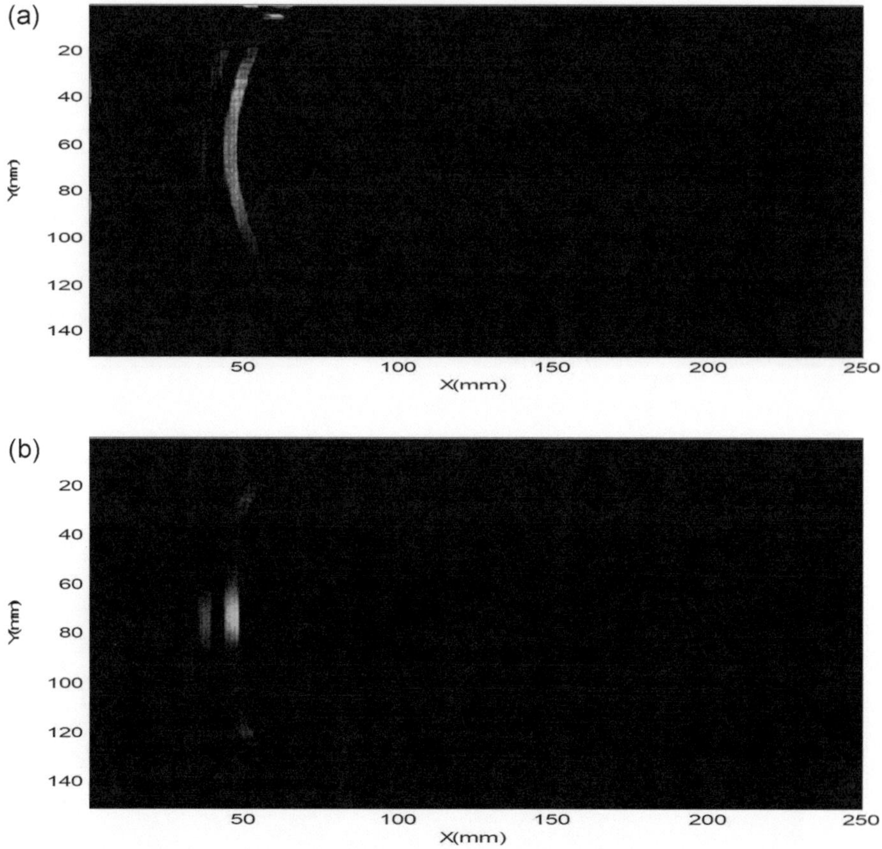

Fig. 4.48 **a** Imaging of a single crack with a center coordinate of (157.5, 64), starting from a time window of 0.2375 ms. **b** Imaging of two cracks with center coordinates of (157.5, 64) and (157.5, 86) respectively, starting from a time window of 0.2375 ms

the positions of the two cracks are close, the wavefronts of different sound waves will overlap at a specific point. If the phases of the intersecting sound waves are not exactly opposite, their vibrations will add up and enhance at the intersection. That is, at these positions, the corresponding particle vibration speed will increase, forming a pseudo-crack.

On the other hand, in Fig. 4.46, the shape of the bright spot corresponding to the real crack also has a significant difference from the actual shape of the crack (the area inside the black solid line). To avoid the mutual influence between cracks, a single crack with a center coordinate of (157.5, 64) is analyzed separately. Comparing Fig. 4.46 with Fig. 4.44, it can be found that for cracks of the same actual position and size, their imaging shapes are significantly different. When a $\Delta t = 6.25 \times 10^{-4}$ ms, $t_R = 0.2375$ ms time window is added to both, the imaging is as shown in Fig. 4.48a, b respectively.

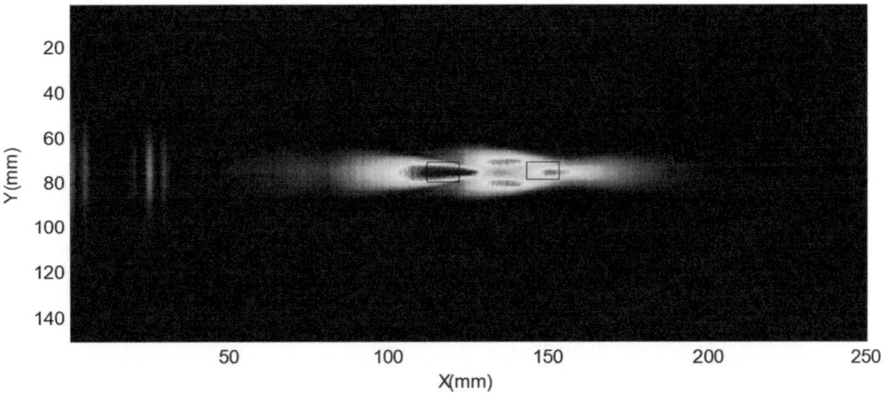

Fig. 4.49 Imaging of two cracks with center coordinates of (117.5, 75) and (147.5, 75) respectively

Observing the wavefront of the imaging after adding the time window, in Fig. 4.48b, the wavefront of the higher harmonics converges at the crack at position (157.5, 64), then splits into two parts propagating along the *y* axis and diagonally; while in Fig. 4.48a, the wavefront does not split. Therefore, the reason for the difference in imaging shape between Figs. 4.44 and 4.46 is that due to the influence of nearby cracks, the propagation and convergence of high-order nonlinear harmonics have changed.

Figure 4.49 shows the imaging of two cracks with center coordinates of (117.5, 75) and (147.5, 75) imaging results of the two cracks. It can be seen that although the error between the position of the maximum value of the speed distribution and the center position of the crack is small, two bright spots appear outside the actual crack position. The reason for the appearance of these bright spots is that during the time reversal process, the high-order harmonics that will converge to the right crack intersect and superimpose with the high-order harmonics that will converge to the left crack.

Furthermore, as shown in Fig. 4.50, when there are two cracks, one crack may mask the nonlinearity of the other crack, so one crack appears brighter in the image, while the other is darker. In the process of judging the cracks, the darker imaged crack is easy to be missed.

4.8.2.4 Nonlinear Elastic Wave Spectroscopy—Time Reversal—Nonlinear Elastic Wave Spectroscopy (NEWS-TR-NEWS)

As previously mentioned, the maximum value imaging method can extract the center position of the cracks in the solid sample. However, when detecting real cracks, there are two possibilities: (1) The position judged to have cracks in the image does not actually have cracks, i.e., false cracks appear; (2) Real cracks exist in the darker areas

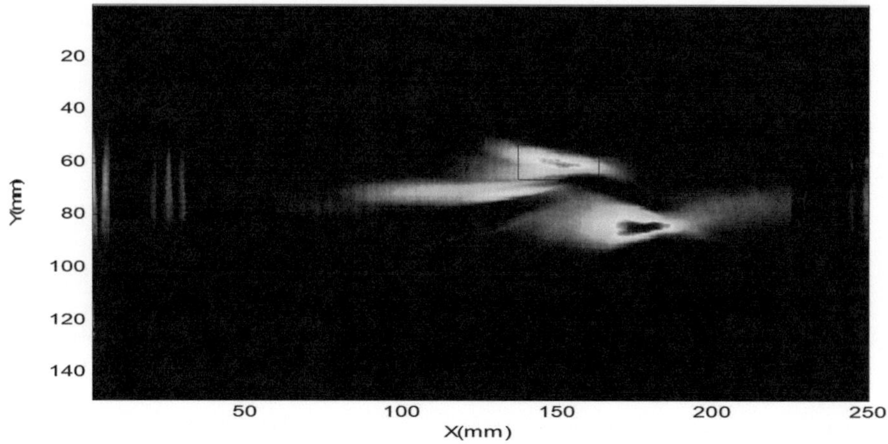

Fig. 4.50 Imaging of two cracks with center coordinates of (147.5, 61.5) and (177.5, 83.5) respectively

of the image, which are easy to be missed when judged. To distinguish false cracks, this section uses the NEWS-TR-NEWS method shown in Fig. 4.53, that is, a NEWS process is performed again after the original NEWS-TR process.

According to the NEWS-TR-NEWS method in Fig. 4.51, during the NEWS-TR process, all crack positions (including false cracks) are first determined and recorded. In the added forward NEWS stage, the velocity of the particles at these recorded positions, $v_x(x, y; t)$, is scanned. To implement this process, two forward stages need to be added, and the impact reversal filtering method is used to obtain the nonlinear signal at the corresponding position. For a specific position in the solid sample, $x = x_k$, $y = y_k$, the obtained nonlinear signal is the time series $v_x(t)$, and its maximum value is extracted to obtain $\max(v_x(t))$. When the abscissa is the same and the receiver scans in the y-axis direction, the corresponding position $M = \max(v_x(t))$ is obtained, which can distinguish whether the crack at this position really exists: if the crack really exists, the nonlinear signal is generated at the crack and continues to spread in the subsequent propagation process. Therefore, when the nonlinear sound wave is generated at the crack and has not spread, the nonlinear vibration is the maximum value, that is, the one-dimensional array $M = \max(v_x(t))$ image will produce a peak. Similarly, the darker cracks can also be distinguished by the added forward NEWS process to avoid missing. This section will specifically analyze the maximum value imaging and false crack identification of the cracks in the sample.

According to the above analysis, when there are two cracks in the solid material sample, the maximum value imaging method with a time window can effectively image two cracks with the same center position abscissa and distinguish false cracks in the image. However, for multiple cracks with the same ordinate and darker crack imaging, a new method needs to be proposed for judgment.

For the two crack distributions in Fig. 4.49, in the added forward NEWS process after the original NEWS-TR process, at $x = 120$ mm, $x = 135$ mm, and $x = 150$ mm,

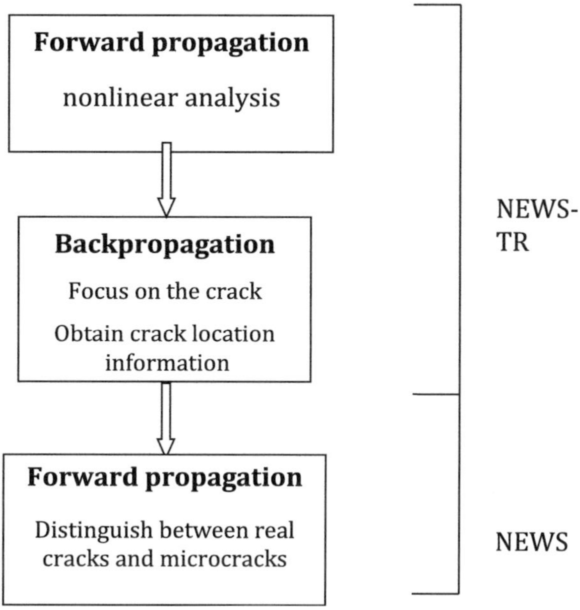

Fig. 4.51 NEWS-TR-NEWS

the y-axis direction is scanned respectively, and the speed distribution is shown in Fig. 4.52. According to the actual position of the crack, peaks appear at $x = 120$ mm and $x = 150$ mm, and the speed peak at $x = 150$ mm is greater than the speed peak at $x = 135$ mm, so it can be judged that in Fig. 4.49, x = The bright spot at 150 mm is the imaging of a real crack, not caused by the propagation of higher harmonics generated previously.

On the other hand, if the two bright spots between the two real cracks in Fig. 4.49 are real cracks, the velocity distribution at $x = 135$ mm should have a double-peak waveform. However, the velocity waveform scanned at $x = 135$ mm in Fig. 4.52 does not show a double peak. Its peak is caused by the propagation of higher harmonics generated at $x = 120$ mm, so these two bright spots can be judged as false cracks.

For the case where the x-coordinates of the two cracks are the same, when the center coordinates are distributed as shown in Fig. 4.46, scans are performed along the y axis at $x = 148$ mm and $x = 158$ mm in the newly added forward process, and the maximum velocity distribution is shown in Fig. 4.53a, b. In Fig. 4.53b, because the two cracks are real, the waveform is a clear double peak; while in Fig. 4.53a, there is no obvious peak and the vibration speed is very small, so the bright spot in the corresponding position in Fig. 4.46 can be judged as a false crack.

In addition, in Fig. 4.50, the crack imaging result at the center at (147.5, 61.5) is obviously dimmer than the crack at (177.5, 83.5), making it difficult to directly judge its authenticity. According to the NEWS-TR-NEWS method, scans are performed along the y axis at $x = 178$ mm and $x = 150$ mm, and the results are shown in Fig. 4.54. Since there are obvious peaks at both places, it can be judged that both are real cracks, thus avoiding the omission of cracks in darker imaging areas.

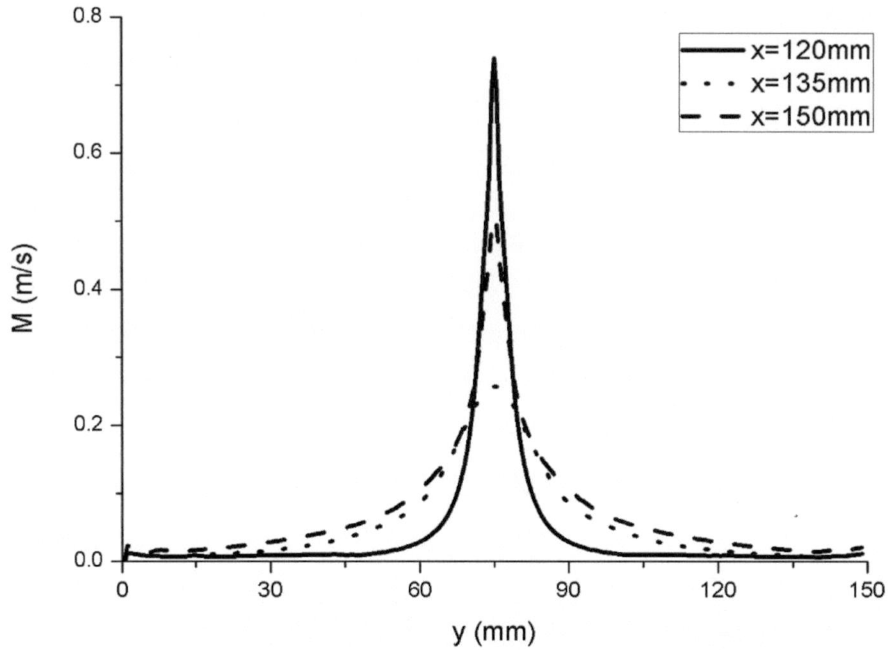

Fig. 4.52 The maximum velocity distribution along the y axis at $x = 120$ mm, $x = 135$ mm, and $x = 150$ mm when the center positions of the two cracks are (117.5, 75) and (147.5, 75)

It should be noted that even in the case of only one defect, if the spatial sampling rate is insufficient, the time-reversed sound field will produce sidelobes, which will also lead to the generation of illusions. As in the ultrasonic linear array, if the size of a single transducer unit is too large, sidelobes will be produced during focusing. The generation of sidelobes is unrelated to the object to be imaged and is caused by the transducer array used, which is different from the false cracks discussed in this section. The false cracks discussed in this section are caused by: during the time reversal process, the multiple high-order harmonic wavefronts that are about to focus on the original multiple defects meet and overlap during propagation, causing instantaneous large vibrations, and then captured and imaged using the maximum value imaging method. Therefore, the false cracks considered in this section are mainly caused by the influence of multiple cracks being too close to the focusing of higher harmonics.

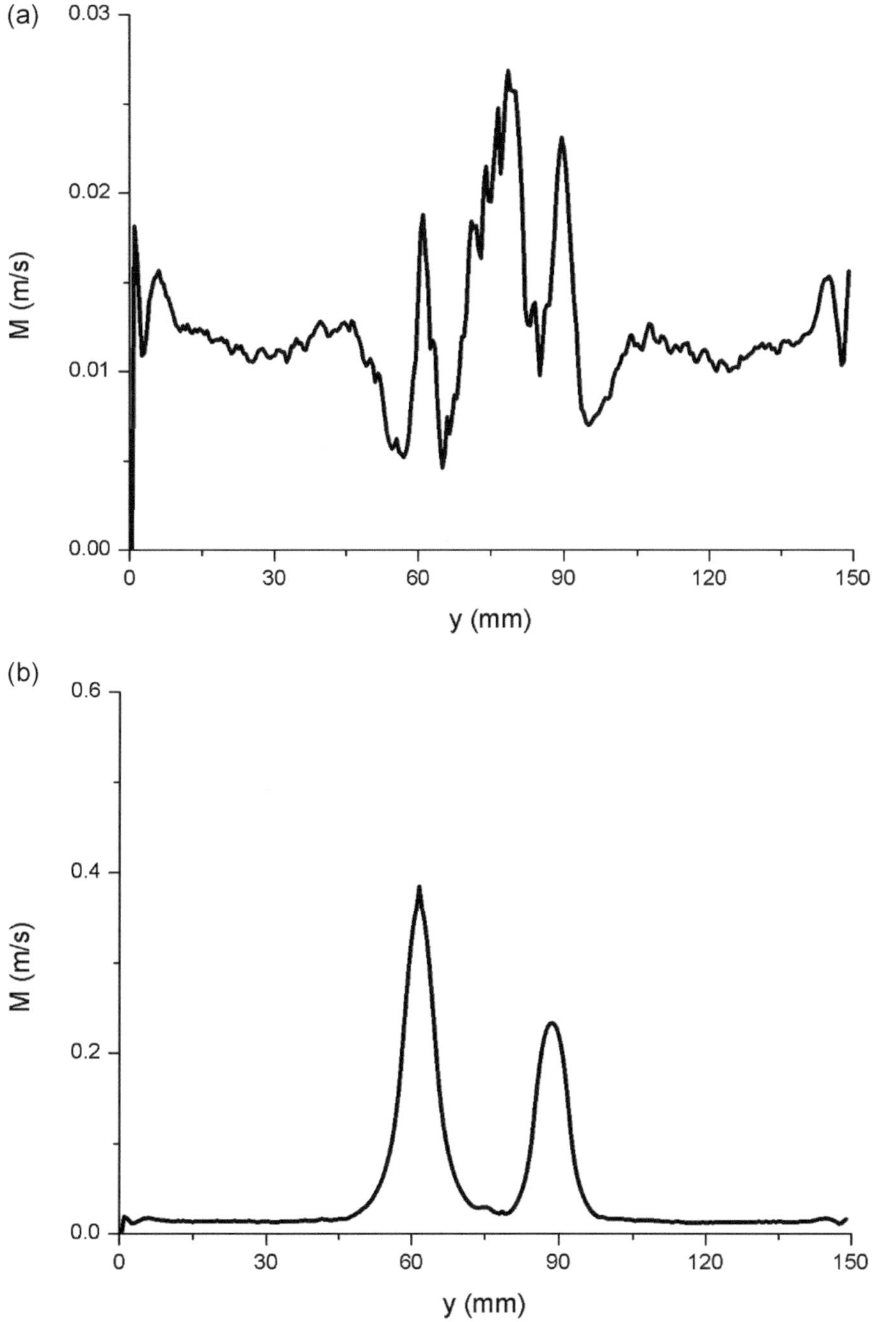

Fig. 4.53 **a** The maximum velocity distribution along the y axis at $x = 148$ mm when the center positions of the two cracks are (157.5, 64) and (157.5, 86). **b** The maximum velocity distribution along the y axis at $x = 158$ mm when the center positions of the two cracks are (157.5, 64) and (157.5, 86)

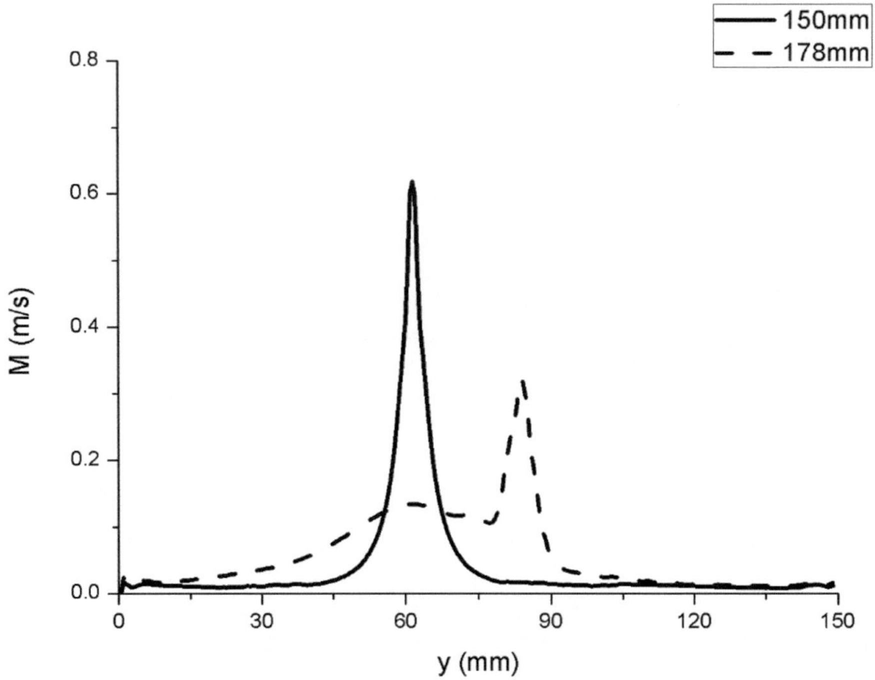

Fig. 4.54 The maximum velocity distribution along the y axis at $x = 178$ mm and $x = 150$ mm when the center positions of the two cracks are (147.5, 61.5) and (177.5, 83.5)

References

1. Shah AA, Ribakov Y, Hirose S. Nondestructive evaluation of damaged concrete using nonlinear ultrasonics. Mater Design. 2009;30:775–82.
2. Zhenggan Z, Haiwei F. Research progress of ultrasonic guided wave detection technology. Nondestruct Test. 2006;28(2):57–63.
3. Liu Z. Guided wave technology in ultrasonic nondestructive testing. Nondestruct Test. 1999;21(8):375–76.
4. Zhenggan Z, Dong W. Development of air-coupled ultrasonic nondestructive testing technology. J Mechan Eng. 2008;44(6):10–4.
5. Xiaozhou L, Jinlin Z, Chang Y, et al. Ultrasonic study of harmonic characteristics of nonlinear meso-elastic solid materials such as rocks. Adv Phys. 2006;26(3):386–9.
6. Solodov I, Wackerl J, Pfleider K, Busse G. Nonlinear self-modulation and subharmonic acoustic spectroscopy for damage detection and location. Appl Phys Lett. 2004;84(26):5386–88.
7. Du G, Zhu Z, Gong X. Acoustic fundamentals. Nanjing: Nanjing University Press; 2001.
8. Li D, Gao G, Dong J. Review of the current status of the application of nonlinear acoustics and time-reversed acoustics in material defect identification. J Mechan Eng. 2009;45(1):1–8.
9. Campos-Pozuelo C, Vanhille C, Juan A, et al. Nonlinear elastic behavior and ultrasonic fatigue of metals, vol. 3. Universality of Nonclassical Nonlinearity, 2006. p. 443–65.
10. Nazarov VE, Kolpakov AB, Radostin AV. Amplitude dependent internal friction and generation of harmonics in granite resonator. Acoust Phys. 2009;55(1):100–7.
11. Debaditya DD, Hoon SH, Kent AK, et al. A nonlinear acoustic technique for crack detection in metallic structures. Struct Health Monit. 2009;8(3):251–62.

12. Ostrovsky L, Johnson PA. Dynamic nonlinear elasticity in geomaterials. La Rivista Del Nuovo Cimento. 2008;24(7):1–46.
13. Preisach F. Uber die magnetische Nachwirkung. "About the magnetic after effect" Z. Phys. 1935;94:277–302.
14. Mayergoyz ID. Hysteresis models from the mathematical and control theory points of view. J Appl Phys. 1985;57:3803–5.
15. McCall KR, Guyer RA. Equation of state and wave propagation in hysteretic nonlinear materials. J Geophys Res. 1994;99:23887–97.
16. McCall KR, Guyer RA. A new theoretical paradigm to describe hysteresis, discrete memory and nonlinear elastic wave propagation in rock. Nonlin Process Geophys. 1996; 89–101.
17. Van Den Abeele K. Multi-mode nonlinear resonance ultrasound spectroscopy for defect imaging: an analytical approach for the one-dimensional case. J Acoust Soc Am. 2007;122:73–90.
18. Van Den Abeele K. Resonant bar simulations in media with localized damage. Ultrasonics. 2004;42:1017–24.
19. Vanaverbeke, Van Den Abeele K. Two-dimensional modeling of wave propagation in materials with hysteretic nonlinearity. J Acoust Soc Am. 2007;122:58–72.
20. Donskoy D, Sutin A, Ekimov A. Nonlinear acoustic interaction on contact interfaces and its use for nondestructive testing. NDT&E Int. 2001;34:231–8.
21. Graves RW. Simulating seismic wave propagation in 3D elastic media using staggered–grid finite differences. Bull Seismol Soc Am. 1996;86(4):1091–106.
22. Fellinger P, Marklein R, Langenberg KJ, et al. Numerical modeling of elastic wave propagation and scattering with EFIT-elastodynamic finite integration technique. Wave Motion. 1995;21:47–66.
23. Helbig K, Rasolofosaon PNJ. A new theoretical paradigm to describe hysteresis and nonlinear elasticity in arbitrary anisotropic rocks. In: Anisotropy fractures. Converted waves and case studies. 2000, p. 383–98.
24. Fuyuki M, Matsumoto Y. Finite difference analysis of Rayleigh wave scattering at a trench. Bull Seismol Soc Am. 1980;70(6):2051–69.
25. Fink M. Time reversal of ultrasonic fields—Part I: basic principles. IEEE Trans Ultrason Ferroelectr Freq Control. 1992;39(5):555–66.
26. Goursolle T, Callé S, Santos SD. A two-dimensional pseudospectral model for time reversal and nonlinear elastic wave spectroscopy. J Acoust Soc Am. 2007;122(6):3220–30.
27. Ulrich TJ, Johnson PA, Guyer RA. Interaction dynamics of elastic waves with a complex nonlinear scatterer through the use of a time reversal mirror. Phys Rev Lett. 2007; 98(10).
28. Cassereau D, Wu F, Fink M. Limits of self-focusing using closed time-reversed cavities and mirrors—theory and experiment. In: Proceeding IEEE ultrasonic symposium; 1990. p. 1613–18.
29. Cassereau D, Fink M. Time-reversal of ultrasonic fields-Part III: theory of the closed time-reversal cavity. IEEE Trans Ultrason Ferroelectr Freq Control. 1992;39(5):579–92.
30. Anderson BE, Griffa M, Larmat C, Ulrich TJ, Johnson PA. Time reversal. Acoust Today. 2008;4:5–15.
31. McCall KR, Guyer RA. Equation of state and wave propagation in hysteretic nonlinear elastic materials. J Geophys Res. 1994;99:23887–97.
32. Gliozzi AS, Griffa M, Scalerandi M. Efficiency of time-reversed acoustics for nonlinear damage detection in solids. J Acoust Soc Am. 2006;120:2506–17.
33. Simpson DH, Chin CT, Burns PN. Pulse inversion Doppler: a new method for detecting nonlinear echoes from microbubble contrast agents. IEEE Trans Ultrason Ferroelectr Freq Control. 1999;46:372–82.

Chapter 5
Nonlinear Theory and Experimental Study of Sound Waves in Concrete

As is well known, concrete, especially high-performance concrete, is a complex substance. Due to the widespread use of concrete, non-destructive testing of concrete has received widespread attention. There are many methods for non-destructive testing of concrete, such as ultrasonic pulse method, vibration method, and acoustic emission technology method. Ultrasonic technology is an important method for testing concrete. However, so far, these methods are all based on ultrasonic linear features. For example, Schubert conducted numerical simulations of ultrasound propagation in concrete [1]. Popovics and Rose summarized the methods for testing concrete [2]. Etsuzo studied the relationship between sound speed and the water content of concrete [3]. In addition, strength is generally considered an important characteristic of concrete. Cracks are an important reason affecting the strength of concrete, and the strength of concrete is related to the cracks in it [4], but traditional methods cannot provide a reliable method for testing concrete. Therefore, there is a strong need to provide a more accurate method to test the strength of concrete.

Due to the acoustic nonlinearity in solid media, the nonlinear phenomena in solids are very meaningful. The feasibility of this method is based on the relationship between acoustic characteristics and cracks. Research results show that porous media, such as rocks and concrete, are highly nonlinear media [5]. Concrete is between solid and liquid, and experimental results show that concrete is a non-classical hysteresis elastic material. Entahar et al. [6] studied the fast and slow dynamic characteristics in concrete. Lacouture et al. [7] used linear and nonlinear methods to study the characteristics of concrete. The drying of cement affecting the hardness of concrete is an important research topic [8], however, the nonlinear characteristics of dried concrete are affected by cracks to a certain extent [9].

The mechanism of concrete sulfate erosion can be divided into physical erosion and chemical erosion. There are mainly three views on the mechanism of physical erosion, summarized by Thaulow and Sahu [10] as: the pressure growth theory caused by the solid phase growth during the conversion of anhydrous sulfate crystals

X. Liu, *Nonlinear Sound Waves in Solids*,
Advances in Acoustics, https://doi.org/10.1007/978-981-96-5253-2_5

to hydrated sulfate crystals, the crystalline water pressure theory, and the salt crystallization pressure theory. The mechanism of chemical erosion is mainly the external diffusion of sulfate ions into the concrete or the sulfate ions carried by the concrete itself, which react with the main components in the concrete such as calcium silicate hydrate, calcium hydroxide, and calcium aluminate hydrate. The main reaction products are gypsum, ettringite, etc., which generate enough stress to cause concrete damage.

Cody [11] and others compared the expansion of concrete under three conditions in a sodium sulfate solution: continuous soaking, wet-dry cycling, and freeze–thaw cycling. The results showed that the maximum was in wet-dry cycling, followed by freeze–thaw cycling, and the minimum was in continuous soaking. Among the 5 current American concrete sulfate erosion destruction test methods proposed by the American Materials and Testing Association, the wet-dry cycle alternating method is also one of them [12]. Qiao et al. [13] found that when studying the dynamic modulus of elasticity of fly ash concrete in a sulfate environment, after the wet-dry cycle in drinking water, the dynamic modulus of elasticity of the concrete decreased after the cycle, indicating that the wet-dry cycle caused damage to the concrete, and the concrete in the wet-dry cycle environment was more severely eroded by sulfate and salt crystallization than the concrete in the persistent wet environment.

The mix ratio of concrete has a great impact on the strength of structural concrete, and the mix ratio of concrete includes many aspects, such as water-cement ratio, aggregate gradation, and the addition ratio of fly ash or polypropylene fiber materials, etc. The water-cement ratio is a very important factor affecting the strength of concrete. The size of the water-cement ratio determines the fluidity of the slurry, determines whether the slurry can evenly wrap the aggregate without flowing, thereby affecting the strength. If the water-cement ratio is too small, the fluidity of the slurry is too low to wrap the aggregate. If the water-cement ratio is too large, the fluidity of the slurry increases, the cement slurry begins to drip down, and the cement wrapping aggregate will be uneven, forming a dense lower part of the specimen with a low porosity, and the upper part is just the accumulation of aggregate, lacking cementation, and the overall strength is very low. Literature [14] Tests were conducted on the strength of ordinary concrete without the addition of water-reducing agents under different water-cement ratios. It was concluded that the strength of the concrete is highest when the water-cement ratio is 0.4. The coarse aggregate particles in the concrete are bonded together by hardened cement slurry. According to the concrete structure model, it can be analyzed that the strength of the concrete mainly has the following characteristics: First, the strength of the coarse aggregate determines the main strength of the concrete skeleton. Coarse aggregate is the material that occupies the most volume in concrete, and it is bonded together by cement slurry to form the skeleton of the concrete. Coarse aggregate is also the main load-bearing material. Second, the mechanical properties of the cement slurry shell have a great impact on the strength of the concrete. When the concrete is under stress, there is a tendency for the particles of the coarse aggregate to slide along the interface, and the cement slurry shell can play a restraining and limiting role where the particles contact each other.

Therefore, the microscopic morphology of the coarse aggregate-slurry shell interface has a great impact on the interface strength. Therefore, the larger the bonding area between the coarse aggregate particles, the more bonding points, the greater the overall strength of the concrete. It can be seen from the above that optimizing the concrete mix ratio and increasing the number of cement slurry bonding points between coarse aggregate particles is the key to improving the strength of concrete. This chapter studies the nonlinear propagation of sound waves in concrete under different degrees of damage, different water content, sulfate erosion, and different mix ratios.

5.1 Nonlinear Theoretical Model of Sound Wave Propagation in Concrete

The Poisson's ratio of concrete is between 0.14 and 0.17, which is much higher than that of ordinary metals. Therefore, the transverse vibration caused by the longitudinal wave excitation of the concrete rod can be ignored. When the longitudinal wave propagates along the length direction of the concrete rod, the sound wave equation can be written as:

$$\rho U_{tt} = \sigma_x(\varepsilon) \tag{5.1}$$

U is the vibration displacement of the concrete particles. ρis the density of the concrete. σ is stress, ε is strain. t is time.

In non-classical nonlinear theory, there is a hysteresis relationship between stress and strain, as shown in Fig. 5.1. The basic law of stress and strain when the least two parameters method is used to describe the hysteresis relationship in (5.2).

$$\sigma(\varepsilon) = \begin{cases} E\left(\varepsilon - \frac{\gamma_1}{2}\varepsilon^2\right), \varepsilon > 0, \varepsilon_t > 0; \\ E\left(1 - \frac{\gamma_1}{2}\varepsilon_0\right), \varepsilon > 0, \varepsilon_t < 0 \\ E\left(\varepsilon + \frac{\gamma_2}{2}\varepsilon^2\right), \varepsilon < 0, \varepsilon_t < 0 \\ E\left(1 - \frac{\gamma_2}{2}\varepsilon_0\right), \varepsilon\langle 0, \varepsilon_t\rangle 0 \end{cases} \tag{5.2}$$

ε_0 is the strain amplitude, γ_1, γ_2 is the small change of Hooke's law, $\gamma_1 \neq \gamma_2$ reflects the asymmetry of compression waves and expansion waves.

Under external excitation, the wave equation of the longitudinal wave in the concrete rod $\rho U_{tt} = \sigma_x(\varepsilon)$ becomes

$$U_{tt} - c^2 U_{\mathrm{xx}} + \mu U_{\mathrm{ttt}} = c^2 F(U) \tag{5.3}$$

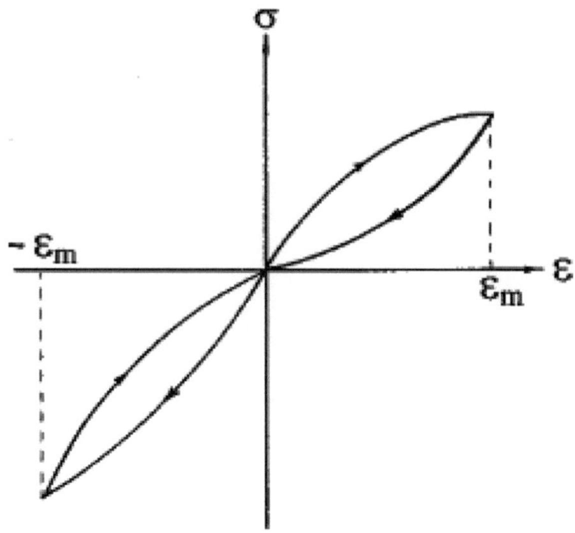

Fig. 5.1 Hysteresis phenomenon of stress and strain relationship

$$F(U) = \begin{cases} -\gamma_1 U_x U_{xx};\ U_x > 0,\ U_{xt} > 0 \\ -\frac{\gamma_1}{2}(\varepsilon_0 U_{xx} + \varepsilon_{0x} U_x);\ U_x > 0,\ U_{xt} < 0 \\ \gamma_2 U_x U_{xx};\ U_x < 0,\ U_{xt} < 0 \\ -\frac{\gamma_2}{2}(\varepsilon_0 U_{xx} + \varepsilon_{0x} U_x);\ U_x \langle 0,\ U_{xt} \rangle 0 \end{cases} \tag{5.4}$$

U is the displacement of the particles, c is the sound speed of the longitudinal wave, μU_{tt} is the quantity caused by linear viscosity.

During the experiment, the boundary conditions of the concrete rod are considered to be a rod that is subjected to a harmonic external force at one end and is free at the other end. The boundary conditions can be written as

$$\begin{aligned} U(0, t) &= A_0 \cos \omega t, \\ U(L, t) &= 0 \end{aligned} \tag{5.5}$$

For the convenience of analysis, let

$$V(x, t) = U(x, t) - A_0 \cos \omega t (1 + \sin \mathrm{k}_n x) \tag{5.6}$$

$k_n = 2\pi f_n / c$ is the wave number of the resonant rod at the nth mode resonance frequency. At this time, the boundary condition is

$$V(0, t) = V_x(L, t) = 0 \tag{5.7}$$

Considering the nth mode, substituting (5.7) into (5.3), we get

$$V_{\mathrm{tt}} - c^2 V_{\mathrm{xx}} = c^2 F(V) - \mu V_{\mathrm{ttt}} + A_0 \left\{ \omega^2 + \left(\omega^2 + \omega_n^2 \right) \sin \mathrm{k}_n x \right\} \cos \omega t \tag{5.8}$$

The right side of the above equation includes nonlinear, attenuation, and external force effects. When these three items are very small, the solution of the above equation can be written as

$$V^{(1)} = V_0 \cos[(\omega_n + \delta)t + \Psi] \sin k_n + W(x, t) \tag{5.9}$$

Substituting (5.8) into (5.7), we get

$$\begin{aligned} W_{\mathrm{tt}} - c^2 W_{\mathrm{xx}} = {} & c^2 F\left(V^{(1)}\right) + A_0\left\{\omega^2 + \left(\omega^2 - \omega_n^2\right) \sin \mathrm{k}_n x\right\} \cos \omega t \\ & + \mu_t \omega_n^3 V_0 \sin[(\omega_n + \delta)t + \psi] \sin \mathrm{k}_n x \\ & + 2\omega_n \delta V_0 \cos[(\omega_n + \delta)t + \psi] \sin \mathrm{k}_n x \end{aligned} \tag{5.10}$$

The above equation also satisfies the boundary condition (5.7). To ensure the time-domain invariance of the solution of the above equation, the right side should also satisfy the orthogonality of the characteristic function. For the nonlinear term $F\left(V^{(1)}\right)$, the Fourier analysis is as follows

$$\begin{aligned} V^{(1)} &= V_0 \cos[(\omega_n + \delta)t + \phi] \sin k_n x \\ V^{(1)} &= V_0 \cos \Theta \sin k_n x \\ V_x^{(1)} &= V_0 k_n \cos \Theta \cos k_n x \\ V_{xx}^{(1)} &= -V_0 k_n^2 \cos \Theta \sin k_n x \\ V_x^{(1)} V_{xx}^{(1)} &= -V_0^2 k_n^3 \cos^2 \Theta \sin k_n x \cos k_n x \quad ((\omega_n + \delta)t + \phi = \Theta) \end{aligned} \tag{5.11}$$

$$\begin{aligned} \varepsilon &= \frac{\partial V}{\partial x} = V_0 k_n \cos \Theta \cos k_n x \\ \varepsilon_0 &= V_0 k_n \cos k_n x \\ \varepsilon_{0x} &= \frac{\partial \varepsilon_0}{\partial x} = -V_0 k_n^2 \sin k_n x \end{aligned} \tag{5.12}$$

On the length of the concrete resonant rod, according to kl = Pi/2, the length L is divided into N = 2n − 1 sections.

(1) When

$$\Theta \in \left(-\frac{\pi}{2}, 0\right), \quad F = \gamma_1 V_0^2 k_n^3 \cos^2 \Theta \sin k_n x \cos k_n x \tag{5.13}$$

$$\frac{1}{\pi L} \int_{-\frac{\pi}{2}}^{0} \int_{0}^{L} \gamma_1 V_0^2 k_n^3 \cos^3 \Theta \sin^2 k_n x \cos k_n x d\Theta dx$$

$$= V_0^2 k_n^3 * \frac{1}{\pi} \int_{-\frac{\pi}{2}}^{0} \gamma_1 \cos^3 \Theta d\Theta * \frac{1}{L} \int_0^L \sin^2 k_n x \cos k_n x dx$$

$$= V_0^2 k_n^3 * \frac{2}{3\pi} \gamma_1 * \frac{2}{3\pi} \tag{5.14}$$

$$\frac{1}{\pi} \int_{-\frac{\pi}{2}}^{0} \gamma_1 \cos^3 \Theta d\Theta = \frac{2}{3\pi} \gamma_1 \tag{5.15}$$

Let $k_n L' = \frac{\pi}{2}$, since $k_n L = \frac{\pi}{2}(2n-1)$, therefore $L = (2n-1) \times L'$.

$$\frac{1}{L} \int_0^L \sin^2 k_n x \cos k_n x dx = \frac{1}{k_n L} \int_0^L \sin^2 k_n x d \sin k_n x$$

$$= \frac{(2n-1)}{k_n L} \int_0^{L'} \sin^2 k_n x d \sin k_n x \tag{5.16}$$

$$= \frac{(2n-1)}{k_n L} * \frac{1}{3} * \sin^3 k_n L'$$

$$= \frac{1}{k_n L} * \frac{1}{3} = \frac{2}{3\pi}$$

$$\frac{1}{\pi L} \gamma_1 V_0^2 k_n^3 \int_{-\frac{\pi}{2}}^{0} \cos^2 \Theta \sin \Theta d\Theta \int_0^L \sin^2 k_n x \cos k_n x dx$$

$$= V_0^2 k_n^3 * \gamma_1 * \frac{1}{\pi} \int_{-\frac{\pi}{2}}^{0} \cos^2 \Theta \sin \Theta d\Theta * \frac{1}{L} \int_0^L \sin^2 k_n x \cos k_n x dx$$

$$= V_0^2 k_n^3 \gamma_1 * -\frac{1}{3\pi} * \frac{2}{3\pi}$$

$$\frac{1}{\pi} \int_{-\frac{\pi}{2}}^{0} \cos^2 \Theta \sin \Theta d\Theta = -\frac{1}{3\pi} \tag{5.17}$$

$$\frac{1}{L} \int_0^L \sin^2 k_n x \cos k_n x dx = \frac{2}{3\pi} \tag{5.18}$$

(2) When

$$\Theta \in \left(0, \frac{\pi}{2}\right), \quad F\left(V^{(1)}\right) = \gamma_1 V_0^2 k_n^3 \cos\Theta \sin k_n x \cos k_n x \tag{5.19}$$

$$\frac{1}{\pi L}\int_0^{\pi/2}\int_0^L \gamma_1 V_0^2 k_n^3 \cos^2\Theta \sin^2 k_n x \cos k_n x d\Theta dx$$

$$= V_0^2 k_n^3 * \frac{1}{\pi}\int_0^{\frac{\pi}{2}} \gamma_1 \cos^2\Theta d\Theta * \frac{1}{L}\int_0^L \sin^2 k_n x \cos k_n x dx$$

$$= V_0^2 k_n^3 * \frac{2}{3\pi} * \frac{\gamma_1}{4}$$

$$\frac{1}{L}\int_0^L \sin^2 k_n x \cos k_n x dx = \frac{2}{3\pi}$$

$$\frac{1}{\pi}\int_0^{\pi/2} \gamma_1 \cos^2\Theta d\Theta = \frac{\gamma_1}{4} a \tag{5.20}$$

$$\frac{1}{\pi L}\int_0^{\pi/2}\int_0^L \gamma_1 V_0^2 k_n^3 \cos\Theta \sin\Theta \sin^2 k_n x \cos k_n x d\Theta dx$$

$$= V_0^2 k_n^3 * \frac{1}{\pi}\int_0^{\frac{\pi}{2}} \gamma_1 \cos\Theta \sin\Theta d\Theta * \frac{1}{L}\int_0^L \sin^2 k_n x \cos k_n x dx$$

$$= V_0^2 k_n^3 * \frac{2}{3\pi} * \frac{\gamma_1}{2\pi}$$

$$\frac{1}{\pi}\int_0^{\pi/2} \gamma_1 \cos\Theta \sin\Theta d\Theta = \frac{\gamma_1}{2\pi} \tag{5.21}$$

$$\frac{1}{L}\int_0^L \sin^2 k_n x \cos k_n x dx = \frac{2}{3\pi} \tag{5.22}$$

(3) When

$$\Theta \in \left(\frac{\pi}{2}, \pi\right), \quad F\left(V^{(1)}\right) = \gamma_2 V_x^{(1)} V_{xx}^{(1)} = -\gamma_2 V_0^2 k_n^3 \cos^2\Theta \sin k_n x \cos k_n x \tag{5.23}$$

$$
\begin{aligned}
& -\frac{1}{\pi L}\gamma_2 V_0^2 k_n^3 \int\limits_{\frac{\pi}{2}}^{\pi} \cos^3\Theta d\Theta \int\limits_0^L \sin^2 k_n x \cos k_n x dx \\
& = -V_0^2 k_n^3 * \frac{1}{L}\int\limits_0^L \sin^2 k_n x \cos k_n x dx * \frac{1}{\pi}\int\limits_{\frac{\pi}{2}}^{\pi} \gamma_2 \cos^3\Theta d\Theta \\
& = V_0^2 k_n^3 * \frac{2}{3\pi} * \frac{2}{3\pi}\gamma_2
\end{aligned} \tag{5.24}
$$

$$
\begin{aligned}
& \frac{1}{\pi}\int\limits_{\frac{\pi}{2}}^{\pi} \gamma_2 \cos^3\Theta d\Theta = -\frac{2}{3\pi}\gamma_2 \\
& \frac{1}{L}\int\limits_0^L \sin^2 k_n x \cos k_n x dx = \frac{2}{3\pi}
\end{aligned} \tag{5.25}
$$

$$
\begin{aligned}
& -\frac{1}{\pi L}\gamma_2 V_0^2 k_n^3 \int\limits_{\frac{\pi}{2}}^{\pi} \cos^2\Theta \sin\Theta d\Theta \int\limits_0^L \sin^2 k_n x \cos k_n x dx \\
& = -V_0^2 k_n^3 * \frac{1}{L}\int\limits_0^L \sin^2 k_n x \cos k_n x dx * \frac{1}{\pi}\int\limits_{\frac{\pi}{2}}^{\pi} \gamma_2 \cos^2\Theta \sin\Theta d\Theta \\
& = V_0^2 k_n^3 * \frac{2}{3\pi} * -\frac{1}{3\pi}\gamma_2
\end{aligned} \tag{5.26}
$$

$$
\begin{aligned}
& \frac{1}{\pi}\int\limits_{\frac{\pi}{2}}^{\pi} \gamma_2 \cos^2\Theta \sin\Theta d\Theta = \frac{1}{3\pi}\gamma_2 \\
& \frac{1}{L}\int\limits_0^L \sin^2 k_n x \cos k_n x dx = \frac{2}{3\pi}
\end{aligned}
$$

(4) When

$$
\Theta \in \left(\pi, \frac{3\pi}{2}\right), \quad F\left(V^{(1)}\right) = \gamma_2 V_0^2 k_n^3 \cos\Theta \sin k_n x \cos k_n x \tag{5.27}
$$

$$
\frac{1}{\pi L}\int\limits_{\pi}^{3\pi/2}\int\limits_0^L \gamma_2 V_0^2 k_n^3 \cos^2\Theta \sin^2 k_n x \cos k_n x d\Theta dx
$$

$$= V_0^2 k_n^3 * \frac{1}{\pi}\int\limits_{\pi}^{\frac{3\pi}{2}} \gamma_2 \cos^2 \Theta d\Theta * \frac{1}{L}\int\limits_0^L \sin^2 k_n x \cos k_n x dx$$

$$= V_0^2 k_n^3 * \frac{2}{3\pi} * \frac{\gamma_2}{4} \tag{5.28}$$

$$\frac{1}{L}\int\limits_0^L \sin^2 k_n x \cos k_n x dx = \frac{2}{3\pi}$$

$$\frac{1}{\pi}\int\limits_{\pi}^{\frac{3\pi}{2}} \gamma_2 \cos^2 \Theta d\Theta = \frac{\gamma_2}{4}$$

$$\frac{1}{\pi L}\int\limits_{\pi}^{3\pi/2}\int\limits_0^L \gamma_2 V_0^2 k_n^3 \cos\Theta \sin\Theta \sin^2 k_n x \cos k_n x d\Theta dx$$

$$= V_0^2 k_n^3 * \frac{1}{\pi}\int\limits_{\pi}^{\frac{3\pi}{2}} \gamma_2 \cos\Theta \sin\Theta d\Theta * \frac{1}{L}\int\limits_0^L \sin^2 k_n x \cos k_n x dx$$

$$= V_0^2 k_n^3 * \frac{2}{3\pi} * \frac{\gamma_2}{2\pi} \tag{5.29}$$

$$\begin{aligned} &\frac{1}{L}\int\limits_0^L \sin^2 k_n x \cos k_n x dx = \frac{2}{3\pi} \\ &\frac{1}{\pi}\int\limits_{\pi}^{3\pi/2} \gamma_2 \cos\Theta \sin\Theta d\Theta = \frac{\gamma_2}{2\pi} \end{aligned} \tag{5.30}$$

Combining the four cases, we can get the Fourier analysis of the nonlinear term $F(V^1)$ in Eq. (5.10) as

$$\begin{aligned} &\cos: V_0^2 k_n^3 \frac{(8+3\pi)(\gamma_1+\gamma_2)}{18\pi^2} = V_0^2 k_n^3 \frac{2}{3\pi} a_1 \\ &\sin: V_0^2 k_n^3 \frac{(\gamma_1+\gamma_2)}{9\pi^2} = V_0^2 k_n^3 \frac{2}{3\pi} b_1 \\ &a_1 = \frac{(8+3\pi)(\gamma_1+\gamma_2)}{12\pi}, b_1 = \frac{(\gamma_1+\gamma_2)}{6\pi} \end{aligned} \tag{5.31}$$

The Fourier analysis of in Eq. (5.10) is

$$\frac{1}{\pi L} \int_0^{2\pi \int_n \begin{pmatrix} \cos\Theta \\ \sin\Theta \end{pmatrix}} \int_0^{L \int} A_0 \cos\omega t\left\{\omega^2 + \left(\omega^2 - \omega_n^2\right)\sin k_n x\right\} \sin$$

$$= \frac{A_0}{\pi}\frac{1}{L}\int_0^L \left\{\omega^2 + \left(\omega^2 - \omega_n^2\right)\sin k_n x\right\}\sin k_n x dx * \int_0^{2\pi} \cos\omega t \begin{pmatrix} \cos\Theta \\ \sin\Theta \end{pmatrix} d\Theta *$$

$$= \frac{A_0}{\pi} * \frac{\omega_n^2}{\frac{(2n-1)\pi}{2}} * \pi \begin{pmatrix} \cos\psi \\ \sin\psi \end{pmatrix}$$

$$= \frac{A_0 \omega_n^2}{\pi(n - 1/2)} \begin{pmatrix} \cos\psi \\ \sin\psi \end{pmatrix} \tag{5.32}$$

$$\int_0^{2\pi} \cos\omega t \begin{pmatrix} \cos\Theta \\ \sin\Theta \end{pmatrix} d\Theta$$

$$= \int_0^{2\pi} \cos(\Theta - \psi) \begin{pmatrix} \cos\Theta \\ \sin\Theta \end{pmatrix} d\Theta(\Theta = \omega t + \psi)$$

$$= \pi * \begin{pmatrix} \cos\psi \\ \sin\psi \end{pmatrix} \tag{5.33}$$

$$\frac{1}{L}\int_0^L \left\{\omega^2 + \left(\omega^2 - \omega_n^2\right)\sin k_n x\right\}\sin k_n x dx$$

$$= \frac{1}{L}\int_0^L \left\{\omega_n^2\right\}\sin k_n x dx \ (\omega = \omega_n)$$

$$= \frac{(2n-1)\omega_n^2}{k_n L} * \int_0^{\frac{\pi}{2}} \sin\alpha d\alpha; \alpha = k_n x \in \left(0, (2n-1)\frac{\pi}{2}\right),$$

$$= \frac{\omega_n^2}{\pi/2} \tag{5.34}$$

The Fourier analysis of the term $\mu_1 \omega_n^2 V_0 \sin\Theta \sin k_n x$ in Eq. (5.10) is

$$\frac{1}{\pi L} \int_0^{2\pi \int_n \begin{pmatrix} \cos\Theta \\ \sin\Theta \end{pmatrix}} \int_0^{L \int} \mu_1 \omega_n^3 V_0 \sin\Theta \sin k_n x \sin$$

$$= \frac{\mu_1\omega_n^3 V_0}{\pi} * \frac{1}{2} * \begin{pmatrix} 0 \\ \pi \end{pmatrix} = \begin{pmatrix} 0 \\ \frac{\mu_1\omega_n^3 V_0}{2} \end{pmatrix} \tag{5.35}$$

$$\frac{1}{L}\int_0^L \sin k_n x \sin k_n x dx$$

$$= \frac{(2n-1)}{k_n L}\int_0^{\frac{\pi}{2}} \sin\alpha \sin\alpha d\alpha \quad \alpha = k_n x \in \left(0, (2\text{n}-1)\frac{\pi}{2}\right),$$

$$= \frac{(2n-1)}{\frac{(2n-1)\pi}{2}} * \frac{\pi}{4} = \frac{1}{2}$$

$$\int_0^{2\pi} \sin\Theta \begin{pmatrix} \cos\Theta \\ \sin\Theta \end{pmatrix} d\Theta = \begin{pmatrix} 0 \\ \pi \end{pmatrix} \tag{5.36}$$

In Eq. (5.10), the Fourier analysis of the term $2\omega_n\delta V_0 \cos\Theta \sin k_n x$ is

$$\frac{1}{\pi L}\int_0^{2\pi \int_n \begin{pmatrix} \cos\Theta \\ \sin\Theta \end{pmatrix}} \int_0^{L\int} 2\omega_n\delta V_0 \cos\Theta \sin k_n x \sin$$

$$= \frac{2\omega_n\delta V_0}{\pi} * \frac{1}{2} * \begin{pmatrix} \pi \\ 0 \end{pmatrix} = \begin{pmatrix} \omega_n\delta V_0 \\ 0 \end{pmatrix} \tag{5.37}$$

$$\frac{1}{L}\int_0^L \sin k_n x \sin k_n x dx = \frac{1}{2}$$

$$\int_0^{2\pi} \cos\Theta \begin{pmatrix} \cos\Theta \\ \sin\Theta \end{pmatrix} d\Theta = \begin{pmatrix} \pi \\ 0 \end{pmatrix} \tag{5.38}$$

Let the Fourier analysis on the right side of Eq. (5.10) be zero, thus obtaining the equation about amplitude and phase:

$$\begin{aligned} \frac{2a_1 V_0^2\omega_n^2}{3\pi c} + \delta V_0 &= -\frac{A_0\omega_n}{\pi\left(n-\frac{1}{2}\right)}\cos\psi \\ \frac{2b_1 V_0^2\omega_n^2}{3\pi c} + \frac{\mu_1\omega_n^2 V_0}{2} &= -\frac{A_0\omega_n}{\pi(n-1/2)}\sin\psi \end{aligned} \tag{5.39}$$

The solution to the above system of equations is

$$V_0 = \frac{A_0\left(\frac{c}{L}\right)}{[(\delta - \delta_N)^2 + (\mu_1 + \mu_N)^2\omega_n^2/4]^{\frac{1}{2}}}$$
$$\delta_N = -2(n - 1/2)a_1 V_0 \omega_n / 3L,\ \mu_N = 4(n - 1/2)b_1 V_0 / 3L\omega_n \tag{5.40}$$

$$\frac{1}{\pi L}\int_0^{2\pi L}\int_0^{L} F\left(V^{(1)}\right)\sin \mathrm{k}_n x\begin{pmatrix}\cos\Theta\\ \sin\Theta\end{pmatrix}\mathrm{d}x\mathrm{d}\Theta = \mathrm{V}_0^2 k_n^3 \frac{2}{3\pi}\begin{pmatrix}a_1\\ b_1\end{pmatrix} \tag{5.41}$$

$$a_1 = \frac{(8 + 3\pi)(\gamma_1 + \gamma_2)}{12\pi},\ b_1 = \frac{(\gamma_1 + \gamma_2)}{6\pi}$$
$$\Theta = (\omega_n + \delta)t + \psi$$

Let the Fourier analysis of (5.41) be zero, thus obtaining the equation about amplitude and phase

$$\begin{aligned}\frac{2a_1 V_0^2 \omega_n^2}{3\pi c} + \delta V_0 &= -\frac{A_0 \omega_n}{\pi(n - 1/2)}\cos\psi\\ \frac{2b_1 V_0^2 \omega_n^2}{3\pi c} + \frac{\mu_1 \omega_n^2 V_0}{2} &= -\frac{A_0 \omega_n}{\pi(n - 1/2)}\sin\psi\end{aligned} \tag{5.42}$$

Solve the above system of equations to get

$$V_0 = \frac{A_0(c/L)}{\left[(\delta - \delta_N)^2 + (\mu_1 + \mu_N)^2\omega_n^2/4\right]^{1/2}} \tag{5.43}$$

$$\delta_N = -2(n - 1/2)a_1 V_0 \omega_n / 3L,\ \mu_N = 4(n - 1/2)b_1 V_0 / 3L\omega_n$$
$$\delta_N\text{: nonlinear frequency shift, } \mu_N\text{: nonlinear loss factor} \tag{5.44}$$

When considering the generation of higher harmonics, analyze the harmonic problem through perturbation method. Obtain the displacement equation of the second harmonic at this frequency

$$V_{\mathrm{tt}}^{(2)} - c^2 V_{\mathrm{xx}}^{(2)} + \mu_2 V_{\mathrm{ttt}}^{(2)} = \frac{V_0^2 k_n^3 c^2}{4}\sin 2\mathrm{k}_n x(a_2\cos 2\Theta + \mathrm{b}_2 \sin 2\Theta) \tag{5.45}$$

$$a_2 = \frac{(8 + 3\pi)(\gamma_1 - \gamma_2)}{24\pi},\ b_2 = \frac{(\gamma_1 - \gamma_2)}{6\pi} \tag{5.46}$$

The solution of the second harmonic can be written as

$$V^{(2)} = V_2(x)\sin(2\Theta + \varphi) \tag{5.47}$$

Similar to the analysis of the first harmonic, the equation of amplitude and phase under the second harmonic can be obtained (assuming $\delta = 0$)

$$\left(V_{2xx}+4k_n^2V_2\right)\cos\varphi=-\frac{V_0^2k_n^3b_2}{4}\sin 2k_n\text{x} - 8\mu_2\omega_nk_n^2V_2\sin\phi$$
$$\left(V_{2xx}+4k_n^2V_2\right)\sin\varphi=-\frac{V_0^2k_n^3a_2}{4}\sin 2k_n\text{x} + 8\mu_2\omega_nk_n^2V_2\cos\phi \tag{5.48}$$

Let $\mu_2=0$, eliminate $\cos\varphi$, $\sin\varphi$ term, to get

$$V_{2xx}+4k_n^2V_2=\frac{V_0^2k_n^3}{4}\sin 2k_nx\sqrt{a_2^2+b_2^2} \tag{5.49}$$

According to the boundary conditions, solve the above system of equations to get:

$$V_2(x)=\frac{V_0^2k_n}{32}\sqrt{a_2^2+b_2^2}(\sin 2k_nx-2k_nx\cos 2k_nx)$$

At the free end of the rod

$$k_nL=\frac{\pi}{2}(2n-1),\ \sin 2k_nL=0,\ \cos 2k_nL=-1 \tag{5.50}$$

$$V_2(L)=\frac{V_0^2k_n^2L}{16}\sqrt{a_2^2+b_2^2}$$

Similarly, the wave equation under the third harmonic is

$$\begin{aligned}V_{\text{tt}}^{(3)}-c^2V_{\text{xx}}^{(3)}+\mu_3V_{ttt}^{(3)}&=\frac{V_0^2k_n^3c^2}{4}\sin 2\text{k}_nx(a_3\cos 3\Theta+\text{b}_3\sin 3\Theta)\\&\quad-\text{V}_0c\omega_0a_0\left[V_x^{(3)}\cos k_nx\right]_n'\end{aligned} \tag{5.51}$$

$$a_0=\frac{(\pi+4)(\gamma_1+\gamma_2)}{8\pi}$$
$$a_3=\frac{2(\gamma_1+\gamma_2)}{15\pi},\ b_3=\frac{\gamma_1+\gamma_2}{30\pi}$$

Unlike the second harmonic, the frequency of the third harmonic is close to the resonance frequency, so the solution of (5.51) is written as the following general vibration form of the rod:

$$V^3(x,t)=V_3\sin 3k_nx\sin(3\Theta+\chi) \tag{5.52}$$

Substitute (5.52) into (5.51), the displacement amplitude of the third harmonic at the free end of the concrete rod is obtained

$$V_3=\frac{2(n-1/2)}{45L}\frac{V_0^2\omega_n\sqrt{a_3^2+b_3^2}}{\left[9\mu_3^2\omega_n^4+4(1-27a_0/35a_1)^2\delta_N^2\right]^{1/2}} \tag{5.53}$$

The nonlinear parameters γ_1 and γ_2 can be obtained from the following two formulas

$$|\gamma_1 - \gamma_2| = \frac{16 * V_2}{V_0^2 k_n^2 L \sqrt{\left(\frac{8+3\pi}{24\pi}\right)^2 + \left(\frac{1}{6\pi}\right)^2}} \tag{5.54}$$

$$\gamma_1 + \gamma_2 = \frac{45LV_3\sqrt{9u_3^2\omega_n^4 + 4\left(1 + \frac{81(4+\pi)}{70(8+3\pi)}\right)^2 \delta_N^2}}{(2n-1)\omega_n V_0^2 \sqrt{\left(\frac{2}{15\pi}\right)^2 + \left(\frac{1}{30\pi}\right)^2}} \tag{5.55}$$

V_0, V_2, V_3 is the displacement amplitude of the first, second, and third harmonics at the first resonance frequency.

5.2 Experimental Method

The experimental setup is shown in Fig. 5.2. The transmitting transducer is tightly bound to the mass load, and the cement mortar prism sample is placed on the transmitting transducer. The sample is tied to the experimental bracket near the end of the transmitting transducer, using Vaseline as a coupling agent to ensure good coupling between the sample and the transducer. This end can be considered a rigid boundary. An accelerometer is placed on the upper end of the sample, with an aluminum sheet placed on the accelerometer to ensure good coupling between the sample box and the accelerometer. Since the mass of the aluminum sheet and the accelerometer is small, this end can be considered free.

The experimental system is shown in Fig. 5.3. The signal generator excites a single-frequency pulse signal, which is transmitted to the transmitting transducer

Fig. 5.2 Experimental setup

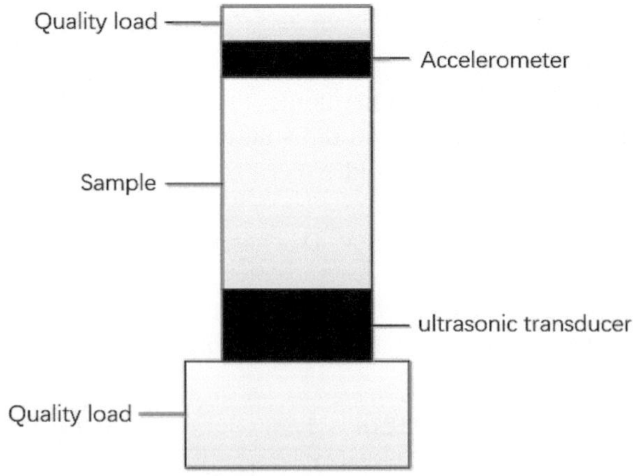

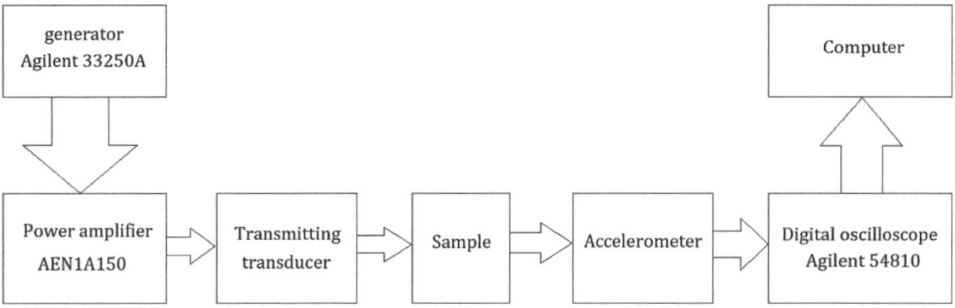

Fig. 5.3 Experimental system

after being amplified by the power amplifier, and the signal is received by the accelerometer. The sensitivity of the accelerometer is 10 mv/g, and the effective bandwidth is from 5 to 100 kHz. The received signal is received by the oscilloscope and collected by LabVIEW. The relationship between acceleration a and the displacement amplitude of the sample end V is:

$$V = \frac{a}{\omega^2} \tag{5.56}$$

5.3 Non-classical Nonlinear Analysis of Concrete Under Different Degrees of Damage

The dimensions of the six cubic concrete test blocks are $8 \times 6 \times 5$ cm^3. We use a nonlinear method to perform non-destructive testing of concrete under the first resonant frequency. Considering the transmission performance of the transmitting transducer and the effective bandwidth of the accelerometer, such a sample size was chosen for the experiment. The mass ratio of cement, sand, and gravel is 1:2.46:4.19, and 1.8 tons of water is injected into each cubic meter of the sample. The well-stirred mud and sand mixture is poured into a prepared wooden box and naturally dried in the sun. Due to the stirring process, a large amount of bubbles will be generated in the mud and sand mixture. In the early stage of drying, the concrete sample is baked under an infrared lamp for an hour, which is very beneficial for degassing.

A week later, the concrete samples are completely dried. A rock grinder from the Department of Geology is used to grind the concrete samples, ensuring that the sizes of the six different concrete samples are consistent and that the smoothness of the acoustic contact surfaces of different samples is consistent. At this time, the measured speed of sound in the concrete sample is 2900 m/s, and the density is 2400 kg/m^3. A 100-ton fatigue testing machine from the Key Laboratory of Concrete Mechanics of the Ministry of Education at Southeast University is used to apply different pressures to the six concrete samples. The type of concrete under this formula is ordinary concrete, and the compressive limit is 21 MPa. The experiment proved that the compressive limit of the concrete samples we made is indeed 21 MPa. Under this pressure, the concrete sample is almost crushed. Pressures of 3.5, 7, 10.5, 14, 17.5, 21 MPa are applied to the six concrete samples in the thickness direction. After being compressed, the two bottom surfaces of the concrete test block in the length direction undergo different degrees of deformation and expansion. The two compressed surfaces in the thickness direction are very flat. Therefore, in the experiment, we still used the thickness direction as the sound propagation direction.

Table 5.1 gives the linear acoustic parameters of the six test blocks, including the speed of sound and size. It also compares the difference between the theoretical calculation value and the measured value of the first resonant frequency. In samples with stronger damage, the speed of sound is smaller. The speed of sound of the first sample is 50% higher than that of the sixth sample, and the first resonant frequency drops from 13.5 to 8.4 kHz. In theoretical calculations, it is assumed that the receiving end is free. In the experiment, although the mass of the accelerometer and the load thin aluminum sheet is very small, it cannot be ignored, so the experimental value of the first resonant frequency is larger than the theoretical value.

The first and third harmonics are close to resonance, so loss must be considered. However, μ_1 and μ_3 are difficult to measure directly. The relationship between nonlinear loss and quality factor is $\mu_n = \left[\omega Q_n\right]^{-1}$, so the nonlinear loss can be indirectly calculated through the quality factor. The quality factor is different in different samples and under different excitation voltages. As shown in Table 5.2, in the sixth sample, the quality factor changes with the change in strain at the receiving end.

Figure 5.4 shows the power spectrum of six samples at the first resonant frequency when the excitation voltage is 300 mv. From Fig. 5.4a, we find that the non-classical nonlinearity of the first test block is the weakest, with the third harmonic weaker than the second harmonic. In samples with greater damage, the non-classical nonlinearity gradually strengthens, the third harmonic gradually strengthens, and even surpasses the second harmonic. In the second sample, the third harmonic is almost as strong as the second harmonic. In the other four samples, the third harmonic is stronger than

Table 5.1 Linear acoustic parameters in six concrete samples

Number	Pressure (MPa)	Length (m)	Delay time (μS)	Speed of longitudinal wave (m/s)	F_1 (KHz) Theoretical value	F_1 (KHz) Measured value
1	3.5	0.0550	19.0	2894	13.2	13.52
2	7.0	0.0545	20.0	2725	12.5	13.45
3	10.5	0.0540	21.5	2547	11.6	11.63
4	14.0	0.0535	24.0	2230	10.4	11.57
5	17.5	0.0610	28.0	2178	8.8	8.88
6	21.0	0.060	30.0	2033	8.3	8.39

Table 5.2 Quality factor under different strains in the sixth sample

Strain amplitude	Q-Factor (− 3 dB)
3.74E−8	66.7
7.64E−8	65.3
1.219E−7	63.1
1.58E−7	61.5
2.12E−7	59.9
2.58E−7	57.1
3.04E−7	55.1
3.50E−7	53.4
3.94E−7	51.1
4.37E−7	49.6

the second harmonic. In the sixth sample, as shown in Fig. 5.4f, the amplitude of the odd harmonics is significantly stronger than the even harmonics, which is a main feature of non-classical nonlinear acoustics.

Figure 5.5 illustrates the linear relationship between the nonlinear frequency shift and the strain amplitude in the first and sixth samples. The experimental results confirm the theoretical results of the linear relationship between the nonlinear frequency shift and the amplitude of the first harmonic, as in Eq. (5.44).

According to classical nonlinear theory, the amplitude of the third harmonic is cubically related to the first harmonic. However, in non-classical nonlinear theory, the amplitude of the third harmonic is squared with the first harmonic. Figure 5.6 shows the relationship between the strain amplitude of the second and third harmonics and the strain amplitude of the first harmonic in the sixth sample. From the slopes of the two curves, we can conclude that, like the second harmonic, in damaged concrete, the third harmonic is squared with the amplitude of the first harmonic. This once again proves that concrete is a material with strong non-classical nonlinearity.

From Table 5.3, we can see that when the excitation voltage is 300 mv, as the degree of damage increases, the nonlinear frequency shift increases from 18 to 603 Hz. The first sample was damaged under a pressure of 3.5 MPa. It is the least damaged, with the fewest cracks. The nonlinear frequency shift and nonlinear parameters are the smallest. In the second, third, and fourth test blocks, the non-classical nonlinear characteristics gradually become significant. The fifth and sixth test blocks are the most severely damaged. Their non-classical nonlinear characteristics are the strongest. If we assume γ_1 is greater than γ_2, γ_1 and γ_2 can be easily calculated. As the degree of damage increases, $|\gamma_1 - \gamma_2|$ and $\gamma_1 + \gamma_2$ increase by about a hundred times. The nonlinear parameters γ_1 and γ_2 increase to 10^8. From Table 5.3, we conclude that compared with linear parameters, non-classical nonlinear parameters more significantly reflect the changes in the degree of concrete damage, demonstrating the superiority of the non-classical nonlinear method.

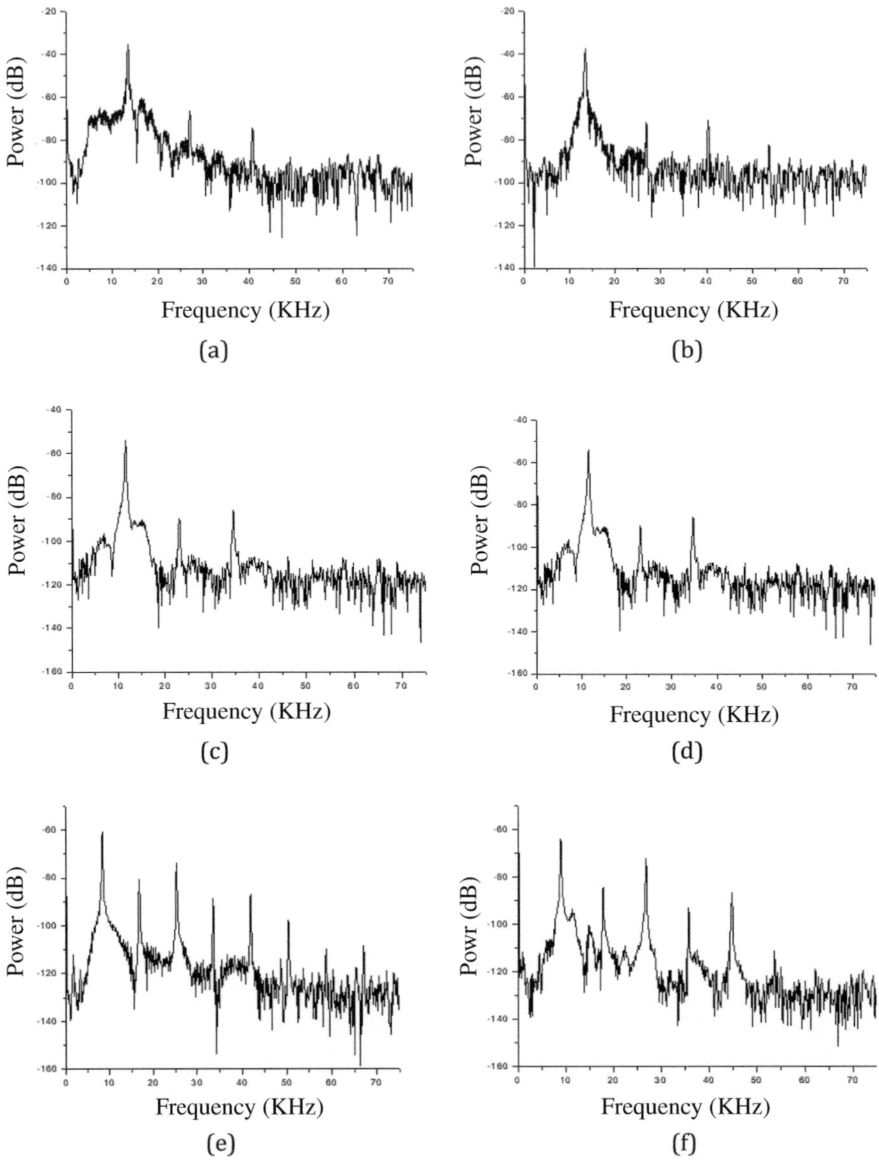

Fig. 5.4 The power spectrum of the free-end acceleration of the first (**a**), second (**b**), third (**c**), fourth (**d**), fifth (**e**), and sixth (**f**) samples

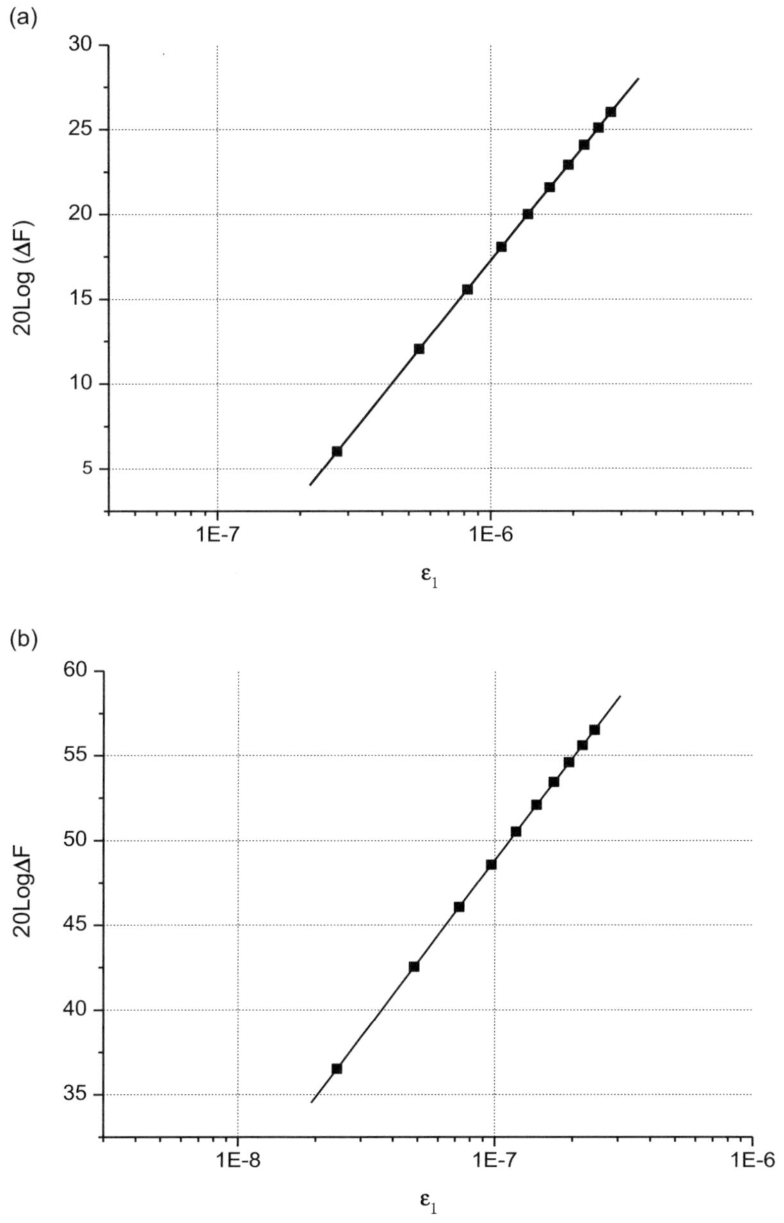

Fig. 5.5 The relationship between the nonlinear frequency shift and the amplitude of the first harmonic in the first (**a**) and sixth (**b**) concrete samples (The solid line is the first fitting curve)

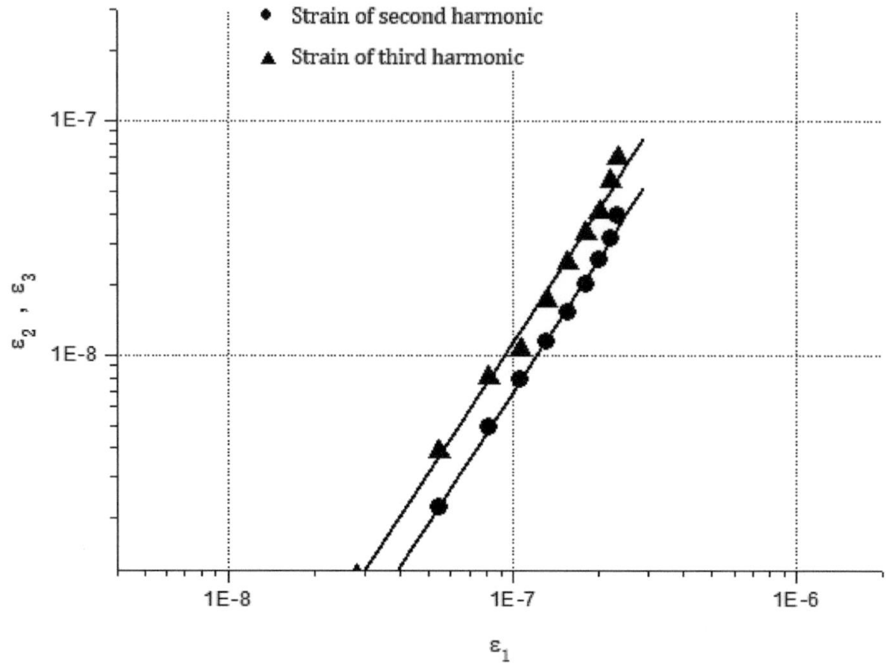

Fig. 5.6 The relationship between the strain of the second and third harmonics and the strain of the first harmonic in the sixth concrete sample (The solid line is the first fitting curve)

Table 5.3 Nonlinear frequency shift and nonlinear parameters in the six samples

Number	Nonlinear frequency shift (Hz)	$\|\gamma_1 - \gamma_2\|$	$\gamma_1 + \gamma_2$	γ_1	γ_2
1	18	2.02E+5	1.56E+6	8.81E+5	6.79E+5
2	56	2.40E+5	1.63E+6	9.35E+6	6.95E+5
3	85	6.9E+5	3.56E+6	2.13E+6	1.44E+6
4	222	1.19E+5	8.70E+6	4.95E+6	3.76E+6
5	450	2.42E+7	2.57E+8	1.46E+8	1.16E+8
6	603	3.04E+7	5.78E+8	3.27E+8	2.74E+8

5.4 Non-classical Nonlinear Changes in Concrete Samples with Different Water Contents

The strength of concrete is an important parameter for evaluating the quality of concrete. The water content of the concrete severely affects the strength of the concrete [15]. By studying the relationship between water content and acoustic nonlinear parameters, this section discusses the method of measuring the water content of concrete using non-classical nonlinear acoustics theory.

In this section, a concrete test block is placed in an oven to bake, continuously changing the water content in the concrete, and measuring parameters such as the sound speed, quality factor, harmonic amplitude, and nonlinear frequency shift of the test block after each evaporation of 1 g of water. All experimental results are measured under the first harmonic frequency of the concrete test block with different water contents. The boundary condition of the concrete test block is one end free and one end rigid. The experimental results show that the nonlinear parameters γ_1 and γ_2 are functions related to water content, and the nonlinear parameter method can provide a method for detecting the water content of a concrete. $\gamma_1 \neq \gamma_2$ reflects the asymmetry of the concrete compression process and expansion process. The relationship between the change of water content γ_1 and γ_2 is opposite, which also reflects the difference between compression vibration and expansion vibration. Moreover, compared with traditional methods, nonlinear frequency shift and nonlinear parameters more sensitively reflect the change in concrete water content.

As shown in Fig. 5.7, as the water content increases, the amplitude of the second harmonic decreases, and the amplitude of the third harmonic increases. The weakening of the second harmonic amplitude indicates that the classical nonlinear characteristics are weakening; the increase of the third harmonic indicates that the non-classical nonlinear characteristics are increasing. In concrete, non-classical nonlinear characteristics are very strong. Although the classical nonlinearity is weakened, the influence of non-classical nonlinearity on the third harmonic is dominant. Therefore, overall, the third harmonic is still enhanced. When the water content is higher than 1.3%, the amplitude of the third harmonic is stronger than the second harmonic, as shown in Fig. 5.8.

When the excitation voltage is 300 mv, the nonlinear frequency shift under different water contents, as shown in Fig. 5.9. When the concrete test block changes from dry to saturated, the nonlinear frequency shift increases from 16 to 115 Hz, increased by five times. Clearly, the nonlinear frequency shift is more sensitive to changes in the water content of concrete than conventional linear parameters, such as sound speed and density.

According to Nazarov's non-classical nonlinear acoustics theory [16, 17], the nonlinear parameters γ_1 and γ_2 can be obtained from Eqs. (5.22)–(5.23)

If we consider $\gamma_1 > \gamma_2$, γ_1 and γ_2 can be obtained from Table 5.4. The measurement results show that the relationship between compressive vibration and expansive vibration is opposite to that of water content. Therefore, the difference between compressive vibration and expansive vibration decreases with the difference in water content. When the concrete changes from dry to water-saturated, the nonlinear parameter γ_2 increases by two times, which can provide a method for detecting the water content of concrete.

At any water content, the quality factor will change with the change of the excitation signal. In fully water-saturated concrete samples, when the excitation voltage increases from 30 to 300 mv, Q increases from 6.1 to 8.5, as shown in Fig. 5.10.

Taking the concrete under water-saturated conditions as an example, we have once again verified the strong non-classical nonlinear acoustic characteristics in concrete.

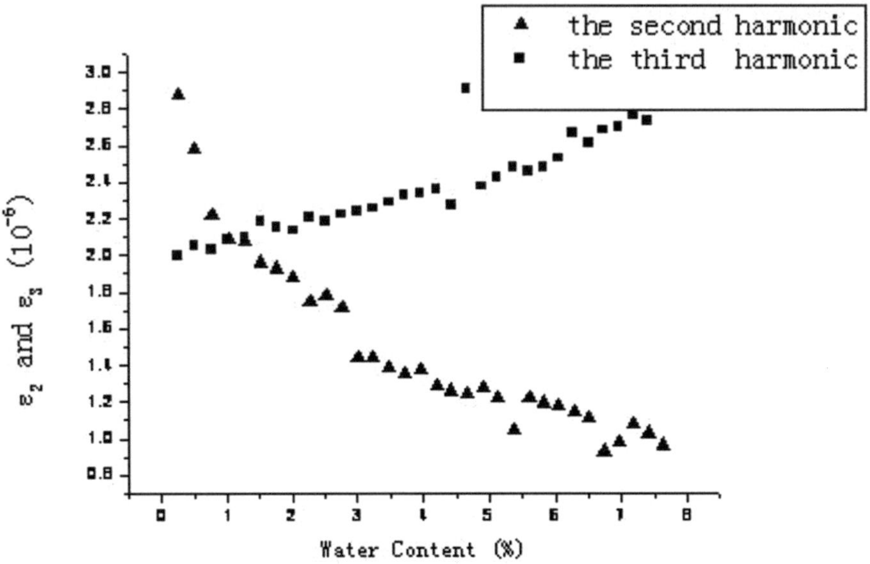

Fig. 5.7 Comparison of the amplitude of the second harmonic and the third harmonic under different water contents

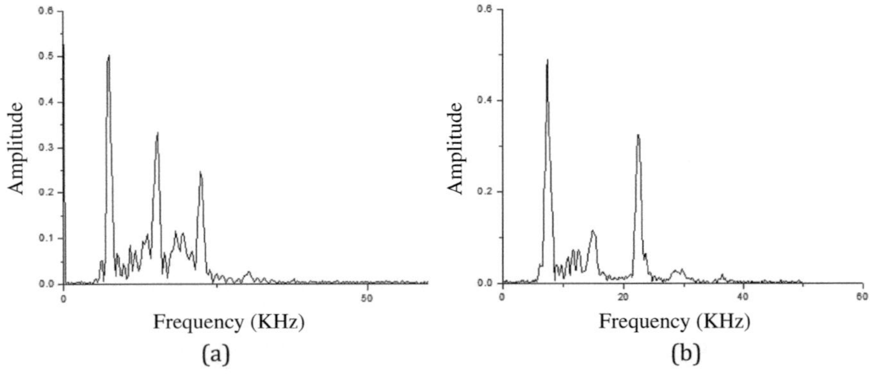

Fig. 5.8 The amplitude spectrum of the acceleration received at the free end of the test block when the water content is 0.26% (**a**) and 7.6% (**b**)

As shown in Fig. 5.11, the nonlinear frequency shift is linearly related to the excitation voltage; as shown in Fig. 5.12, the amplitude of the third harmonic is square related to the excitation voltage, which are important features of non-classical nonlinear acoustics.

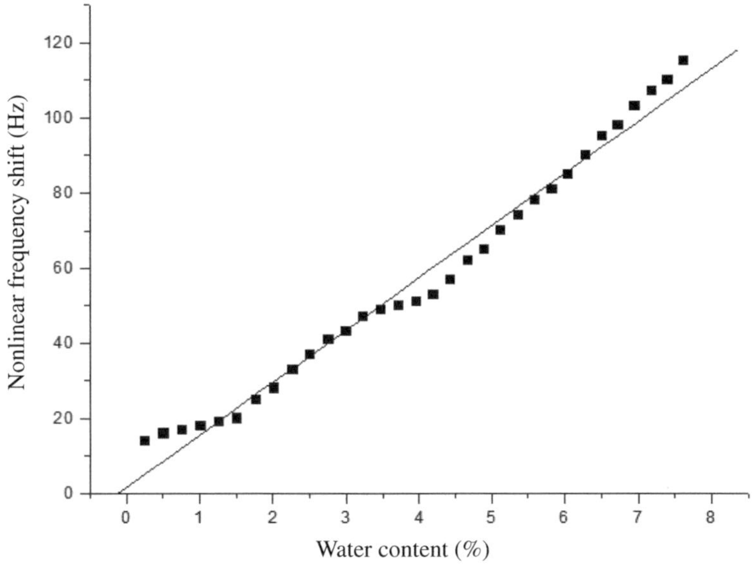

Fig. 5.9 The nonlinear frequency shift at different water contents

Table 5.4 Relationship between nonlinear parameters of test samples and water content

Water content (%)	0.26	3.00	5.37	7.62
γ_1	$1.54 * 10^7$	$1.43 * 10^7$	$1.38 * 10^7$	$1.34 * 10^7$
γ_2	$5.66 * 10^6$	$9.54 * 10^6$	$1.04 * 10^7$	$1.14 * 10^7$

5.5 Changes in Non-classical Nonlinear Effects of Concrete Under Sulfate Erosion

Since the alkaline solution has the most obvious erosion effect on the aggregate in concrete, and the impact of sulfate on concrete is mainly reflected in the erosion of cement mortar, most domestic and foreign research on concrete's resistance to sulfate erosion also uses cement mortar. Therefore, this experiment also chose cement mortar as the sample. In this section, six cubic cement mortar edge column samples were made, marked as samples 1–6, with dimensions of $16 \times 4 \times 4$ cm^3.

The quality ratio of water, cement, and sand is 1:2.16:4.85, with the cement being engineering cement and the sand being international standard sand. The samples were cured under standard conditions, then polished with a rock grinder to ensure that the six samples were of the same size and that each sample had a consistent surface smoothness.

To verify the erosion of concrete by sulfate and the effect of wet-dry cycles, sample 1 was taken as a reference in this experiment, and the remaining five cement mortar

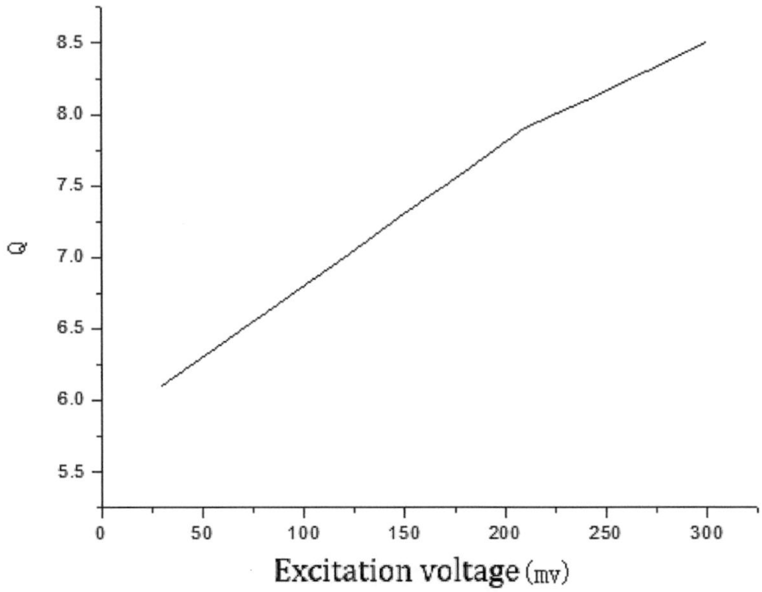

Fig. 5.10 Relationship between Q value and excitation voltage under water-saturated conditions

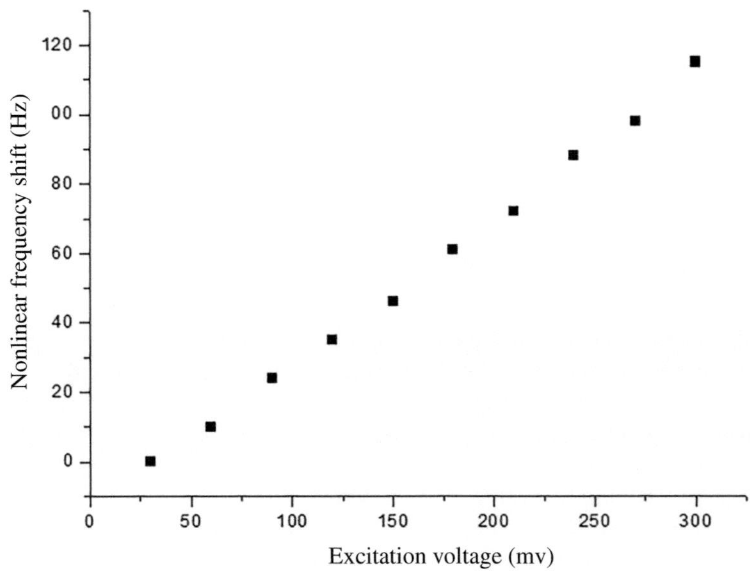

Fig. 5.11 Relationship between nonlinear frequency shift and the excitation voltage under water-saturated conditions

prism samples (samples 2–6) were subjected to different numbers of wet-dry cycles with a 6% sodium sulfate solution by mass fraction. The parameters of each sample were then measured to serve as a basis for judging the degree of sample damage.

According to the Arrhenius equation, for every 10 °C increase in temperature, the speed of a general chemical reaction increases by about 2–3 times. The increase in temperature will lead to an increase in the diffusion of sulfate ions, as well as an increase in ion movement speed and chemical reaction speed. These will lead to an increase in the speed of concrete sulfate erosion, accelerating the destruction of concrete [18]. Lawrence [19] believes that the highest allowable curing temperature to effectively prevent the instability of gypsum structure is between 65 and 70 °C. Considering these factors, the drying temperature in this experiment was set at 65 °C, and the curing temperature was set at 30 °C.

The wet-dry cycle method adopted in this experiment is as follows: at a temperature of 30 °C, the specimens are soaked in a sodium sulfate solution with a concentration of 6% (mass fraction) for 17 h, then drained and dried in an oven at a temperature of 65 °C for 6 h, then cooled for 1 h. This constitutes one wet-dry cycle. After different numbers of wet-dry cycles, the degree of erosion of the cement mortar prism samples varies. This experiment aims to use ultrasonic non-destructive testing methods to detect changes in the degree of erosion. To compare traditional ultrasonic detection methods using sound speed and other linear parameters with non-classical nonlinear methods, this experiment measured the linear parameter sound speed of the samples, studied non-classical nonlinear phenomena (including nonlinear frequency shift, etc.) in the samples, and discussed the possibility of using non-classical nonlinear parameters to characterize the degree of deterioration of cement mortar prism samples caused by sulfate erosion and wet-dry cycles.

First, the size and sound speed and other linear parameters of samples 1 to 6 were measured. Concrete is a non-uniform medium, with a typical longitudinal wave speed between 3500 and 4500 m/s. Assuming the longitudinal wave speed c is 4000 m/s, for an ultrasonic wave with a frequency of 1 MHz, the wavelength is $\lambda = c/f = 4$ mm, which is similar in size to the stones in the concrete, so the scattering of the stones must be considered. To avoid the influence of scattering, a lower frequency, such as 50 kHz, must be used. At this time, the sound wave wavelength is roughly a few centimeters, meeting the spatial resolution requirements for concrete quality control.

The sound speed measurement experiment was conducted using the RSMSYS5 sonic detector and general-purpose operating software developed by the Wuhan Institute of Rock and Soil Mechanics of the Chinese Academy of Sciences. The sound speed of the samples before sulfate erosion was 4417 m/s. Tables 5.4 and 5.5 provide the size, sound speed, and theoretical and experimental values of the first harmonic frequency F_1 after sulfate erosion and wet-dry cycles.

As can be seen from Table 5.5, for cement mortar prism samples affected by different degrees of sulfate erosion and dry–wet cycling, the change in the linear parameter sound speed is not significant, and the degree of change can almost be compared with the measurement error. Therefore, the linear parameter sound speed

Table 5.5 Linear sound parameters of the six samples after erosion

Number	Cycle times	Length (cm)	Sound of speed (m/s)	F1 (Hz) Theoretical	F1 (Hz) Experimental
1	0	15.89	4417 ± 10	13,899	13,840 ± 18
2	1	15.82	4417 ± 10	13,960	13,898 ± 22
3	2	15.85	4413 ± 8	13,921	13,880 ± 25
4	3	15.88	4413 ± 8	13,895	13,859 ± 25
5	4	15.84	4407 ± 10	13,911	13,888 ± 19
6	5	15.89	4407 ± 10	13,867	13,820 ± 28

is not sufficient to characterize the degree of damage to the cement mortar prism samples under the influence of sulfate erosion and dry–wet cycling.

Figure 5.13 is the power spectrum at the first resonance frequency of samples 1 and 6. Among them, Fig. 5.13a–c are the power spectra at the first resonance frequency of sample 1 at excitation amplitudes of 30 mv/60 mv/210 mv respectively, and Fig. 5.13d–f are the power spectra at the first resonance frequency of sample 6 at excitation amplitudes of 30 mv/60 mv/210 mv respectively. In the undamaged sample 1, the second harmonic is greater than the third harmonic. In the slightly damaged sample 6, the second harmonic and the third harmonic are almost equally

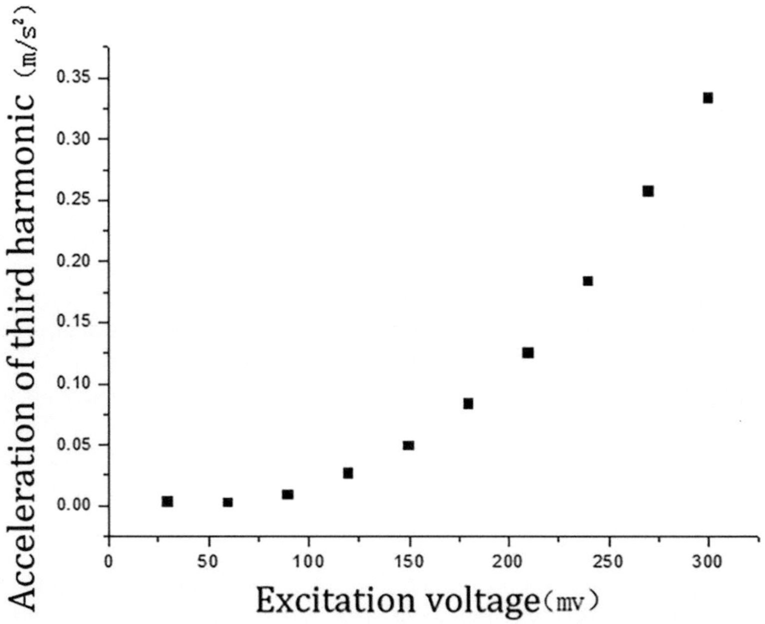

Fig. 5.12 Relationship between the amplitude of the third harmonic and the excitation voltage under water-saturated conditions

strong at 30 mv, and as the excitation amplitude increases, the phenomenon of the third harmonic amplitude being greater than the second harmonic amplitude appears, which is one of the main characteristics of non-classical nonlinearity.

Figure 5.14 shows the change in nonlinear frequency shift in six samples at an excitation voltage of 300 mV. The nonlinear frequency shift reflects the degree of non-classical nonlinearity. Compared with the change in sound speed in the six samples in Table 5.5, the change in nonlinear frequency shift is more obvious and more suitable for characterizing the degree of micro-damage inside the sample.

Assuming $\gamma_1 > \gamma_2$, γ_1 and γ_2 can be easily calculated. Table 5.6 shows that as the number of times the cement mortar prism samples are subjected to sulfate erosion and dry–wet cycling increases, $|\gamma_1 - \gamma_2|$ and $\gamma_1 + \gamma_2$ increase significantly. The non-classical nonlinear parameters γ_1 and γ_2 not only consider the nonlinear frequency shift, but also consider non-classical nonlinear characteristics such as nonlinear loss and harmonic generation, which can reflect the non-classical nonlinear acoustic characteristics of cement mortar prism samples more comprehensively and more obviously than simply using nonlinear frequency shift. Therefore, it is feasible to use the non-classical nonlinear parameters γ_1 and γ_2 to characterize the micro-damage and degradation degree of cement mortar prism samples caused by sulfate erosion and dry–wet cycling. Due to the damage, more micro-cracks are caused, making the nonlinear effect more obvious, thus leading to a significant change in the nonlinear parameters.

5.6 Non-classical Nonlinear Research of Concrete with Different Mix Ratios

The porosity of concrete has a significant impact on the mechanical and acoustic properties of concrete materials. The pores in concrete are divided into three types: the first type is closed pores, the second type is open but not connected pores, and the third type is pores that penetrate and connect through the concrete. Since the first type of pores is the least, we only study the latter two types of pores, which are collectively referred to as effective voids. The porosity referred to in this section refers to the measured effective porosity. The method for determining the effective porosity of concrete materials is as follows:

1. Measure the size of the specimen and calculate its volume, denoted as V;
2. Weigh the mass of the specimen in a water-saturated state, denoted as M;
3. Weigh the mass of the specimen in a dry state, denoted as m;
4. Calculate the effective porosity of the specimen according to formula (2.25), denoted as P:

$$P = \left[\frac{(M - m)}{\rho_w V} \right] \times 100\% \tag{5.57}$$

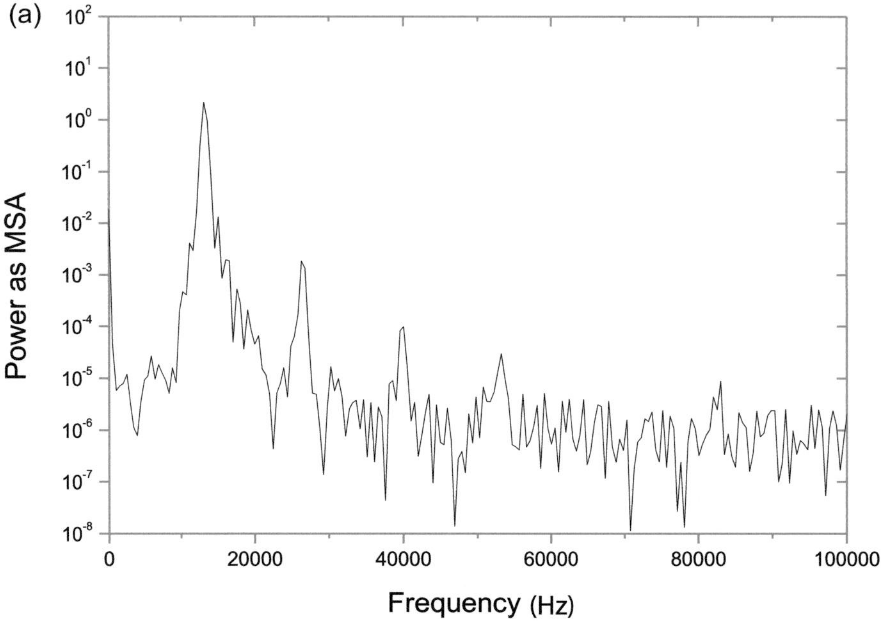

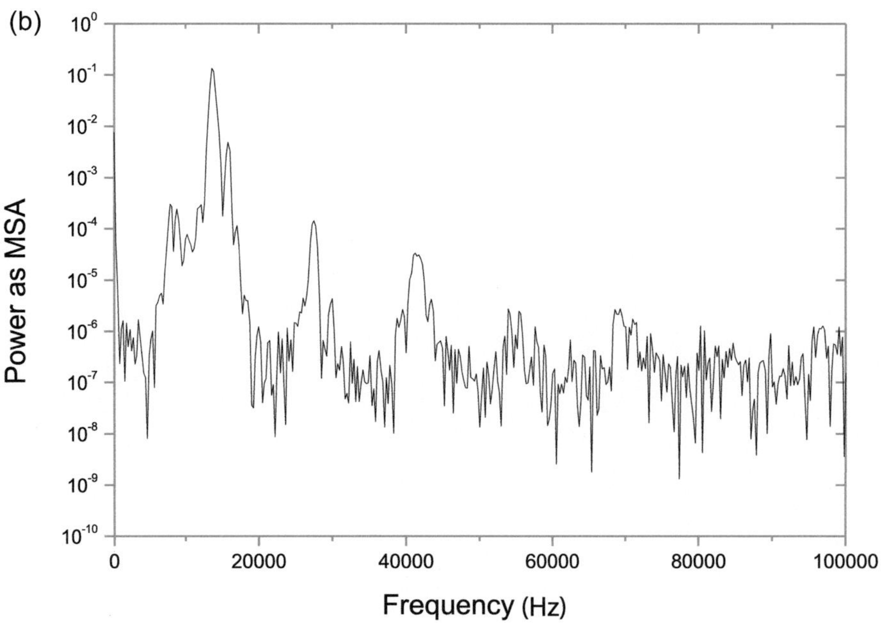

Fig. 5.13 The free-end power spectrum of samples 1 (**a–c**) and 6 (**d–f**) at excitation amplitudes of 30/60/210 mv

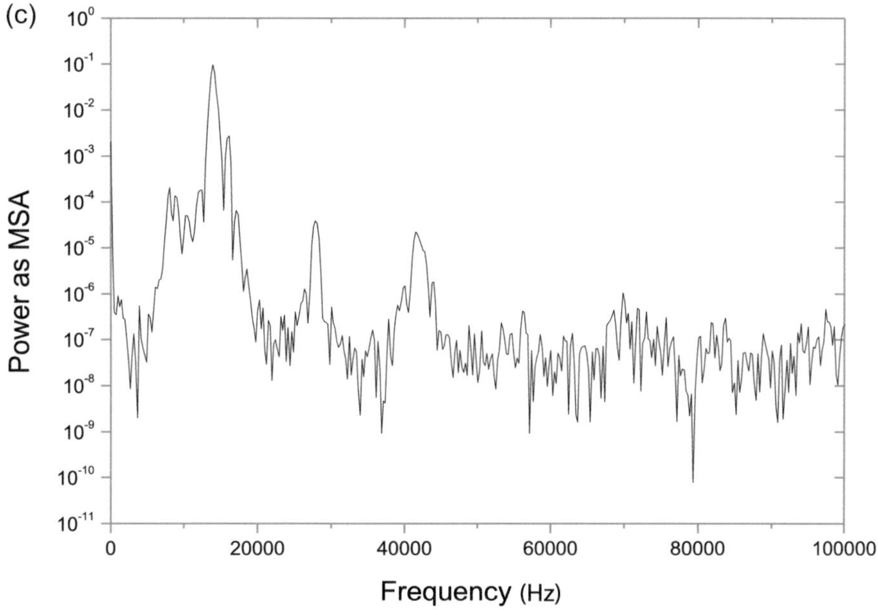

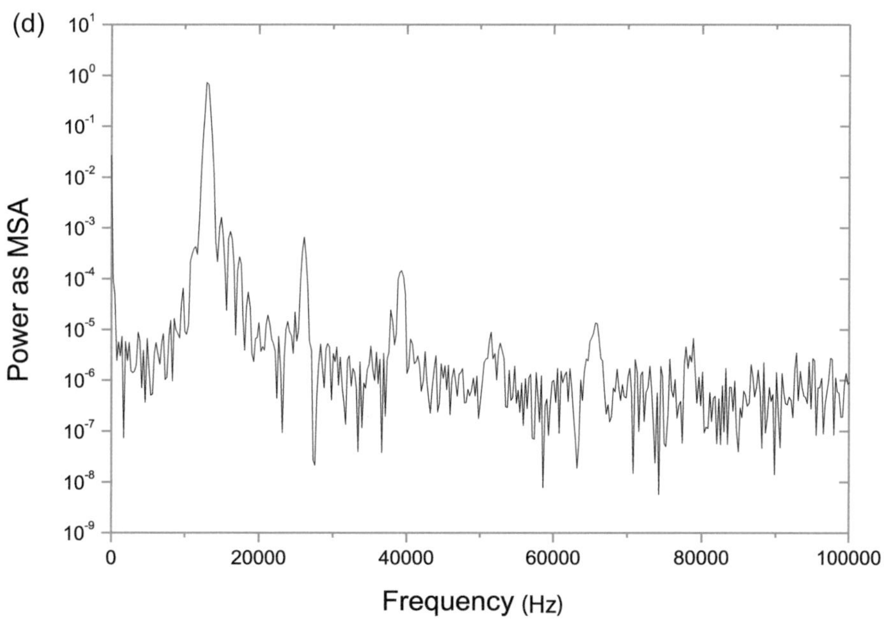

Fig. 5.13 (continued)

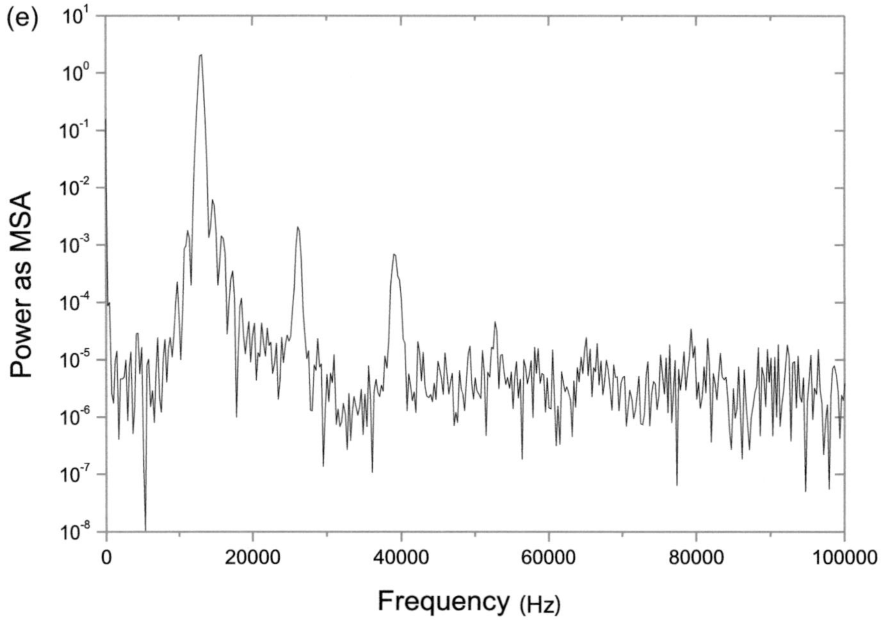

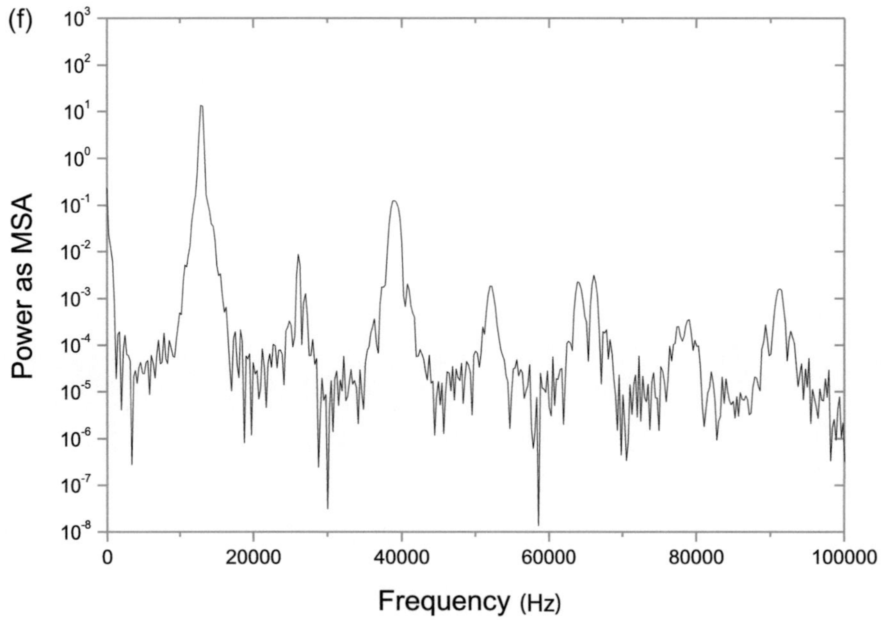

Fig. 5.13 (continued)

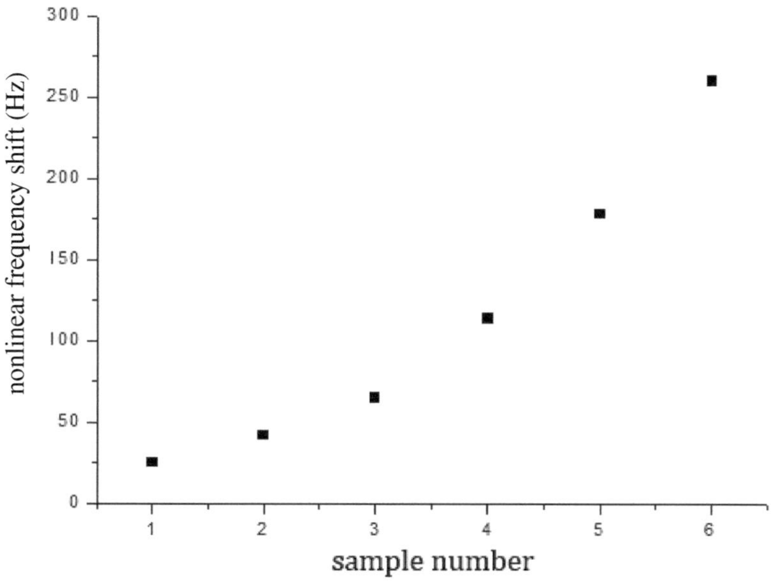

Fig. 5.14 The nonlinear frequency shift for samples 1–6

Table 5.6 Nonlinear frequency shift and nonlinear parameters in six samples

Number	Nonlinear frequency offset (Hz)	$\|\gamma_1 - \gamma_2\|$	$\gamma_1 + \gamma_2$	γ_1	γ_2
1	25	2.27E+05	1.77E+06	9.99E+05	7.72E+05
2	42	2.51E+05	1.98E+06	1.12E+06	8.65E+05
3	65	3.86E+05	3.23E+06	1.81E+06	1.42E+06
4	114	3.50E+05	3.17E+06	1.76E+06	1.41E+06
5	178	3.90E+05	3.85E+06	2.12E+06	1.73E+06
6	260	4.99E+05	4.78E+06	2.64E+06	2.14E+06

where, ρ_w represents the density of water, $\rho_w = 1000\,\text{kg/m}^3$.

As mentioned above, the mix ratio of concrete affects the porosity, and the porosity will affect the strength and mechanical and acoustic characteristics of the concrete material. Therefore, this section mainly studies the ultrasonic characteristics of concrete with different mix ratios, including linear and nonlinear characteristics. Since there are many factors affecting the strength of concrete, such as the water-cement ratio, aggregate gradation, and admixture additives, this experiment will keep the water-cement ratio of each sample the same, all at 0.4, and no admixture additives such as fly ash are added, all using ordinary cement, sand, and crushed stone materials.

This experiment used ordinary Portland cement with a density of 3150 kg/m^3 and a strength grade of 32.5. The aggregate used was sand with a bulk density of 1450 kg/

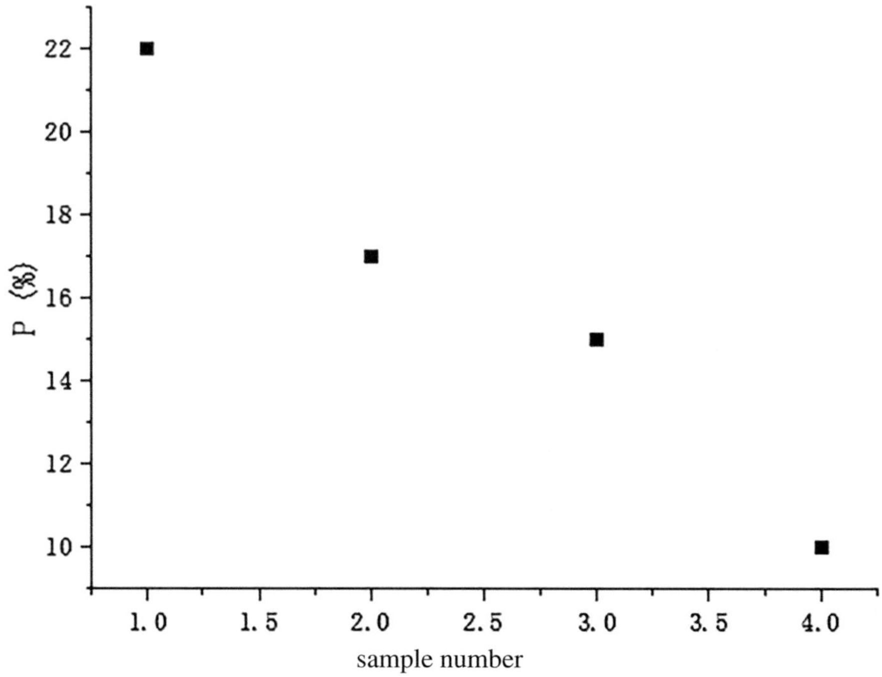

Fig. 5.15 Porosity for samples 1–4

m^3 and a porosity of 42%, and crushed stone with a bulk density of 1520 kg/m^3. This experiment did not use water reducers or other additives.

This section of the experiment had four concrete column samples, labeled A_1–A_4, with dimensions of $16 \times 4 \times 4$ cm^3. The mix ratios of the four samples are shown in Table 5.7, where W represents water, C represents cement, S represents sand, and G represents crushed stone. After mixing the cement and other materials, they were poured and compacted indoors, then dried under natural room temperature conditions. The samples were demolded after a week of curing, and then the surfaces were polished smooth.

Table 5.7 Concrete mix ratios

Sample number	Concrete material ratios			
	W	C	S	G
A_1	0.4	1	1	0
A_2	0.4	1	1	0.5
A_3	0.4	1	1	1
A_4	0.4	1	1	1.5

This experiment had four concrete samples with different mix ratios, labeled A_1–A_4. The porosity of the four concrete samples was first measured. The dimensions of each sample were measured, and the volume of each sample was calculated V_i. The four concrete samples were placed in a 60 °C oven to dry until the mass of the concrete no longer decreased. At this point, the mass of each sample was weighed m_i, recorded as the mass of the dry concrete sample. Then the concrete samples were placed in water, and after they were fully saturated, they were drained and weighed again M_i, recorded as the mass of the water-saturated concrete sample. The porosity of each concrete sample P_i can be calculated according to formula (5.57).

The sound speed and nonlinear frequency shift in each sample were measured, and the relationship between the sound speed and porosity changes in the four samples was compared. The nonlinear parameters were calculated, and non-classical nonlinear phenomena such as harmonic generation were observed.

According to formula (5.57), the calculated porosity is shown in Table 5.8, and the pattern is shown in Fig. 5.15. As can be seen from the figure and table, the porosity decreases monotonically with the increase of coarse aggregate, but the increase of porosity with the increase of coarse aggregate is not a linear relationship. Therefore, the change of porosity with the mix ratio still needs to be determined through a large number of experiments.

The measurement method is the same as in Sect. 5.4. The length and sound speed are shown in Table 5.9, and the variation pattern is shown in Fig. 5.16. The linear parameter sound speed increases monotonically with the increase of the proportion of crushed stone in the concrete mix.

As shown in Fig. 5.17, it is the spectrum of the received signal at the free end of samples A_1–A_4. From the figure, it can be seen that the odd harmonics of samples A_1, A_2 are obvious, while the even harmonics are weak, showing obvious non-classical

Table 5.8 Porosity

Sample number (N)	V (cm^3)	M (g)	m (g)	P (%)
A_1	278	651	590	22
A_2	301	725	675	17
A_3	275	708	668	15
A_4	286	758	730	10

Table 5.9 Linear parameters

Sample number (N)	Length (cm)	Sound speed (m/s)	F_1 (Hz) theoretical value	F_1 (Hz) experimental value
A_1	16.05	3019	14,107	13,703
A_2	16.00	3200	15,000	14,300
A_3	15.75	3351	15,957	14,720
A_4	15.90	3424	16,151	15,160

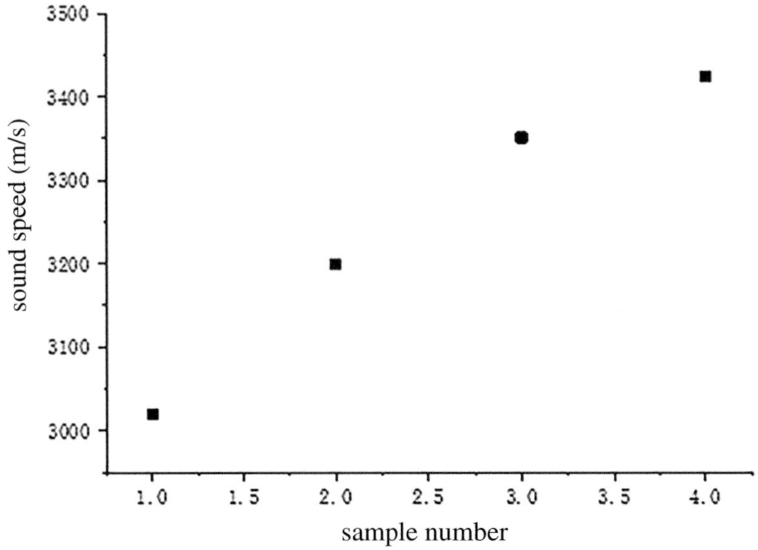

Fig. 5.16 Sound speed for sample 1–4

nonlinear phenomena. However, the second harmonic amplitude of samples A_3, A_4 is slightly higher than the third harmonic amplitude, and the non-classical nonlinear effect is lower than A_1.

The nonlinear frequency shift of the four samples is shown in Fig. 5.18. With the increase of the proportion of crushed stone in the concrete samples, the nonlinear frequency shift of the four samples decreases. Compared with the change trend of harmonics in Fig. 5.17, they all show the weakening of non-classical nonlinear effects.

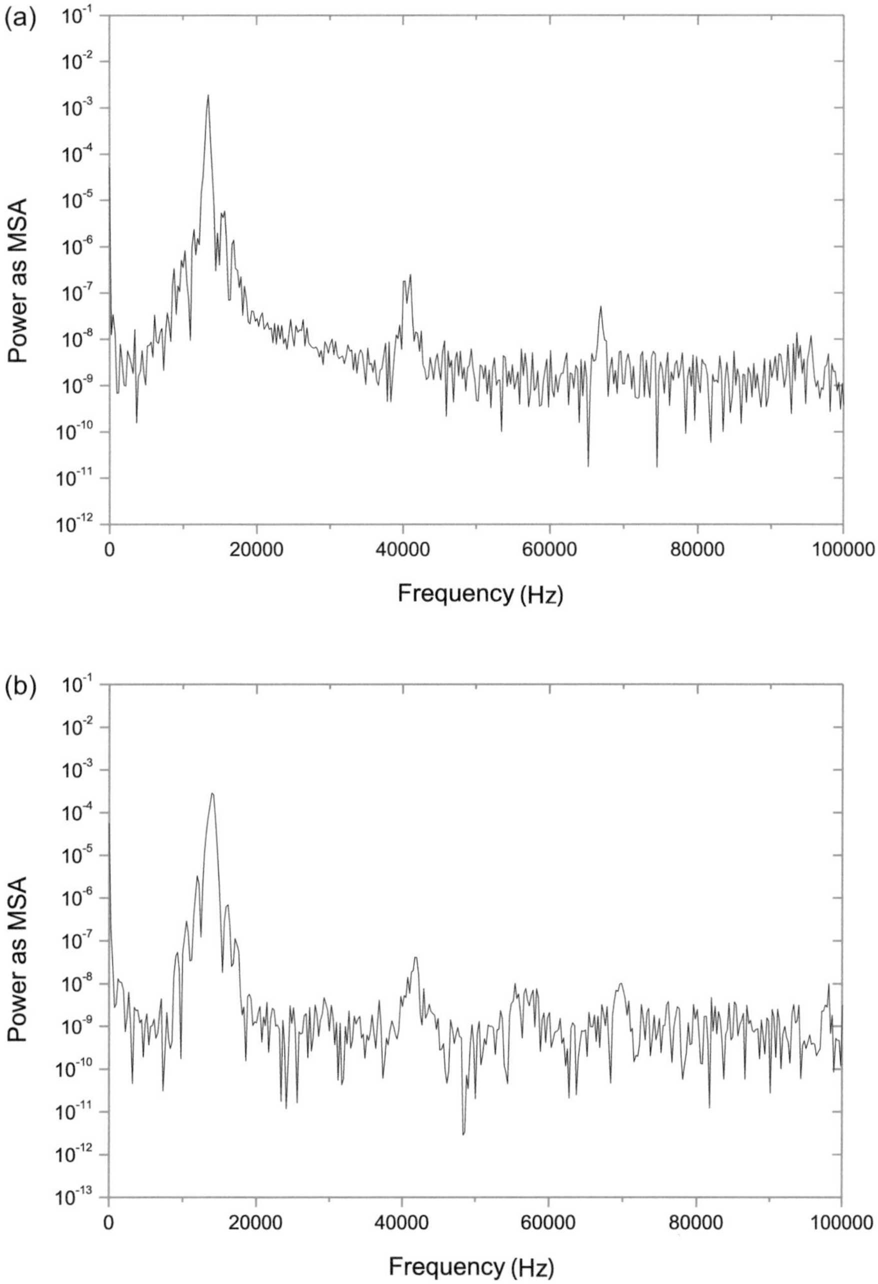

Fig. 5.17 Spectrum of the received signal at the free end of samples A_1–A_4 (**a–d**)

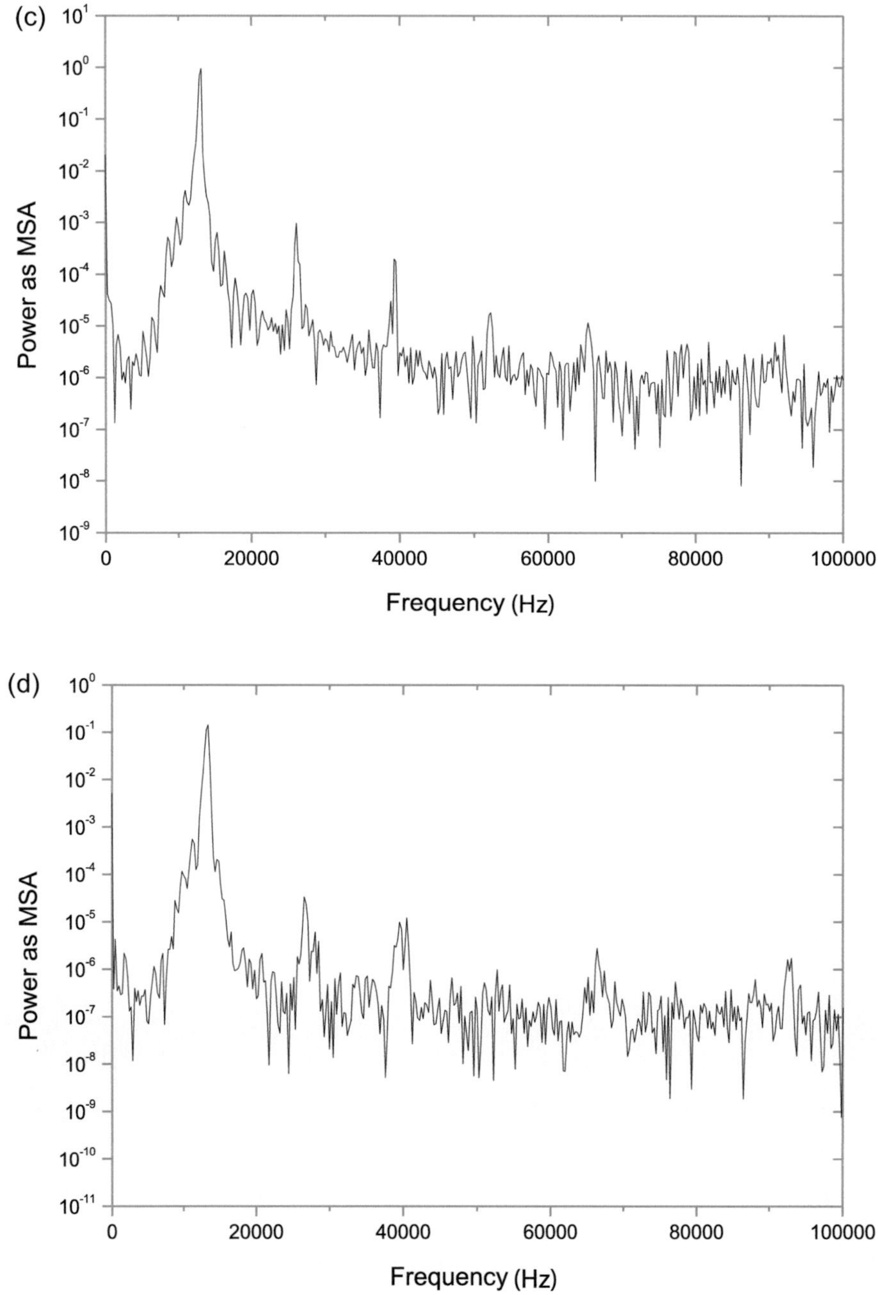

Fig. 5.17 (continued)

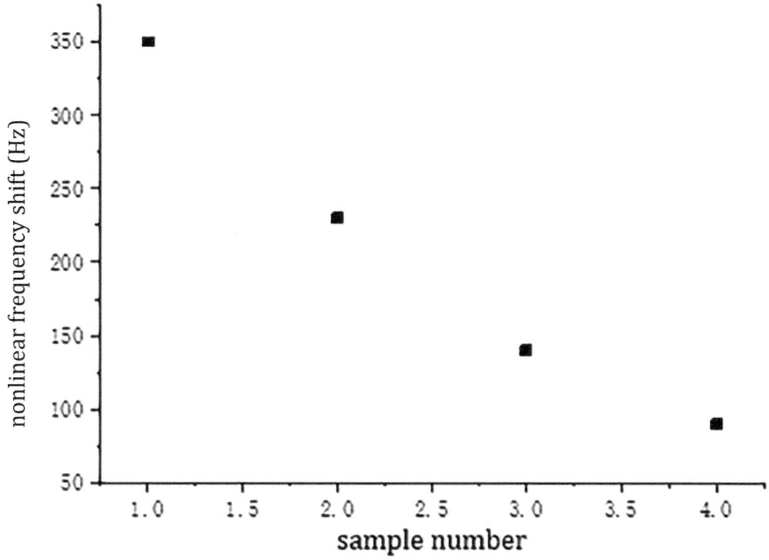

Fig. 5.18 The nonlinear frequency shift for samples 1–4

References

1. Schubert F, Marklein R. Numerical computation of ultrasonic wave propagation in concrete using the elastodynamic finite integration technique (EFIT). IEEE Ultrason Symp. 2002;1(1):799–804.
2. Popovics JS, Rose JL. A survey of developments in ultrasonic NDE of concrete. IEEE Trans Ultrason Ferroelectr Freq Control. 1994;41:140–3.
3. Ohdaira E, Masuzawa N. Water content and its effect on ultrasound propagation in concrete—the possibility of NDE. Ultrasonics. 2000;38:546–52.
4. Terill JM, Richardson JM, Sekby AR. Nonlinear moisture profile and shrinkage in concrete members. Mag Concr Res. 1986;38:220–5.
5. Abeele KV, Visscher JD. Damage assessment in reinforced concrete using spectral and temporal nonlinear vibration techniques. Cem Concr Res. 2000;30:1453–64.
6. Bentahar M, El AH, El GR, Griffa M, Scalerandi M. Hysteretic elasticity in damaged concrete: quantitative analysis of slow and fast dynamics. Phys Rev B. 2006;73:0141161.
7. Lacouture JC, Johnson PA, Cohen-Tenoudji F. Study of critical behavior in concrete during curing by application of dynamic linear and nonlinear means. J Acoust Soc Am. 2003;113:1325.
8. Schutter GD, Taerwe L. General hydration for porland cement and blast furnace slag cememt. Cem Concr Res. 1995;25:593–604.
9. Johnson PA. The new wave in acoustic testing. Mater World. 1999;7:544–6.
10. Niels T, Sadananda S. Mechanism of concrete deterioration due to salt crystallization. Mater Charact. 2004;53:123–8.
11. Cody RD, Cody AM. Reduction of concrete deterioration by ettringite using crystal growth inhibition techniques. Iowa DOT TR-431; 2001.
12. Idom GM, Skaluy J. Rapid test of concrete expansivity due to internal sulfate attack. ACI Mater J. 1993:383–5.
13. Qiao HX, He ZM, Liu CL. Study on dynamic modulus of elasticity of fly ash concrete under the condition of sulfate. Fly Ash Comprehen Util. 2006;1:6–8.

14. Pan X. Study on the preparation method and properties of porous cement concrete pavement materials; 2010.
15. Terill M, Richardson M, Selby AR. Non-linear moisture profiles and shrinkage in concrete members. Mag Concr Res. 1986;137(38):220–5.
16. Nazarov VE, Ostrovskii LA, Soustova IA, et al. Anomalous acoustic nonlinearity in metals. Soviet Phys Acoust. 1988;34(3):284–8.
17. Nazarov VE, Kolpakov AB. Experimental investigations of nonlinear acoustic phenomena in polycrystalline zinc. J Acoust Soc Am. 2000;107(4):1915–21.
18. Cao ZL, Yuan XZ, Xing F. American test methods for evaluating the durability of cement based materials against external sulfate attack. J Shenzhen Univ Sci Eng. 2006;7:201–11.
19. Lawrence CD. Mortar expansion due to delayed ettringite and formation, effect of luring period and temperature. Cem Concr Res. 1995;25:903–14.

Chapter 6
Nonlinear Propagation of Acoustic Waves in Mass-Spring Periodic Structures

It is well known that the path of light is reversible, that is, light can propagate from point a to point b, and it can also propagate from point b to point a. In nature, not only light waves, but all forms of wave propagation show reversible properties, which we call reciprocity. In order to control the directional flow of energy, people have been striving to find ways to rectify various wave motions. The electronic diode is the earliest device to achieve rectification. It was the advent of electronic diodes, represented by silicon crystals, that sparked the electronic industry revolution and ushered us into the information age. In recent years, inspired by the rectification effect of electronic diodes, rectification effects of other forms of waves have also received increasing attention. Research groups around the world have achieved rectification effects on light waves, heat flow, and sound waves by designing some special artificial materials [1–12]. In 2004, Baowen Li from the National University of Singapore theoretically proposed a model that can achieve unidirectional thermal rectification by using a simple harmonic spring to couple two one-dimensional nonlinear crystal oscillator chains [7]. Due to the difference in phonon spectra of two nonlinear oscillator chains, their phonon bands are significantly different at different temperatures. By controlling the overlap degree of the phonon bands in the forward and reverse directions, the one-way flow of heat is achieved. Based on this thermal rectification effect, theoretical models of thermal diodes, thermal triodes, and thermal logic gates have been proposed, thus demonstrating that phonons can also be used to carry and process information. Subsequently, the thermal rectification effect has received great attention, and more theories and experiments have successively supported this new effect [8–10]. This one-way heat flow is realized at the nanoscale, but it was difficult to rectify sound waves at the macro scale at that time. Inspired by the solid band theory, in recent years, people have broken through the limitations of existing natural materials by simulating natural lattices, and obtained new functional materials and devices through band design. In 2009, Cheng Jianjun and others proposed a new mechanism of acoustic rectification [13], they combined the characteristics of nonlinear acoustic materials and linear phonon crystals, and realized the rectification

X. Liu, *Nonlinear Sound Waves in Solids*,
Advances in Acoustics, https://doi.org/10.1007/978-981-96-5253-2_6

of sound waves at the macro scale. When a finite amplitude fundamental frequency sound wave exists in the nonlinear medical ultrasound contrast agent microbubbles, it will generate harmonics, and the frequency of the generated second harmonic just falls within the passband of the phonon crystal, so the second harmonic can propagate in the lattice. If the fundamental frequency sound wave is input in the reverse direction, the frequency of the fundamental wave at this time is within the band gap in the lattice, and it decays exponentially with the increase of distance, that is, it cannot pass through the lattice. In this way, when the same energy sound waves are input in the forward and reverse directions, the ratio of the energy received at the receiving end is significantly different, thus realizing the rectification of sound waves. After the research on sound rectification received widespread attention, research institutions in various countries have conducted more in-depth research on the sound rectification effect. N. Boechler from Caltech and others proposed using pre-pressurized one-dimensional nonlinear particle chains to achieve asymmetric control of sound waves [14], and this mechanism is considered to be applicable to light, heat, and sound and other materials.

6.1 Mechanism of Bragg Scattering Type Phonon Crystal Band Gap Formation

For the band gap characteristics of phonon crystals under the action of Bragg scattering mechanism and its influence rules, there has been a lot of research work [15–20]. When the matrix of the Bragg scattering type phonon crystal is liquid or gas, only longitudinal waves exist in the matrix. Therefore, the band gap originates from the in-phase superposition of reflected waves between adjacent cells, and the center frequency of the first band gap corresponds to the elastic wave wavelength which is about twice the lattice constant. This is similar to Bragg's study of X-ray diffraction behavior in crystals, hence the name Bragg scattering type phonon crystal. When Bragg scattering occurs in solids, because it has both transverse and longitudinal waves, the formation mechanism of the band gap is more complicated. We focus on the analysis of the mechanism of the Bragg scattering type phonon crystal band gap formed by solid scatterers in the solid matrix. The research results show that the wavelength corresponding to the band gap frequency is on the same order of magnitude as the transverse wave wavelength. It is also for this reason that the Bragg scattering type phonon crystal is not conducive to the control of low-frequency sound waves.

For a one-dimensional two-component phonon crystal, we can equate it to a one-dimensional diatomic chain. For a one-dimensional diatomic chain complex lattice, we know that there is a band gap in the band structure, and there is no corresponding vibration propagation mode in the band gap, so the vibration in this frequency range cannot stably propagate in the one-dimensional diatomic chain. In the band structure diagram, the frequency area where no dispersion curve enters for all wave vectors of the irreducible Brillouin zone is called the complete band gap. The bands tangent to the upper and lower edges of the band gap are called band edge modes. Since the group velocity of all band edge modes is zero at the band edge, the eigenfield can reach a very high density at the band edge. And in the band gap, the eigenfield density is zero, therefore, the eigenfield can also reflect the band gap of the solid.

6.1.1 Linear Band Structure of Mass-Spring System

We consider an infinitely long one-dimensional diatomic chain, with periodically arranged different masses and springs of the same stiffness connecting them. According to the Bragg scattering mechanism, there is a band gap in such a system. If the vibration corresponding to the band gap frequency passes through this diatomic chain, then the amplitude of the vibration will decay exponentially in it. If the vibration frequency is within the passband, then the vibration can theoretically pass through this diatomic chain without decay [21, 22]. Suppose we change this diatomic chain structure to the same mass, springs of different stiffness alternately arranged, as shown in Fig. 6.1, there should also be a band gap.

We can write out the $2n$. The motion equation of the $2n + 1$ mass blocks

$$m\frac{d^2\xi_{2n}}{dt^2} = K_1(\xi_{2n-1} - \xi_{2n}) + K_2(\xi_{2n+1} - \xi_{2n}) \tag{6.1}$$

$$m\frac{d^2\xi_{2n+1}}{dt^2} = K_2(\xi_{2n} - \xi_{2n+1}) + K_1(\xi_{2n+2} - \xi_{2n+1}) \tag{6.2}$$

where ξ represents the displacement of vibration, K_1 and K_2 represent the stiffness of the spring, m represents the mass. According to the Bloch periodicity theorem, the displacement has a wave solution, which can be expressed as

$$\xi_{2\mathrm{n}} = Ae^{j(\omega t-2nqa)} \tag{6.3}$$

$$\xi_{2\mathrm{n}\,+\,1} = Be^{j[\omega t-(2n+1)qa]} \tag{6.4}$$

where A and B represent the amplitudes of the $2n$ and $2n+1$ mass blocks respectively, ω, q, a represent the circular frequency of vibration, wave vector, and the length of a period of the mass-spring system respectively. Substituting Eqs. (6.3) and (6.4) into Eqs. (6.1) and (6.2) we get

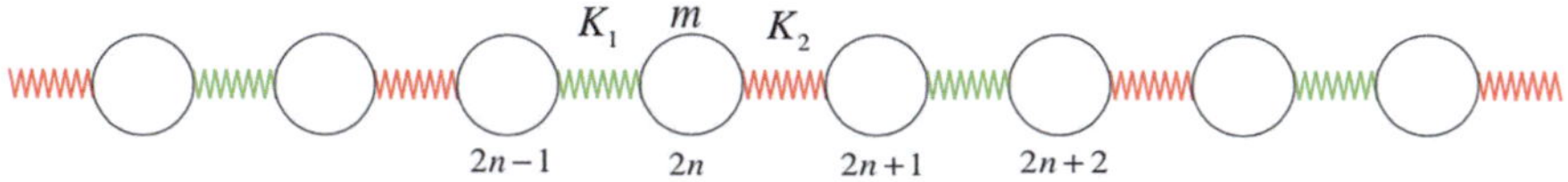

Fig. 6.1 Schematic diagram of a one-dimensional mass-spring system

$$\left(K_1 + K_2 - \omega^2 m\right)A - \left(K_1 e^{jqa} + K_2 e^{-jqa}\right)B = 0 \tag{6.5}$$

$$\left(K_2 e^{jqa} + K_1 e^{-jqa}\right)A - \left(K_1 + K_2 - \omega^2 m\right)B = 0 \tag{6.6}$$

Since the amplitudes A and B cannot both be zero, the determinant of the matrix should be zero, so we have

$$\left|\begin{pmatrix} K_1 + K_2 - \omega^2 m & -\left(K_1 e^{jqa} + K_2 e^{-jqa}\right) \\ K_2 e^{jqa} + K_1 e^{-jqa} & -\left(K_1 + K_2 - \omega^2 m\right) \end{pmatrix}\right| = 0 \tag{6.7}$$

We can then obtain a relationship between the circular frequency ω and the wave vector q

$$\omega^2 = \frac{1}{m}\left[K_1 + K_2 \pm \sqrt{K_1^2 + K_2^2 + 2K_1K_2\cos(2qa)}\right] \tag{6.8}$$

This relationship is the dispersion relation of the mass-spring system. Using the frequency $f = \omega/2\pi$ Substituting the angular frequency, when the mass is set to 1kg, and the spring stiffness are $K_1 = 1000$ N/m, $K_2 = 1200$ N/m, we can obtain the frequency f as a function of the normalized wave vector qa, as shown in Fig. 6.2.

6.1.2 *Linear Simulation of the Mass-Spring System*

There are two different curves in Fig. 6.2a, the upper one is called the optical branch, and the lower one is called the acoustic branch. If the vibration frequency is on the optical branch or the acoustic branch, then the vibration can propagate without attenuation. There is a range of frequencies between the acoustic branch and the optical branch that does not correspond to any curve, indicating that there is no corresponding vibration mode for these frequencies, which is the band gap. When qa tends to $\pi/2$, the upper and lower boundaries of the band gap can be described as $f_u = \frac{1}{2and}$ (if the parameters $K_1 = 1000$ N/m, $K_2 = 1200$ N/m, $m = 1$ kg are substituted into Eq. (6.8), we can get $f_u = 7.80$ Hz, $f_l = 7.12$ Hz). The density of intrinsic fields is very high at the upper and lower boundaries of the bandgap, where resonance can easily occur; for an infinitely long mass-spring system, the state density within the bandgap is zero, so it does not support the propagation of elastic waves. In fact, the number of periodic structures is always finite, and the propagation of elastic waves within the bandgap will decay exponentially. We also simulated the transmission coefficient of this system with finite element software. The finite element simulation software performs excellently in simulating linear and nonlinear structures, and allows us to set material parameters ourselves. During the simulation, we loaded a period of stress on the first mass block of the mass-spring system in the Load module; in the Visualization module, we extracted the displacements of the

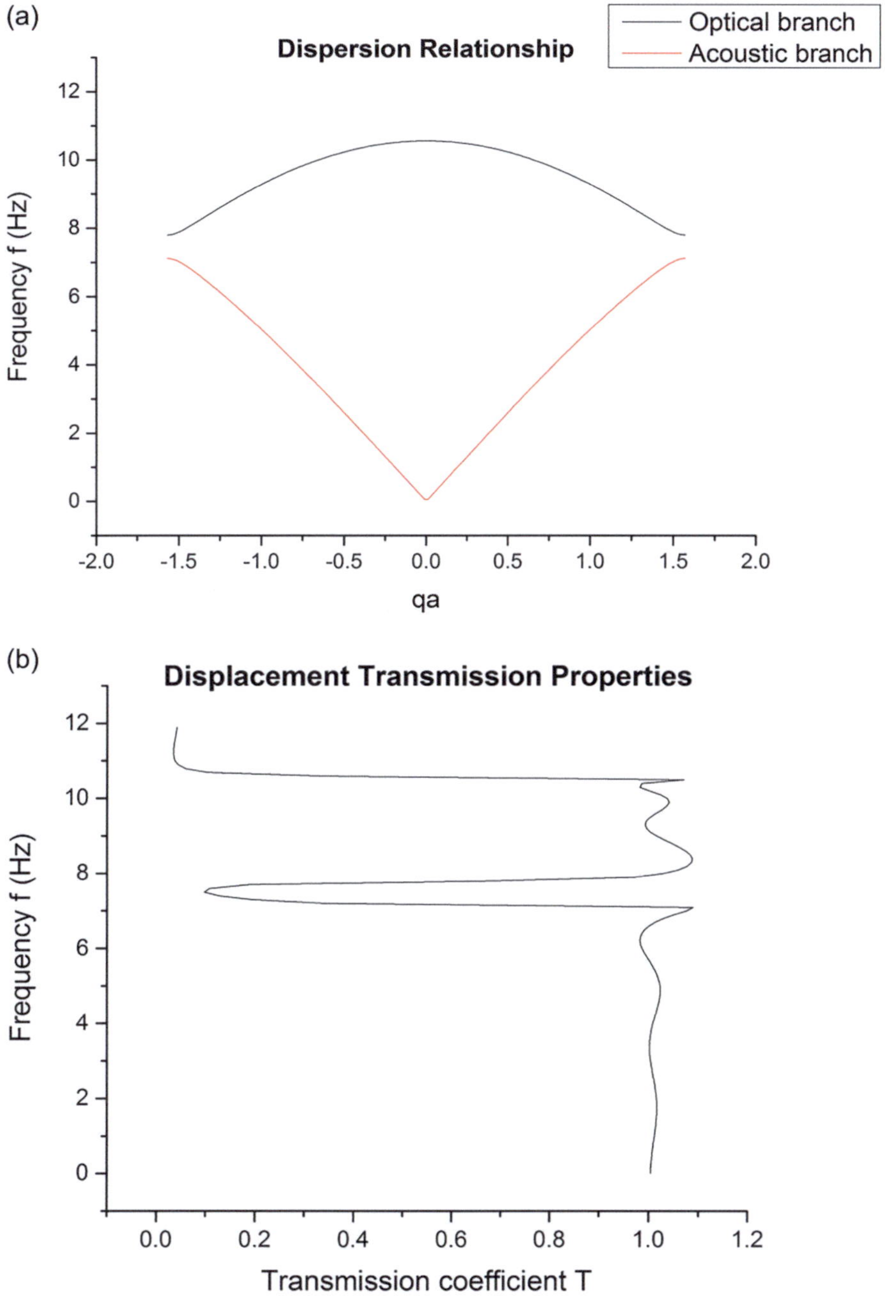

Fig. 6.2 **a** Dispersion relation of the linear infinite mass-spring system ($m = 1\,\text{kg}$, $K_1 = 1000\,\text{N/m}$, $K_2 = 1200\,\text{N/m}$). **b** Simulated displacement transmission coefficient when the number of masses is 11

first and eleventh mass blocks obtained from the simulation. The displacement at this time is in the time domain, and after Fourier transformation (FFT), it is converted to the frequency domain. After division, a transmission coefficient can be obtained, as shown in Fig. 6.2b. The transmission coefficient matches the dispersion curve (Fig. 6.2a) very well: the transmission coefficient is approximately equal to 1 in the optical branch (7.80–10.56 Hz) and the acoustic branch (0.00–7.12 Hz), and it is significantly less than 1 between the two branches (7.12–7.80 Hz). The simulation results show that the propagation of waves within the system's bandgap will be significantly attenuated, while in the transmission band, they can pass through the mass-spring system without attenuation.

6.2 Nonlinear Regulation Based on the Mechanism of Phononic Crystal Bandgap Formation

6.2.1 Nonlinear Analysis Based on the Mass-Spring System

For the linear case of a system with the same mass block and different spring stiffness, when $qa = \pm\pi/2$, according to formula (6.8), the cutoff frequencies at the upper and lower boundaries of the bandgap are respectively $f_u = \frac{1}{2\pi}\sqrt{2K_2/m}$ and $f_l = \frac{1}{2\pi}\sqrt{2K_1/m}$. If the stiffness of the two springs K_1 and K_2 are kept constant, then the cutoff frequencies will not change. Now we replace the linear spring K_1 with a nonlinear spring $K_1 + \varepsilon\Omega\xi^2$ replacing it (where ε represents a small parameter, Ω represents a nonlinear parameter) and the linear spring K_2 and mass m remain unchanged. As the displacement ξ gradually increases, the stiffness of the nonlinear spring $K_1 + \varepsilon\Omega\xi^2$ should also increase, while the stiffness of the linear spring K_2 remains unchanged. From this, it can be concluded that the lower boundary of the bandgap $f_l = \frac{1}{2\pi}\sqrt{2K_1/m}$ should move towards a better frequency, and the upper boundary $f_u = \frac{1}{2\pi}\sqrt{2K_2/m}$ should remain unchanged. In the case of $\Omega = 0$, this represents the linear situation of the dispersion curve; in the case of $\Omega = 1000$, it indicates that the linear spring K_1 is replaced by a hard spring, the stiffness of the hard spring will increase with the increase of amplitude, so the upper boundary of the bandgap does not change while the lower boundary will move up with the increase of amplitude; when $\Omega = -1000$, the linear spring K_1 is replaced by a soft spring, the stiffness of the soft spring decreases with the increase of amplitude, so at this time, the upper boundary of the bandgap remains unchanged with the increase of amplitude, while the lower boundary of the bandgap will move down (These results are based on the condition $K_1 < K_2$).

6.2.2 Nonlinear Simulation of Mass-Spring System

In order to verify whether the above situation will occur after replacing the linear spring K_1 with a nonlinear spring, we used finite element simulation software to simulate the nonlinear system. To prevent strong reflections at the last mass block (n = 11), we increased the number of mass blocks in the Assembly module to 200 (Fig. 6.3). In the Interaction module, we used parameters $K_1 = \left(1000 \pm 100 * \xi^2\right)$ N/m. The connector (connecter) has replaced the linear spring. In the Load module, the amplitude of the input vibration can be determined by the size of the concentrated force, so we control the input amplitude and the stiffness of the nonlinear spring by changing the size of the concentrated force. The Load module gives an excitation to the first mass block, and the Visualization module extracts the time-domain displacement of the first and eleventh mass blocks, which is transformed into frequency through Fourier Transform (FFT), and divided to get the transmission coefficient curve, as shown in Fig. 6.4.

The parameters of the mass-spring system are set as $m = 1\,\text{kg}, K_1 = \left(1000 \pm 100 * \xi^2\right)\,\text{N/m}, K_2 = 1200\,\text{N/m}$. When a small amplitude ($\xi \sim 10^{-1}$ m, stiffness variation is 1 N/m 1/1000 of the linear case) vibration input, the system's nonlinearity can be ignored, so the transmission coefficient obtained at this time is the transmission coefficient under linear conditions. The transmission coefficient values for the optical branch (7.80–10.56 Hz) and acoustic branch (0.00–7.12 Hz) are 1, while the transmission coefficient values within the bandgap (7.12–7.80 Hz) are much less than 1. It is worth noting that the reason why the transmission coefficient curve we obtained from the simulation under linear conditions is so good is because the number of mass blocks we simulated is not 11 but 200 (the mass blocks after the 11th can be regarded as vibration absorption devices), and the simulation time is 10 s, which can greatly reduce the reflection at the 11th mass block. When the amplitude of the input vibration is of the appropriate size ($\xi \sim 1$ m), the system's nonlinearity cannot be ignored. As can be seen from Fig. 6.4, for K_1 Replacing with a hard spring, the upper boundary of the bandgap remains unchanged, while the lower boundary moves towards high frequency as the amplitude increases; if K_1 is replaced with a soft spring, the situation is just the opposite, the lower boundary will move towards low frequency as the amplitude increases. It's like a filter controlled by amplitude. The simulation results are consistent with the previous section's analysis of the upper and

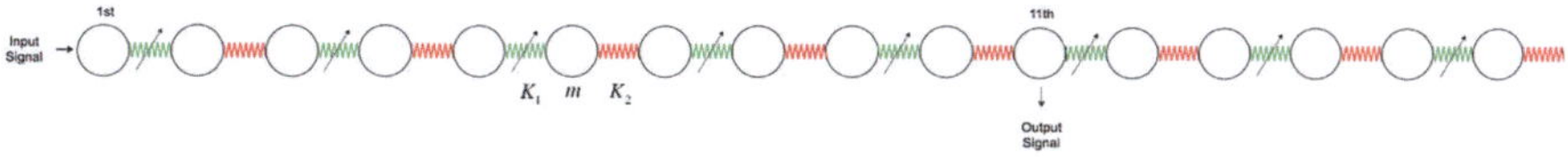

Fig. 6.3 Finite element simulation schematic: an excitation is added to the first mass block, and after the simulation, the time-domain displacement of the first and eleventh mass blocks is extracted, then through Fourier Transform (FFT), divided to get the transmission coefficient curve. To reduce reflection, the number of mass blocks has been increased to 200

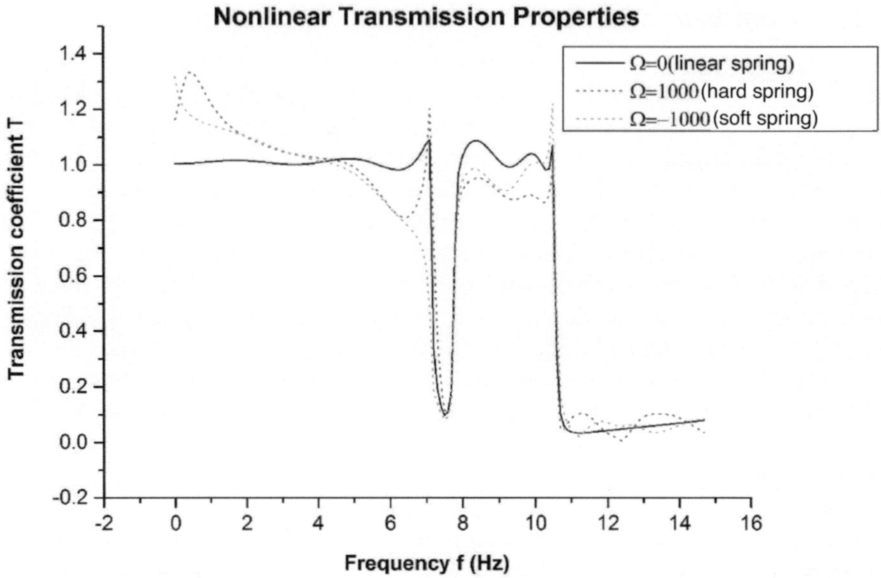

Fig. 6.4 Transmission coefficient curve of the nonlinear mass-spring system

lower boundaries of the bandgap $f_u = \frac{1}{2\pi}\sqrt{2K_2/m}$ and $f_l = \frac{1}{2\pi}\sqrt{2K_1/m}$. In Fig. 6.4, the transmission coefficient of some frequency bands of the acoustic branch is greater than 1, which is caused by the reflection caused by the impedance mismatch due to the existence of nonlinearity.

We also did other simulations to prove this result. In the simulation, the input signal contains two frequencies, one frequency is $f_1 = 7.08$ Hz, which is very close to the cutoff frequency of the acoustic branch (7.12 Hz), and the other frequency is $f_2 = 3.18$ Hz on the acoustic branch, but it is far from the cutoff frequency. Figure 6.5 is the schematic diagram of the amplitude-dependent filter. The parameters of the mass spring system are m = 1, $K_1 = (1000 - 100 * \xi^2)$ N/m, $K_2 = 1200$ N/m, m = 1, $K_1 = (1000 - 100 * \xi^2)$ N/m, $K_2 = 1200$ N/m, for the first case, the amplitude is $A_0 \approx 0.2$, both frequency components of the elastic wave can pass through the system without significant attenuation (Fig. 6.6), because both frequencies are located on the acoustic branch. As the amplitude gradually increases to $A_0 \approx 2.0$, because the nonlinear spring at this time is a soft spring, the stiffness decreases with the increase of amplitude, and the cutoff frequency of the acoustic branch will also decrease, so the frequency at $f_1 = 7.08$ Hz may be in the band gap, then the frequency component at $f_1 = 7.08$ Hz cannot pass through the mass-spring system while the other frequency component at $f_2 = 3.18$ Hz can pass through the system, as shown in Fig. 6.7.

It can be imagined that when K_1 is replaced with a hard spring, when the input signal also contains two frequency components $f_1 = 7.16$ Hz (very close to the acoustic branch) and $f_2 = 3.18$ Hz (in the band gap), the system will behave differently. 3.18 Hz $f_2 = 3.18$ Hz (in the middle of the acoustic branch). Under the small

Fig. 6.5 Schematic of the amplitude-dependent frequency filter

amplitude excitation $A_0 \approx 0.2$, the frequency components within the bandgap should be attenuated, while the frequency components on the acoustic branch will not be attenuated; when the amplitude increases to $A_0 \approx 2$, the frequency $f_1 = 7.16$ Hz falls into the passband of the acoustic branch due to the upward shift of the lower boundary of the bandgap, and at this time, both frequency components will not have significant attenuation, as shown in the simulation results of Figs. 6.8 and 6.9. In Fig. 6.8, the simulation results show that the frequency of $f_1 = 7.16$ Hz does not have a particularly large attenuation, which is because the bandgap has narrowed compared to the case of the soft spring. Due to the optical branch boundary resonance and nonlinearity, extra frequencies appear in the figure.

6.3 Non-reciprocal Propagation Mechanism

As nonlinear optical isolators become a research hotspot [23–25], the concept of acoustic metamaterials has been proposed for several decades. These pioneering works on acoustic metamaterials have attracted much attention and may bring revolutionary changes to existing acoustic technologies in various important fields. Linear periodic structures exhibit many interesting phenomena, while nonlinearities may present even more peculiar propagation characteristics. Sievers and Takeno discovered the phenomenon of localization by studying pure anharmonic crystals [26]. Vakakis et al. showed us that during localization, the response is confined to a finite subdomain, while the rest of the system oscillates at a relatively lower amplitude [27]. Daraio and others experimentally demonstrated that a one-dimensional nonlinear phononic crystal made of viscoelastic polytetrafluoroethylene (PTFE) chains, like stainless steel elastic beads, can have a large wave speed by changing the preload [28]. There are also some mathematical methods for studying nonlinear systems. Some perturbation techniques are used for the propagation of waves in nonlinear continuous systems [29] and discrete systems. In 1995, a multiscale analysis method was used to study the nonlinear oscillators of infinite single-ring chains [30]. In 2001, a perturbation analysis method was also used for weakly nonlinear single-atom chains and found the boundary frequency [31].

People have always been interested in the unidirectional manipulation of acoustics. However, due to the limitations of the reciprocity principle, the unidirectional manipulation of sound waves has always been a challenge [32, 33]. In 2009, Cheng Jianchun and others first used the coupling of nonlinear materials and acoustic superlattices to break the barrier of the reciprocity principle and realized the acoustic diode

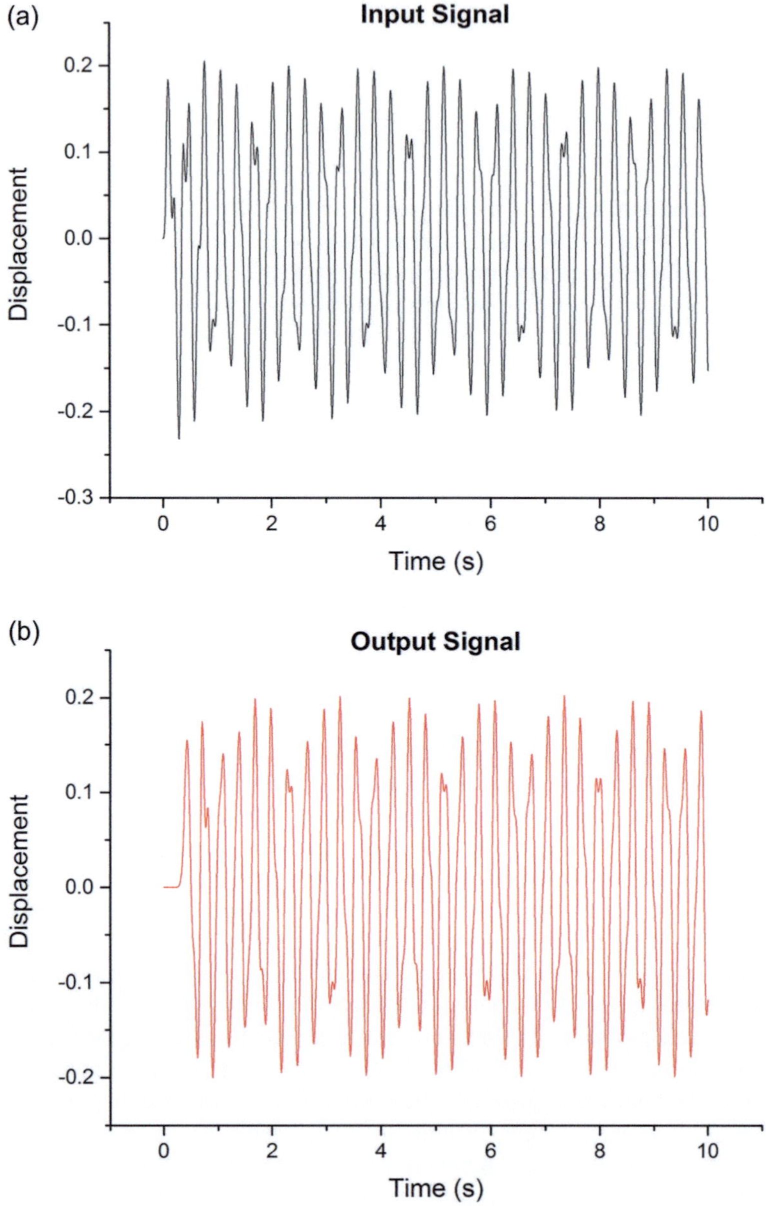

Fig. 6.6 In the case of a soft spring, the input signal contains two frequencies $f_1 = 7.08$ Hz and $f_2 = 3.18$ Hz, amplitude $A_0 \approx 0.2$, the above figure shows the time-domain signals of the input and output **a** $f_1 = 7.08$ Hz, **b** $f_2 = 3.18$ Hz; the below figure shows the frequency-domain signals of the input and output, **c** $f_1 = 7.08$ Hz, **d** $f_2 = 3.18$ Hz. When the amplitude is small, both frequencies of the signal can pass through the system smoothly without significant attenuation

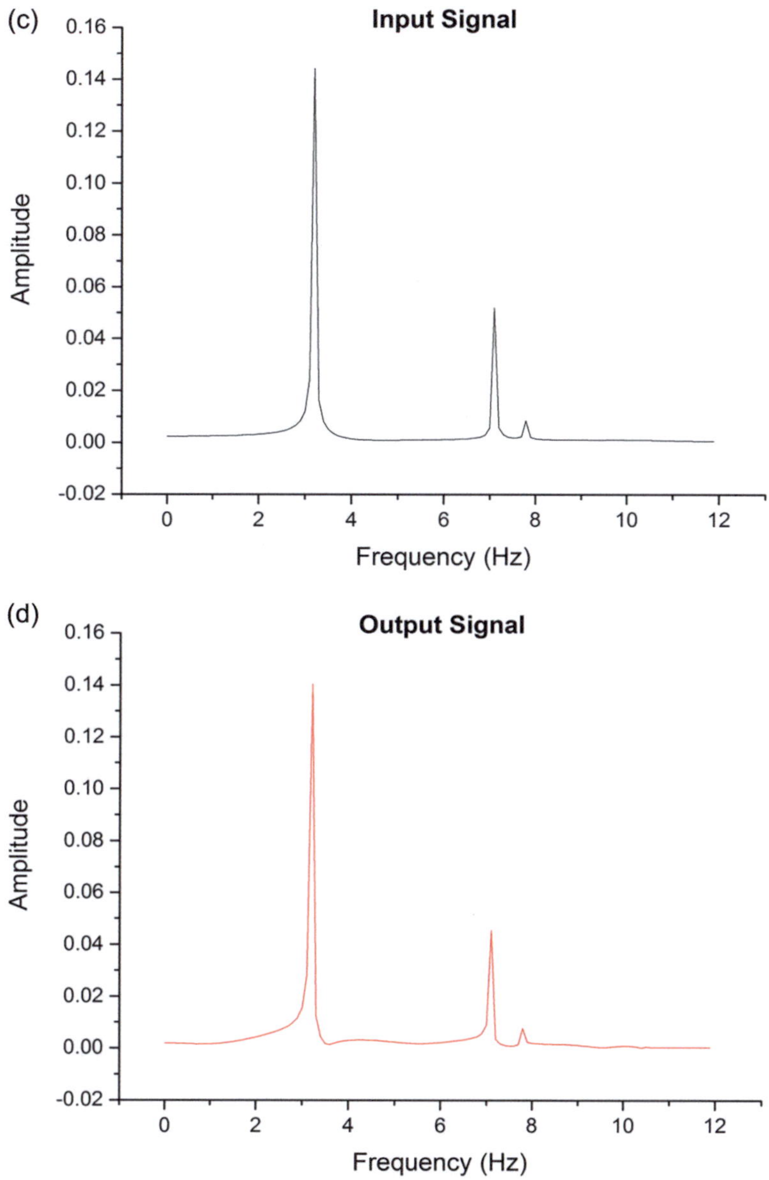

Fig. 6.6 (continued)

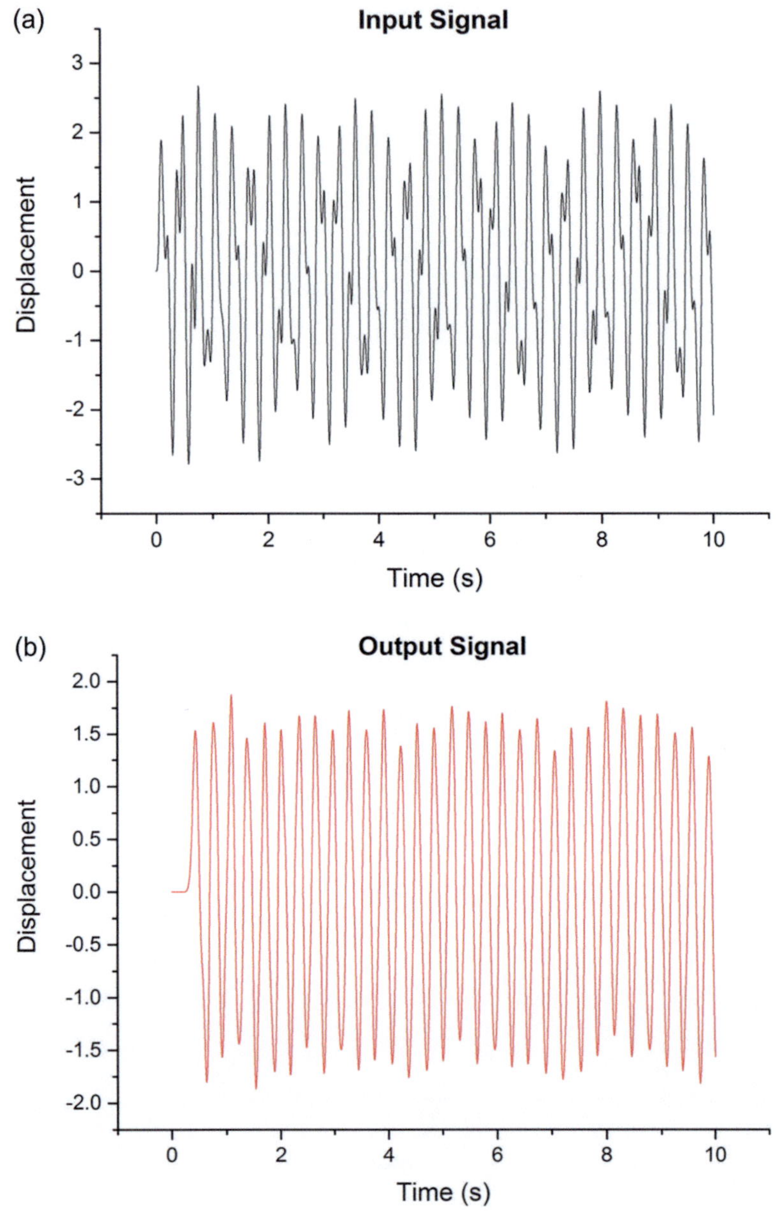

Fig. 6.7 In the case of a soft spring, the input signal contains two frequencies $f_1 = 7.08\,\text{Hz}$ and $f_2 = 3.18\,\text{Hz}$, amplitude $A_0 \approx 2$, the above figure shows the time-domain signals of input and output (**a**, **b**); the figure below shows the frequency-domain signals of input and output (**c**, **d**). When the amplitude is large enough, the signal's $f_1 = 7.08\,\text{Hz}$ frequency will have a significant attenuation when passing through the system, while the attenuation of the frequency $f_2 = 3.18\,\text{Hz}$ is not significant when passing through the system

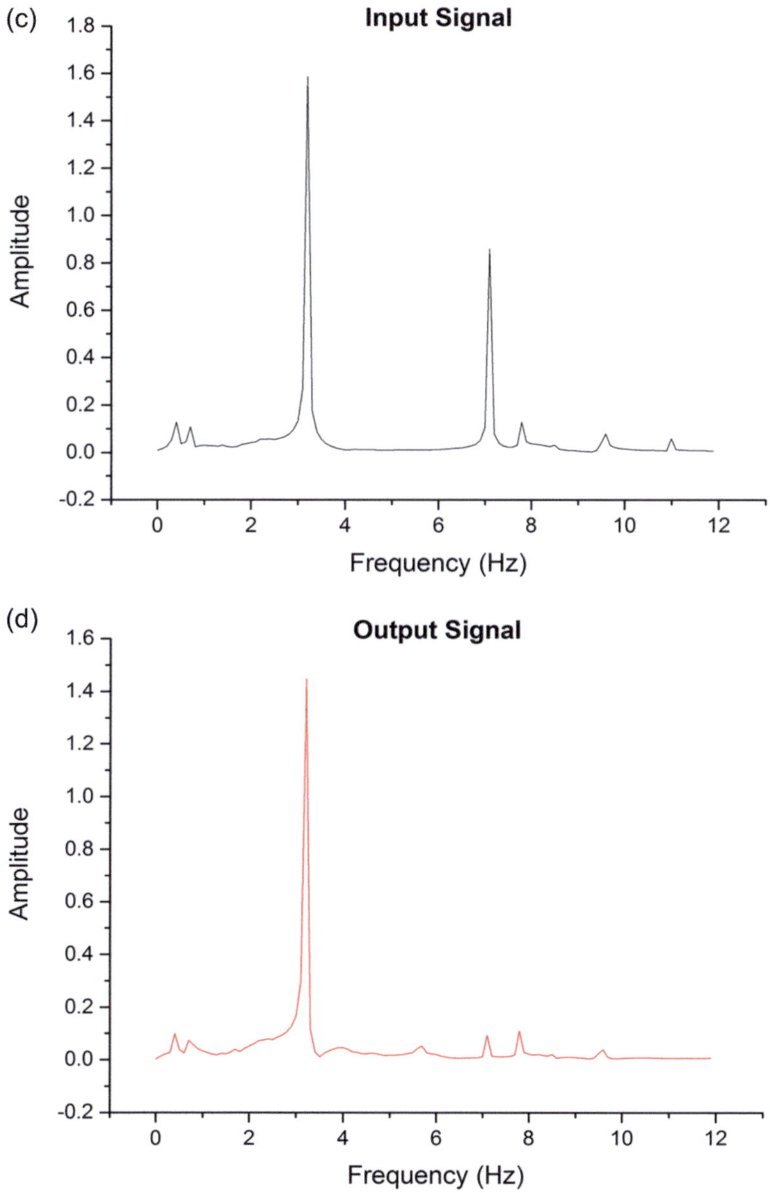

Fig. 6.7 (continued)

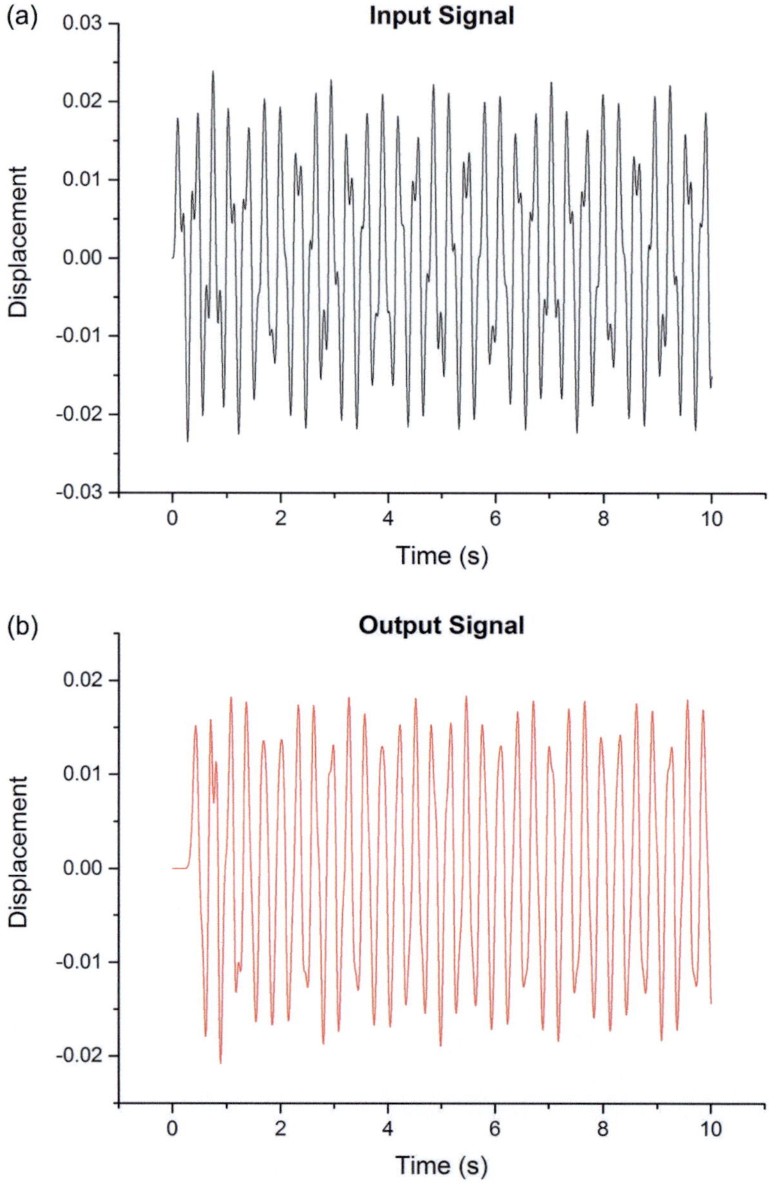

Fig. 6.8 In the case of a hard spring, the input signal contains frequencies $f_1 = 7.16\,\text{Hz}$ and $f_2 = 3.18\,\text{Hz}$. With an amplitude $A_0 \approx 0.2$, In the above figure, the top graph represents the time-domain signals of input and output, while the bottom graph represents the frequency-domain signals. When the amplitude is small, the component of the signal with a frequency of $f_1 = 7.16\,\text{Hz}$ experiences a significant attenuation when passing through the system, while the component with a frequency of $f_2 = 3.18\,\text{Hz}$ does not experience a noticeable attenuation when passing through the system

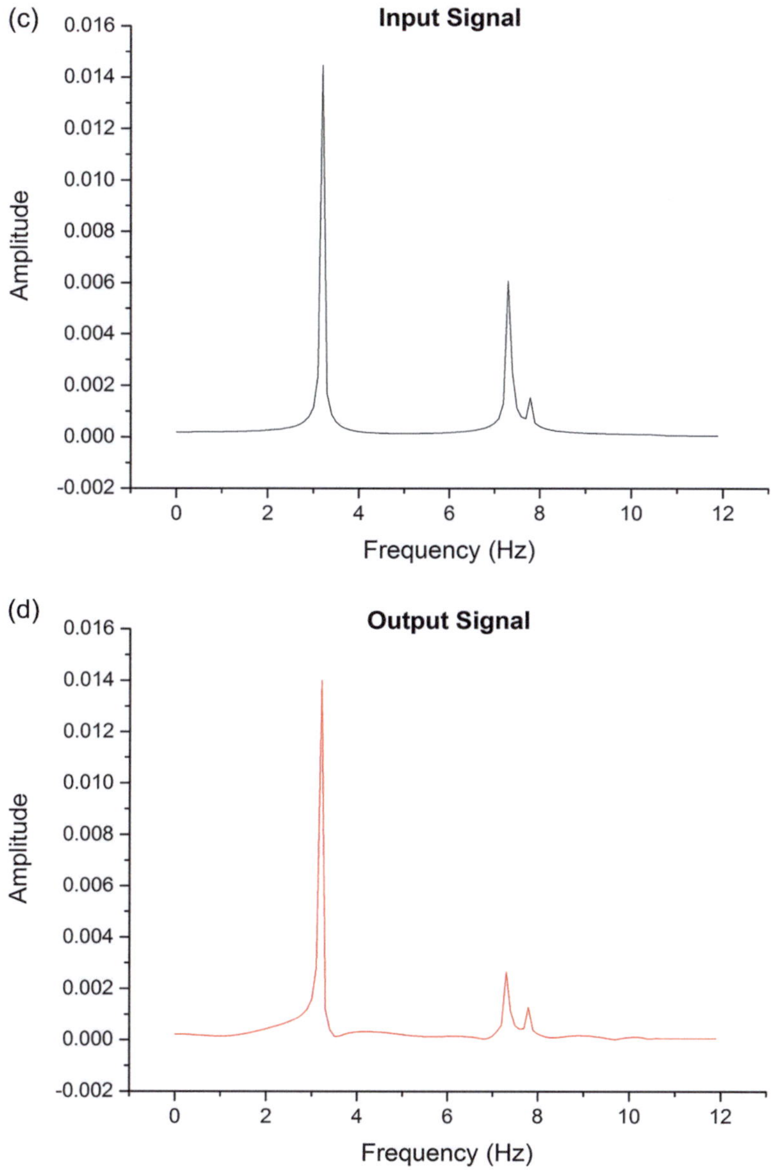

Fig. 6.8 (continued)

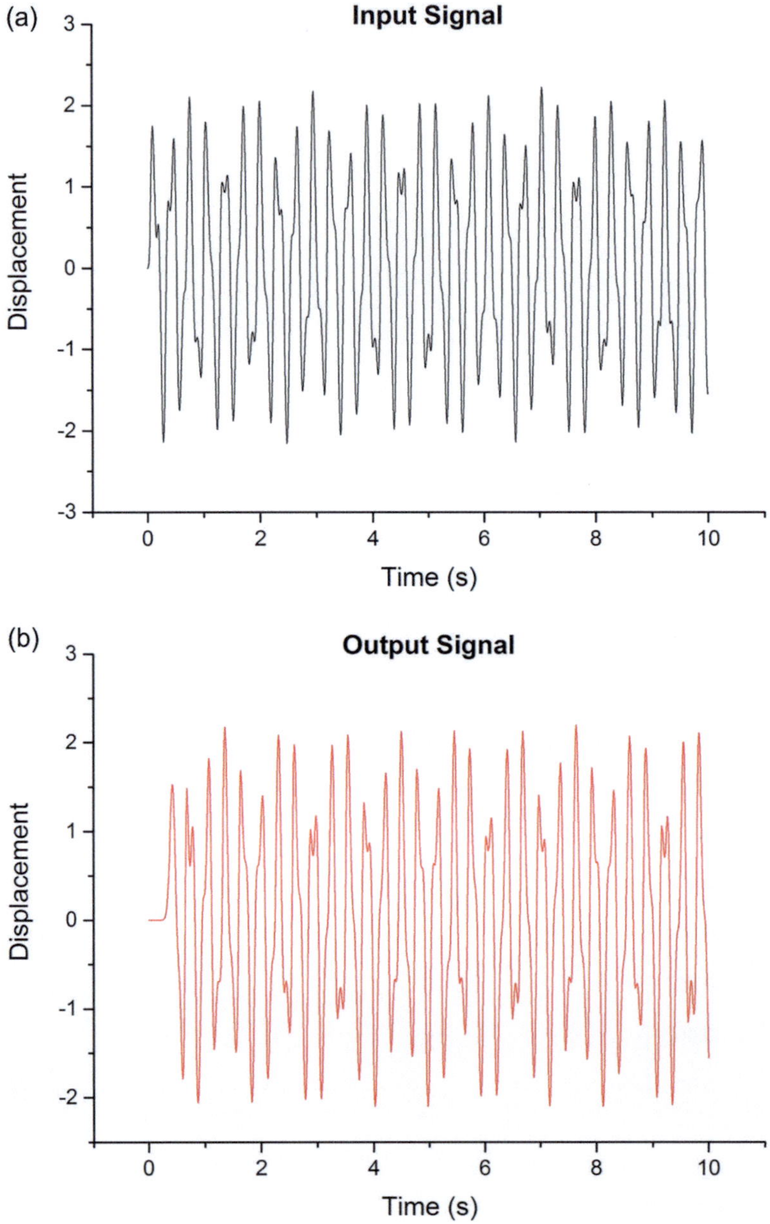

Fig. 6.9 In the case of a hard spring, the input signal contains frequencies $f_1 = 7.16\,\text{Hz}$ and $f_2 = 3.18\,\text{Hz}$. At an amplitude of $A_0 \approx 2$, the top graph represents the time-domain signals of input and output, while the bottom graph represents the frequency-domain signals. When the amplitude is large, both frequencies in the signal can pass through the system without significant attenuation

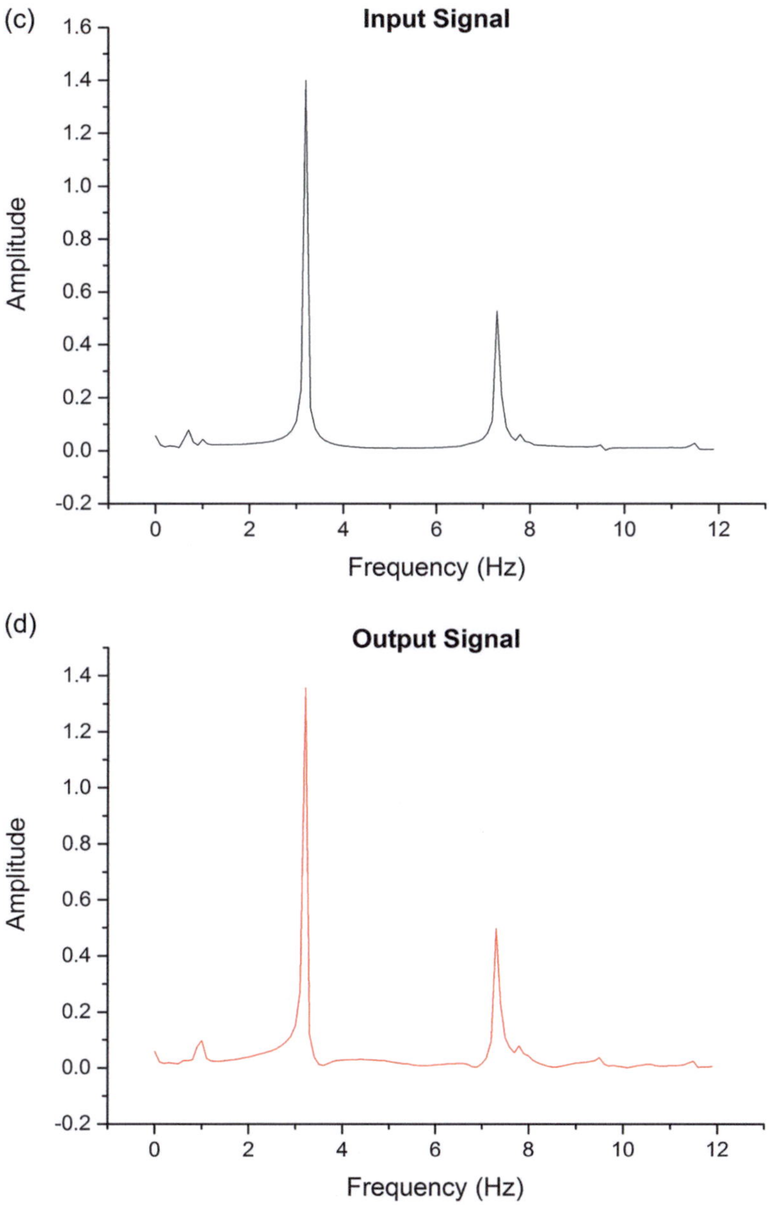

Fig. 6.9 (continued)

[13]. In 2011, N. Boechler and G. Theocharis used frequency filters and asymmetric combinations to produce bifurcations to obtain a rectification ratio greater than 10^4 [34]. Alù and colleagues realized that acoustic isolation can be achieved by breaking spatiotemporal symmetry through external flow [35]. Popa and others proposed a more compact structure of active acoustic metamaterials to achieve unidirectional transmission [36]. Cheng Jianchun and others designed a broadband near-perfect acoustic diode using acoustic nonlinear materials and nonlinear electronic components in a compensatory manner [37]. In 2015, T. Devaux studied the experimental characterization of single-mode elastic waveguides obtained by coupling selection layers and conversion layers [38], the results show that some acoustic metamaterials cannot play a key role in circuits like electronic components. The emergence of nonlinear acoustic metamaterials is that linear materials cannot break reciprocity [39–43]. We have designed an acoustic material with a rectifying effect composed of nonlinear periodic structures and linear periodic structures.

6.3.1 Principle of Asymmetric Structure

We have designed a device that includes a nonlinear periodic structure and a linear periodic structure as shown in Fig. 6.10. We introduce a nonlinear structure that depends on the input amplitude to break the reciprocity principle. Theoretical analysis shows that as long as the structural parameters are selected reasonably, the transmission spectrum of the nonlinear structure significantly depends on the amplitude response, which can lead to asymmetry in the transmission of sound energy. As the amplitude or direction changes, the sound energy flow through the system will be different. Next, we did a simulation to confirm our prediction.

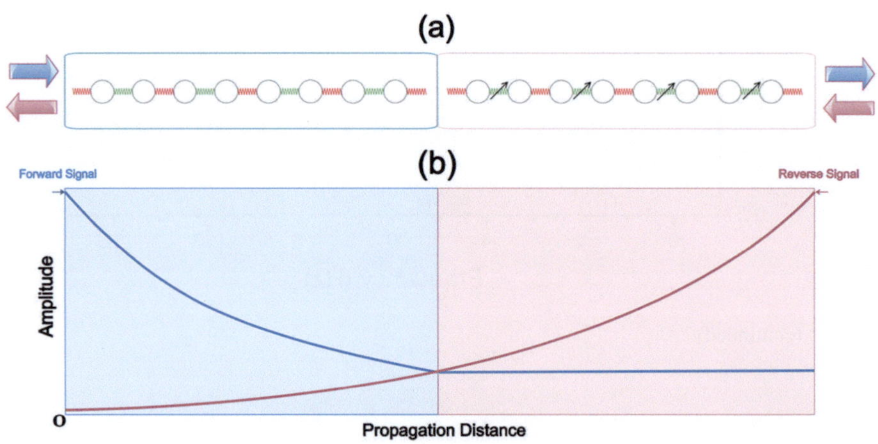

Fig. 6.10 **a** Schematic of the asymmetric structure. **b** Amplitude distribution of elastic waves in two opposite directions

Figure 6.10 illustrates how asymmetric transmission occurs in our designed structure. As shown in Fig. 6.10a, the positive direction of the device is the direction in which the incident wave propagates from the linear system at the left end to the nonlinear system on the right, and the reverse direction is just the opposite. In Fig. 6.10b, for the linear periodic structure, $K_1 = 850\,\text{N/m}$, $K_2 = 1200\,\text{N/m}$, $m = 1\,\text{kg}$, $N = 15$, the lower boundary (acoustic branch cutoff frequency) and upper boundary (optical branch cutoff frequency) of the bandgap are $f_1 = 6.56\,\text{Hz}$, $\text{f}_1 = 6.56$ Hz and $f_2 = 7.80\,\text{Hz}$, and remain unchanged. For the nonlinear periodic structure, $K_1 = \left(1000 - 100\xi^2\right)\text{N/m}$, $K_2 = 1200\,\text{N/m}$, $m = 1\,\text{kg}, N = 27$, the upper boundary of the bandgap remains at $f_2 = 7.80\,\text{Hz}$ unchanged while the lower boundary moves to a lower frequency from 7.12 Hz as the amplitude increases. When the input amplitude is small enough, the nonlinear medium can be approximated as a quasi-linear system. As we gradually increase the strength of the input signal, the medium will gradually exhibit strong nonlinear effects. This provides the possibility for acoustic waves to propagate asymmetrically from different directions.

In order to achieve non-reciprocal transmission, it is very necessary to select an appropriate incident frequency and a sufficiently large amplitude. The selected incident frequency needs to be within the bandgap of the linear periodic structure and also within the passband of the quasi-linear periodic structure. We analyze the situations where the incident wave passes through the non-reciprocal structure in the forward and reverse directions, as shown in Fig. 6.11. When a large amplitude sound wave ($f = 7.08\,\text{Hz}$) is incident from the left end into the linear periodic structure, because the frequency is within the bandgap of the linear periodic structure, the amplitude will decay with the propagation distance (Fig. 6.11a). Then, the elastic wave that has decayed for the first time arrives at the nonlinear periodic structure. Due to the quasi-linear property of the nonlinear material, the frequency of the elastic wave is within the passband of the nonlinear structure (0.00–7.12 Hz), then it undergoes a second attenuation (Fig. 6.11b). For the case of sound waves incident from the right end, when the large amplitude sound wave first passes through the nonlinear periodic structure, the strong nonlinearity generated causes the sound wave frequency (7.08 Hz) to fall within the bandgap of the nonlinear structure, causing attenuation at this time (Fig. 6.11c). Then, the attenuated elastic wave undergoes a second attenuation when it passes through the linear periodic structure (Fig. 6.11d). The result is: when the square wave and the reverse direction input sound wave are exactly the same, the output amplitude is completely different, which is due to the basic mechanism of non-reciprocal materials.

6.3.2 Non-reciprocal Structure Simulation

In order to prove the mechanism of our designed non-reciprocal device, we simulate this structure. During the simulation, we increase the number of mass blocks to 300, and set the simulation time to 15 s to minimize reflection as much as possible. For the linear mass-spring system, the acoustic branch cutoff frequency and optical

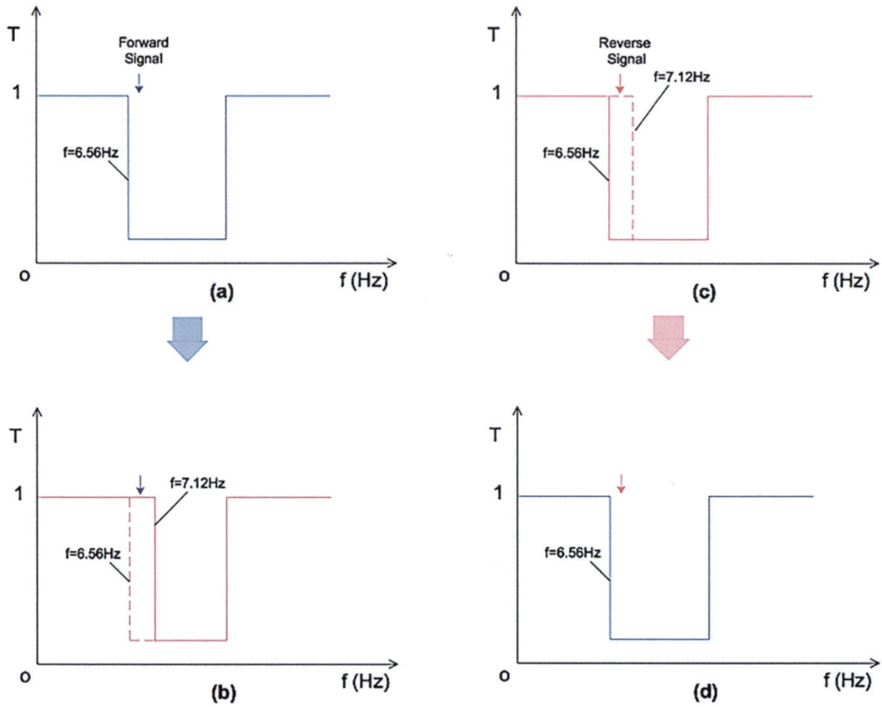

Fig. 6.11 Schematic diagram of incident waves exciting the non-reciprocal structure from opposite directions

branch cutoff frequency are $f = 6.56$ Hz and $f = 7.80$ Hz respectively, and remain constant. For the nonlinear mass-spring system, the optical branch medium frequency remains at $f = 7.80$ Hz while the cutoff frequency of the acoustic branch moves to a lower frequency as the amplitude increases from $f = 7.12$ Hz. The frequency of the incident wave should satisfy the relationship 6.56 Hz $< f <$ 7.12 Hz. During the simulation, in the Load module, the first mass block is given a periodic excitation with a frequency of $f = 7.80$ Hz, and the amplitude of the force is 1600 N. In order to better understand the physical picture of the vibration process, we extracted the vibration conditions of the first mass block (N = 1), the mass block connecting the linear structure and the nonlinear structure in the middle (N = 15), and the last mass block (N = 41), as shown in Fig. 6.12. For the forward case, the initial amplitude is very large, and when it reaches the connecting mass block (N = 15), the amplitude will decrease to a certain value so that the subsequent nonlinear structure exhibits quasi-linear properties, so that the final mass block (N = 41) has an amplitude almost identical to the connecting mass block (N = 15). For the reverse case, the incident wave with the same amplitude first hits the nonlinear periodic structure. Due to the strong nonlinearity, the frequency of the incident wave is in the bandgap, so the amplitude of the elastic wave has a significant attenuation when it reaches the connecting mass block (N = 15). Then it comes to the linear periodic structure and experiences a second attenuation. Figure 6.13a extracts the amplitude of every odd mass block from the simulation, which corresponds exactly to the principle diagram of the non-reciprocal structure in Fig. 6.10b. From the simulation, it can be seen that when the input amplitude is the same, the output amplitude is very different. Figure 6.13b reveals that the forward output amplitude is much larger than the reverse output amplitude (the amplitude ratio is an order of magnitude). It can also be seen that the output signal at the frequency of $f = 7.08$ Hz generates additional frequency components, which is due to the resonance phenomenon caused by the high modal density at the cutoff frequency of the optical branch.

6.4 Propagation of Sound Waves in Nonlinear Metamaterials

6.4.1 Formation Mechanism and Characteristics of the Bandgap of Localized Resonance Type Phononic Crystals

In 2000, Professor Zhengyou Liu proposed the concept of localized resonance type phononic crystals in Science [44], thus opening up a new field of phononic crystal research. The physical mechanism of the bandgap formation of localized resonance type phononic crystals is different from Bragg scattering, and is determined by the strong resonance characteristics of the localized resonance unit. Metamaterials have a novel characteristic that natural materials do not possess due to their localized resonance mechanism. In the case of acoustic metamaterials, this new characteristic directly comes from the frequency-limited behavior of the two related parameters, mass density and bulk modulus. Below we will focus on introducing the dynamic mass density related to metamaterials.

The impact of these negative effective mass and negative bulk modulus characteristics on the motion of sound waves in acoustic metamaterials (AM) is fascinating, as these negative properties do not exist in any naturally occurring material. This has led to many research activities aimed at developing new designs and technologies based on AM [45]. Acoustic metamaterials are divided into three overlapping types: phononic crystals, intrinsic acoustic metamaterials, and inertial acoustic metamaterials. The concept of phononic crystals predates the current wave of research on acoustic metamaterials (AM). The bandgap caused by Bragg scattering due to atomic-scale periodic impedance mismatch arrangement [46], wave redirection and focusing [47]. Fok et al. divided acoustic metamaterials into intrinsic acoustic metamaterials and inertial acoustic metamaterials based on whether the size of the microstructure is fully coupled or decoupled with the manipulated wavelength scale [45]. In intrinsic acoustic metamaterials, the phase velocity, which includes

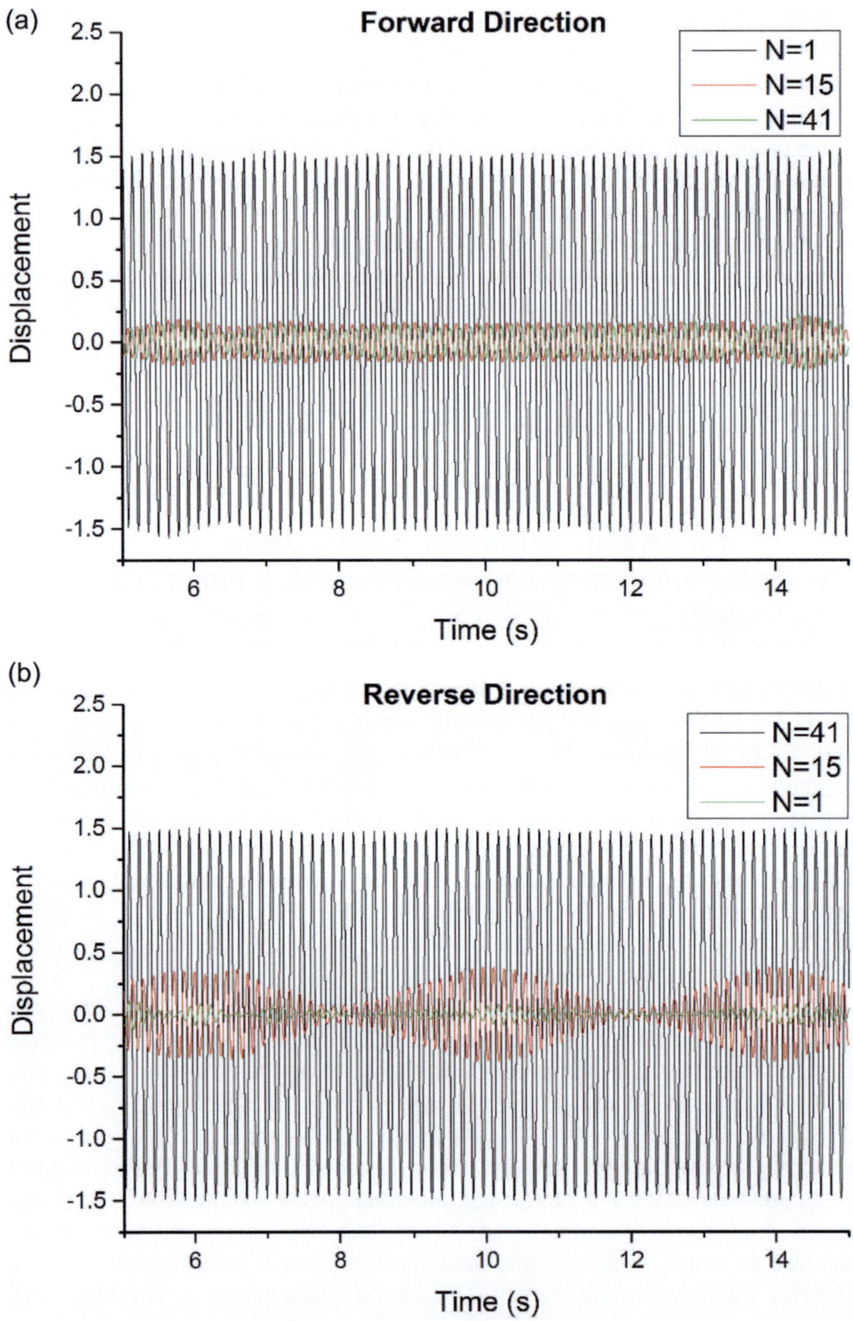

Fig. 6.12 Simulation results of the input and output signals of the non-reciprocal structure. **a** Forward direction. **b** Reverse direction

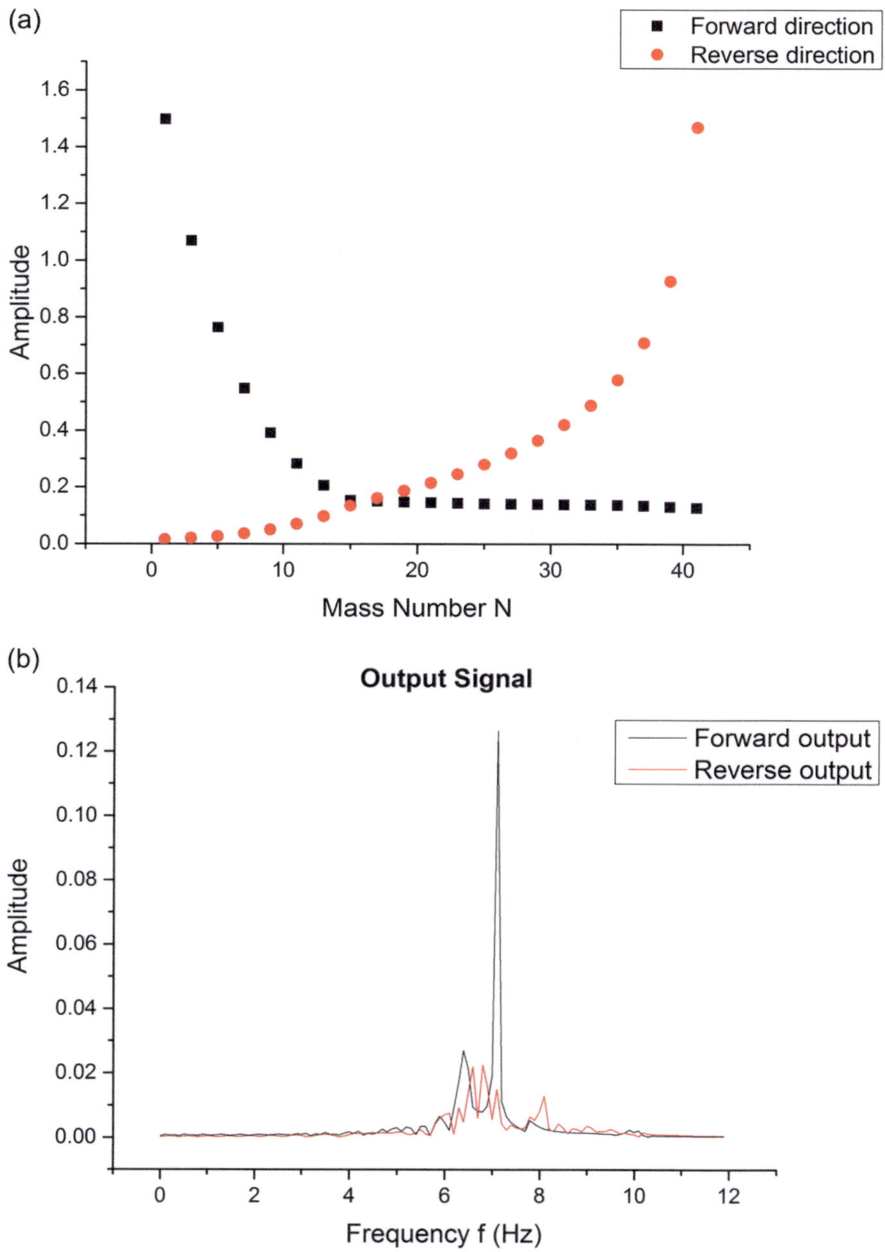

Fig. 6.13 Time-domain simulation results of the amplitude distribution of the elastic wave in the non-reciprocal structure (**a**) and the comparison of the output in the frequency domain in both directions (**b**)

phase, is much lower than the phase velocity in the bulk material, leading to localized resonance in the adjustable frequency range [48, 49]. For inertial acoustic metamaterials [50, 51], the energy of sound waves can be localized using engineered microstructures, such as mass-spring-damper resonators, without affecting the host material. Bandgaps caused by periodicity and resonance can also be interactively enhanced [52]. Lu et al. speculated that further reducing the size of the microstructure would lead to an increase in the acoustic frequency of the metamaterial structure, eventually overlapping with the electromagnetic counterpart of the material, which could cause unusual interactions [53] Combining acoustics and electromagnetic resonance, new microstructures can be designed. We know that for Bragg scattering type phononic crystals, the low-frequency dispersion curve can be considered linear, that is, low-frequency elastic waves exhibit the characteristics of long-wave traveling waves. The presence of localized oscillators changes the original dispersion curve of the system and splits it at its resonance frequency, leading to the generation of localized resonance band gaps. It can be seen that the localized resonance band gap is the result of the coupling of the long-wave traveling wave characteristics in the matrix and the resonance characteristics of the periodically distributed localized oscillators. Localized resonance type phononic crystals and Bragg scattering type phononic crystals have obvious differences, mainly: (1) The band gap frequency is much lower than the Bragg band gap of the same crystal size, achieving small size control of large wavelengths. Usually, the lattice constant of the Bragg scattering type phononic crystal is equivalent to half of the wavelength corresponding to the band gap frequency in the matrix material. Therefore, in order to obtain a sound wave band gap near 1000 Hz in water, the lattice size of the Bragg scattering type phononic crystal must reach 0.75 m. Plus, several periods are needed to form effective attenuation in the band gap frequency segment, so to use the band gap to achieve sound insulation around 1000 H requires several meters of Bragg scattering type phononic crystal, which is obviously impractical. The lattice constant of the localized resonance type phononic crystal is two orders of magnitude smaller than the wavelength corresponding to the band gap frequency in the matrix material. To achieve 1000 Hz underwater sound insulation only requires a thickness of the centimeter level, which is completely acceptable. Therefore, localized resonance type phononic crystals have opened up a new path for the application of phononic crystals in low-frequency sound wave and vibration control. (2) There are flat bands in the band structure, and there is a localized resonance phenomenon in the internal wave field. (3) The band gap is determined by the localized resonance characteristics of individual scatterers, regardless of their arrangement. For localized resonance type phononic crystals, the width and position of the band gap are mainly determined by the localized resonance characteristics of the scatterers, regardless of their lattice form. Under the same filling rate, using the same scatterers to adopt simple cubic, face-centered cubic, body-centered cubic, hexagonal close-packed and diamond structure lattice forms to obtain three-dimensional localized resonance type phononic crystals, their first band gaps are almost exactly the same. (4) The band gap width increases with the increase of the filling rate. For Bragg scattering type phononic crystals, the band gap width first increases and then decreases with the increase of the filling rate, while for localized resonance type phononic crystals, the larger the filling rate, the wider the band gap, and the band gap width increases monotonically with the increase of the filling rate.

6.4.2 Nonlinear Research Based on the Band Gap Formation Mechanism of Localized Resonance Type Phononic Crystals

Considering a one-dimensional lattice composed of the same mass-in-mass units, as shown in Fig. 6.14. The motion equation of the jth unit cell is:

$$m_1^{(j)}\frac{d^2u_2^{(j)}}{dt^2}+k_1\left(2u_1^{(j)}-u_1^{(j-1)}-u_1^{(j+1)}\right)+k_2\left(u_1^{(j)}-u_2^{(j)}\right)=0 \tag{6.9}$$

$$m_2^{(j)}\frac{d^2u_2^{(j)}}{dt^2}+k_2\left(u_2^{(j)}-u_1^{(j)}\right)=0 \tag{6.10}$$

where, $u_1^{(j)}$ and $u_2^{(j)}$ represent the displacements of the vibrations of mass block 1 and mass block 2 of the jth cell, respectively. Consider a regular single-atom lattice (Fig. 6.14b) to represent this mass-in-mass structure. The dispersion relations of this mass-in-mass structure and the single-atom lattice can be represented as

$$\cos ql = 1-\frac{\delta}{2\theta}\frac{\left(\frac{\omega}{\omega_0}\right)^2\left[\left(\frac{\omega}{\omega_0}\right)^2-(1+\theta)\right]}{\left(\frac{\omega}{\omega_0}\right)^2-1} \tag{6.11}$$

$$\cos ql = 1-\frac{\delta}{2\theta}\frac{(1+\theta)m_{eff}}{m_{st}}\left(\frac{\omega}{\omega_0}\right)^2 \tag{6.12}$$

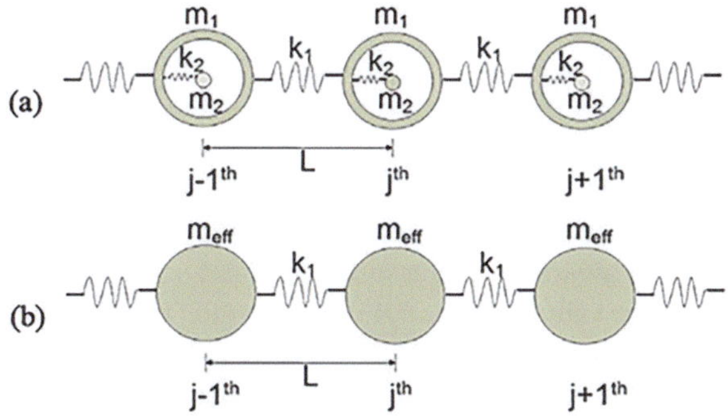

Fig. 6.14 **a** Infinite length mass-in-mass lattice structure. **b** Corresponding equivalent lattice system

where $\theta = \frac{m_2}{m_1}, \delta = \frac{k_2}{k_1}, \omega_0^2 = \frac{k_2}{m_2}, m_{st} = m_1 + m_2$ [54]. In order to make the two dynamic lattice systems equivalent, the effective mass of the single-atom lattice is chosen to replace the mass-in-mass structure, so the two systems can be perfectly matched. In other words, when Eqs. (6.11) and (6.12) are completely equivalent, the relationship Eq. (6.13) needs to be satisfied.

$$\frac{m_{eff}}{m_{st}} = 1 + \left(\frac{\theta}{1+\theta}\right)\frac{(\frac{\omega}{\omega_0})^2}{1-(\frac{\omega}{\omega_0})^2} \tag{6.13}$$

From Eq. (6.13), it can be seen that when the frequency ω approaches the localized resonance frequency ω_0, negative mass and density will appear. According to the equation, negative effective mass and density occur in the interval

$$1 < \left(\frac{\omega}{\omega_0}\right)^2 < 1 + \theta \tag{6.14}$$

From Eq. (6.14), it can be seen that by adjusting the ratio of the two masses θ, the frequency interval of negative equivalent mass can be easily controlled. One extreme case is when $\theta = \frac{m_2}{m_1} = 0$, the situation is the same as that of a single atomic chain. Another extreme case is $\theta = \frac{m_2}{m_1} = \infty$, which will result in negative equivalent mass in all frequency intervals greater than the resonance frequency. However, for a certain material, the mass ratio $\theta = \frac{m_2}{m_1}$ is fixed, so the mass ratio cannot be arbitrarily adjusted in one material. In order to adjust the frequency interval of negative equivalent mass to achieve the purpose of adjusting the position and width of the bandgap, we introduce nonlinearity.

Since both the upper and lower boundaries of the bandgap are related to ω_0, we can change the upper and lower boundaries by changing the working frequency ω_0. At this time, we only need to make k_2 a nonlinear spring. By $\omega_0 = \sqrt{\frac{k_2}{m_2}}$ It is known that when k_2 is a hard spring, the stiffness increases with the increase of amplitude, ω_0 will also move upward with the increase of amplitude; when k_2 is a soft spring, the situation is reversed, ω_0 moves downward with the increase of amplitude. To verify our conjecture, we used finite element simulation software to perform linear and nonlinear simulations. In the linear simulation process, we set $m_1 = 0.18$ kg, $m_2 = 0.245$ kg, $k_1 = 10{,}000$ N/m, $k_2 = 1000$ N/m. We can get $\omega_0 = 63.89$ Rad/s, the corresponding frequency is $f_0 = 10.17$ Hz, $\theta = \frac{m_2}{m_1} = 1.36$, the upper boundary of the effective mass with a negative bandgap is $f_1 = 15.62$ Hz. The simulation unit has a total of 50, and the first and eighth mass blocks m_1 are used to make the transmission coefficient curve. We used the method of perfect matching layer to avoid too much reflection, and set damping in k_1 after the 15th unit, with damping values of 0.1, 0.4, 0.9, 1.6, etc. The simulation results of the transmission coefficient curve are shown in Fig. 6.15.

As can be seen from Fig. 6.15, the simulation results are in good agreement with the theoretical bandgap boundaries (10.17–15.62 Hz). In the corresponding bandgap frequency range, the transmission coefficient is significantly lower than the

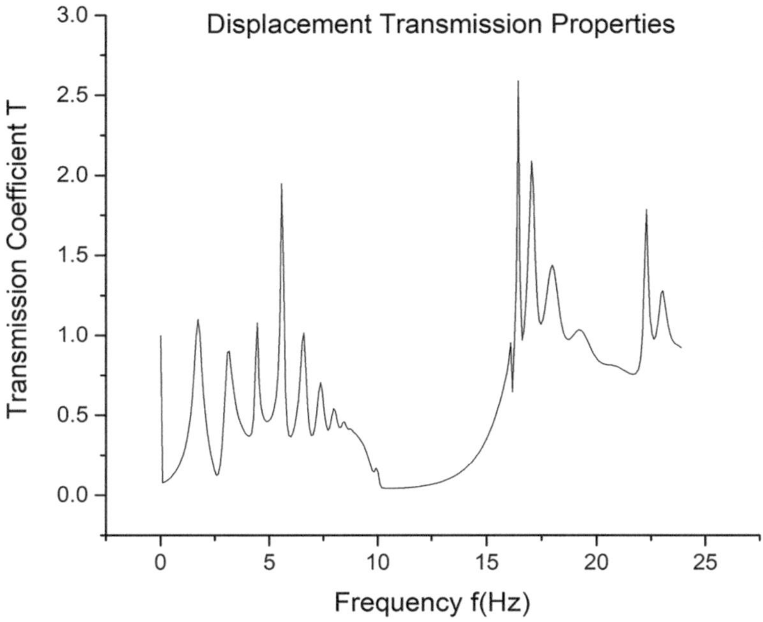

Fig. 6.15 Simulation diagram of the transmission curve of the linear mass-in-mass structure

transmission coefficient of other frequency ranges. And the transmission coefficient near the lower boundary of the bandgap is lower than the transmission coefficient at the upper boundary of the bandgap, this is because the resonance frequency of the internal mass-spring system is exactly at the lower boundary of the bandgap, and the absorption of vibration should be strongest at this frequency. We also simulated the transmission coefficient of sound waves passing through 10 units. As shown by the red curve in Fig. 6.16. The results show that the bandgap transmission coefficient value of passing through 10 units is less than the bandgap transmission coefficient value of passing through 8 units. This indicates that the more units passed, the more the vibration attenuates, which is consistent with the actual situation.

Because the formation of the bandgap is due to the absorption of sound energy by the resonance brought by the negative mass, the intensity of the absorption is related to the damping of k_2. So next, we simulated the transmission coefficient curve under different damping conditions. Set the damping of k_2 to 0.5, the other steps are exactly the same as the linear simulation steps. The simulation results are shown in Fig. 6.17. The results show that when there is damping, the transmission coefficient of the bandgap is less than when there is no damping, this is because when there is damping, the energy absorption of the vibration in the bandgap is more intense.

In addition, we also conducted a nonlinear simulation of this acoustic metamaterial. In the simulation, we set the linear spring k_2 Replace with a nonlinear connector (connector), the parameters of the nonlinear connector are set to $F =$

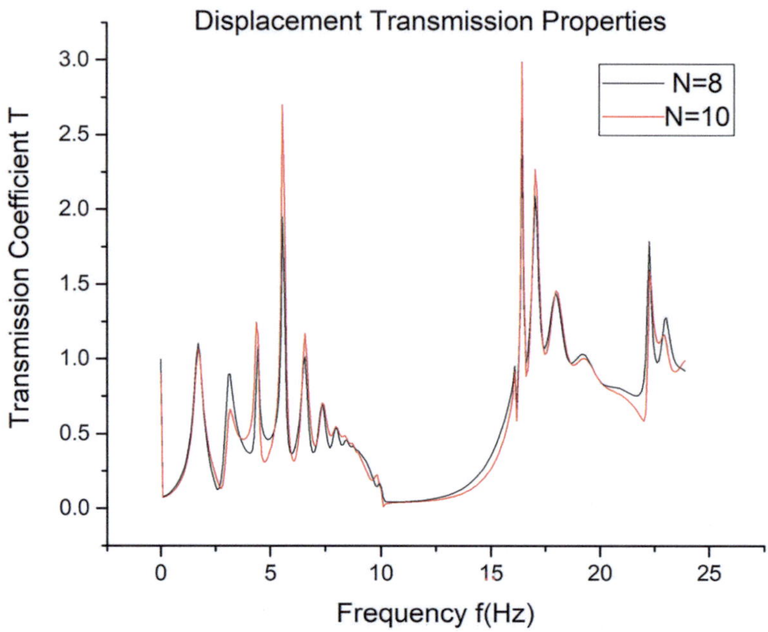

Fig. 6.16 Simulation of linear transmission coefficient with different unit numbers

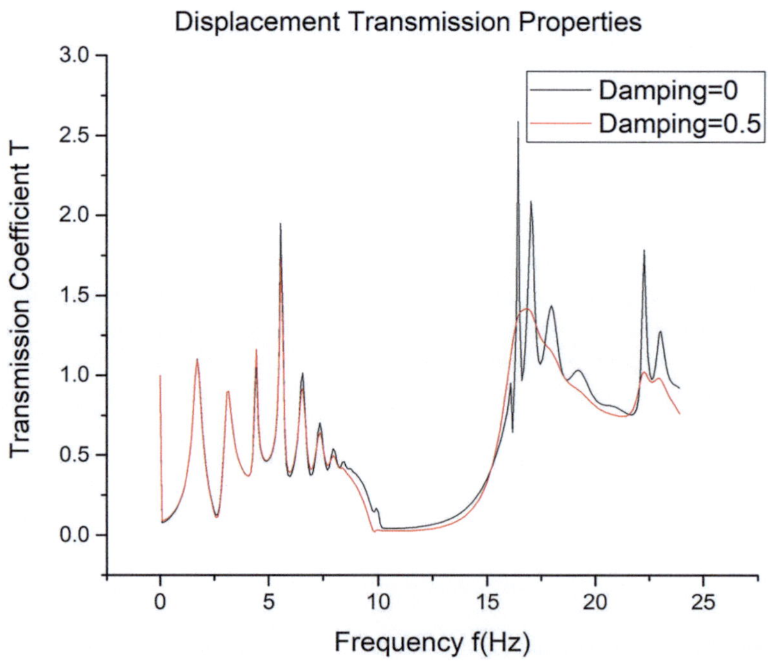

Fig. 6.17 Simulation of the transmission coefficient curve under different damping conditions

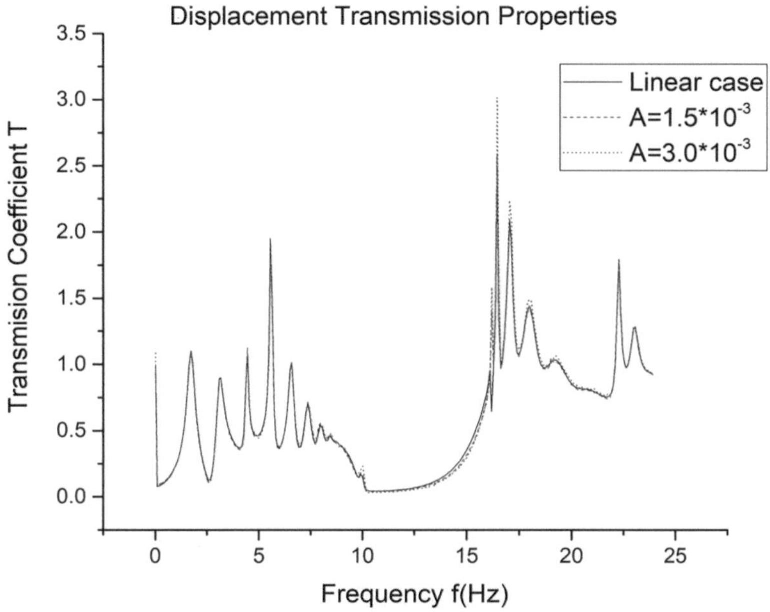

Fig. 6.18 Simulation of nonlinear transmission coefficient of hard spring

$1000 * x + 2{,}000{,}000 * x^3$ (hard spring). When the amplitude is small ($x < 10^{-3}$ m), we can consider the connector as a linear spring; when the amplitude reaches a suitable value, the nonlinear effect cannot be ignored at this time, and the bandgap will move according to the above analysis. The simulation results show that at small amplitudes, the transmission coefficient curve is the same as the linear case, and the bandgap interval is 10.17–15.62 Hz. When the amplitude (A) reaches 10^{-3} m, the nonlinear effect cannot be ignored, and the bandgap moves towards high frequency. And within a certain amplitude range, the larger the amplitude, the more the upper and lower boundaries of the bandgap move towards high frequency, as shown in Fig. 6.18. Similarly, we also simulated the transmission coefficient under the condition of a soft spring. That is, replace k_2 with a nonlinear connector (connector), the parameters of the nonlinear connector are set to $F = 1000 * x - 2{,}000{,}000 * x^3$. In this case, as the amplitude increases, the stiffness of the nonlinear connector will gradually decrease, the resonance frequency of m $_2$ will also decrease, and the bandgap will move towards low frequency. The simulation results are shown in Fig. 6.19. It can be concluded that after replacing k_2 with a nonlinear spring, we can control the position of the bandgap of the acoustic metamaterial structure by controlling the amplitude.

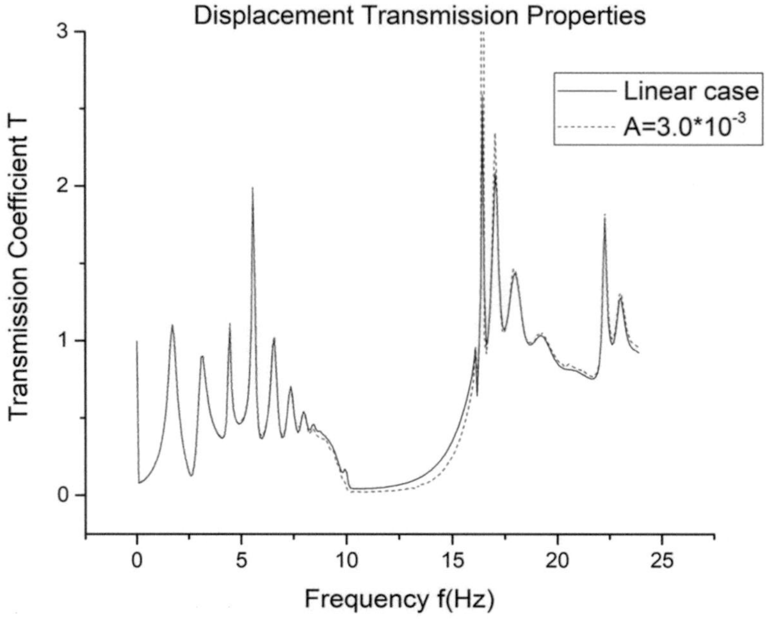

Fig. 6.19 Simulation of nonlinear transmission coefficient of soft spring

References

1. Wang Z, Chong YD, Joannopoulos JD, et al. Reflection-free one-way edge modes in a gyromagnetic photonic crystal. Phys Rev Lett. 2008;100:013905.
2. Haldane FD, Raghu S. Possible realization of directional optical waveguides in photonic crystals with broken time-reversal symmetry. Phys Rev Lett. 2008;100:013904.
3. He C, Chen XL, Lu MH, et al. Tunable one-way cross-waveguide splitter based on gyromagnetic photonic crystal. Appl Phys Lett. 2010;96:111111.
4. Serebryannikov AE. One-way diffraction effects in photonic crystal gratings made of isotropic materials. Phys Rev B. 2009;80:155117.
5. Serebryannikov AE, Ozbay E. Unidirectional transmission in non-symmetric gratings containing metallic layers. Opt Exp. 2009;17:13335.
6. Yu Z, Fan S. Complete optical isolation created by indirect interband photonic transitions. Nat Photon. 2009;3:91–4.
7. Li BW, Wang L, Casati G. Thermal diode: rectification of heat flux. Phys Rev Lett. 2004;93:184301.
8. Li BW, Lan J, Wang L. Interface thermal resistance between dissimilar anharmonic lattices. Phys Rev Lett. 2005;95:104302.
9. Chang CW, Okawa D, Majumdar A, et al. Solid-state thermal rectifier. Science. 2006;314:1121.
10. Li BW, Wang L. Phononics gets hot. Phys World. 2008;21:27.
11. Kobayashi W, Teraoka Y, Terasaki I. An oxide thermal rectifier. Appl Phys Lett. 2009;95:171905.
12. Nesterenko V, Daraio CE, Herbold H, et al. Anomalous wave reflection at the interface of two strongly nonlinear granular media. Phys Rev Lett. 2005;95:158702.
13. Liang BB, Yuan B, Cheng JC. Acoustic diode: rectification of acoustic energy flux in one-dimensional systems. Phys Rev Lett. 2009;103:104301.

14. Boechler N, Theocharis G, Daraio C. Bifurcation-based acoustic switching and rectification. Nat Mater. 2011;10:665–8.
15. Sigalas MM, Economou EN. Elastic and acoustic wave band structure. J Sound Vib. 1992;158(2):377–82.
16. Kushwaha MS, Halevi P, Dobrzynski L, et al. Acoustic band structure of periodic elastic composites. Phys Rev Lett. 1993;71(13):2022–5.
17. Dowling JP. Sonic band structure in fluids exhibiting periodic density variations. J Acoust Soc Am. 1992;91(5):2539–43.
18. Sigalas MM, Economou EN. Elastic-wave propagation through disorderd and/or absorptive layered systems. Phys Rev B. 1995;51(5):2780–9.
19. Kafesaki M, Sigalas MM, Economou EN. Elastic wave band gaps in 3-D periodic polymer matrix composites. Solid State Commun. 1995;96(5):285–9.
20. Vasseur JO, deymier PA, Frantziskonic G, et al. Experimental evidence for media. J Phys Condens Matter 1998;10(27):6051–64.
21. Mead DJ. Wave propagation in continuous periodic structures: research contributions from Southampton. J Sound Vib. 1996;190(3):495–524.
22. Narisetti RK, Leamy MJ, Ruzzene M. A perturbation approach for predicting wave propagation in one-dimensional nonlinear periodic structures. J Vib Acoust. 2010;132(3):031001.
23. Nesterenko VF, Daraio C, Herbold EB, et al. Anomalous wave reflection at the interface of two strongly nonlinear granular media. Phys Rev Lett. 2005;95:158702.
24. Fan L, Wang J, Varghese LT, et al. An all-silicon passive optical diode. Science; 2012:1214383.
25. Chang L, Jiang X, Hua S, et al. Parity-time symmetry and variable optical isolation in active-passive-coupled microresonators. Nat Photon. 2014;8:524–9.
26. Zhu SL, Dreyer TM, Liebler R, Preminger GM, Zhong P, et al. Reduction of tissue injury in shock-wave lithotripsy by using an acoustic diode. Ultrasound Med Biol. 2004;30:675.
27. Peng B, Ozdemir SK, Lei F, Monifi F, Gianfreda M, Long GL, Fan S, Nori F, Bender CM, Yang L, et al. Parity–time-symmetric whispering-gallery microcavities. Nat Phys. 2014;10:394–8.
28. Takeno S, Kisoda K, Sievers AJ. Intrinsic localized vibrational modes in anharmonic crystals. Progr Theor Phys Suppl. 1988;94:242–69.
29. Vakakis AF, King ME, Pearlstein AJ. Forced localization in a periodic chain of non-linear oscillators. Int J Nonlin Mechan. 1994;29(3):429–47.
30. Daraio C, Nesterenko VF, Herbold EB, et al. Tunability of solitary wave properties in one-dimensional strongly nonlinear phononic crystals. Phys Rev E. 2006;73(2):026610.
31. Nayfeh AH, Kelly SG. Nonlinear propagation of waves induced by an infinite vibrating cylinder. J Phys Suppl. 1979;40:8–13.
32. Vakakis AF, King ME. Nonlinear wave transmission in a monocoupled elastic periodic system. J Acoust Soc Am. 1995;98(3):1534–46.
33. Chakraborty G, Malik AK. Dynamics of weakly non-linear periodic chain. Int J Non-Linear Mech. 2001;36(2):375–89.
34. Boechler N, Theocharis G, Daraio C. Bifurcation-based acoustic switching and rectification. Nat Mater. 2011;10(9):665–68.
35. Fleury R, Sounas DL, Sieck CF, et al. Sound isolation and giant linear nonreciprocity in a compact acoustic circulator. Science. 2014;343:516–9.
36. Popa B-I, Cummer SA. Non-reciprocal and highly nonlinear active acoustic metamaterials. Nat Commun. 2014;5:3398.
37. Gu ZM, Hu J, Liang B. Broadband non-reciprocal transmission of sound with invariant frequency. Sci Rep. 2015;6:19824.
38. Devaux T, Tournat V, Richoux O, et al. Asymmetric acoustic propagation of wave packets via the self-demodulation effect. Physics Review Letters. 2015;115:234301.
39. Li Y, Liang B, Gu ZM, et al. Unidirectional acoustic transmission through a prism with near-zero refractive index. Appl Phys Lett. 2013;103:053505.
40. Yuan B, Liang B, Tao JC, et al. Broadband directional acoustic waveguide with high efficiency. Appl Phys Lett. 2012;101:043503.

41. Li Y, Tu J, Liang B, et al. Unidirectional acoustic transmission based on source pattern reconstruction. J Appl Phys. 2012;112:064504.
42. Maznev AA, Every AG, Wright OB. Reciprocity in reflection and transmission: what is a 'phonon diode'? Wave Motion. 2013;50:776–84.
43. Hamilton MF, Blackstock DT. Nonlinear-Wave propagation in a fluid layer. In: Nonlinear acoustics, 12th ISNA; 1990, p. 321–326.
44. Liu ZY, Zhang X, Mao Y, et al. Locally resonant sonic materials. Science. 2000;289:1734–6.
45. Fok L, Ambati M, Zhang X. Acoustic metamaterials. MRS Bull. 2008;33:931–4.
46. Yang SX, Page JH, Liu ZY, et al. Focusing of sound in a 3D phononic crystal. Phys Rev Lett. 2004;93(2):024301.
47. Celli P, Gonella S. Low-frequency spatial wave manipulation via phononic crystals with relaxed cell symmetry. J Appl Phys. 2014;115:103502.
48. Li J, Chan CT. Double-negative acoustic metamaterial. Phys Rev E. 2004;70(5):055602.
49. Ding YQ, Liu ZY, Qiu CY, et al. Metamaterial with simultaneously negative bulk modulus and mass density. Phys Rev Lett. 2007;99(9):093904.
50. Fang N, Xi D, Xu J, et al. Ultrasonic metamaterial with negative modulus. Nat Mater. 2006;5(6):452–56.
51. Ao XY, Chan CT. Far-field image magnification for acoustic waves using anisotropic acoustic metamaterials. Phys Rev E. 2008;77(2):025601.
52. Raghavan L, Phani AS. Local resonance bandgaps in periodic media: theory and experiment. J Acoust Soc Am. 2013;134(3):1950–9.
53. Lu M, Feng L, Chen Y. Phononic crystals and acoustic metamaterials. Mater Today. 2009;12(12):34–42.
54. Huang HH, Sun CT, Huang G. On the negative effective mass density in acoustic metamaterials. Int J Eng Sci. 2009;47:610–7.

Chapter 7
Reverse Excitation and Enhancement of Second Harmonics in Metamaterials

Nonlinear acoustic metamaterials are considered one of the most important future research directions in the field of metamaterials [1, 2]. Current applications of nonlinear metamaterials in electromagnetic waves include coordination [3, 4], parameter down-conversion [5], generation of higher harmonics [6–8], etc. Due to the similarity between sound waves and electromagnetic waves, nonlinear acoustic metamaterials have become a current research hotspot [9, 10]. In acoustics, the nonlinearity of metamaterials cannot be ignored, especially for pipe-type or resonant unit-containing acoustic metamaterials. In 1992, Sugimoto first conducted a nonlinear analysis of metamaterials periodically paralleled with Helmholtz resonators in ordinary waveguides [11], and found that the appearance of shock waves can be effectively suppressed by reasonably adjusting the structural parameters of the Helmholtz resonators. Later, in 2007, Richoux et al. studied the acoustic transmission characteristics of this type of metamaterial [12], and found that as the sound pressure level of the input sound wave increases, the resonance frequency of the Helmholtz resonator shifts towards the low frequency direction. In 1995, Bradley et al. used nonlinear Bloch theory to analyze the nonlinear higher harmonics of metamaterials with periodic parallel unit structures on ordinary waveguides [13]. In 2014, Fan et al. constructed a double-negative acoustic metamaterial using a thin film and side holes, and analyzed the shift rule of the two characteristic frequencies of this metamaterial under nonlinear conditions [14]. Subsequently, in 2016, Li et al. first conducted a nonlinear analysis of the transmission coefficient and scattering coefficient of the double-negative metamaterial based on the thin film and Helmholtz resonator [9]. Recently, Lan et al. theoretically obtained the specific expressions of the nonlinear equivalent bulk modulus and acoustic impedance of the single-negative metamaterial based on the Helmholtz resonator using the nonlinear perturbation method [10].

X. Liu, *Nonlinear Sound Waves in Solids*,
Advances in Acoustics, https://doi.org/10.1007/978-981-96-5253-2_7

7.1 Second Harmonic Reverse Excitation Caused by Acoustic Negative Refraction Materials

As known from Chap. 4, when sound waves propagate in nonlinear media, they will excite higher harmonics, which can be used in industry to image cracks in materials. Usually, the amplitude of the second harmonic increases with the increase of the fundamental wave. However, due to factors such as material attenuation, the higher harmonics are very small. Moreover, the use of harmonics to detect cracks cannot be performed inside the object, which limits its application prospects. This section theoretically predicts the reverse excitation of the second harmonic in acoustic metamaterials, that is, the direction of sound energy propagation can change without relying on reflection, so we can obtain the second harmonic at the sound source. At the same time, when the metamaterial is long enough, the size of the second harmonic will be independent of the length of the metamaterial, and if a strongly nonlinear medium is used as the base material, a large second harmonic can be obtained.

Figure 7.1 is a schematic diagram of the structure of an acoustic metamaterial, composed of a series of sub-wavelength scale vibrating thin plates and small tubes with holes. Its period constant is d, the cross-sectional area of the periodic hole tube is $S_0 = \pi a^2$ (where a is the radius of the hole tube), and the length of the hole tube is l. The mass of the periodic vibrating plate is M_m, and its force is C_m. If its period constant satisfies $d < \lambda/5$ (where λ is the wavelength of the sound wave), then this structure can be considered uniform.

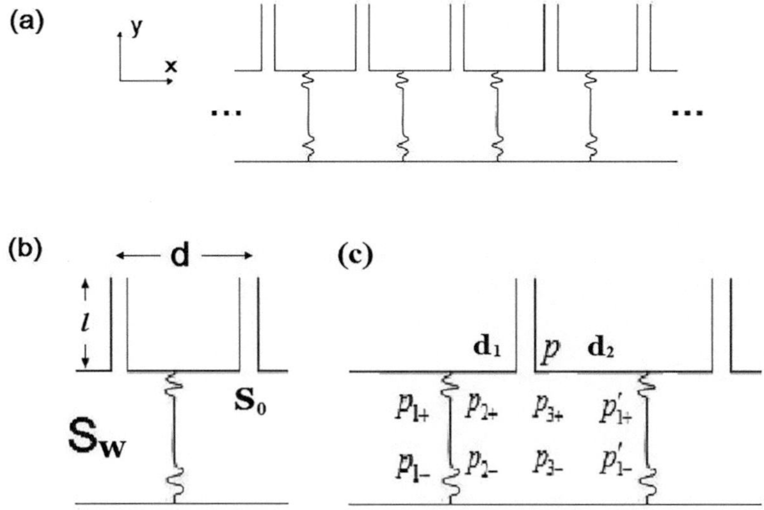

Fig. 7.1 **a** Schematic diagram of the metamaterial structure formed after embedding periodic vibrating thin plates and periodic hole tubes in a one-dimensional sound waveguide. **b** Schematic diagram of a single cell of the metamaterial. **c** Schematic diagram of sound pressure distribution analysis in the metamaterial

Therefore, its momentum equation and continuity equation can be written as

$$S_w \frac{\partial p}{\partial x} + \rho \frac{dU_1}{dt} + \frac{M_m}{d}\frac{dv_1}{dt} + \frac{1}{C_m d}\int_{-\infty}^{t} v_1 dt = 0 \tag{7.1}$$

$$S_w \frac{\partial \rho}{\partial t} + \frac{\partial}{\partial x}(\rho U_1) + \frac{\rho U_2}{d} = 0 \tag{7.2}$$

where, $\frac{M_m}{d}$ and $C_m d$ are the mass and force per unit length respectively, $U_1 = v_1 S_w$ is the volume velocity in the waveguide, $U_2 = v_2 S_0$ is the volumetric velocity in the perforated tube. According to Newton's law of motion, the momentum equation in the perforated tube can be written as

$$pS_0 = \rho l' S_0 \frac{dv_2}{dt} + \eta v_2 \tag{7.3}$$

where $l' = l + 1.46a$ is the effective length of the perforated tube, η is the dissipation coefficient, which can be measured by acoustic methods. If the condition $l \ll \lambda$ is met, the perforated tube on the waveguide can be considered a lumped parameter system, which satisfies the condition: $\frac{\partial v^2}{\partial y} = 0$, therefore $\frac{dv_2}{dt} = \frac{\partial v_2}{\partial t} + v_2 \frac{\partial v_2}{\partial y} = \frac{\partial v_2}{\partial t}$, Eq. (7.3) can be rewritten as

$$pS_0 = \rho l' S_0 \frac{\partial v_2}{\partial t} + \eta v_2 \tag{7.4}$$

By using Eqs. (7.2) and (7.4) to eliminate v_2, we can get

$$\left(\eta \frac{\partial}{\partial t} + \rho l' S_0 \frac{\partial^2}{\partial t^2}\right)\rho + \left(\eta \frac{\partial}{\partial x} + \rho l' S_0 \frac{\partial^2}{\partial x \partial t}\right)(\rho v_1) + \frac{\rho S_0^2}{S_w d} p = 0 \tag{7.5}$$

Rewrite Eq. (7.1) as

$$\frac{\partial p}{\partial x} + \rho \frac{dv_1}{dt} + \frac{M_m}{S_w d}\frac{dv_1}{dt} + \frac{1}{C_m S_w d}\int_{-\infty}^{t} v_1 dt = 0 \tag{7.6}$$

Under the condition of linear approximation, Eqs. (7.5) and (7.6) can be simplified as

$$\left(\eta \frac{\partial}{\partial t} + \rho_0 l' S_0 \frac{\partial^2}{\partial t^2}\right)\rho' + \rho_0\left(\eta \frac{\partial}{\partial x} + \rho_0 l' S_0 \frac{\partial^2}{\partial x \partial t}\right) v_1 + \frac{\rho_0 S_0^2}{S_w d} p = 0 \tag{7.7}$$

$$\frac{\partial^2 p}{\partial x \partial t} + \rho_0 \frac{\partial^2 v_1}{\partial t^2} + \frac{M_m}{S_w d} \frac{\partial^2 v_1}{\partial t^2} + \frac{1}{C_m S_w d} v_1 = 0 \tag{7.8}$$

Similarly, under the condition of linear approximation, its constitutive equation can be written as

$$\frac{p}{\rho'} = c_0^2 = \frac{\kappa}{\rho_0} \tag{7.9}$$

where κ is its bulk modulus under adiabatic conditions, c_0 is the speed of sound under perturbation approximation.

For a single frequency wave, we assume that the sound pressure and particle vibration velocity in the new structure can be represented as: $p = p(x)e^{j\omega t}$ and $v_1 = v_1(x)e^{j\omega t}$, where ω is the angular frequency of the sound wave. Using Eq. (7.9), in the frequency domain, we can rewrite Eqs. (7.7) and (7.8) in the following form

$$\frac{\frac{j\omega\eta - \omega^2 \rho_0 l' S_0}{\kappa} + \frac{S_0^2}{S_w d}}{\eta + j\omega\rho_0 l' S_0} p(x) + \frac{\partial}{\partial x} v_1(x) = 0 \tag{7.10}$$

$$\frac{\partial p(x)}{\partial x} + \frac{\frac{1}{C_m S_w d} - \omega^2 \left(\rho_0 + \frac{M_m}{S_w d}\right)}{j\omega} v_1(x) = 0 \tag{7.11}$$

where j is the imaginary unit. Under linearization approximation, using Eqs. (7.9), (7.1) and (7.2) can be simplified in the frequency domain as

$$\frac{\partial p}{\partial x} + j\omega\rho v = 0 \tag{7.12}$$

$$\frac{j\omega}{\kappa} p + \frac{\partial}{\partial x} v = 0 \tag{7.13}$$

By comparing Eqs. (7.6), (7.11) with Eqs. (7.12), (7.13), we can obtain the expressions for the equivalent bulk modulus $\kappa_{eff}(\omega)$ and the equivalent density $\rho_{eff}(\omega)$

$$\kappa_{eff}^{-1}(\omega) = \kappa^{-1}\left(1 - \frac{F}{\omega^2 + i\Gamma\omega}\right) \tag{7.14}$$

$$\rho_{eff}(\omega) = \rho_0 \left(1 - \frac{\omega_0^2 - \omega^2}{G\omega^2}\right) \tag{7.15}$$

where, $F = \frac{S_0 \kappa}{\rho_0 S_w d l'}$, $\Gamma = \frac{\eta}{\rho_0 l' S_0}$, $\omega_0^2 = \frac{1}{M_m C_m}$, $G = \frac{\rho_0 S_w d}{M_m}$. From Eqs. (7.10) and (7.11) we can get

$$\frac{\partial^2 p(x)}{\partial x^2} + k^2(\omega) p(x) = 0 \tag{7.16}$$

Here, $k(\omega) = \sqrt{\frac{\omega^2 \rho_{eff}(\omega)}{\kappa_{eff}(\omega)}}$ is the wave number of the sound wave in this metamaterial. The solution of Eq. (7.16) is

$$p = p_i e^{j(\omega t - kx)} + p_r e^{j(\omega t + kx)} \tag{7.17}$$

We can get the expression of the wave Poynting vector in this structure

$$\begin{aligned} Pv &= \mathrm{Re}\big[p^* v_1\big] = \mathrm{Re}\big[p^*(x)e^{-j\omega t} v_1(x)e^{j\omega t}\big] \\ &= \mathrm{Re}\big[p^*(x)v_1(x)\big] = \mathrm{Re}\left[j\frac{p^*(x)\frac{\partial p(x)}{\partial x}}{\omega \rho_{eff}}\right] \end{aligned} \tag{7.18}$$

For the wave $p = p_i e^{j(\omega t - kx)}$ with phase propagating along the x axis in the positive direction, Eq. (7.18) can be expressed as

$$Pv = \mathrm{Re}\left[\frac{k|p_i|^2}{\omega \rho_{eff}}\right] \tag{7.19}$$

We can see that if κ_{eff} and ρ_{eff} are both positive, then we can get $k > 0$ and $Pv > 0$, which means that the phase propagation direction and energy propagation direction of the sound wave in this structure are both towards the positive direction of the x axis, and the metamaterial shows right-handed properties at this time. If κ_{eff} and ρ_{eff} are both negative, then $k > 0$ but $k > 0$, which means that the phase of the sound wave propagates in the positive direction, but the sound energy propagates in the negative direction, and the metamaterial shows left-handed properties at this time. If the signs of κ_{eff} and ρ_{eff} are opposite, that is, one is positive and the other is negative, then k is an imaginary number, $Pv \to 0$, so the sound wave decays rapidly in the x direction and cannot propagate in this metamaterial.

In numerical calculations, we use the following parameters: $\rho_0 = 998$ m/s, $c_0 = 1483$ m/s, $\eta = 5 \times 10^{-3}$ N s/m, $a = 1$ mm, $S_0 = 3.14$ mm^2, $l = 1$ mm, $l' = 2.46$ mm, $d = 8$ mm, $S_w = 16$ mm^2, $M_m = 0.1$ g, $C_m = 2.5 \times 10^{-7}$ m/N. Figure 7.2a, b are the relationships between the equivalent volume modulus κ_{eff} and equivalent density ρ_{eff} of the metamaterial and frequency. We can find that there are two critical frequencies in this system, $f_{VP} = 21.092$ KHz and $f_{SH} = 23.575$ KHz, which are determined by the vibrating thin plate and the periodically perforated small tube. From Fig. 7.2, we can find that when.

$f < f_{SH}$, $\mathrm{Re}(\kappa_{eff}) < 0$, and the equivalent volume modulus is negative at this time. When $f < f_{VP}$, $\rho_{eff} < 0$, and the equivalent density is negative at this time. Therefore, when $f < f_{VP}$, the equivalent volume modulus and equivalent density of this structure are both negative, and the metamaterial shows left-handed properties at this time, and sound waves can propagate in this structure. When $f_{VP} < f < f_{SH}$, that is, the equivalent volume modulus is negative, but the equivalent density is positive, sound waves cannot propagate in this metamaterial. When $f > f_{SH}$, that is, the equivalent volume modulus and equivalent density are both positive, the metamaterial shows

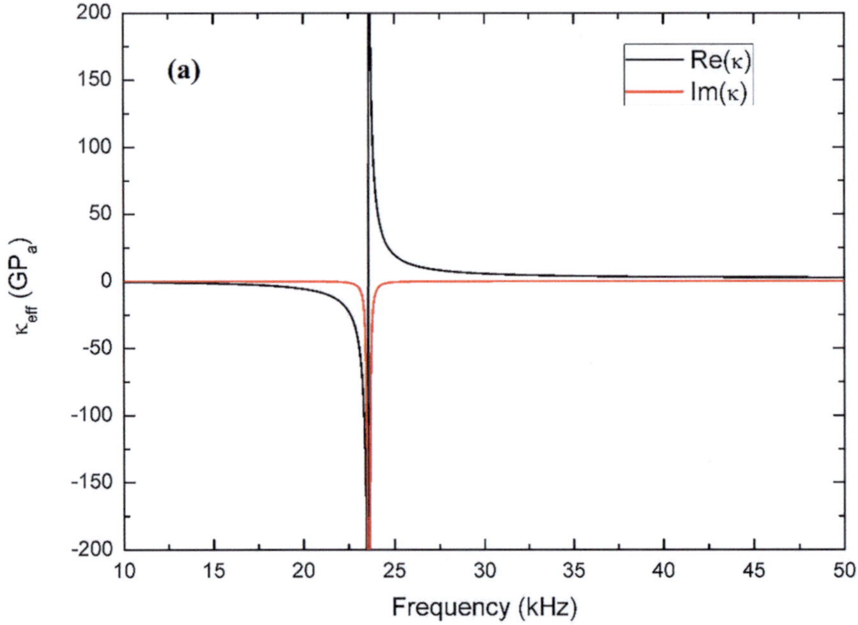

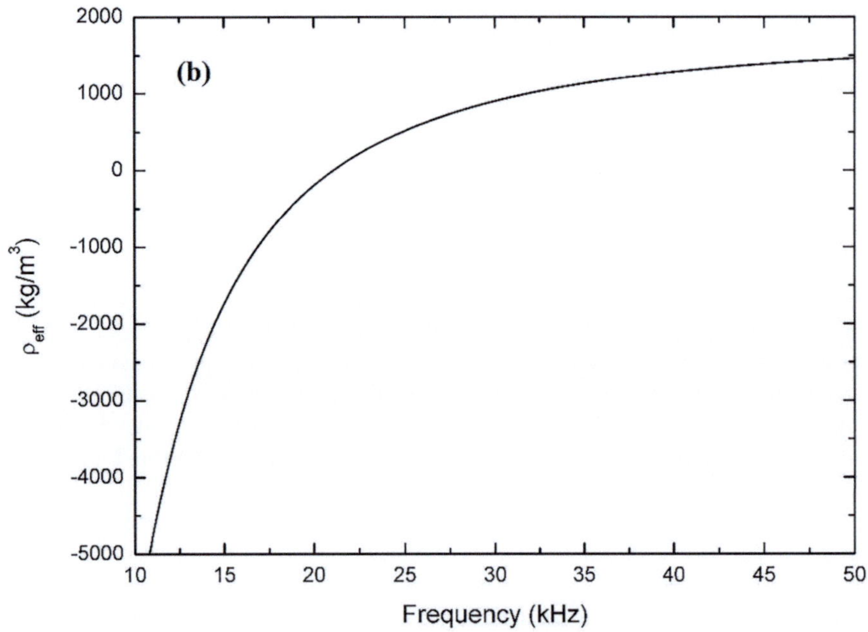

Fig. 7.2 **a** The relationship between the equivalent volume modulus and frequency, where the black line represents the relationship between the real part of the equivalent volume modulus Re(κ_{eff}) and frequency, and the red line represents the relationship between the imaginary part of the equivalent volume modulus and frequency. **b** The relationship between the equivalent mass and frequency

right-handed properties. Figure 7.3 is the dispersion curve of this metamaterial. It is not difficult to find from Fig. 7.3a that when $f < f_{VP}$, the slope of the dispersion curve is negative, so the phase propagation direction of the sound wave is opposite to the energy propagation direction, showing left-handed properties. When $f_{VP} < f < f_{SH}$, we can find that $\mathrm{Re}(k) \to 0$, and the imaginary part of the wave number $\mathrm{Im}(k)$ is very large in this frequency range, as shown in Fig. 7.3b. Therefore, the sound wave will decay rapidly in this case and cannot propagate. When $f > f_{SH}$, the slope of the dispersion curve is positive, so the phase propagation direction of the sound wave is the same as the energy propagation direction, and the material shows right-handed properties at this time. From Fig. 7.3a, b, we can also find that when $f < f_{VP}$, $\mathrm{Re}(k) > 0$ but $\mathrm{Im}(k) < 0$, this means that the sound pressure increases with the increase of the phase. This phenomenon does not actually violate the law of conservation of energy. This is because in this frequency band, the direction of phase propagation of the sound wave is opposite to the direction of sound energy propagation, so the increase of sound pressure with the increase of phase means that the sound pressure decreases along the direction of energy propagation. This phenomenon is consistent with the law of conservation of energy, indicating that this metamaterial shows left-handed properties within the frequency band.

To verify the correctness of the uniform medium assumption used in our calculations, we consider the multiple reflection effects of sound waves in this structure, and use the transfer matrix method to strictly calculate its dispersion curve. The analysis schematic is shown in Fig. 7.1c, where we assume $\mathrm{d}_1 = 0.003$ m, $\mathrm{d}_2 = 0.005$ m and satisfy the condition $\mathrm{d} = \mathrm{d}_1 + \mathrm{d}_2$. Therefore, we can get:

$$p_{1+}S_w + p_{1-}S_w = p_{2+}S_w + p_{2-}S_w + \frac{1}{C_m}(x_{2+} + x_{2-}) + M_m(a_{2+} + a_{2-}) \quad (7.20)$$

$$v_{1+} + v_{1-} = v_{2+} + v_{2-} \quad (7.21)$$

$$p_{2+}e^{-jkd_1} + p_{2-}e^{jkd_1} = p_{3+} + p_{3-} = p \quad (7.22)$$

$$S_w v_{2+}e^{-jkd_1} + S_w v_{2-}e^{jkd_1} = S_w v_{3+} + S_w v_{3-} + S_0 v \quad (7.23)$$

$$p_{3+}e^{-jkd_2} + p_{3-}e^{jkd_2} = p'_{1+} + p'_{1-} \quad (7.24)$$

$$v_{3+}e^{-jkd_2} + v_{3-}e^{jkd_2} = v'_{1+} + v'_{1-} \quad (7.25)$$

Here, p_+ and v_+ represent the sound pressure and particle velocity of the sound wave propagating in the positive direction of the x axis, p_- and v_- represent the sound pressure and particle velocity of the sound wave propagating in the negative direction of the x axis. The subscripts 1, 2, 3 represent the positions in front of the vibrating thin plate, behind the vibrating thin plate, and behind the opening, as shown in Fig. 7.1c. The displacement and acceleration satisfy the relationship: $x_{2+} = \int v_{2+}dt = \frac{v_{2+}}{j\omega}$,

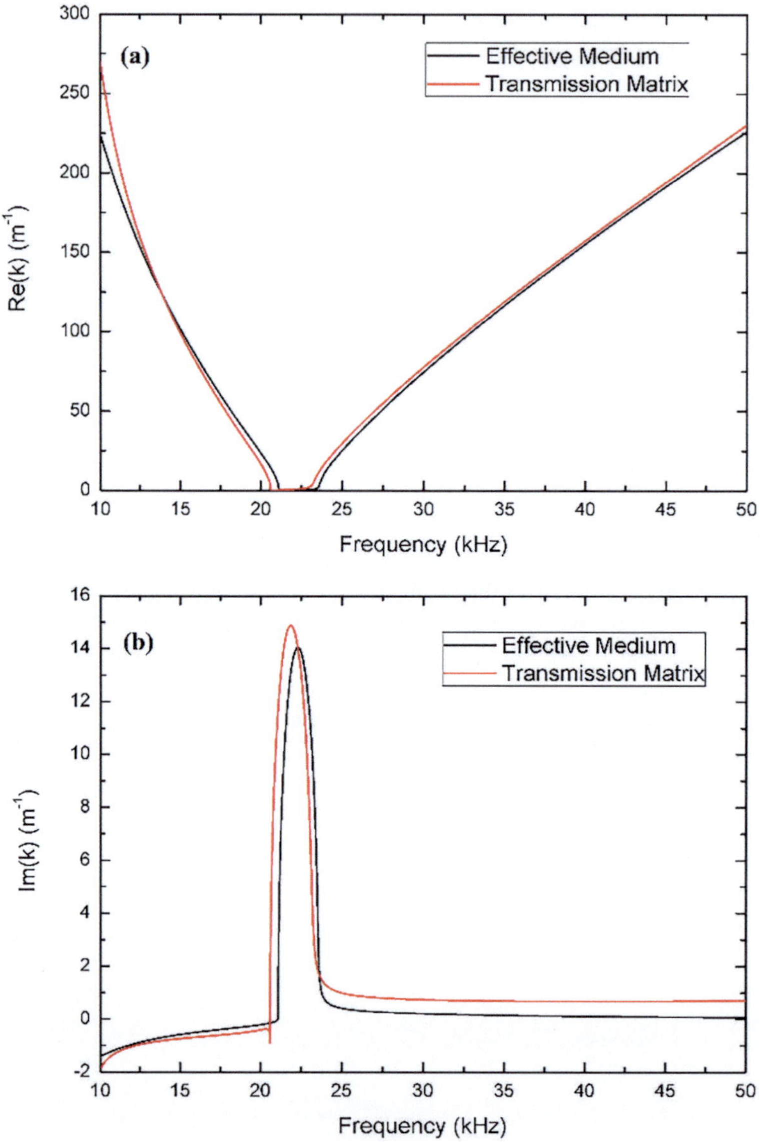

Fig. 7.3 **a** The dispersion curve of the real part of the wave number in the metamaterial Re(k) from 10 to 50 kHz. The black line is calculated by the equivalent medium method, and the red line is calculated by the transfer matrix method. **b** The dispersion curve of the imaginary part of the wave number in the metamaterial Im(k) from 10 to 50 kHz. The black line is calculated by the equivalent medium method, and the red line is calculated by the transfer matrix method. **c** The relationship between the sound intensity transmission coefficient and frequency in this metamaterial with a length of 0.2 m. The black line, red line, and blue line respectively represent the transmission coefficients under the conditions of both periodic vibrating thin plates and perforated small tubes, only perforated small tubes, and only periodic vibrating thin plates

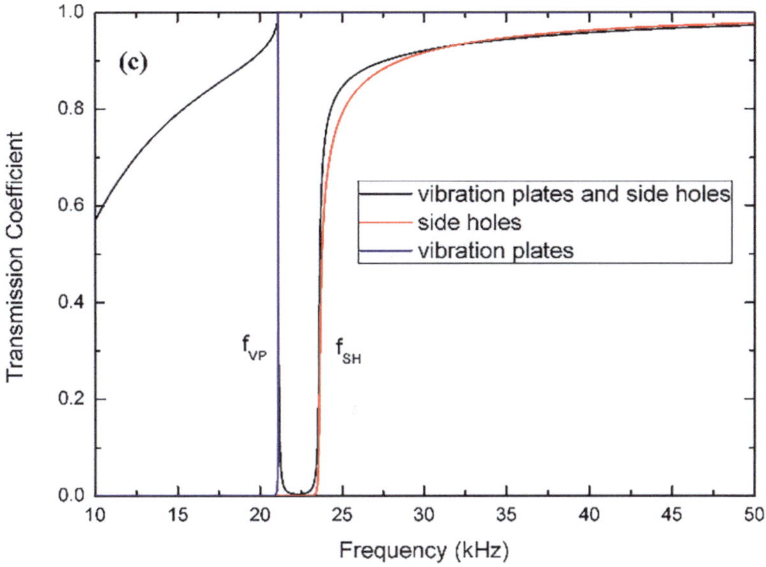

Fig. 7.3 (continued)

$x_{2-} = \int v_{2-}dt = \frac{v_{2-}}{j\omega}$, $a_{2+} = \frac{dv_{2+}}{dt} = j\omega v_{2+}$, $a_{2-} = \frac{dv_{2-}}{dt} = j\omega v_{2-}$, $v_{1+} = \frac{p_{1+}}{\rho_0 c_0}$, $v_{1-} = -\frac{p_{1-}}{\rho_0 c_0}$, $v_{2+} = \frac{p_{2+}}{\rho_0 c_0}$, $v_{2-} = -\frac{p_{2-}}{\rho_0 c_0}$, $v_{3+} = \frac{p_{3+}}{\rho_0 c_0}$, $v_{3-} = -\frac{p_{3-}}{\rho_0 c_0}$, $v'_{1+} = \frac{p'_{1+}}{\rho_0 c_0}$, $v'_{1-} = -\frac{p'_{1-}}{\rho_0 c_0}$, $v = \frac{p}{Z_a}$. Here $Z_a = j\omega M_a + R_a$ is the acoustic impedance rate of the perforated small tube at the connection of the waveguide tube, where $M_a = \rho_0 l' S_0$, and $R_a = R_{a1} + R_{a2}$, which includes two parts, namely $R_{a1} = l\pi a\sqrt{2\eta'\omega\rho_0}$ caused by the viscous properties of the medium, and $R_{a2} = 0.785\rho_0 c_0 k^2 a^4$ caused by the sound radiation of the perforated small tube. $\eta' = 9.98 \times 10^{-4}$ N s/m^2 is the viscosity coefficient in water. The velocity variable v After elimination, we can obtain:

$$p_{1+} + p_{1-} = \left(1 + \frac{1}{j\omega C_m S_w \rho_0 c_0} + \frac{j\omega M_m}{S_w \rho_0 c_0}\right)p_{2+} + \left(1 - \frac{1}{j\omega C_m S_w \rho_0 c_0} - \frac{j\omega M_m}{S_w \rho_0 c_0}\right)p_{2-} \tag{7.26}$$

$$p_{1+} - p_{1-} = p_{2+} - p_{2-} \tag{7.27}$$

$$p_{2+}e^{-jkd_1} + p_{2-}e^{jkd_1} = p_{3+} + p_{3-} \tag{7.28}$$

$$p_{2+}e^{-jkd_1} - p_{2-}e^{jkd_1} = \left(\frac{S_0\rho_0 c_0}{S_w Z_a} + 1\right)p_{3+} + \left(\frac{S_0\rho_0 c_0}{S_w Z_a} - 1\right)p_{3-} \tag{7.29}$$

$$p_{3+}e^{-jkd_2} + p_{3-}e^{jkd_2} = p'_{1+} + p'_{1-} \tag{7.30}$$

$$p_{3+}e^{-jkd_2} + p_{3-}e^{jkd_2} = p'_{1+} + p'_{1-} \tag{7.31}$$

Using the matrix method, we can write the above six equations as:

$$\begin{bmatrix} 1 & 1 \\ 1 & -1 \end{bmatrix}\begin{bmatrix} p_{1+} \\ p_{1-} \end{bmatrix} = \begin{bmatrix} 1 + \frac{1}{j\omega C_m S_w \rho_0 c_0} + \frac{j\omega M_m}{S_w \rho_0 c_0} & 1 - \frac{1}{j\omega C_m S_w \rho_0 c_0} - \frac{j\omega M_m}{S_w \rho_0 c_0} \\ 1 & -1 \end{bmatrix}\begin{bmatrix} p_{2+} \\ p_{2-} \end{bmatrix} \tag{7.32}$$

$$\begin{bmatrix} e^{-jkd_1} & e^{jkd_1} \\ e^{-jkd_1} & -e^{jkd_1} \end{bmatrix}\begin{bmatrix} p_{2+} \\ p_{2-} \end{bmatrix} = \begin{bmatrix} 1 & 1 \\ \frac{S_0\rho_0 c_0}{S_w Z_a} + 1 & \frac{S_0\rho_0 c_0}{S_w Z_a} - 1 \end{bmatrix}\begin{bmatrix} p_{3+} \\ p_{3-} \end{bmatrix} \tag{7.33}$$

$$\begin{bmatrix} e^{-jkd_2} & e^{jkd_2} \\ e^{-jkd_2} & -e^{jkd_2} \end{bmatrix}\begin{bmatrix} p_{3+} \\ p_{3-} \end{bmatrix} = \begin{bmatrix} 1 & 1 \\ 1 & -1 \end{bmatrix}\begin{bmatrix} p'_{1+} \\ p_{1-}' \end{bmatrix} \tag{7.34}$$

Let Set $\Sigma = \begin{bmatrix} 1 & 1 \\ 1 & -1 \end{bmatrix}$, $\Omega = \begin{bmatrix} 1 + \frac{1}{j\omega C_m S_w \rho_0 c_0} + \frac{j\omega M_m}{S_w \rho_0 c_0} & 1 - \frac{1}{j\omega C_m S_w \rho_0 c_0} - \frac{j\omega M_m}{S_w \rho_0 c_0} \\ 1 & -1 \end{bmatrix}$, $\Lambda = \begin{bmatrix} e^{-jkd_1} & e^{jkd_1} \\ e^{-jkd_1} & -e^{jkd_1} \end{bmatrix}$, $\Theta = \begin{bmatrix} 1 & 1 \\ \frac{S_0\rho_0 c_0}{S_w Z_a} + 1 & \frac{S_0\rho_0 c_0}{S_w Z_a} - 1 \end{bmatrix}$, $\Xi = \begin{bmatrix} e^{-jkd_2} & e^{jkd_2} \\ e^{-jkd_2} & -e^{jkd_2} \end{bmatrix}$, we can obtain:

$$\Sigma\begin{bmatrix} p_{1+} \\ p_{1-} \end{bmatrix} = \Omega\begin{bmatrix} p_{2+} \\ p_{2-} \end{bmatrix} \tag{7.35}$$

$$\Lambda\begin{bmatrix} p_{2+} \\ p_{2-} \end{bmatrix} = \Theta\begin{bmatrix} p_{3+} \\ p_{3-} \end{bmatrix} \tag{7.36}$$

$$\Xi\begin{bmatrix} p_{3+} \\ p_{3-} \end{bmatrix} = \Sigma\begin{bmatrix} p'_{1+} \\ p'_{1-} \end{bmatrix} \tag{7.37}$$

Finally, we can obtain:

$$\begin{bmatrix} p'_{1+} \\ p'_{1-} \end{bmatrix} = \Sigma^{-1}\Xi\Theta^{-1}\Lambda\Omega^{-1}\Sigma\begin{bmatrix} p_{1+} \\ p_{1-} \end{bmatrix} \tag{7.38}$$

Let $T = \Sigma^{-1}\Xi\Theta^{-1}\Lambda\Omega^{-1}\Sigma$, we can obtain the relationship between the sound pressure of the previous cell and the sound pressure of the next cell:

$$\begin{bmatrix} p'_{1+} \\ p'_{1-} \end{bmatrix} = T\begin{bmatrix} p_{1+} \\ p_{1-} \end{bmatrix} \tag{7.39}$$

According to Bloch's theorem, we have:

$$\begin{bmatrix} p'_{1+} \\ p'_{1-} \end{bmatrix} = e^{-j\alpha d} \begin{bmatrix} p_{1+} \\ p_{1-} \end{bmatrix} \tag{7.40}$$

According to Eqs. (7.39) and (7.40), we can obtain

$$\left(e^{-j\alpha d} I - T\right) \begin{bmatrix} p_{3+} \\ p_{3-} \end{bmatrix} = \begin{bmatrix} 0 \\ 0 \end{bmatrix} \tag{7.41}$$

Here, I is the unit matrix, we can get the equation

$$\left| e^{-j\alpha d} I - T \right| = 0 \tag{7.42}$$

Solve this equation, we can get the Bloch wave vector α, its real part and imaginary part are represented by red lines in Fig. 7.3a, b.

The consistency of the red line and the black line in Fig. 7.3a, b proves the correctness of our calculation using the assumption of uniform medium. Figure 7.3c shows the relationship between the sound intensity transmission coefficient and frequency in this kind of metamaterial with a length of 0.2 m. We can find that when $f_{VP} < f < f_{SH}$, sound waves cannot propagate in this kind of metamaterial. When $f < f_{VP}$ and $f > f_{SH}$, sound waves can propagate in this kind of metamaterial. This kind of metamaterial can realize left-handed properties in a wide frequency band range, and has the structure of periodic vibrating thin plates and perforated tubes. The boundaries of the sound transmission curves of the structure match well with the boundaries of the transmission curves of the structure with only periodic vibrating thin plates and the structure with only periodic perforated tubes.

Now let's explain why when $f < f_{VP}$, this structure can have both negative volume modulus and negative density. When sound waves propagate in traditional sound waveguides, the phase of the particle vibration speed is ahead of the phase of the sound pressure gradient $\frac{\pi}{2}$. When a series of vibrating thin plates are introduced, it is equivalent to introducing a series of sound capacitors, so the phase of the particle speed lags behind the phase of the sound pressure gradient $\frac{\pi}{2}$. Compared with before the introduction of vibrating thin plates, the phase of the particle vibration speed relative to the phase of the sound pressure gradient changes by π, so we get a negative equivalent mass. Similarly, when sound waves propagate in traditional sound waveguides, the phase of the sound pressure is ahead of the phase of the divergence of the particle speed $\frac{\pi}{2}$. When a series of parallel perforated small tubes are introduced, it is equivalent to introducing a series of sound masses, so the phase of the sound pressure will lag behind the phase of the divergence of the particle speed. $\frac{\pi}{2}$ The phase of the sound pressure relative to the phase of the particle velocity divergence has changed compared to before the introduction of the perforated tube π, so we obtained a negative volume modulus in the low frequency situation. This is the physical mechanism by which this metamaterial can exhibit both negative equivalent mass and negative volume modulus when $f < f_{VP}$.

Under nonlinear conditions, we expand the equation of state under adiabatic conditions by Taylor series and retain the second-order term, we can obtain:

$$P = P_0 + A\left(\frac{\rho - \rho_0}{\rho_0}\right) + \frac{1}{2}B\left(\frac{\rho - \rho_0}{\rho_0}\right)^2 \tag{7.43}$$

Here, $A = \rho_0\left(\frac{\partial P}{\partial \rho}\right)_{s,\rho_0} = \rho_0 c_0^2 = \kappa$, $B = \rho_0^2\left(\frac{\partial^2 P}{\partial \rho^2}\right)_{s,\rho_0} = \rho_0^2\left(\frac{\partial c^2}{\partial \rho}\right)_{s,\rho_0}$. We can then obtain the relationship between density ρ and sound pressure p under the second-order approximation condition.

$$\rho = \rho_0 + \frac{\rho_0}{\kappa}p - \frac{\rho_0 B}{2\kappa^3}p^2 \tag{7.44}$$

Therefore, we can obtain the relationship of the density change.

$$\rho' = \rho - \rho_0 = \frac{\rho_0}{\kappa}p - \frac{\rho_0 B}{2\kappa^3}p^2 \tag{7.45}$$

Equation (7.5) is the sound wave equation of this metamaterial without linear approximation. After substituting Eq. (7.44) into Eq. (7.5), we can get:

$$\begin{aligned}
&\left[\eta\frac{\partial}{\partial t} + \left(\rho_0 + \frac{\rho_0}{\kappa}p - \frac{\rho_0 B}{2\kappa^3}p^2\right)l'S_0\frac{\partial^2}{\partial t^2}\right]\left(\rho_0 + \frac{\rho_0}{\kappa}p - \frac{\rho_0 B}{2\kappa^3}p^2\right)\\
&+\left[\eta\frac{\partial}{\partial x} + \left(\rho_0 + \frac{\rho_0}{\kappa}p - \frac{\rho_0 B}{2\kappa^3}p^2\right)l'S_0\frac{\partial^2}{\partial x\partial t}\right]\\
&\left[\left(\rho_0 + \frac{\rho_0}{\kappa}p - \frac{\rho_0 B}{2\kappa^3}p^2\right)v_1\right] + \frac{\left(\rho_0 + \frac{\rho_0}{\kappa}p - \frac{\rho_0 B}{2\kappa^3}p^2\right)S_0^2}{S_w d}p = 0
\end{aligned} \tag{7.46}$$

We first simplify the first part of Eq. (7.46). Due to the existence of the differential operator, the static density ρ_0 is 0 under the action of the differential operator, so $\left(\rho_0 + \frac{\rho_0}{\kappa}p - \frac{\rho_0 B}{2\kappa^3}p^2\right)$ can be simplified to $\left(\frac{\rho_0}{\kappa}p - \frac{\rho_0 B}{2\kappa^3}p^2\right)$. And in our calculation process, we only consider the second-order approximation, we ignore the $\frac{\rho_0 B}{2\kappa^3}p^2$ term in the first bracket of the first term, therefore, we can get:

$$\begin{aligned}
&\left[\eta\frac{\partial}{\partial t} + \left(\rho_0 + \frac{\rho_0}{\kappa}p - \frac{\rho_0 B}{2\kappa^3}p^2\right)l'S_0\frac{\partial^2}{\partial t^2}\right]\\
&\left(\rho_0 + \frac{\rho_0}{\kappa}p - \frac{\rho_0 B}{2\kappa^3}p^2\right)\\
&\approx \left[\eta\frac{\partial}{\partial t} + \left(\rho_0 + \frac{\rho_0}{\kappa}p\right)l'S_0\frac{\partial^2}{\partial t^2}\right]\left(\frac{\rho_0}{\kappa}p - \frac{\rho_0 B}{2\kappa^3}p^2\right)
\end{aligned} \tag{7.47}$$

Expand the above formula, and ignore the third-order term $\frac{\rho_0}{\kappa}pl'S_0\frac{\partial^2}{\partial t^2}\frac{\rho_0 B}{2\kappa^3}p^2$, we can get the simplified expression of the first term of Eq. (7.46).

$$\begin{aligned}
&\left(\eta\frac{\partial}{\partial t}+\rho_0 l'S_0\frac{\partial^2}{\partial t^2}\right)\left(\frac{\rho_0}{\kappa}p\right)-\left(\eta\frac{\partial}{\partial t}+\rho_0 l'S_0\frac{\partial^2}{\partial t^2}\right)\\
&\left(\frac{\rho_0 B}{2\kappa^3}p^2\right)+\frac{\rho_0}{\kappa}pl'S_0\frac{\partial^2}{\partial t^2}\frac{\rho_0}{\kappa}p\\
&=\rho_0\left(\frac{\rho_0 l'S_0}{\kappa}\frac{\partial^2}{\partial t^2}+\frac{\eta}{\kappa}\frac{\partial}{\partial t}\right)p\\
&-\rho_0\frac{B}{2\kappa^3}\frac{\partial}{\partial t}\left(\eta+\rho_0 l'S_0\frac{\partial}{\partial t}\right)p^2+\rho_0\frac{\rho_0 l'S_0}{\kappa^2}p\left(\frac{\partial^2}{\partial t^2}p\right)
\end{aligned}\tag{7.48}$$

Similarly, we simplify the second term of Eq. (7.46). Since the last part of the second term is multiplied by v_1, and v_1 is a first-order term, and we only retain the second-order approximation in the end, so we can ignore $\left(\rho_0+\frac{\rho_0}{\kappa}p-\frac{\rho_0 B}{2\kappa^3}p^2\right)v_1$ in $\frac{\rho_0 B}{2\kappa^3}p^2$, so the second part becomes:

$$\begin{aligned}
&\left[\eta\frac{\partial}{\partial x}+\left(\rho_0+\frac{\rho_0}{\kappa}p-\frac{\rho_0 B}{2\kappa^3}p^2\right)l'S_0\frac{\partial^2}{\partial x\partial t}\right]\\
&\left[\left(\rho_0+\frac{\rho_0}{\kappa}p-\frac{\rho_0 B}{2\kappa^3}p^2\right)v_1\right]\\
&\approx\left[\eta\frac{\partial}{\partial x}+\left(\rho_0+\frac{\rho_0}{\kappa}p\right)l'S_0\frac{\partial^2}{\partial x\partial t}\right]\left[\left(\rho_0+\frac{\rho_0}{\kappa}p\right)v_1\right]
\end{aligned}\tag{7.49}$$

Expand it to the second-order approximation, we can get:

$$\begin{aligned}
&\left(\eta\frac{\partial}{\partial x}+\rho_0 l'S_0\frac{\partial^2}{\partial x\partial t}\right)(\rho_0 v_1)+\left(\eta\frac{\partial}{\partial x}+\rho_0 l'S_0\frac{\partial^2}{\partial x\partial t}\right)\\
&\left(\frac{\rho_0}{\kappa}pv_1\right)+\frac{\rho_0}{\kappa}pl'S_0\frac{\partial^2}{\partial x\partial t}(\rho_0 v_1)
\end{aligned}\tag{7.50}$$

In the approximation process, we use $\frac{\rho'}{\rho_0}$ to replace $\frac{v_1}{c_0}$, and according to Eq. (7.45), we can get:

$$v_1=\frac{c_0}{\kappa}p-\frac{c_0 B}{2\kappa^3}p^2\tag{7.51}$$

Substitute Eq. (7.51) into Eq. (7.50), and retain the second-order term, we can get the simplified second part of Eq. (7.46).

$$\left(\eta\frac{\partial}{\partial x}+\rho_0 l'S_0\frac{\partial^2}{\partial x\partial t}\right)(\rho_0 v_1)$$

$$
\begin{aligned}
&+\left(\eta\frac{\partial}{\partial x}+\rho_0 l' S_0\frac{\partial^2}{\partial x\partial t}\right)\left(\frac{\rho_0 c_0}{\kappa^2}p^2\right)\\
&+\frac{\rho_0}{\kappa}pl' S_0\frac{\partial^2}{\partial x\partial t}\left(\rho_0\frac{c_0}{\kappa}p\right)\\
&=\rho_0\left(\eta+\rho_0 l' S_0\frac{\partial}{\partial t}\right)\left(\frac{\partial}{\partial x}v_1\right)\\
&+\rho_0\frac{c_0}{\kappa^2}\frac{\partial}{\partial x}\left(\eta+\rho_0 l' S_0\frac{\partial}{\partial t}\right)p^2\\
&+\rho_0\frac{\rho_0 l' S_0}{\kappa^2}p\left(c_0\frac{\partial^2}{\partial x\partial t}p\right)
\end{aligned}
\tag{7.52}
$$

The third part of Eq. (7.46) can be simplified to:

$$
\frac{\left(\rho_0+\frac{\rho_0}{\kappa}p-\frac{\rho_0 B}{2\kappa^3}p^2\right)S_0^2}{S_w d}p=\rho_0\frac{S_0^2}{S_w d}p+\rho_0\frac{S_0^2}{\kappa S_w d}p^2 \tag{7.53}
$$

Add the first three terms (7.48), (7.52) and (7.53) of Eq. (7.46), and then divide by ρ_0, we can obtain the expression of Eq. (7.5) under the second-order approximation

$$
\begin{aligned}
&\left(\frac{\rho_0 l' S_0}{\kappa}\frac{\partial^2}{\partial t^2}+\frac{\eta}{\kappa}\frac{\partial}{\partial t}+\frac{S_0^2}{S_w d}\right)p+\left(\eta+\rho_0 l' S_0\frac{\partial}{\partial t}\right)\frac{\partial}{\partial x}v_1\\
&+\left[\left(\frac{c_0}{\kappa^2}\frac{\partial}{\partial x}-\frac{B}{2\kappa^3}\frac{\partial}{\partial t}\right)\left(\eta+\rho_0 l' S_0\frac{\partial}{\partial t}\right)+\frac{S_0^2}{\kappa S_w d}\right]p^2\\
&+\frac{\rho_0 l' S_0}{\kappa^2}p\left(c_0\frac{\partial}{\partial x}+\frac{\partial}{\partial t}\right)\frac{\partial}{\partial t}p=0
\end{aligned}
\tag{7.54}
$$

Define the operator: $\hat{H}=\frac{\rho_0 l' S_0}{\kappa}\frac{\partial^2}{\partial t^2}+\frac{\eta}{\kappa}\frac{\partial}{\partial t}+\frac{S_0^2}{S_w d}$ and $\hat{I}=\left(\eta+\rho_0 l' S_0\frac{\partial}{\partial t}\right)\frac{\partial}{\partial x}$, and define the nonlinear term: $\psi=\left[\left(\frac{c_0}{\kappa^2}\frac{\partial}{\partial x}-\frac{B}{2\kappa^3}\frac{\partial}{\partial t}\right)\left(\eta+\rho_0 l' S_0\frac{\partial}{\partial t}\right)+\frac{S_0^2}{\kappa S_w d}\right]p^2+\frac{\rho_0 l' S_0}{\kappa^2}p\left(c_0\frac{\partial}{\partial x}+\frac{\partial}{\partial t}\right)\frac{\partial}{\partial t}p$, we can simply represent Eq. (7.54) with operators as

$$
\hat{H}p+\hat{I}v_1+\psi=0 \tag{7.55}
$$

Similarly, applying the total differential operator $\frac{d}{dt}$ to Eq. (7.8), we can get

$$
\frac{d}{dt}\left(\frac{\partial p}{\partial x}\right)+\frac{d}{dt}\left(\rho\frac{dv_1}{dt}\right)+\frac{M_m}{S_w d}\frac{d^2 v_1}{dt^2}+\frac{1}{C_m S_w d}v_1=0 \tag{7.56}
$$

We know

$$
\frac{d}{dt}=\frac{\partial}{\partial t}+v_1\frac{\partial}{\partial x} \tag{7.57}
$$

Therefore, the first term of Eq. (7.56) can be transformed into

$$\frac{d}{dt}\left(\frac{\partial p}{\partial x}\right)=\left(\frac{\partial}{\partial t}+v_1\frac{\partial}{\partial x}\right)\left(\frac{\partial p}{\partial x}\right)=\frac{\partial^2}{\partial x\partial t}p+v_1\frac{\partial^2 p}{\partial x^2} \tag{7.58}$$

Substituting Eq. (7.51) into Eq. (7.58) and retaining the second-order term, we can get

$$\frac{d}{dt}\left(\frac{\partial p}{\partial x}\right)=\frac{\partial^2}{\partial x\partial t}p+\frac{c_0}{\kappa}p\frac{\partial^2}{\partial x^2}p \tag{7.59}$$

The second and third terms of Eq. (7.56) can be transformed into

$$\begin{aligned}\frac{d}{dt}\left(\rho\frac{dv_1}{dt}\right)+\frac{M_m}{S_w d}\frac{d^2v_1}{dt^2}&=\frac{d\rho}{dt}\frac{dv_1}{dt}+\rho\frac{d^2v_1}{dt^2}\\&+\frac{M_m}{S_w d}\frac{d^2v_1}{dt^2}=\frac{d\rho}{dt}\frac{dv_1}{dt}+\left(\rho+\frac{M_m}{S_w d}\right)\frac{d^2v_1}{dt^2}\end{aligned} \tag{7.60}$$

According to Eq. (7.57), Eq. (7.60) can be transformed into

$$\begin{aligned}&\frac{d}{dt}\left(\rho\frac{dv_1}{dt}\right)+\frac{M_m}{S_w d}\frac{d^2v_1}{dt^2}\\&=\left[\left(\frac{\partial}{\partial t}+v_1\frac{\partial}{\partial x}\right)\rho\right]\cdot\left[\left(\frac{\partial}{\partial t}+v_1\frac{\partial}{\partial x}\right)v_1\right]\\&+\left(\rho+\frac{M_m}{S_w d}\right)\left[\left(\frac{\partial}{\partial t}+v_1\frac{\partial}{\partial x}\right)^2v_1\right]\end{aligned} \tag{7.61}$$

Substituting Eq. (7.44) into Eq. (7.61) and expanding the operator $\left(\frac{\partial}{\partial t}+v_1\frac{\partial}{\partial x}\right)^2$, we can get

$$\begin{aligned}&\frac{d}{dt}\left(\rho\frac{dv_1}{dt}\right)+\frac{M_m}{S_w d}\frac{d^2v_1}{dt^2}\\&=\left[\left(\frac{\partial}{\partial t}+v_1\frac{\partial}{\partial x}\right)\left(\rho_0+\frac{\rho_0}{\kappa}p-\frac{\rho_0 B}{2\kappa^3}p^2\right)\right]\cdot\left[\left(\frac{\partial}{\partial t}+v_1\frac{\partial}{\partial x}\right)v_1\right]\\&+\left(\rho_0+\frac{\rho_0}{\kappa}p-\frac{\rho_0 B}{2\kappa^3}p^2+\frac{M_m}{S_w d}\right)\\&\left\{\left[\frac{\partial^2}{\partial t^2}+\frac{\partial}{\partial t}\left(v_1\frac{\partial}{\partial x}\right)+v_1\frac{\partial^2}{\partial x\partial t}+v_1\frac{\partial}{\partial x}\left(v_1\frac{\partial}{\partial x}\right)\right]v_1\right\}\end{aligned} \tag{7.62}$$

Since only the second-order term is retained, ignore $\frac{\rho_0 B}{2\kappa^3}p^2$ term and $v_1\frac{\partial}{\partial x}\left(v_1\frac{\partial}{\partial x}\right)$ term, and note that $\frac{\partial}{\partial t}\left(v_1\frac{\partial}{\partial x}\right)=\frac{\partial v_1}{\partial t}\cdot\frac{\partial}{\partial x}+v_1\frac{\partial^2}{\partial t\partial x}$, so we can get

$$\frac{d}{dt}\left(\rho\frac{dv_1}{dt}\right)+\frac{M_m}{S_w d}\frac{d^2v_1}{dt^2}=\left[\left(\frac{\partial}{\partial t}+v_1\frac{\partial}{\partial x}\right)\left(\rho_0+\frac{\rho_0}{\kappa}p\right)\right]$$
$$\cdot\left[\left(\frac{\partial}{\partial t}+v_1\frac{\partial}{\partial x}\right)v_1\right]+\left(\rho_0+\frac{\rho_0}{\kappa}p+\frac{M_m}{S_w d}\right)$$
$$\left\{\left[\frac{\partial^2}{\partial t^2}+\frac{\partial v_1}{\partial t}\cdot\frac{\partial}{\partial x}+v_1\frac{\partial^2}{\partial t\partial x}+v_1\frac{\partial^2}{\partial x\partial t}\right]v_1\right\} \tag{7.63}$$

Expand Eq. (7.63) and retain the second-order term, we can get

$$\frac{d}{dt}\left(\rho\frac{dv_1}{dt}\right)+\frac{M_m}{S_w d}\frac{d^2v_1}{dt^2}=\frac{\partial}{\partial t}\left(\frac{\rho_0}{\kappa}p\right)$$
$$\cdot\frac{\partial v_1}{\partial t}+\left(\rho_0+\frac{M_m}{S_w d}\right)\frac{\partial^2 v_1}{\partial t^2}+\left(\rho_0+\frac{M_m}{S_w d}\right)$$
$$\left(\frac{\partial v_1}{\partial t}\cdot\frac{\partial v_1}{\partial x}+2v_1\frac{\partial^2 v_1}{\partial t\partial x}\right)+\frac{\rho_0}{\kappa}p\frac{\partial^2 v_1}{\partial t^2} \tag{7.64}$$

Substituting Eq. (7.51) into Eq. (7.64) and retaining the second-order term, we can get

$$\frac{d}{dt}\left(\rho\frac{dv_1}{dt}\right)+\frac{M_m}{S_w d}\frac{d^2v_1}{dt^2}=\frac{\rho_0 c_0}{\kappa^2}\left(\frac{\partial p}{\partial t}\right)^2$$
$$+\left(\rho_0+\frac{M_m}{S_w d}\right)\frac{\partial^2 v_1}{\partial t^2}+\left(\frac{1}{\kappa}+\frac{M_m}{\rho_0 S_w d\kappa}\right)\frac{\partial p}{\partial t}$$
$$\cdot\frac{\partial p}{\partial x}+p\left(\frac{2}{\kappa}+\frac{2M_m}{\rho_0 S_w d\kappa}\right)\left(\frac{\partial^2 p}{\partial t\partial x}\right)+\frac{\rho_0 c_0}{\kappa^2}p\frac{\partial^2 p}{\partial t^2} \tag{7.65}$$

The fourth term of Eq. (7.56) does not need to be simplified. Combining Eq. (7.59), (7.65) and the fourth term of Eq. (7.56), Eq. (7.56) can be simplified as

$$\frac{\partial^2}{\partial x\partial t}p+\left[\left(\rho_0+\frac{M_m}{S_w d}\right)\frac{\partial^2}{\partial t^2}+\frac{1}{C_m S_w d}\right]v_1$$
$$+p\left[\frac{c_0}{\kappa}\frac{\partial^2}{\partial x^2}+\frac{\rho_0 c_0}{\kappa^2}\frac{\partial^2}{\partial t^2}+\left(\frac{2}{\kappa}+\frac{2M_m}{\rho_0 S_w d\kappa}\right)\frac{\partial^2}{\partial x\partial t}\right]p$$
$$+\left(\frac{1}{\kappa}+\frac{M_m}{\rho_0 S_w d\kappa}\right)\frac{\partial p}{\partial x}\frac{\partial p}{\partial t}+\frac{\rho_0 c_0}{\kappa^2}\left(\frac{\partial p}{\partial t}\right)^2=0 \tag{7.66}$$

Define the operator $\hat{J}=\frac{\partial^2}{\partial x\partial t}$ and $\hat{K}=\left(\rho_0+\frac{M_m}{S_w d}\right)\frac{\partial^2}{\partial t^2}+\frac{1}{C_m S_w d}$, and define the nonlinear term: $\varphi=p\left[\frac{c_0}{\kappa}\frac{\partial^2}{\partial x^2}+\frac{\rho_0 c_0}{\kappa^2}\frac{\partial^2}{\partial t^2}+\left(\frac{2}{\kappa}+\frac{2M_m}{\rho_0 S_w d\kappa}\right)\frac{\partial^2}{\partial x\partial t}\right]p+\left(\frac{1}{\kappa}+\frac{M_m}{\rho_0 S_w d\kappa}\right)\frac{\partial p}{\partial x}\frac{\partial p}{\partial t}+\frac{\rho_0 c_0}{\kappa^2}\left(\frac{\partial p}{\partial t}\right)^2$, Eq. (7.66) can be represented by the operator method as

$$\hat{J}p + \hat{K}v_1 + \varphi = 0 \tag{7.67}$$

Applying the operator $\hat{K}$ to Eq. (7.67), we can get

$$\hat{K}\hat{H}p + \hat{K}\hat{I}v_1 + \hat{K}\psi = 0 \tag{7.68}$$

Applying the operator $\hat{I}$ to Eq. (7.55), we can get

$$\hat{I}\hat{J}p + \hat{I}\hat{K}v_1 + \hat{I}\varphi = 0 \tag{7.69}$$

Since the operator satisfies the commutativity $\hat{K}\hat{I} = \hat{I}\hat{K}$, subtracting Eq. (7.69) from Eq. (7.78), we can get

$$\left(\hat{K}\hat{H} - \hat{I}\hat{J}\right)p = \hat{I}\varphi - \hat{K}\psi \tag{7.70}$$

This is the acoustic equation in this metamaterial, preserved to the second-order approximation. Let $p = p_1 + p_2$, where p_1 is the first-order approximate solution of the equation, p_2 is the second-order approximate solution of the equation, satisfying $p_1 >> p_2$, we can then obtain the first-order approximate equation

$$\left(\hat{K}\hat{H} - \hat{I}\hat{J}\right)p_1 = 0 \tag{7.71}$$

and the second-order approximate equation

$$\left(\hat{K}\hat{H} - \hat{I}\hat{J}\right)p_2 = \hat{I}\phi_1 - \hat{K}\psi_1 \tag{7.72}$$

Here $\psi_1 = \left[\left(\frac{c_0}{\kappa^2}\frac{\partial}{\partial x} - \frac{B}{2\kappa^3}\frac{\partial}{\partial t}\right)\left(\eta + \rho_0 l' S_0 \frac{\partial}{\partial t}\right) + \frac{S_0^2}{\kappa S_w d}\right]p_1^2 + \frac{\rho_0 l' S_0}{\kappa^2} p_1 \left(c_0 \frac{\partial}{\partial x} + \frac{\partial}{\partial t}\right)\frac{\partial}{\partial t} p_1$ and $\varphi_1 = p_1\left[\frac{c_0}{\kappa}\frac{\partial^2}{\partial x^2} + \frac{\rho_0 c_0}{\kappa^2}\frac{\partial^2}{\partial t^2} + \left(\frac{2}{\kappa} + \frac{2M_m}{\rho_0 S_w d\kappa}\right)\frac{\partial^2}{\partial x \partial t}\right]p_1 + \left(\frac{1}{\kappa} + \frac{M_m}{\rho_0 S_w d\kappa}\right)\frac{\partial p_1}{\partial x}\frac{\partial p_1}{\partial t} + \frac{\rho_0 c_0}{\kappa^2}\left(\frac{\partial p_1}{\partial t}\right)^2$. Under linear approximation, the solution of Eq. (7.77) is Eq. (7.17). If the sound-absorbing material is placed at the end of this metamaterial, without causing reflection, and only the sound waves propagating along the x axis need to be considered, then $p_1 = \mathrm{Re}[e^{j(\omega t + kx)}]$, where $k = k(\omega) = a - bj$. Here, the sign of the wave vector k is taken as positive, because in the left-hand frequency band, the sound energy radiates in the positive direction of the x axis, so the direction of phase propagation is in the negative direction of the x axis.

Under linear approximation, $p_1 = \mathrm{Re}[p_0 e^{j(\omega t + kx)}] = p_0 e^{bx}\cos(\omega t + ax)$, we can get $p_1^2 = p_0^2 e^{2bx}\frac{1+\cos 2(\omega t + ax)}{2}$. Since $\frac{1}{2}p_0^2 e^{2bx}$ is a non-propagating term, this term does not contribute to the propagation of sound waves, so we only consider the propagating term in the nonlinear equation $\frac{1}{2}p_0^2 e^{2bx}\cos(\omega t + ax)$. Therefore, $p_1^2 = \frac{1}{2}p_0^2 e^{2bx}\cos 2(\omega t + ax) = \mathrm{Re}\left[\frac{1}{2}p_0^2 e^{j(2\omega t + 2kx)}\right]$. So the first term of ψ_1 can be transformed into

$$\left[\left(\frac{c_0}{\kappa^2}\frac{\partial}{\partial x}-\frac{B}{2\kappa^3}\frac{\partial}{\partial t}\right)\left(\eta+\rho_0 l' S_0\frac{\partial}{\partial t}\right)+\frac{S_0^2}{\kappa S_w d}\right]p_1^2$$
$$=\left[\left(\frac{c_0}{\kappa^2}\frac{\partial}{\partial x}-\frac{B}{2\kappa^3}\frac{\partial}{\partial t}\right)\left(\eta+\rho_0 l' S_0\frac{\partial}{\partial t}\right)+\frac{S_0^2}{\kappa S_w d}\right]\mathrm{Re}\left[\frac{1}{2}p_0^2 e^{j(2\omega t+2kx)}\right]$$
$$=\mathrm{Re}\left\{\left[\left(\frac{c_0}{\kappa^2}\frac{\partial}{\partial x}-\frac{B}{2\kappa^3}\frac{\partial}{\partial t}\right)\left(\eta+\rho_0 l' S_0\frac{\partial}{\partial t}\right)+\frac{S_0^2}{\kappa S_w d}\right]\left[\frac{1}{2}p_0^2 e^{j(2\omega t+2kx)}\right]\right\} \quad (7.73)$$

For a monochromatic wave, and noting that the operator here acts on the second harmonic, we can use $2j\omega$ and $2jk$ to replace the operators $\frac{\partial}{\partial t}$ and $\frac{\partial}{\partial x}$ respectively.

Similarly, the second term of ψ_1 can be transformed into:

$$\frac{\rho_0 l' S_0}{\kappa^2}p_1\left(c_0\frac{\partial}{\partial x}+\frac{\partial}{\partial t}\right)\frac{\partial}{\partial t}p_1$$
$$=\frac{\rho_0 l' S_0}{\kappa^2}\mathrm{Re}[p_0 e^{j(\omega t+kx)}]\times\left\{\left(c_0\frac{\partial}{\partial x}+\frac{\partial}{\partial t}\right)\frac{\partial}{\partial t}\mathrm{Re}\left[p_0 e^{j(\omega t+kx)}\right]\right\}$$
$$=\frac{\rho_0 l' S_0}{\kappa^2}\mathrm{Re}\left[p_0 e^{j(\omega t+kx)}\right]\times\mathrm{Re}\left\{\left(c_0\frac{\partial}{\partial x}+\frac{\partial}{\partial t}\right)\frac{\partial}{\partial t}\left[p_0 e^{j(\omega t+kx)}\right]\right\} \quad (7.74)$$

For a monochromatic wave, and noting that the operator here acts on the fundamental wave, we can use $j\omega$ and jk to replace the operators $\frac{\partial}{\partial t}$ and $\frac{\partial}{\partial x}$ respectively. Therefore, ψ_1 can be simplified to:

$$\psi_1=\mathrm{Re}\left\{\left[\left(2jk\frac{c_0}{\kappa^2}-2j\omega\frac{B}{2\kappa^3}\right)(\eta+2j\omega\rho_0 l' S_0)+\frac{S_0^2}{\kappa S_w d}\right]\left[\frac{1}{2}p_0^2 e^{j(2\omega t+2kx)}\right]\right\}$$
$$+\frac{\rho_0 l' S_0}{\kappa^2}\mathrm{Re}\left[p_0 e^{j(\omega t+kx)}\right]\times\mathrm{Re}\left\{(jkc_0+j\omega)j\omega\left[p_0 e^{j(\omega t+kx)}\right]\right\}$$
$$=\mathrm{Re}\left\{\left[\left(2jk\frac{c_0}{\kappa^2}-2j\omega\frac{B}{2\kappa^3}\right)(\eta+2j\omega\rho_0 l' S_0)+\frac{S_0^2}{\kappa S_w d}\right]\left[\frac{1}{2}p_0^2 e^{j(2\omega t+2kx)}\right]\right\}-\frac{\rho_0 l' S_0}{2\kappa^2}\omega p_0^2 e^{2bx}$$
$$[(ac_0+\omega)\cos 2(\omega t+ax)+bc_0\sin 2(\omega t+ax)+(ac_0+\omega)] \quad (7.75)$$

Similarly, since the non-propagating term does not affect the propagation of sound waves, this part is not considered, so the second part of ψ_1 can be simplified to

$$-\frac{\rho_0 l' S_0}{2\kappa^2}\omega p_0^2 e^{2bx}[(ac_0+\omega)\cos 2(\omega t+ax)+bc_0\sin 2(\omega t+ax)]$$

$$= -\frac{\rho_0 l' S_0}{2\kappa^2}\omega \mathrm{Re}\left[(kc_0 + \omega)p_0^2 e^{j(2\omega t + 2kx)}\right] \tag{7.76}$$

Therefore, we can get:

$$\psi_1 = \mathrm{Re}\left\{\left[-\frac{\rho_0 l' S_0}{2\kappa^2}\omega(kc_0 + \omega) + j\left(k\frac{c_0}{\kappa^2} - \omega\frac{B}{2\kappa^3}\right)(\eta + 2j\omega\rho_0 l' S_0) + \frac{S_0^2}{2\kappa S_w d}\right]p_0^2 e^{j(2\omega t + 2kx)}\right\} \tag{7.77}$$

When the operator $\hat{K}$ acts on ψ_1, $2j\omega$ and $2jk$ can be used to replace the operator $\frac{\partial}{\partial t}$ and $\frac{\partial}{\partial x}$, therefore:

$$\hat{K}\psi_1 = \mathrm{Re}\left\{\left[\frac{1}{C_m S_w d} - 4\omega^2\left(\rho_0 + \frac{M_m}{S_w d}\right)\right] \times \left[-\frac{\rho_0 l' S_0}{2\kappa^2}\omega(kc_0 + \omega) + j\left(k\frac{c_0}{\kappa^2} - \omega\frac{B}{2\kappa^3}\right)(\eta + 2j\omega\rho_0 l' S_0) + \frac{S_0^2}{2\kappa S_w d}\right] p_0^2 e^{j(2\omega t + 2kx)}\right\} \tag{7.78}$$

Similarly,

$$\begin{aligned}\varphi_1 &= p_1\left[\frac{c_0}{\kappa}\frac{\partial^2}{\partial x^2} + \frac{\rho_0 c_0}{\kappa^2}\frac{\partial^2}{\partial t^2} + \left(\frac{2}{\kappa} + \frac{2M_m}{\rho_0 S_w d\kappa}\right)\frac{\partial^2}{\partial x \partial t}\right]p_1 \\ &\quad + \left(\frac{1}{\kappa} + \frac{M_m}{\rho_0 S_w d\kappa}\right)\frac{\partial p_1}{\partial x}\frac{\partial p_1}{\partial t} + \frac{\rho_0 c_0}{\kappa^2}\left(\frac{\partial p_1}{\partial t}\right)^2 \\ &= \mathrm{Re}\left[p_0 e^{j(\omega t + kx)}\right] \times \mathrm{Re}\left\{\left[\frac{c_0}{\kappa}\frac{\partial^2}{\partial x^2} + \frac{\rho_0 c_0}{\kappa^2}\frac{\partial^2}{\partial t^2} + \left(\frac{2}{\kappa} + \frac{2M_m}{\rho_0 S_w d\kappa}\right)\frac{\partial^2}{\partial x \partial t}\right]p_0 e^{j(\omega t + kx)}\right\} \\ &\quad + \left(\frac{1}{\kappa} + \frac{M_m}{\rho_0 S_w d\kappa}\right)\mathrm{Re}\left[\frac{\partial}{\partial x}p_0 e^{j(\omega t + kx)}\right] \times \mathrm{Re}\left[\frac{\partial}{\partial t}p_0 e^{j(\omega t + kx)}\right] \\ &\quad + \frac{\rho_0 c_0}{\kappa^2}\mathrm{Re}\left[\left(\frac{\partial}{\partial t}p_0 e^{j(\omega t + kx)}\right)\right]^2 \end{aligned} \tag{7.79}$$

Here, $j\omega$ and jk can be used to replace the operators $\frac{\partial}{\partial t}$ and $\frac{\partial}{\partial x}$, therefore

$$\begin{aligned}\varphi_1 &= \mathrm{Re}\left[p_0 e^{j(\omega t + kx)}\right] \\ &\quad \times \mathrm{Re}\left\{-k^2\frac{c_0}{\kappa} - \omega^2\frac{\rho_0 c_0}{\kappa^2} - \omega k\left(\frac{2}{\kappa} + \frac{2M_m}{\rho_0 S_w d\kappa}\right)]p_0 e^{j(\omega t + kx)}\right\} \\ &\quad + \left(\frac{1}{\kappa} + \frac{M_m}{\rho_0 S_w d\kappa}\right)\mathrm{Re}\left[jkp_0 e^{j(\omega t + kx)}\right]\end{aligned}$$

$$\times \mathrm{Re}\left[j\omega p_0 e^{j(\omega t+kx)}\right] + \frac{\rho_0 c_0}{\kappa^2}\mathrm{Re}\left[\left(j\omega p_0 e^{j(\omega t+kx)}\right)\right]^2 \tag{7.80}$$

Only retaining the transmission terms, through complex calculations, we can obtain

$$\varphi_1 = \mathrm{Re}\left\{\left[-k^2\frac{c_0}{\kappa} - 2\omega^2\frac{\rho_0 c_0}{\kappa^2} - 3\omega k\left(\frac{1}{\kappa} + \frac{M_m}{\rho_0 S_w d\kappa}\right)\right]\left[\frac{1}{2}p_0^2 e^{j(2\omega t+2kx)}\right]\right\} \tag{7.81}$$

Similarly, we can obtain

$$\hat{I}\phi_1 = \mathrm{Re}\left\{-jk\left(\eta + 2j\omega\rho_0 l' S_0\right)\times\left[k^2\frac{c_0}{\kappa} + 2\omega^2\frac{\rho_0 c_0}{\kappa^2} + 3\omega k\left(\frac{1}{\kappa} + \frac{M_m}{\rho_0 S_w d\kappa}\right)\right]p_0^2 e^{j(2\omega t+2kx)}\right\} \tag{7.82}$$

According to Eqs. (7.78) and (7.82), we can obtain

$$\hat{I}\phi_1 - \hat{K}\psi_1 = \mathrm{Re}\left[\sigma p_0^2 e^{j(2\omega t+2kx)}\right] \tag{7.83}$$

Here:

$$\begin{aligned}\sigma = &- jk\left(\eta + 2j\omega\rho_0 l' S_0\right)\\ &\times\left[k^2\frac{c_0}{\kappa} + 2\omega^2\frac{\rho_0 c_0}{\kappa^2} + 3\omega k\left(\frac{1}{\kappa} + \frac{M_m}{\rho_0 S_w d\kappa}\right)\right]\\ &-\left[\frac{1}{C_m S_w d} - 4\omega^2\left(\rho_0 + \frac{M_m}{S_w d}\right)\right]\\ &\times\left[-\frac{\rho_0 l' S_0}{2\kappa^2}\omega(kc_0 + \omega) + j\left(k\frac{c_0}{\kappa^2} - \omega\frac{B}{2\kappa^3}\right)\left(\eta + 2j\omega\rho_0 l' S_0\right) + \frac{S_0^2}{2\kappa S_w d}\right]\end{aligned} \tag{7.84}$$

Since $\hat{H}, \hat{I}, \hat{J}$ and $\hat{K}$ are all linear operators, Eqs. (7.61) and (7.84) can be expressed as

$$\left(\hat{K}\hat{H} - \hat{I}\hat{J}\right)p_2' = \sigma p_0^2 e^{j(2\omega t+2kx)} \tag{7.85}$$

Here $p_2 = \mathrm{Re}[p_2']$, for convenience, the p_2' in Eq. (7.85) is written as p_2, and using $2j\omega$ to replace the operator $\frac{\partial}{\partial t}$, we can obtain

$$\left\{\frac{\partial^2}{\partial x^2} - \frac{\left[\frac{1}{C_m S_w d} - 4\omega^2\left(\rho_0 + \frac{M_m}{S_w d}\right)\right]\times\left(\frac{S_0^2}{S_w d} + 2j\omega\frac{\eta}{\kappa} - 4\omega^2\frac{\rho_0 l' S_0}{\kappa}\right)}{2j\omega\left(\eta + 2j\omega\rho_0 l' S_0\right)}\right\}p_2$$

$$= \frac{\sigma}{-2j\omega(\eta + 2j\omega\rho_0 l' S_0)} p_0^2 e^{j(2\omega t + 2kx)} \tag{7.86}$$

Comparing Eqs. (7.10), (7.11), (7.14) and (7.15), we can obtain

$$\frac{\partial^2 p_2}{\partial x^2} + k^2(2\omega) p_2 = \frac{\sigma}{-2j\omega(\eta + 2j\omega\rho_0 l' S_0)} p_0^2 e^{j(2\omega t + 2kx)} \tag{7.87}$$

The general solution of Eq. (7.87) is

$$p_2 = \frac{1}{k^2(2\omega) - (2k)^2} \times \frac{\sigma}{-2j\omega(\eta + 2j\omega\rho_0 l' S_0)} p_0^2 e^{j(2\omega t + 2k(\omega)x)} + p_{21} e^{j(2\omega t - k(2\omega)x)} + p_{22} e^{j(2\omega t + k(2\omega)x)} \tag{7.88}$$

Here p_{21} and p_{22} are determined by the boundary conditions. Consider the following situation: the fundamental wave of this metamaterial is left-handed while the second harmonic is right-handed. Therefore, the phase velocity of the fundamental wave is opposite to the direction of propagation of the group velocity, while the phase velocity of the second harmonic is the same as the direction of propagation of the group velocity. If we assume that the expression of the fundamental wave is $p_0 \cos(\omega t \pm kx)$, then the second harmonic must be proportional to $p_0^2 \cos(2\omega t \pm 2kx)$. That is, the direction of the phase velocity of the fundamental wave is consistent with the direction of the phase velocity of the harmonic. When the sound source radiates sound waves outward, the direction of propagation of the group velocity of the fundamental wave is in the positive direction of the x axis. Since the metamaterial shows a left-handed effect in this frequency band, the direction of propagation of the phase velocity of the fundamental wave is in the negative direction of the x axis. Since the direction of the phase velocity of the fundamental wave is consistent with the direction of the phase velocity of the second harmonic, the direction of the phase velocity of the second harmonic is in the negative direction of the x axis. Through previous analysis, in the double frequency, the equivalent volume modulus and equivalent density of the metamaterial are both positive ($\kappa_{eff}(2\omega) > 0$, $\rho_{eff}(2\omega) > 0$), belongs to the right-handed property, therefore the energy propagation direction of the second harmonic is consistent with the phase propagation direction of the second harmonic, heading towards the negative direction of the x axis. That is to say, in this metamaterial, the reverse propagation characteristic of the second harmonic can be realized. Therefore, it can be considered that in this metamaterial, the energy propagation direction of the fundamental wave is opposite to the propagation direction of the second harmonic, which is a special property of the metamaterial. If sound-absorbing material is used at the end of the metamaterial, there is no reflection wave of the fundamental wave, and because the phase propagation direction of the fundamental wave is opposite to the harmonic, it will not excite the second harmonic propagating in the positive direction of the x axis, that is to say $p_{21} = 0$. Assuming the length of the metamaterial is L_0, and the origin is set at $x = 0$, then because the second harmonic is propagating in the negative direction of the x axis, and the generation of the second harmonic is due to the accumulation effect, the sound pressure of the second harmonic at $x = L_0$ is zero. Therefore, we can obtain

$$p_{22} = \frac{1}{k^2(2\omega) - (2k)^2} \times \frac{\sigma}{2j\omega(\eta + 2j\omega\rho_0 l'S_0)} p_0^2 e^{j[2k(\omega) - k(2\omega)]L_0} \tag{7.89}$$

Then, the second-order approximate solution can be obtained

$$p_2 = \frac{1}{k^2(2\omega) - (2k)^2} \times \frac{\sigma}{-2j\omega(\eta + 2j\omega\rho_0 l'S_0)} p_0^2 e^{j(2\omega t + 2k(\omega)x)} + p_{22} e^{j(2\omega t + k(2\omega)x)} \tag{7.90}$$

where, p_{22} is determined by Eq. (7.89). If the fundamental frequency is $f = 17.575$ kHz, the length of the metamaterial is $L_0 = 0.2$ m, the nonlinear parameter of water is $\frac{B}{A} = 5.2$, through calculation, the expression of the second harmonic is:

$$\begin{aligned} p_2 = {} & 10^{-10} p_0^2 \\ & \times \left[e^{-0.779x}(6.1262 + j795.924) e^{j(220854t + 117.976x)} \right. \\ & \left. + e^{0.142x}(-6.01471 - j662.011) e^{j(220854t + 117.983x)} \right] \end{aligned} \tag{7.91}$$

From Eq. (7.91), it can be found that the phase propagation direction of the second harmonic is along the negative direction of the x axis. Since the second harmonic is in the right-handed property band, the sound energy propagation direction is the same as the fundamental direction, also along the negative direction of the x axis. The relationship between the amplitude of the second harmonic and the change of position can be seen from Fig. 7.4a. At $x = 0.2$ m, the amplitude of the second harmonic is 0, and the amplitude of the second harmonic gradually increases in the reverse direction along the x axis (the propagation direction of the second harmonic is from right to left, that is, the negative direction of the x axis, so the amplitude of the second harmonic increases from right to left). Figure 7.4b shows the distribution of the amplitude of the second harmonic when $L_0 = 5$ m. It can also be found that because the second harmonic is accumulated in reverse, at $x = 5$ m, the amplitude of the second harmonic is 0. And it can be found that from right to left (along the negative direction of the x axis), the amplitude of the second harmonic gradually increases, and then the speed of increase gradually becomes faster. This is because the energy of the second harmonic is propagating in the negative direction of the x axis, while the energy of the fundamental wave is propagating in the positive direction of the x axis. Due to the attenuation factor, the fundamental wave decreases with the increase of propagation distance, so the contribution of the fundamental wave to the second harmonic is very small in the distance, so the growth speed of the second harmonic is relatively slow in the distance. But near the sound source, because the attenuation effect is not very significant, the fundamental wave is relatively large, which will make a large contribution to the second harmonic, so the second harmonic grows very fast near the sound source. The phase and sound intensity distribution of the fundamental wave and the second harmonic are shown in Fig. 7.5. From Fig. 7.5a, b it can be found that the phase propagation direction of the fundamental wave is consistent with the phase propagation direction of the second harmonic, and from Fig. 7.5c, d it can be found that the sound energy propagation direction of the fundamental wave is towards the positive direction of the x axis, while the energy propagation direction of the second harmonic is towards the negative direction of the x axis. Figure 7.6 shows the length of the metamaterial and the second harmonic at $x =$ The relationship of the amplitude at 0 position can be observed from the graph. Initially, the amplitude of the second harmonic increases with the length of the metamaterial. However, when the length of the metamaterial is long enough, the amplitude of the second harmonic gradually approaches a constant. The mechanism behind this is that under the effect of attenuation, the fundamental wave will slowly approach zero after propagating a sufficient distance, thus its contribution to the excitation of the second harmonic becomes smaller and smaller, until it approaches zero. Figure 7.7 shows the relationship between the amplitude of the second harmonic and the acoustic nonlinear parameter. It can be observed that the amplitude of the second harmonic is basically proportional to the nonlinear parameter. Therefore, if a strong nonlinear medium is added to this structure, the second harmonic at the sound source can become very large, and when the length of the metamaterial is long enough, the amplitude of the second harmonic will not decrease due to the increase in propagation distance. This is due to the benefits brought by the reverse excitation of the second harmonic. In traditional media, the second harmonic is excited and propagates in the positive direction of the x axis. The second harmonic first increases, then under the effect of attenuation, it gradually decreases with the increase in propagation distance. However, in this metamaterial, because the second harmonic is excited and propagates in the negative direction of the x axis, the fundamental wave gradually decreases with the increase in propagation distance, and its contribution to the second harmonic becomes smaller and smaller, until it is ignored. Therefore, the second harmonic will not decrease with the increase in propagation distance. This conclusion can be clearly seen from Fig. 7.6. Figure 7.8 shows the relationship between the amplitude of the second harmonic and the amplitude of the fundamental wave. It can be observed that the amplitude of the second harmonic is proportional to the square of the amplitude of the fundamental wave. Therefore, we can effectively increase the amplitude of the second harmonic by increasing the nonlinear parameter of the medium and the amplitude of the fundamental wave. Therefore, our research will have a certain impact on the application of ultrasonic harmonics.

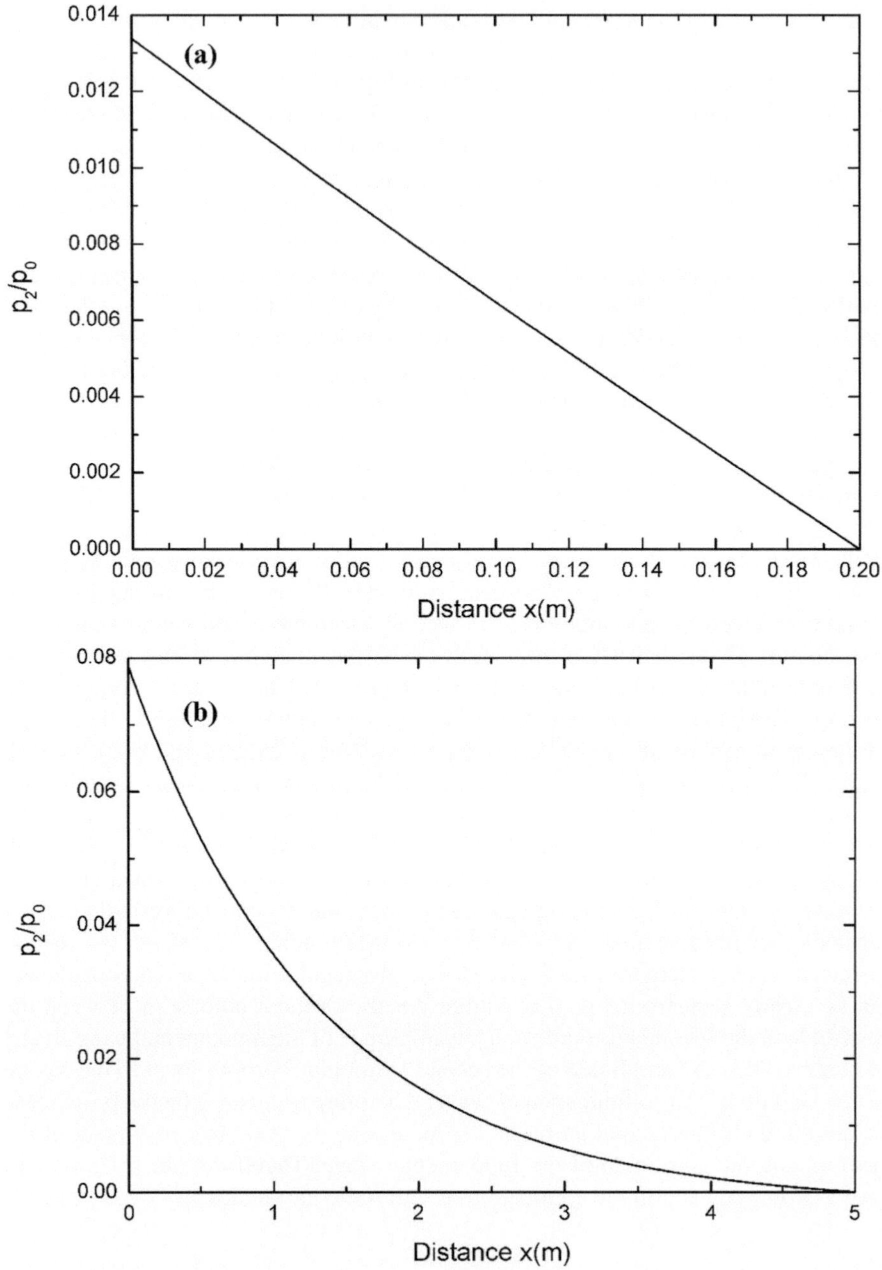

Fig. 7.4 **a** The distribution of the amplitude of the second harmonic when $L_0 = 0.2\,\text{m}$. **b** The distribution of the amplitude of the second harmonic when $L_0 = 5\,\text{m}$. Here, the fundamental wave is uniformly set to $p_0 = 1.0\,\text{MPa}$

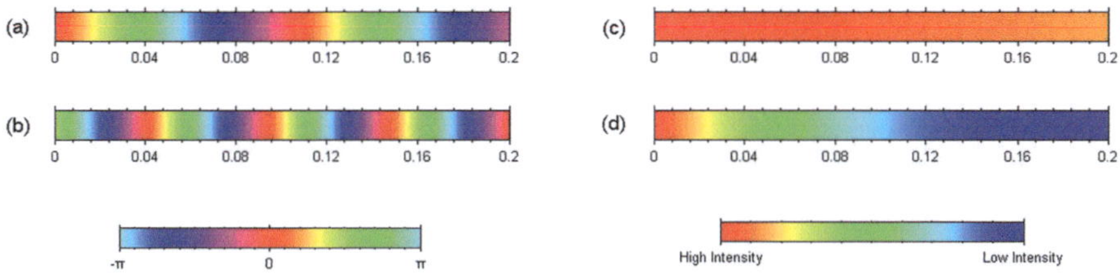

Fig. 7.5 **a** The phase distribution of the fundamental wave in the metamaterial. **b** The phase distribution of the second harmonic in the metamaterial. **c** The sound intensity distribution of the fundamental wave in the metamaterial. **d** The sound intensity distribution of the second harmonic in the metamaterial. Here, the amplitude of the fundamental wave is set to $p_0 = 1.0\,\text{MPa}$

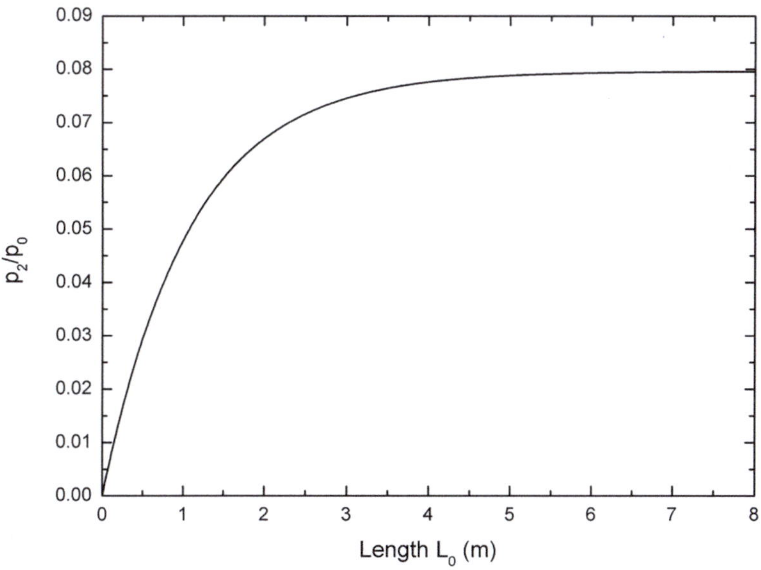

Fig. 7.6 The relationship between the length of the metamaterial L_0 and the amplitude of the second harmonic at the origin. Here, the amplitude of the fundamental wave is set to $p_0 = 1.0\,\text{MPa}$

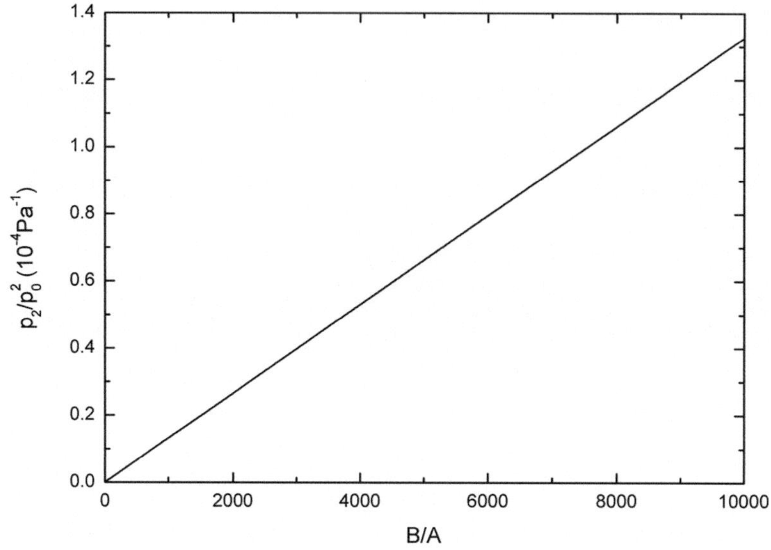

Fig. 7.7 The relationship between the amplitude of the second harmonic and the nonlinear parameter

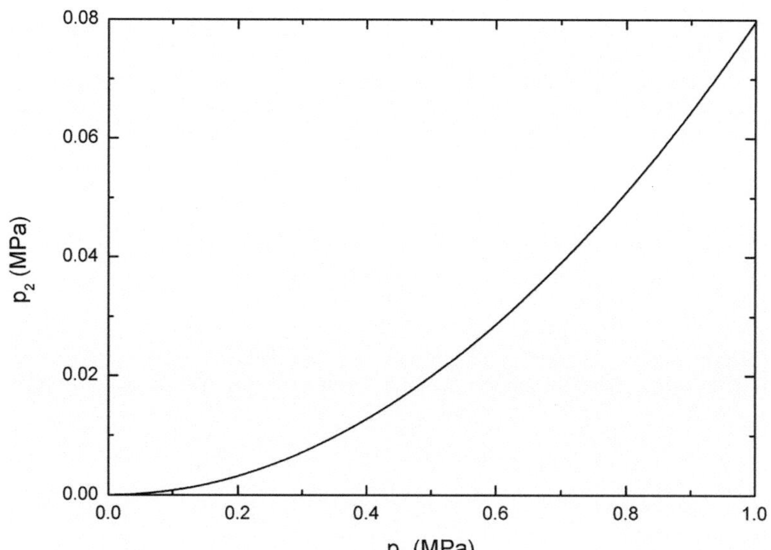

Fig. 7.8 The relationship between the amplitude of the second harmonic and the amplitude of the fundamental wave

7.2 Quasi-Phase Matching Enhances the Second Harmonic of Dispersion Metamaterials

In the past decade, metamaterials have exhibited many novel properties, such as achieving special phenomena such as negative refraction, invisibility cloaks, and super-resolution imaging that breaks the diffraction limit, attracting the research interest of many scientists [15–21]. However, all these properties are based on linear conditions. After the linear characteristics of metamaterials have been well studied, researchers have extended the research of metamaterials to the nonlinear field, and have discovered many previously undiscovered new phenomena, such as reverse second harmonic excitation, reverse second harmonic localization, etc. [22–24]. To realize the above phenomena, a higher second harmonic conversion efficiency is required. However, physically, to ensure that the energy stored in the medium is positive, all metamaterials will inevitably produce dispersion and dissipation properties [25], which means that there is a certain difficulty in obtaining high-efficiency second harmonic conversion efficiency in metamaterials. Phase matching is an effective method to achieve high-efficiency second harmonic excitation. Its principle is to find a special frequency point that satisfies the momentum conservation condition [26], to achieve high-efficiency second harmonic excitation. However, only certain specific frequency points can satisfy the phase matching condition, which will greatly reduce the application value of nonlinear metamaterials. Quasi-phase matching methods are widely used in the electromagnetic wave and light wave fields to achieve high-efficiency forward second harmonic excitation [27], which brings us some thoughts: Are there other methods that can achieve wide-frequency reverse second harmonic excitation in acoustic metamaterials? Can the quasi-phase matching method be applied to nonlinear acoustic metamaterials?

7.2.1 Phase Matching

In the study of nonlinear metamaterials, the excitation characteristics of reverse second harmonics have been discovered in optics [23, 24], these properties have many potential application values. In acoustics, the excitation of reverse second harmonics can be used for imaging biological tissues, and the development of a mature and feasible method to obtain efficient reverse second harmonic excitation will greatly promote the application of nonlinear construction materials. However, media with negative nonlinear parameters are not common in acoustics, so the traditional quasi-phase matching method widely used in optics is difficult to apply in the field of acoustics. In this section, we propose a new quasi-phase matching method using nonlinear complementary materials to achieve efficient reverse second harmonic excitation. First, let's discuss how to implement the phase matching method using nonlinear complementary materials.

The concept of complementary materials was proposed with the research of metamaterials. Under linear conditions, one of the most important applications of metamaterials is the perfect lens proposed by Pendry [15]. Complementary materials can "eliminate" their corresponding spaces, the effect is like inserting a completely blank area in space, no effect is produced in this area, and the phase does not change after the wave passes through this area, so this characteristic can be used to realize invisibility cloaks. The most important function of complementary materials is to realize the coordinate translation effect: when a beam passes through normal media, its wave path continuously increases, and when it passes through negative refractive media, its wave path will continuously decrease. After connecting these two media, the phase increase produced by a beam passing through normal media can be cancelled by negative refractive media, so this area looks like it does not exist [18]. As shown in Fig. 7.9a, consider a sound source located on the left $x = -L$ excites a sound wave, which passes through two consecutive media, the densities of these two media are ρ_0 and $-\rho_0$, the volume modulus are κ_0 and $-\kappa_0$, and the lengths are L and d. When the sound wave propagates from left to right, when it passes through the first normal medium, its sound path first increases nL, where n is the refractive index of the first medium, and then passes through the second negative refractive medium, its sound path decreases nd. The final effect is as if the sound wave has passed through a normal medium with a length of $L - d$. If we set $L = d$, the phase of the sound wave at $x = -L$ and $x = d$ is exactly the same. If the attenuation of the sound wave can be ignored, it looks like the sound wave has passed through a completely blank space. Now, we replace the previous medium with a second-order nonlinear medium, their characteristics at the fundamental frequency are the same as described above. But at twice the frequency, their densities become $-\rho_0$ and ρ_0, and their volume modulus become $-\kappa_0$ and κ_0, as shown in Fig. 7.9b. The material on the left shows right-handed properties at the angular frequency ω, but shows left-handed properties at the angular frequency 2ω. The material on the right shows left-handed properties at the angular frequency ω, but shows right-handed properties at the angular frequency 2ω. Therefore, we can still call this pair of materials complementary materials. According to the analysis in the first section of this chapter, we know that reverse propagating second harmonics can be excited in these materials [28, 29]. Now let's consider whether the second harmonic will be cancelled or enhanced when it propagates in this pair of nonlinear complementary materials. We assume $L = d$, first, we consider the second harmonic excited by the material on the left; after being excited by the infinitesimal element located at the x coordinate, the size of the second harmonic propagating to $x = -L$ is: $A\cos[2\omega(t + t') - 2k_1x]dx$, where A is a proportionality constant, its value is proportional to the nonlinear parameter and the square of the sound pressure amplitude, $t' = (x + L)/c_2$, $k_1 = \omega/c_1$ and $k_2 = 2\omega/c_2$ respectively represent the wave numbers of sound waves in the left medium, at the fundamental frequency and the second harmonic frequency, c_1 and c_2 respectively represent the speed of sound at the fundamental frequency and the second harmonic frequency. Although the wave vector direction of the second harmonic is positive in this metamaterial, we take the sign of t' as positive because its energy flow direction is towards the negative x axis [23], so its acoustic energy is accumulated in the negative direction.

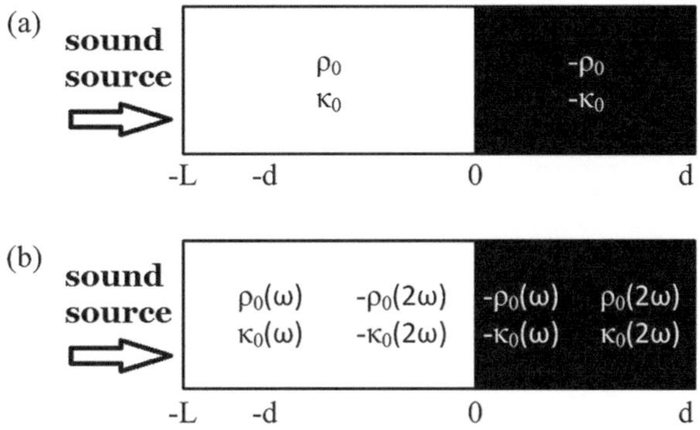

Fig. 7.9 **a** A pair of linear complementary materials. **b** A pair of nonlinear complementary materials

For simplicity, we use the quasi-linear approximation, assuming that this structure is relatively short, so that A is a constant. In our assumed material, the phase matching condition is satisfied, so we can superpose the second harmonics excited by infinitesimal elements everywhere and propagated to $x = -L$, and obtain the sound pressure of the second harmonic at $x = -L$: $P_{2\omega} = A\cos(2\omega t + 2k_1 L)L$. Now, we consider the second harmonic excited by the material on the right. Due to the complementary material effect, the phase of the fundamental wave at x is equal to the phase at $-x$, so the phase of the second harmonic excited by the fundamental wave at x should also be equal to the phase at $-x$. In such a material, the second harmonic is excited in the opposite direction, so the energy of the second harmonic is accumulated in the negative direction. When the second harmonic excited at x propagates to $-x$, its phase does not change due to the complementary effect of this pair of materials at the second harmonic. Finally, the sound pressure of the second harmonic propagated to $x = -L$ excited by the two segments of material is superimposed, and it is found that its value is $2P_{2\omega}$ instead of 0. Therefore, if nonlinear effects are considered, second harmonics will be excited, and this pair of complementary materials can no longer be considered as blank. For this reason, we can use this property to achieve quasi-phase matching methods in fluid media to enhance the excitation of the second harmonic.

7.2.2 Quasi-Phase Matching

In phase-matching materials, the phase of the second harmonic excited by any infinitesimal element and propagated to $x = -L$ is the same, as shown in Fig. 7.10a. The small arrows in the figure represent the phase of the second harmonic excited at x and propagated to $x = -L$. All the small arrows pointing in the same direction mean the enhancement of the second harmonic, as shown in Fig. 7.10b, the whole large arrow represents the total second harmonic after superposition. Now, we will extend the analysis process to the case of phase mismatch. As shown in Fig. 7.10c, the material parameters of the right medium at the fundamental frequency ω are ρ_1 and κ_1, and the material parameters at the second harmonic frequency 2ω are $-\rho_2$ and $-\kappa_2$.

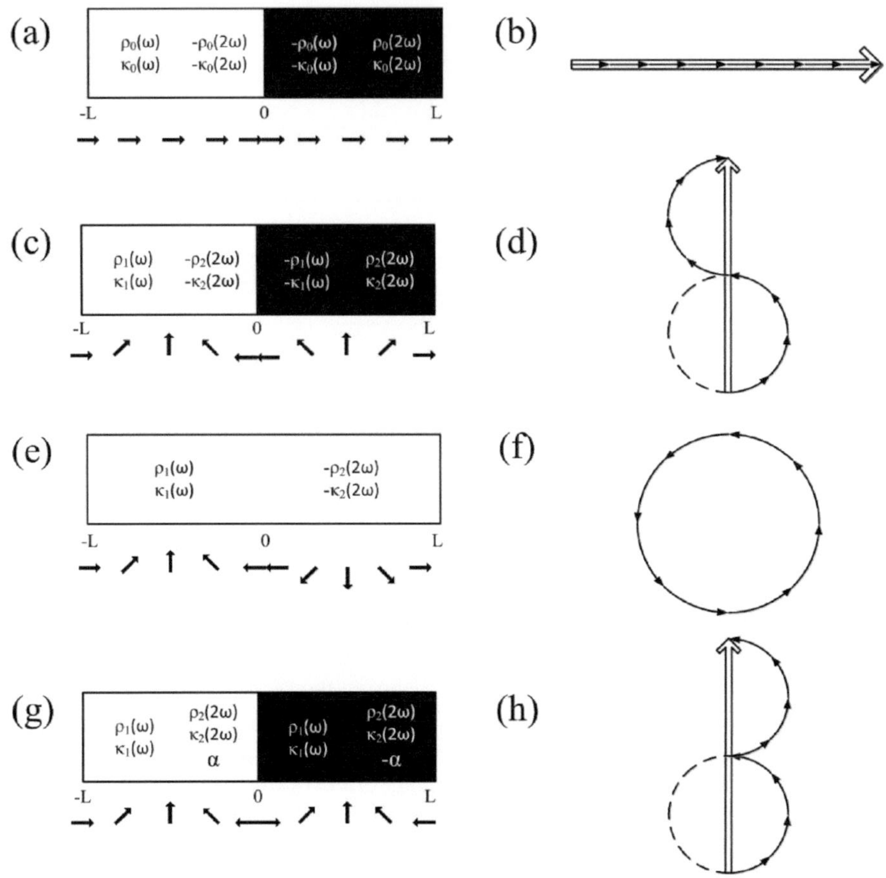

Fig. 7.10 **a** A pair of dispersionless nonlinear complementary materials and their **b** second harmonic. **c** A pair of dispersive nonlinear complementary materials and their **d** second harmonic. **e** Normal dispersive ordinary medium and its **f** second harmonic. **g** A pair of media connected together with nonlinear polarization and their **h** second harmonic

Due to dispersion effects, the phases of the second harmonics excited by infinitesimal elements at different places and propagated to $x = -L$ are different. We define the coherence length as $l_0 = \pi/|k_2 - 2k_1|$, if the material is converted to its complementary material after the sound wave passes through a length of l_0, then the quasi-phase matching condition can be achieved. Let $x = -L$, we can get the phase of the second harmonic excited by infinitesimal elements at different places and propagated to $x = -L$, and represent it with the direction of small arrows, as shown in Fig. 7.10c. If we only consider one period, by superposing these infinitesimal elements, we can get the second harmonic at $x = -L$ $P_{2\omega} = 4A\cos(2\omega t + 2k_1 l_0) l_0/\pi$. By vectorially adding the small arrows in Fig. 7.10c, we can obtain Fig. 7.10d. The length of the large arrow in the figure represents the amplitude of the second harmonic, and its direction represents the phase of the second harmonic. If we cascade the structure of n periods, the total length is $2nl_0$, then the total size of the second harmonic obtained at the sound source is: $P_{2\omega} = 4A\cos(2\omega t + 2k_1 l_0)nl_0/\pi$. If we use ordinary media instead of complementary media, as shown in Fig. 7.10e, then when the sound wave passes through two coherent lengths, all the excited second harmonics form a closed circle, as shown in Fig. 7.10f, at this time the superimposed second harmonic is 0. If we extend the structure in Fig. 7.10e and extend its length to $2nl_0$, the size of the second harmonic it excites is still 0. In order to excite stronger second harmonics, our method is to periodically cascade each group of nonlinear complementary materials in the order of ABABABAB..., as shown in Fig. 7.11a. Similarly, we can calculate the second harmonic excited by the traditional quasi-phase matching method, as shown in Fig. 7.10g. The parameters of the left material at the fundamental frequency ω are ρ_1 and κ_1, and the parameters at the double frequency 2ω are ρ_2 and κ_2, the nonlinear parameters of the left material and the right material are α and $-\alpha$ respectively. In this case, a forward second harmonic will be excited, and at $x = -L$, the size of the second harmonic produced by superposition is $P_{2\omega} = 4A\cos(2\omega t + 2k_1 l_0)l_0/\pi$, as shown in Fig. 7.10h, the amplitude of its second harmonic is equal to the size of the second harmonic obtained by our proposed method of using complementary materials to achieve quasi-phase matching. The difference between our method and the traditional quasi-phase matching method is [30, 31]: Our method is to introduce nonlinear complementary materials to correct the phase difference caused by dispersion (the generation of phase difference makes the second harmonic superimposed into a circle, resulting in the inability to excite the second harmonic), so there is no need for a medium with 180° antiparallel polarization and negative nonlinear parameters.

7.2.3 Difference Between Phase Matching in Metamaterials and Ordinary Phase Matching

In order to compare our proposed quasi-phase matching method and the traditional quasi-phase matching method, we let the period of the structure in Fig. 7.11a be l_0, so the length of each unit is $l_0/2$. First, introduce nonlinear complementary materials into this structure and apply our quasi-phase matching method. When the sound wave passes through a structure with a length of $2l_0$, the size of the second harmonic at the origin becomes $P_{2\omega} = 4\sqrt{2}A\cos(2\omega t + 2k_1 l_0)l_0/\pi$, as shown in Fig. 7.11b. If the traditional quasi-phase matching method is applied, we need to introduce a pair of media with 180° antiparallel polarization into A and B in the structure shown in Fig. 7.11a. When the sound wave passes through a structure with a length of $2l_0$, the size of the second harmonic at the origin becomes 0, as shown in Fig. 7.11c. If we define the normalization coefficient $\eta = |P_{2\omega}|/AL$ to represent the excitation efficiency of the second harmonic, where $P_{2\omega}$ is the amplitude of the second harmonic at the origin, L is the total length of the sound wave passed. As mentioned earlier, if

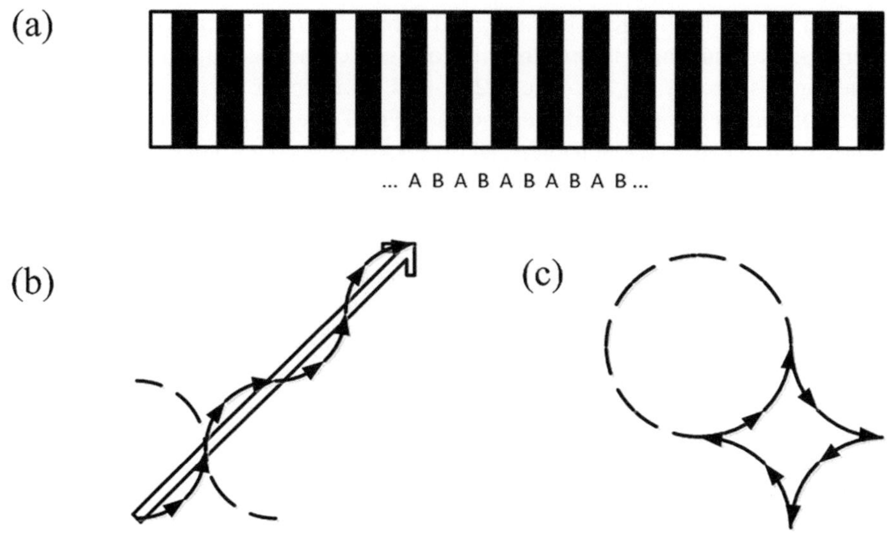

Fig. 7.11 **a** Schematic diagram of a structure formed by cascading a pair of materials with complementary properties or materials with 180° antiparallel polarization in the order of ABABABAB…. **b** The second harmonic produced by the structure formed by cascading complementary media when $L = \frac{l_0}{2}$. **c** The second harmonic produced by the structure formed by cascading materials with 180° antiparallel polarization when $L = \frac{l_0}{2}$

we set the period of the structure to $2l_0$, the efficiency of our quasi-phase matching method and the traditional quasi-phase matching method is the same, both equal to 2/π. If we set the period of the structure to l_0, as mentioned earlier, the efficiency of our quasi-phase matching method is $2\sqrt{2}/\pi$, while the efficiency of the traditional quasi-phase matching method is 0. Figure 7.12 shows the relationship between the second harmonic excitation efficiency and the structural period *D*. For the traditional quasi-phase matching method, the second harmonic excitation efficiency is:

$$
\begin{aligned}
\eta &= f\left(\frac{D}{2l_0}\right) = f(l) \\
&= \frac{\sqrt{\left(\int_0^L sign\left(\sin\frac{2\pi x}{l}\right)\cos 2\pi x dx\right)^2 + \left(\int_0^L sign\left(\sin\frac{2\pi x}{l}\right)\sin 2\pi x dx\right)^2}}{L}
\end{aligned}
\tag{7.92}
$$

where the sign function represents the phase sign of the second harmonic relative to the period length. *L* is the total length of the structure, in our calculations, we take $L = 50$, the first and second parts of the integral represent the contributions generated by the real and imaginary parts of the second harmonic respectively. For our proposed quasi-phase matching method using complementary materials, we have:

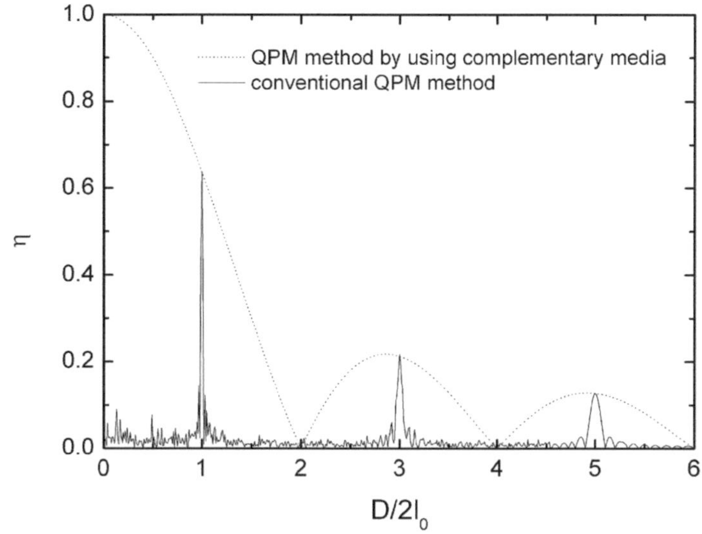

Fig. 7.12 The relationship between the second harmonic excitation efficiency and the structural period D relative to $2l_0$. The solid line represents the traditional quasi-phase matching method, and the dashed line represents our proposed quasi-phase matching method

$$\eta = f\left(\frac{D}{2l_0}\right) = f(l) = \left|\frac{\sin\frac{\pi l}{2}}{\frac{\pi l}{2}}\right| \tag{7.93}$$

From Fig. 7.12, we can see that our proposed quasi-phase matching method can excite the second harmonic over a wide frequency band. The physical mechanism of the traditional quasi-phase matching method is coordinate translation, while the physical mechanism of our proposed quasi-phase matching method is coordinate inversion, as shown in Fig. 7.10c, d, g, h. When using our quasi-phase matching method, the phase of the second harmonic excited at x and transmitted to the origin is consistent with the phase of the second harmonic excited at $-x$ and transmitted to the origin, as shown in Fig. 7.10c. However, when using the traditional quasi-phase matching method, the phase of the second harmonic excited at x and transmitted to the origin is consistent with the phase of the second harmonic excited at $x - L$ and transmitted to the origin, as shown by the direction of the small arrow in Fig. 7.10g. When the period of the structure is less than $2l_0$, most of the second harmonics excited at x and $-x$ cancel each other out, so in this case, the second harmonic cannot be effectively excited, which is why our proposed quasi-phase matching method can excite the second harmonic over a very wide frequency band. In addition, we can find that when using our proposed quasi-phase matching method, if we can reduce the period of the structure, we can obtain a larger second harmonic. However, it should be noted that the period for implementing the quasi-phase matching method cannot be too small, because the lattice period for implementing the double negative material must be an order of magnitude smaller than the wavelength in the medium, so the

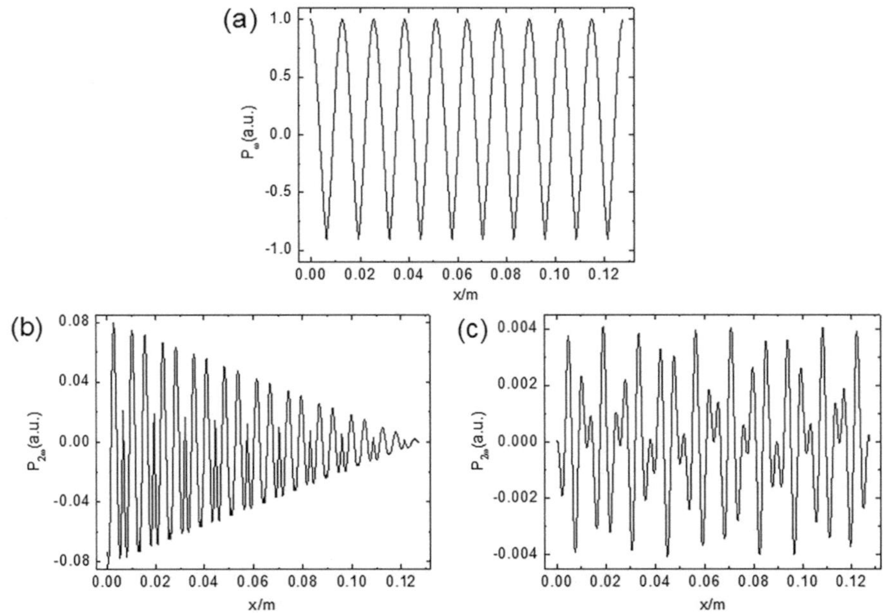

Fig. 7.13 **a** Fundamental wave in a structure cascaded with periodic nonlinear complementary media. **b** The sound pressure distribution of the backward second harmonic excited by using nonlinear complementary materials to achieve quasi-phase matching. **c** The sound pressure distribution of the second harmonic generated without using the quasi-phase matching method

period for implementing the quasi-phase matching must be an order of magnitude larger than the lattice period for implementing the double negative material, in order to satisfy the assumption of a uniform medium.

In order to demonstrate the effectiveness of our proposed quasi-phase matching method, we performed numerical calculations. The parameters are set as: $\rho_1 = 998\,\text{kg/m}^3$ and $\kappa_1 = 2.19\,\text{GPa}$, $\rho_2 = 1594\,\text{kg/m}^3$ and $\kappa_2 = 1.40\,\text{GPa}, f = 0.1$ MHz, as shown in Fig. 7.10c. Therefore, the sound speed, wave number, and coherence length can be calculated as: $c_1 = 1481$ m/s, $c_2 = 938$ m/s, $k_1 = 424\text{ m}^{-1}$, $k_2 = 1340\text{ m}^{-1}$ and $l_0 = 7.39$ mm.

Assuming the coefficient $A = 1\,\text{N/m}^3$, the period length of the structure is $2l_0 = 12.78$ mm, the total length of the material after cascading is $20l_0 = 127.8$ mm. Figure 7.13a shows the fundamental wave in the structure formed by cascading periodic nonlinear complementary media. Figure 7.13b shows the sound pressure distribution of the reverse second harmonic excited by using nonlinear complementary materials to achieve quasi-phase matching. Figure 7.13c shows the sound pressure distribution of the second harmonic generated without using the quasi-phase matching method. Comparing these three figures, it can be found that the quasi-phase matching method implemented using complementary materials can obtain a large second harmonic excitation, which is two orders of magnitude larger than the second harmonic excited without using the quasi-phase matching method.

Our method can also be applied to the situation of forward excitation of the second harmonic. When applied, one medium only needs to have positive refraction at both the fundamental frequency and the double frequency, while its complementary material needs to have negative refraction at both the fundamental frequency and the double frequency. Moreover, even if perfect phase matching conditions cannot be achieved at the fundamental frequency and the double frequency, the phase difference caused by the fundamental wave and the harmonic in the propagation process can be corrected with complementary materials, and the quasi-phase matching method is still effective. However, at this time, the right-hand material should be nonlinear, and the complementary left-hand material needs to be linear.

Obviously, all actual metamaterials have dissipation, but its impact on the excitation of high-order harmonics is very small [32], so in our model, the effect of attenuation is not considered. The quasi-phase matching method we proposed is relatively simple and easy to implement in experiments, and the experimental implementation is crucial for the application of this quasi-phase matching method. It has been reported that periodic structures can be used to achieve simultaneous double negative properties and double positive properties under different frequency bands [33]. Moreover, if the structure is filled with a medium with strong nonlinearity, such as porous materials or ultrasound contrast agents [34–36], a very large second harmonic can be obtained.

References

1. Zheludev NI. The Road ahead for metamaterials. Science. 2010;328(5978):582–3.
2. Wegener M. Metamaterials beyond optics. Science. 2013;342(6161):939–40.
3. Shadrivov IV. Tunable transmission and harmonic generation in nonlinear metamaterials. Appl Phys Lett. 2008;93(16):161903.
4. Fan YC, Wei ZY, Han J, Liu XG, Li HQ. Nonlinear properties of meta-dimer comprised of coupled ring resonators. J Phys D Appl Phys. 2011;44(42):425303.
5. Kozyrev AB. Nonlinear left-handed transmission line metamaterials. J Phys D Appl Phys. 2008;41(17):173001.
6. Klein MW, Wegener M, Feth N, Linden S. Experiments on second- and third-harmonic generation from magnetic metamaterials. Opt Express. 2007;15(8):5238–47.
7. Poutrina E. Analysis of nonlinear electromagnetic metamaterials. New J Phys. 2010;12(9):93010.
8. Rose A, Huang D, Smith DR. Demonstration of nonlinear magnetoelectric coupling in metamaterials. Appl Phys Lett. 2012;101(5):51103.
9. Li YF, Lan J, Li BS, Liu XZ, Zhang JS. Nonlinear effects in an acoustic metamaterial with simultaneous negative modulus and density. J Appl Phys. 2016;120(14):145105.
10. Lan J, Li YF, Yu HY, Li BS, Liu XZ. Nonlinear effects in acoustic metamaterial based on a cylindrical pipe with ordered Helmholtz resonators. Phys Lett A. 2017;381(13):1111–7.
11. Sugimoto N. Propagation of nonlinear acoustic waves in a tunnel with an array of Helmholtz resonators. J Fluid Mech. 1992;244(244):55–78.
12. Richoux O, Tournat V, Le VST. Acoustic wave dispersion in a one-dimensional lattice of nonlinear resonant scatterers. Phys Rev E. 2007;75(2):26615.

13. Bradley CE. Time-harmonic acoustic Bloch wave propagation in periodic waveguides part III Nonlinear effects. J Acoust Soc Am. 1995;96(3):1854–62.
14. Fan L, Chen Z, Deng YC, Ding J, Ge H, Zhang SY, Yang YT, Zhang H. Nonlinear effects in a metamaterial with double negativity. Appl Phys Lett. 2014;105(4):228–582.
15. Pendry JB. Negative refraction makes a perfect lens. Phys Rev Lett. 2000;85:3966–9.
16. Zhu J, Christensen J, Jung J, Martin-Moreno L, Yin X, Fok L, Zhang X, Garcia-Vidal FJ. A holey-structured metamaterial for acoustic deep-subwavelength imaging. Nat Phys. 2011;7:52–5.
17. Pendry JB, Schurig D, Smith DR. Controlling electromagnetic fields. Science. 2006;312:1780–2.
18. Lai Y, Chen HY, Zhang ZQ, Chan CT. Complementary media invisibility cloak that cloaks objects at a distance outside the cloaking shell. Phys Rev Lett. 2009;102:093901.
19. Pendry JB. Time reversal and negative refraction. Science. 2008;322:71–3.
20. Schurig D, Mock JJ, Justice BJ, et al. Metamaterial electromagnetic cloak at microwave frequencies. Science. 2006;314:977–80.
21. Ergin T, Stenger N, Brenner P, et al. Three-dimensional invisibility cloak at optical wavelengths. Science. 2010;328:337–9.
22. Ciraci C, Centeno E. Focusing of second-harmonic signals with nonlinear metamaterial lenses: a biphotonic microscopy approach. Phys Rev Lett. 2009;103:063901.
23. Shadrivov IV, Zharov AA, Kivshar YS. Second-harmonic generation in nonlinear left-handed metamaterials. J Opt Soc Am B. 2006;23(3):529–34.
24. Klein MW, Enkrich C, Wegener M, et al. Second-harmonic generation from magnetic metamaterials. Science. 2006;313(5786):502–4.
25. Smith DR, Kroll N. Negative refractive index in left-handed materials. Phys Rev Lett. 2000;85(14):2933–6.
26. Shankar R. Principles of quantum mechanics. Springer; 1994. Chap. 11, p. 279–304.
27. Rose A, Smith DR. Broadly tunable quasi-phase-matching in nonlinear metamaterials. Phys Rev A. 2011;84:013823.
28. Centeno E, Ciraci C. Theory of backward second-harmonic localization in nonlinear left-handed media. Phys Rev B. 2008;78:235101.
29. Centeno E, Felbacq D, Cassagne D. All-angle phase matching condition and backward second-harmonic localization in nonlinear photonic crystals. Phys Rev Lett. 2007;98:263903.
30. Armstrong JA, Bloembergen N, Ducuing J, et al. Interactions between Light waves in a nonlinear dielectric. Phys Rev. 1962;127:1918–39.
31. Zhu SN, Zhu YY, Ming NB. Quasi-phase-matched third-harmonic generation in a quasi-periodic optical superlattice. Science. 1997;278:843–6.
32. Shadrivov IV, Zharov AA, Zharova NA, et al. Nonlinear left-handed metamaterials. Radio Sci. 2005;40:RS3S90.
33. Lee SH, Park CM, Seo YM, et al. Composite acoustic medium with simultaneously negative density and modulus. Phys Rev Lett. 2010;104:054301.
34. Wu JR, Zhu ZM, Du GH. Nonlinear behavior of a liquid containing uniform bubbles: comparison between theory and experiments. Ultrasound Med Biol. 1995;21:545–52.
35. Quan L, Liu XZ, Gong XF. Quasi-phase matched backward second-harmonic generation by complementary media in nonlinear metamaterials. J Acoust Soc Am. 2012;132:2852–2586.
36. Quan L, Liu XZ, Gong XF. Nonlinear acoustic metamaterial: realization of a backward traveling second-harmonic. J Acoust Soc Am. 2016;139(6):3373.

MIX
Papier aus verantwortungsvollen Quellen
Paper from responsible sources
FSC® C105338

If you have any concerns about our products,
you can contact us on
ProductSafety@springernature.com

In case Publisher is established outside the EU,
the EU authorized representative is:
Springer Nature Customer Service Center GmbH
Europaplatz 3, 69115 Heidelberg, Germany

Printed by Libri Plureos GmbH
in Hamburg, Germany